ESO ASTROPHYSICS SYMPOSIA
European Southern Observatory

Series Editor: Philippe Crane

Philippe Crane (Ed.)

The Light Element Abundances

Proceedings of an ESO/EIPC Workshop
Held in Marciana Marina, Isola d'Elba,
21–26 May 1994

 Springer

Volume and Series Editor

Philippe Crane
European Southern Observatory
Karl-Schwarzschild-Strasse 2
D-85748 Garching

ISBN 978-3-662-22501-1   ISBN 978-3-540-49169-9 (eBook)
DOI 10.10078/978-3-540-49169-9

CIP data applied for

Typesetting: Camera ready by author/editor
SPIN: 10476960   42/3142-543210 - Printed on acid-free paper

Preface

The topic of these proceedings, the abundance of the light elements, is distinguished in several ways. First, it covers a very broad range of astrophysical problems from primordial nuclear synthesis through galactic evolution to stellar astrophysics. All of these areas are inter-related in their effect on our ability to determine the element abundances. Second, the observational aspects of this topic are still tractable by single observers. Many of the observational results presented here have been obtained by individuals or small collaborations. Third, the theoretical work in this field is strongly tied to observations. Finally this topic is one where ESO member state astronomers are actively driving both the observations and theory.

These proceedings gather the papers presented at the ESO/EIPC Workshop on The Light Element Abundances held at Marciana Marina, Isola d' Elba during May 22 – 28, 1994. Together, these papers summarize the current state of our understanding of this topic. In several cases, new results or theoretical developments are included. The topic is being actively developed and, of particular interest to the ESO community, many new ideas on how to exploit 8 meter class telesopes can be developed from the material in these pages.

The Elba International Physics Center (EIPC) co-hosted the workshop. ESO has previously co-hosted with EIPC two workshops on Elba. The relative isolation, beautiful surroundings, and good facilities insured fruitful interactions and a successful workshop.

Acknowledgements It is a great pleasure to acknowledge the input and encouragement of the scientific organizing committee: R. Ferlet, F. Ferrini, D.L. Lambert, P. Molaro, P. Nissen, B. Pagel, L. Pasquini, and G. Steigman. Without C. Stoffer, this workshop would not have happened. Without Antonella Sapere, it would have been chaotic. Without J. Faulkner, it would have been dull.

Garching, January 1995   Philippe Crane

Contents

List of Participants

Name	Institution
ABIA, Carlos	Universidad de Granada, Dept. Fisica Teorica y del Cosmos cabia@ugr.es
AHARPOUR, Nancy	Osservatorio Astrofisico di Arcetri nancy@arcetri.astro.it
ALIMI, Jean-Michel	Observatoire de Paris, LAEC, Meudon alimi@mesiob.obspm.fr
AUDOUZE, Jean	Institut d'Astrophysique, Paris
BALACHANDRAN, Suchitra	The Ohio State University, Dept. of Astronomy suchitra@payne.mps.ohio-state.edu
BANIA, Thomas	Boston University, Astronomy Dept. bania@inanna.bu.edu
BECKMAN, John	Instituto de Astrofisica de Canarias jeb@iac.es
BEERS, Timothy	Michigan State University, Dept. of Physics & Astronomy beers@msupa.pa.ms.edu
BOESGAARD, Ann Merchant	Institute for Astronomy, Univ. of Hawaii boes@galileo.ifa.hawaii.edu
BURBIDGE, Geoffrey	CASS, University of California, San Diego btravell@ucsd.edu
CASSE, Michel	C.E.A. - Service d'Astrophysique, DAPNIA, Gif-sur-Yvette iapobs::casse
CAYREL DE STROBEL, Giusa	Observatoire de Paris, Meudon gcayrel@mesioa.obspm.circe.fr
CHARBONNEL, Corinne	Observatoire Midi-Pyrenees, Toulouse corinne@obs-mip.fr
CHUVENKOV, Vladimir	Rostov University, Dept. of Space Physics cosmos@physfac.rostov-na-donu.su
CRANE, Philippe	ESO, Garching crane@eso.org
D'ANTONA, Francesca	Osservatorio Astronomico di Roma franca@astrmp.astro.it
D'ODORICO, Sandro	ESO, Garching sdodoric@eso.org
DELIYANNIS, Constantine	Institute for Astronomy, Univ. of Hawaii con@galileo.ifa.hawaii.edu
DUNCAN, Douglas	Dept. of Astronomy & Astrophysics University of Chicago duncan@oddjob.uchicago.edu
FAULKNER, John	University of California, Lick Observatory johnf@lick.ucsc.edu
FAVATA, Fabio	ESA - ESTEC Fabio.Favata@astro.estec.esa.nl
FERLET, Roger	Institut d'Astrophysique, Paris ferlet@iap.fr

FERRINI, Federico   Universita di Pisa, Dip. di Fisica
                           federico@astrpi.difi.unipi.it
GALLI, Daniele   Osservatorio Astrofisico di Arcetri
                           galli@arcetri.astro.it
GLUKHOV, Alexander   Rostov University, Dept. of Space Physics
                           cosmos@physfac.rostov-na-donu.su
GUERRERO,   Osservatorio Astronomico di Brera
  Gianantonio   guerrero@astmim.mi.astro.it
HOYLE, Fred   102 Admirals Walk, West Cliff
                           Bournemouth BH2 5HF, U.K
JAKOBSEN, Peter   ESA - ESTEC
                           pjakobse@astro.estec.esa.nl
KHERSONSKY, Valery   Dept. of Physics & Astronomy University of Pittsburgh
                           vkk@phyast.pitt.edu
KISELMAN, Dan   NORDITA, Copenhagen
                           dan@nordita.dk
KLOCHKOVA,   Special Astrophysical Obs.,
  Valentina   Russian Academy of Sciences
                           vkloch@sao.stavropol.su
KUNTH, Daniel   Institut d'Astrophysique, Paris
                           kunth@iap.fr
LE BRE, Agnes   Universite Montpellier II
                           lebre@graal.univ-montp2.fr
LEMOINE, Martin   Institut d'Astrophysique, Paris
                           lemoine@iap.fr
LINSKY, Jeffrey   University of Colorado, JILA
                           jlinsky@jila.colorado.edu
LUDKE, Everton   NRAL, Jodrell Bank
                           el@jb.man.ac.uk
LUDWIG, Hans-Gunter   MPI für Astrophysik, Garching
                           hgl@mpa-garching.mpg.de
MAGAZZU, Antonio   Osservatorio Astrofisico di Catania
                           antonio@ct.astro.it
MATTEUCCI, Francesca   Osservatorio Astronomico di Trieste
                           matteucci@atlantis.oat.ts.astro.it
MOLARO, Paolo   Osservatorio Astronomico di Trieste
                           molaro@astrts.astro.it
MOLLA, Mercedes   Universidad Autonoma de Madrid
                           mercedes@astro1.ft.uam.es
MOWLAVI, Nami   Universite Libre de Bruxelles, IAA
                           nmowlavi@astrohp1.ulb.ac.be
NISSEN, Poul   Inst. of Physics & Astronomy University of Aarhus
                           pen@obs.aau.dk
OLIVE, Keith   School of Physics & Astronomy Univ. of Minnesota
                           olive@mnhep.hep.umn.edu
PAGEL, Bernard   NORDITA, Copenhagen
                           pagel@nbivax.nbi.dk
PALLAVICINI, Roberto   Oservatorio Astrofisico di Arcetri
                           pallavic@arcetri.astro.it

PANCHUK, Vladimir   Special Astrophysical Obs.,
Russian Academy of Sciences
panchuk@sao.stavropol.su

PASQUINI, Luca   ESO, La Silla
lpasquin@eso.org

PEIMBERT, Manuel   UNAM, Mexico
peimbert@astroscu.unam.mx

PILYUGIN, Leonid   Main Astronomical Obs.,
Ukrainian Acad. of Sci., Kiev
pilyugin@gao.kiev.ua

PLEZ, Bertrand   Niels Bohr Institute, Copenhagen
plez@snbivax.nbi.dk

PRIMAS, Francesca   Universita di Trieste, Dip. di Astronomia
astrts::primas

RANDICH, Sofia   MPI für extraterrestrische Physik
msr@mpeu01.rosat.mpe-garching.mpg.de

RAUSCHER, Thomas   Institut für Kernchemie Universität Mainz
rauscher@vkcmzd.chemie.uni-mainz.de

REBOLO, Rafael   Instituto de Astrofisica de Canarias
rrl@iac.es

REEVES, Hubert   CNRS - CEN Saclay, Gif-sur-Yvette

RITOSSA, Claudio   Universita di Bologna, Dip. di Astronomia
ritossa@alma02.cineca.it

ROOD, Robert   University of Virginia, Dept. of Astronomy
rtr@ninkasi.astro.virginia.edu

RUSSELL, Stephen   Dublin Institute for Advanced Studies
SR@cp.dias.ie

RYAN, Sean   Anglo Australian Observatory
sgr@aaoepp2.aao.gov.au

SANDRELLI, Stefano   Universita di Bologna, Dip. di Astronomia
sandrelli@alma02.cineca.it

SATO, Katsuhiko   University of Tokyo, Dept. of Physics
sato@yayoi.phys.s.u-tokyo.ac.jp

SCHRAMM, David   University of Chicago, Dept. of Physics
dns@oddjob.uchicago.edu

SERNA, Arturo   Observatoire de Paris, LAEC, Meudon
serna@mesiob.obspm.fr

SODERBLOM, Dave   STScI, Baltimore
drs@stsci.edu

SPITE, Francois   Observatoire de Paris, Meudon
spitef@memaga.obspm.fr

STEIGMAN, Gary   Dept. of Physics, Ohio State Univ.
steigman@mps.ohio-state.edu

STRANIERO, Oscar   Osservatorio Astronomico Collurania, Teramo
straniero@astrte.te.astro.it

SURAN, Doru Marian   Romanian Acad. of Sci., Astron. Inst., Bucarest
masuran@imar.ro

TAGLIAFERRI,   Osservatorio Astronomico di Brera
Gianpiero   gtagliaf@astmim.mi.astro.it

TERLEVICH, Elena   University of Cambridge, Inst. of Astronomy
                   et@mail.ast.cam.ac.uk
TORRES-PEIMBERT,   UNAM, Mexico
 Silvia   silvia@astroscu.unam.mx
TOSI, Monica   Osservatorio Astronomico di Bologna
                   tosi@alma02.cineca.it
TRAAT, Peeter   Institute of Astrophysics & Atmospheric
                   Physics,Tartu, Estonia
                   traat@obs.ee
TURNSHEK, David   Dept. of Physics & Astronomy, University of Pittsburgh
                   turnshek@vm2.cis.pitt.edu
VANGIONI FLAM,   Instiut d'Astrophysique, Paris
 Elisabeth   flam@iap.fr
VAUCLAIR, Sylvie   Observatoire Midi-Pyrenees, Toulouse
                   svcr@obs-mip.fr
WIERLING, August   Universität Rostock
                   august@darss.mpg.uni-rostock.de
WILSON, Tom   MPI für Radioastronomie, Bonn
                   p073twi@mpifr-bonn.mpg.de

Part I

Light Elements in the Early Universe

Early Nucleosynthesis: The Present Status

Jean Audouze

Institut d'Astrophysique de Paris, CNRS, 98 bis boulevard Arago, 75014 Paris, France

Abstract. Standard models of Big Bang nucleosynthesis (BBN) are those which are the best suited to account for the production of D, ^{3}He, ^{4}He and ^{7}Li, the abundances of which range over about 10 decades. These models are briefly reviewed and compared to more complex ones like those which take into account any possible inhomogeneities resulting from the Quark–Hadron phase transition. Recent abundance determinations (in particular the D abundance in a large redshift extragalactic systems) do not modify our appraisal of the success of the standard BBN models. Should they be confirmed, they would lead to more stringent limits on the baryonic density of the Universe which could be close to that of the "visible matter". Furthermore, specific schemes of galactic evolution should be invoked to account for a possibly large galactic destruction of D unconnected with a large ^{3}He variation.

1. Introduction

The nucleosynthesis of the lightest elements (D, ^{3}He, ^{4}He and ^{7}Li) is a most fascinating problem which has been studied with some success for many years. The list of the scientists who have been involved in such analyses is quite long and starts with the names of G. Gamow, J. Peebles, R. Wagoner, H. Sato, W.A. Fowler, F. Hoyle... The reasons of such an interest are many fold.

(i) in the frame of the most straightforward Big Bang models, the comparison between the nucleosynthesis calculations and the relevant observations provided valuable hints on two important cosmological (physical) parameters : first the predicted baryonic density which is limited to 1–10% of the critical density and second the maximum number of neutrion families which is three. One should note that this limitation was obtained on astrophysical grounds before its confirmation by the relevant particle physics experiments such as the Z^o width undertaken at CERN. Moreover the predicted three lepton families would correspond to the three quark families and give confidence in the Grand Unification schemes. This strengthens the view of D. Sciama who said in 1983 that "Standard Big Bang nucleosynthesis is a triumph for cosmology".

(ii) the second reason is indeed that this problem does not lead to cumbersome calculations. The nucleosynthetic network is most simple. The relevant nuclear reaction rates are know. The physical conditions can be easily modeled. By contrast as we will discuss below, the abundance determinations are most complex and the new developments made through astronomical observations have direct consequences on the evolution of this field.

The purpose of this short text is to show that the simplest standard Big Bang Nucleosynthesis (BBN) scenarios are still today the most successful in the reconciling currents observations with the theoritical predictions. Considerations of more complex models in order to alleviate the strict limits on the baryonic density are not as convincing although they provide interesting constraints on various aspects of particle physics.

A few new observations especially those regarding the D abundance determinations outside of the Solar System may have direct impact on the future of such comparisons with the model predictions. In particular, two groups who observed the H Lyman lines on a high redshift quasar line of sight (Carswell et al. 1994, Songaila et al. 1994) propose that the D abundance can be as high as D/H∼2-3.10⁻⁴. As we argued in previous communications (see eg. Audouze,1993) this fact does not jeopardize the simple BBN models but modifies the predicted baryonic density (see also Vangioni–Flam and Cassé,1994). Should this observation be confirmed, it could lead to a reconsideration of the current models of galactic evolution.

2.   The Standard Big Bang Nucleosynthesis Model

This model has been considered by many authors over more than three decades : Peebles and Wagoner in the sixties, Schramm, Steigman and coworkers since the end of the seventies,. the french group (Reeves, Delbourgo–Salvador, Vangioni–Flam and myself) at the same epoch and many others like Yahil, Beaudet... The hypotheses are straightforward : one considers a homogeneous and isotropic universe the expansion of which is described by the general relativity and which has had at a given epoch a temperature $T > 10^{10}$ K. When the Universe expands, it cools. Its equation of state, its expansion rate and subsequently the outcome of the nucleosynthesis processes, depend on three basic physical parameters :

  i) the neutron lifetime
 ii) the baryonic density
iii) the number of flavors of relativistic particles (leptons).

From the above hypotheses the rate of expansion $\frac{1}{R}\frac{dR}{dt}$ (where R is the scale factor) depends on the total energy density of relativistic particles ρ_R :

$$\frac{1}{R}\frac{dR}{dt} \, \alpha \, \sqrt{\rho_R}$$

and $\rho_R = \dfrac{g_{\text{eff}}}{2}\, \rho_\gamma$ (ρ_γ being the radiative density)

where g_{eff} is the effective number of relativistic degrees of freedom

$$g_{\text{eff}} = \frac{43}{4}\left[1 + \frac{7}{43}\,(N'_\nu - 3)\right],$$

where N_ν is the number of neutrino flavors (see e.g. Boesgaard and Steigman 1985).

The actual expansion time scale t_{exp} is $t_{exp(sec)} = 2.4\, g_{eff}^{-\frac{1}{2}}\, T_{(Mev)}^{-2}$.

If N_ν increases as g_{eff}, t_{exp} decreases, which increases the temperature at which neutrons start to decouple. The net effect is an increase of the neutron proton ratio and an corresponding increase of the Helium density : $\Delta Y \approx 0.014\, (N_\nu - 3)$. Each new neutrino flavour would increase Y by $\approx 1\%$. Several groups (Walker et al. 1991, Smith et al. 1993, Kerman and Krauss 1994) have published the results of their calculations which are in fair agreement. Usually, the estimated abundances are related to the baryonic density (expressed through the parameter η_{10} as:

$$\eta_{10} = 278\, h_o^2 \left(\frac{T_\odot}{2.74}\right)^{-3} \Omega_B$$

with $h_o^2 = (\frac{H}{100})^2$ ranging from 0.25 to 1. From the comparison between the current observations (see below) and the calculations, one deduces a fairly narrow range for $\eta_{10} \sim 3 - 4$ which corresponds to $0.01 \leq \Omega_B \leq 0.04$ i.e. a baryonic density significantly lower than the critical value.

3. Non-Standard Models: A Few Remarks

Several attempts have been performed to alleviate this last statement. Can one build any BBN model predicting the proper abundances of D, He, Li and being consistent with higher values of Ω_B ? Among these "non standard" scenarios, let me quote the models with late decaying particles (Audouze et al. 1985, Salati et al. 1987); those implying neutron degeneracy (David and Reeves 1979, Olive et al. 1991) and the nucleosynthetic models taking into account possible inhomogeneities induced by the quark–hadron transition phase. Many articles including some of ours have considered such hypothesis. This scheme could have been especially attractive:

i)if the Quark–Hadron phase transition is first order, these inhomogeneities are a natural consequence of this process. Dense proton rich zones would have coexisted with low density neutron rich ones.

ii) dense zones would have produced low D, high ^4He abundances while dilute ones would have led to the reverse.

This hypothesis have been found to be unsuccessful because each zone would overproduce ^7Li as shown eg. by Reeves et al. 1988. Higher η values ($5 < \eta < 7$ leading to a doubling of the predicted baryonic density: $0.02 \leq \Omega_B \leq 0.1$) can only be found in a very restricted set of parameters such that the average size of the bubble would be ~ 100 m with a contrast parameter R<100 (see eg. Kurki–Suonio et al. 1990).

In the case of the neutron degeneracy models, Olive et al. (1991) showed after David and Reeves (1979) that η_{10} could be as large as ~ 300 leading to $\Omega \sim 1$, but with an extremely narrow range of electronic and muonic neutrino degeneracy $\xi_e \sim 1.5$ and $\xi_{\mu\tau} \sim 38$ where ξ_e and $\xi_{\mu\tau}$ are such that $\xi_\nu = \frac{\mu_\nu}{T_\nu}$ and μ_ν being the neutrino chemical potential. This extremely limited possibility to reconcile the predicted light element abundances with $\Omega_B = 1$ does exist but at the expense of very strict constraints on the neutrino degeneracy.

4. New Developments on the D Abundance Determination

Many publications have been devoted to reviewing the abundance determinations of the lightest elements. I shall concentrate only on a few points. Progress in the understanding of the BBN problem has always been achieved through new abundance determinations. In 1972–1973 the "cosmic" D abundance was determined from observations of the solar wind and the nearby interstellar medium. This lead to one of the first coherent scheme to account for BBN (see eg. Reeves et al. 1973). Starting from ~ 1979, the primordial ^4He abundance was determined from the ^4He features in the spectra of blue compact galaxies. This was followed by a large number of papers including those fixing a limit of three different lepton families. In 1983, the ^7Li early abundance in pop. II stars was determined by the Spites and confirmed the standard BBN models.

Table 1. "Current" abundance determinations

	"primordial"	"solar"	"present"
			10^{-5} [3]
D/H	$(1.9\text{-}2.5)10^{-4}$ [1]	$(2.6\pm 1)10^{-5}$ [2]	$(1.5^{+0.07}_{-0.08}10^{-5}$ [4]
$\frac{^3He}{H}$	?	$(1.6\pm0.3)10^{-5}$ [2]	$\sim 2.5\ 10^{-5}$ [5]
Y	0.228 ± 0.005 [6]	$0.27\text{-}0.275$ [7]	$0.27\text{-}0.29$ [8]
$\frac{^7Li}{H}$	$(1\text{-}2.3)10^{-10}$ [9]		10^{-9} [9]

[1] Songaila et al. 1994 – Carswell et al. 1994
[2] Geiss 1993
[3] Lemoine et al. 1994
[4] Linsky et al. 1993
[5] Balser et al. 1994
[6] Pagel et al. 1992
[7] Turck–Chièze et al. 1993
[8] Peimbert et al. 1992
[9] Spite et al. 1993

In recent months, a large debate has been raised by the D abundance determination in large redshift QSO lines of sight. Two groups Songaila et al. 1994 and Carswell et al. 1994, have reported a very large D abundance on the Q 0014±813 line of sight such that D/H~3.10^{-4}.

Table 1 summarizes the "current" light element abundance determinations (see also Wilson and Rood 1994) and taking this high D/H value at its face value. It should be noted here that several colleagues express their doubt on this abundance determination not on the quality of the measurement but mainly on the possible confusion of a D line with an H line having a redshift different from that of the principal H lines. We should therefore be extremely cautious about its reliability. Nevertheless given this word of caution, it is still exciting to comment on the consequences of a high D abundance on the standard BBN models.

5. The Consequences of a High Deuterium Abundance

Although it may not be possible to confirm entirely or to refute the high D abundance determinations in the near future (see eg. Linsky, this conference), it is worth to analyzing the implications of such determinations. First, as noticed by all the specialists in the field the $(\frac{D}{H} \sim 3.10^{-4})_{prim}$ hypothesis is not ruled out by the other abundance determinations in the frame of the standard BBN model. As argued in several of my previous papers and by Vangioni–Flam and Cassé, 1995, this large "primordial abundance" of D would imply a very low Ω_B or baryonic density parameter η_{10}. η_{10} would range from 1.4-2 instead of 3-4, if H=50 km s^{-1} Mpc^{-1} 0.02$\leq \Omega_B \leq$0.03; if H=100 km s^{-1} Mpc^{-1} $5.10^{-3} \leq \Omega_B \leq 8.10^{-3}$. In that case the resulting Ω_B would be quite close to Ω_L the density of visible matter. Most of the baryons should be visible. This would imply that the fraction of baryonic dark matter is insignificant compared to the non baryonic one. This could be checked by current studies concerning the matter nature and content in the Universe.

Another consequence lies in the discussion of galactic evolution models. Deuterium has to be destroyed by factors larger than 10-20 during the galactic history. This destruction should not lead to any overproduction of ^3He and metals Z. Moreover, the models should be consistent with the observed stellar luminosity distributions. Some models (e.g. Vangioni–Flam and Audouze 1988, Steigman and Tosi 1992, Vangioni–Flam et al. 1994 and Vangioni–Flam and Cassé 1994) have been proposed which satisfy this constraint.

The first attempt of Vangioni–Flam and Audouze(1988) was based on the prompt initial enrichment hypothesis proposed a long time ago by Truran and Cameron(1971). With a constant rate of star formation (SFR) for $t < t_o$, decreasing afterwards, the D destruction was predicted consistently with the ^3He and Z abundance but the model fails due to the constraint imposed by the G dwarf metallicity distribution. On the other hand, this last constraint was overcome with a bimodal SFR, in a close box model. In this model, the overall metallicity Z was unfortunately excessive.

In the most recent analysis Vangioni–Flam and Cassé 1994 propose a bimodal SFR together with the occurence of an early galactic wind which would sweep up the overabundant Z. The adoption of such constraining hypotheses would account for the possible occurence of a large primordial D abundance.

6. In Summary

Regarding the nucleosynthesis of D, ^3He, ^4He and ^7Li four types of reasoning can be adopted :

a) Big Bang models will prove to be inappropriate as proposed by Hoyle (this conference) together with Burbidge and Narlikar (see eg. Hoyle et al. 1993).

b) Big Bang schemes apply but not the standard ones. The only, very contrived but interesting, model in that respect is the one developed by David and Reeves 1979 and subsequently by Olive et al. 1991 who considered the possibility of neutrino degeneracy. It seems more appropriate to use the D, ^3He, ^4He and ^7Li abundance determinations to constrain some aspect of particle physics like the Quark–Hadron phase transition or some particle characteristics.

c) The new D abundance determination in high redshift QSO lines of sight has to be challenged and abandoned (Steigman, Linsky this conference). In that case we are left with the current BBN scenario $\eta_{10} \sim 3$ and $\Omega_B \sim 0.01$-0.04.

d) It $(D/H)_{prim}$ is as high as 3.10^{-4}, it implies that Ω_B has be very low close to the visible density limit. Moreover the galactic evolution is a bit more complex in order to reconcile this high abundance with the constraints on ^3He and Z abundances such as that on stellar metallicities and luminosities.

To sum up excitement in that field comes mainly from new abundance determinations. This conference and proceedings are proving it clearly!

This paper has been supported by PICS N°114 of CNRS.

References

Audouze, J., Lindley, D. and Silk, J., (1985) : ApJ (letters) **293**, L53

Audouze, J., (1993) : in Origin and Evolution of the Elements, eds. N. Prantzos, E.Vangioni–Flam and M.Cassé, Cambridge U. Press, pp.1

Balser, D.S., Bania, T.M., Brockway, C.J., Rood, R.T. and Wilson, T.L., (1994) : ApJ. in press

Boesgaard, A.M. and Steigman, G., (1985) : Ann. Rev. Astron. Astroph., **23**, 319

Carswell, R.F., Rauch, M., Weymann, R.J., Cook, A.J. and Webb, J.K., (1994) : MN-RAS, **268**, L1

David, Y. and Reeves, H., (1980) : Phil. Trans. R. Soc. London, **A 296**, 415

Geiss, J., (1993) : in Origin and Evolution of the Elements, ed. N. Prantzos, E. Vangioni–Flam and M. Cassé (Cambridge Univ. Press), pp.89

Hoyle, F., Burbidge, G. and Narlikar, J.V., (1993) : ApJ., **410**, 437

Kerman, P.J. and Krauss, L.M., (1994) : Phys. Rev. Letters, **72**, 3309

Kurki–Suonio, R., Matzner, K., Olive, K.A., Schramm, D.N. (1990) : ApJ, **353**, 406

Lemoine, M., Vidal–Madjar, A., Ferlet, R., Bertin, P., Gry, C. and Lallement, R., (1994) : Astron. Astroph. in press

Linsky, J.L., Brown, A., Gayley, K., Diplas, A., Savage, B.D., Ayres, T.R ., Landsman, W., Shore, S.W. and Heap, S.R., (1993) : ApJ., **402**, 694

Olive, K.A., Thomas, D., Schramm, D.N., Walker, T.P., (1991) : Phys. Lett. B, **265**, 239

Pagel, B.E.J., Simonson, E.A., Terlevich, R.J., Edmunds, M.G., (1992) : MNRAS, **255**, 325

Peimbert, M., Torres–Peimbert, S., Ruiz, M.T., (1992) : Rev. Mex. Astron. Astrof., **25**, 155

Reeves, H., Audouze, J., Fowler, W.A. and Schramm, D.N., (1973) : ApJ. **179**, 909

Reeves, H., Delbourgo–Salvador, P., Salati, P. and Audouze, J., (1988) : Eur. J. Phys. **9**, 179

Salati, P., Delbourgo–Salvador, P. and Audouze, J., (1987) : Astron. Astroph. **173**, 1

Smith, M.S., Kawano, L.H. and Malaney, R.A., (1993) : ApJ. Suppl. **85**, 219

Songaila, A., Cowie, L.L., Hogan, C. and Rugers, M., (1994) : Nature, **368**, 599

Spite, F., and Spite, M., (1993) : in Origin and Evolution of the Elements, eds. N. Prantzos, E. Vangioni–Flam and M. Cassé (Cambridge Univ. Press) pp. 201

Steigman, G. and Tosi, M., (1992) : ApJ., **401**, 150

Truran, J.W. and Cameron, A.G.U., (1971) : Astroph. Space Sci., **14**, 179

Turck–Chièze, S., Lopes, I., (1993) : ApJ., **408**, 347

Vangioni–Flam, E. and Audouze, J., (1988) : Astron. Astroph. **193**, 81

Vangioni–Flam, E. and Cassé, M., (1995) : ApJ., in press

Vangioni–Flam, E., Olive, K.A. and Prantzos, N., 1994, ApJ., **427**, 618

Walker, T.P., Steigman, G., Schramm, D.N., Olive, K.A. and Kang, H.S., (1991) : ApJ., **376**, 51

Wilson, T.L., Rood, R.T., (1994) : Ann. Rev. Astron. Astroph. in press.

Big Bang Nucleosynthesis: Consistency or Crisis?

Gary Steigman[1]

[1] Departments of Physics and Astronomy, The Ohio State University, Columbus, OH 43210, USA

Abstract. The early hot, dense, expanding Universe was a primordial reactor in which the light nuclides D, ^{3}He, ^{4}He and ^{7}Li were synthesized in astrophysically interesting abundances. The challenge to the standard hot big bang model (Big Bang Nucleosynthesis $\equiv$ BBN) is the comparison between the observed and predicted abundances, the latter which depend only on the universal abundance of nucleons. The current status of observations is reviewed and the inferred primordial abundances are used to confront BBN. This comparison suggests consistency for BBN for a narrow range in the nucleon abundance but, looming on the horizon are some potential crises which will be outlined.

1 Introduction

Observations of an expanding Universe filled with black body radiation lead naturally to the inference that the early Universe was dense and hot and evolved through an epoch in which the entire Universe was a Primordial Nuclear Reactor. During the first $\sim$ thousand seconds the light elements D, ^{3}He, ^{4}He and ^{7}Li are synthesized in measurable abundances which range from $\sim 10^{-10}$ (for Li/H) to $\sim 10^{-5}$ (for D/H and ^{3}He/H) to $\sim 10^{-1}$ (for ^{4}He/H) (for a review and references see Boesgaard & Steigman 1985; for more recent results and references see Walker et al. 1991 (WSSOK)). The predicted abundances depend on the nucleon density, conveniently measured by the "nucleon abundance", the nucleon-to-photon ratio $\eta \equiv n_N/n_\gamma$ ($\eta_{10} \equiv 10^{10}\eta$; for $T_\gamma = 2.726$K, $n_\gamma = 411$ cm^{-3}). Thus, BBN provides a test of the consistency of the hot big bang model and a probe of cosmology (e.g., of the universal density of nucleons). Specifically, is there a value (or a range of values) of η such that all the predicted abundances are consistent with the inferred primordial abundances derived from the observational data? Further, if there is consistency, is the inferred nucleon density (based on processes which occurred during the first $\sim 10^3$ sec. of the Universe) consistent with that observed at present (when the Universe is $\sim$10 Gyr old)?

According to WSSOK, both questions are answered in the affirmative with $2.8 \lesssim \eta_{10} \lesssim 4.0$. The nucleon density parameter ($\Omega_N \equiv \rho_N/\rho_{CRIT}$) is related to the nucleon abundance and the Hubble parameter ($h_{50} \equiv H_0/50 kms^{-1}Mpc^{-1}$) by,

$$\Omega_N h_{50}^2 = 0.015\, \eta_{10}, \tag{1}$$

(for $T_\gamma = 2.726 \pm 0.010(2\sigma)$, the coefficient in (1) varies from 0.0145 to 0.0148). Thus, for $2.8 \lesssim \eta_{10} \lesssim 4.0$, $0.04 \lesssim \Omega_N h_{50}^2 \lesssim 0.06$, which leads to the conclusion that there are dark baryons ($\Omega_N > \Omega_{LUM}$) but, not all dark matter is baryonic ($\Omega_N < \Omega_{DYN}$).

The physics of BBN is, by now, well understood; for overviews see Boesgaard & Steigman (1985) and Smith, Kawano & Malaney (1993). It is, however, worth emphasizing that Primordial Alchemy is conventional physics. For example, the timescales are long ($\sim 10^{-1} - 10^3$ sec.) and the temperatures (thermal energies) are low (kT$\sim$ 10 keV - 1 MeV) . Although the early Universe is dense, it is dilute on the scale of nuclear physics during the epoch of BBN. For example, for T $\lesssim 1/2$ MeV, the internucleon separation is $\gtrsim 10^6$ fermis. Thus, collective and/or many body effects are entirely negligible. The nucleon reaction network is very limited (effectively, $A \leq 7$) and simple. More importantly, the cross sections are measured at lab energies comparable to the thermal energies during BBN. Thus, in stark contrast to stellar nucleosynthesis (where $kT_\star \ll E_{lab}$), large and uncertain extrapolations are not required. Thus, for fixed η, the BBN predicted abundances of D, ^{3}He and ^{7}Li are known to better than ~ 20 % and the ^{4}He mass fraction (Y_{BBN}) is known to $|\delta Y_{BBN}| \lesssim 6 \times 10^4$ (Thomas et al. 1994).

Since the BBN abundances of D, ^{3}He and ^{7}Li vary noticeably with η, those nuclides serve as "baryometers", leading to constraining lower and upper bounds to η (e.g., WSSOK). Y_{BBN} varies little ($\sim$ logarithmically) with η and, thus, serves as the key to testing the consistency of BBN. In the next sections we first survey the observational data on D, ^{3}He and ^{7}Li and derive bounds to their primordial abundances. Next, the predicted and inferred primordial abundances are compared to test for consistency and to bound η. Then, the ^{4}He abundance is studied for consistency – or crisis. Finally, the health of BBN is assessed and possible crises are outlined.

2 Deuterium

BBN is the only source of astrophysical deuterium. Whenever cycled through stars, D is destroyed (burned to ^{3}He, even during pre-main sequence evolution). Thus, the mass fraction (X_2) of primordial deuterium is no smaller than that observed anywhere in the Universe: $X_{2P} \geq X_{2OBS}$.

As with all of the light elements, there is both bad news and good news. The bad news is that, at least until recently (possibly!), deuterium has been observed only locally (in the interstellar medium (ISM) and the solar system). The good news is that the data is accurate.

Geiss (1993) has reanalyzed the solar system D and ^{3}He data. Using Geiss' results, Steigman & Tosi (1994) find

$$X_{2\odot} = 3.6 \pm 1.3 \times 10^{-5}. \qquad (2)$$

Using older Copernicus and IUE data, along with newer HST data (Linsky et al. 1993), Steigman & Tosi (1994) have noted that over a range of two orders of magnitude in HI column density, $(D/H)_{ISM}$ is constant at a value of $1.6 \pm 0.2 \times 10^{-5}$. To determine X_{2ISM} requires knowledge of the H mass fraction in the ISM, for $Y_{ISM} \approx 0.28 \pm 0.02$ and $Z_{ISM} \approx 0.02$, $X_{ISM} \approx 0.70 \pm 0.02$ and,

$$X_{2ISM} = 2.2 \pm 0.3 \times 10^{-5}. \tag{3}$$

It is expected that X_2 should have decreased in the 4.6 Gyr between the formation of the solar system and the present (although as Steigman and Tosi (1992) show, the decrease may be small). The data are marginally consistent with this expectation: $X_{2\odot}/X_{2ISM} = 1.6 \pm 0.6$.

A lower bound to X_{2OBS} leads to a lower bound to X_{2P} which, in turn, leads to an upper bound to η. For $X_{2P} \geq X_{2ISM} \geq 1.7 \times 10^{-5}(2\sigma)$,

$$D : \eta_{10} \leq 9.0 \tag{4}$$

3 Helium-3

When deuterium is cycled through stars it is burned to ^{3}He. ^{3}He burns at a higher temperature than D so that ^{3}He survives in the cooler, outer layers of stars. Furthermore, since hydrogen burning is incomplete in low mass stars, such stars are net sources of ^{3}He. Thus, any primordial ^{3}He is modified by the competition between stellar production and destruction and, therefore, a detailed evolution model – with its attendant uncertainties – is needed to relate the observed and BBN abundances (Steigman & Tosi 1992). However, since *all* stars do burn D to ^{3}He and, *some* ^{3}He does survive stellar processing, the primordial D + ^{3}He may be bounded by the observed D and ^{3}He (Yang et al. 1984 (YTSSO); Dearborn, Schramm & Steigman 1986). The YTSSO analysis, which has recently been updated (Steigman & Tosi 1994), is "generic" in the sense that it should be consistent with any specific model for Galactic chemical evolution. Its predictions do, however, depend on one model specific parameter g_3, the "effective" survival fraction of ^{3}He.

Since the deuterium observations have already been used to bound the primordial D mass fraction from below, here we are interested in using the solar system observations of D and ^{3}He to bound X_{2P} from above. If any net stellar production of ^{3}He is neglected (so that ^{3}He only increases by burning D and decreases by stellar destruction), it can be shown that (YTSSO; Steigman & Tosi 1994)

$$X_{2P} < X_{2P}^{MAX} = \left[1 - \frac{1}{g_3} \left(\frac{y_3}{y_{23}} \right)_P \right] X_{2\odot} + \frac{2/3}{g_3} \left(\frac{y_2}{y_{23}} \right)_P X_{3\odot}. \tag{5}$$

In (5), the primordial D and ^{3}He abundances (by number) are $y_{2P} = (D/H)_P$ and $y_{3P} = (^3He/H)_P$; $y_{23P} = y_{2P} + y_{3P}$; g_3 is the "effective" survival fraction of ^{3}He (which is model dependent). It can be seen from (5) that the higher/lower the primordial/solar system ^{3}He abundances, the more restrictive the upper bound on primordial deuterium.

Of course, since primordial abundances appear on both sides of eq. 5, care must be excersized in finding the bound. One approach is to evaluate both sides of (5) using the *predicted* abundances as a function of η, identifying those values of η for which the inequality is satisfied (Steigman & Tosi 1994). Alternatively, the inequality can be further relaxed by entirely neglecting *any* primordial ^{3}He. Since $y_{3P} > 0$, we may write,

$$X_{2P} < X_{2P}^{MAX} < \left(X_{2P}^{MAX}\right)^0 = X_{2\odot} + \left(\frac{2/3}{g_3}\right)X_{3\odot}. \tag{6}$$

The inequality in (6) may be further reinforced to relate y_{2P} to $y_{2\odot}$ and $y_{3\odot}$ since the hydrogen mass fraction always decreases from its primordial value $(X_{H\odot} < X_{HP})$,

$$y_{2P} < \left(y_{2P}^{MAX}\right)^0 < y_{2\odot} + g_3^{-1}y_{3\odot}. \tag{7}$$

Using the Geiss (1993) solar system abundances and $g_3 > 1/4$ (Dearborn, Schramm & Steigman 1986), Steigman & Tosi (1994) find $X_{2P} < 11 \times 10^{-5}$ $(y_{2P} < 7.4 \times 10^{-5})$ which leads to a *lower* bound to η,

$$D + {}^3\text{He} : \eta_{10} \gtrsim 3.1 \tag{8}$$

Note that if the more restrictive survival fraction $g_3 > 1/2$ (Steigman & Tosi 1992) is used, we would infer $X_{2P} < 7 \times 10^{-5}$ and $\eta_{10} \gtrsim 4$. It should also be noted that 2σ upper bounds to $X_{2\odot}$ and $X_{3\odot}$ are used in reaching these conclusions.

To summarize the progress so far, solar system and interstellar observations of D and ^{3}He have permitted us to bound primordial deuterium from above and below $(1.6 \lesssim 10^5 X_{2P} \lesssim 11)$ which leads to consistent upper and lower bounds on η $(3.1 \lesssim \eta_{10} \lesssim 9.0)$. Next, we turn to the first consistency test of BBN by considering lithium-7.

4   Lithium-7

As with the other light nuclides, the status of lithium observations has good news and bad news. The good news is that lithium is observed, with relatively good statistical accuracy, in dozens and dozens of stars of varying metallicity, mass (or temperature), evolutionary stage, population, etc. Among the bad news, these stars are all in the Galaxy and, therefore, provide a sample which is not necessarily universal. More serious, however, are the essential corrections which are required to go from the observed surface abundances to their unmodified (by stellar evolution) prestellar values and, to account for the production/destruction of lithium in the course of Galactic chemical evolution.

The overwhelming influence of stellar evolution on the stellar surface lithium abundance is reflected in the enormous range of observed values in Population I stars. The Sun is a case in point. Whereas the meteoritic abundance of lithium is $\sim 2 \times 10^{-9}$([Li] $\equiv 12 + \log(\text{Li/H}) = 3.31$), the solar photosphere abundance is smaller by some two orders of magnitude (Grevesse & Anders 1989). There is, however, evidence for a maximum PopI lithium abundance as inferred from observations of the warmest stars in young open clusters (Balachandran 1994), $[Li]_{PopI} = 3.2 \pm 0.2(2\sigma)$. And, further, there is evidence (e.g., Beckman, Robolo & Molaro 1986) that this maximum decreases with decreasing metalicity until, for $[Fe/H] \lesssim -1.3$, the "Spite Plateau" is reached.

The Spites' discovery (Spite & Spite 1982a,b), subsequently confirmed by many observations (e.g., see WSSOK for an overview and references and, see Thorburn 1994 for the latest observations), is that the warmest $(T \gtrsim 5700K)$, most metal-poor stars ($[Fe/H] \lesssim -1.3$) have, with remarkably few exceptions, the *same* lithium abundance: $[Li]_{PopII} \approx 2.1$ (WSSOK; the values from Thorburn (1994) are systematically higher by ~ 0.2 dex). The value of the Spites' discovery cannot be overestimated but, too, caution is advised. On the one hand, the "plateau" in Fe/H (or, where available, in oxygen abundance) suggests that $[Li]_{PopII}$ may provide an estimate of the primordial abundance free from a (significant) correction for Galactic chemical evolution. On the other hand, the temperature plateau suggests that, "what you see is what you get". That is, the surface abundances of lithium in the warmest PopII stars provide a fair sample of the lithium abundances in the gas out of which those stars formed. If, indeed, $[Li]_P \approx [Li]_{PopII} \approx 2.1 \pm 0.2$ (the uncertainty is mainly systematic, the statistical uncertainties are much smaller (WSSOK)), then BBN is constrained significantly; for $(Li/H)_{BBN} \lesssim 2 \times 10^{-10}$, $1.6 \lesssim \eta_{10} \lesssim 4.0$. However, analysis of Thorburn's (1994) extensive data set raises questions about the *flatness* of the lithium temperature/metallicity plateaus.

Furthermore, it is not clear that corrections for chemical evolution are entirely negligible, even for the very old, very metal-poor PopII stars. Lithium-7 (as well as ^6Li) may be produced by $\alpha-\alpha$ fusion reactions in Cosmic Ray Nucleosynthesis (CRN; Steigman & Walker 1992) as well as by the more familiar spallation reactions of p and α on CNO nuclei. Since the spallation reactions require CNO targets (and/or projectiles) whereas the fusion reactions can utilize primordial ^4He, CRN lithium production has a component which is shallower in its metallicity dependence than that of Be and/or B which are only synthesized in spallation reactions. Thus, if $(Be/H)_{PopII} \sim (Fe/H)^\alpha$, $\Delta(^7Li/H)_{\alpha\alpha} \sim (Fe/H)^{\alpha-1}$ and, since current data (Gilmore et al. 1992; Boesgaard & King 1993) suggests $\alpha \approx 1$, $(\Delta y_7)_{\alpha\alpha}$ should be nearly independent of metallicity and, so, will mimic a primordial component ($y_7 \equiv {}^7Li/H$). Thus, even neglecting any early (PopII) stellar production/destruction of ^7Li, the BBN and observed PopII lithium abundances are, in general related by,

$$y_{7OBS} = f_7(y_{7BBN} + (\Delta y_7)_{CRN}), \tag{9}$$

where $f_7(\leq 1)$ is the stellar surface destruction/dilution factor for ^7Li. Although "standard" (i.e., nonrotating) models for the warmest PopII stars suggest $f_7 \approx 1$ (Chaboyer et al. 1992), models with rotation may permit a significant correction ($f_7 \gtrsim 0.1 - 0.2$; Pinsonneault, Deliyannis & Demarque 1992; Charbonel & Vauclair 1992). The observations of the much more fragile ^6Li in two PopII stars (Smith, Lambert & Nissen 1992; Hobbs & Thorburn 1994) suggests that $f_7 \approx 1$ but this important issue remains unresolved at present. Thus, although the PopII stellar data appears consistent with $[\text{Li}]_{BBN} \lesssim 2.3$, it is unclear that the much higher bound $[\text{Li}]_{BBN} \lesssim 3.0$ (Pinsonneault, Deliyannis & Demarque 1992) can be entirely excluded.

Fortunately, another – independent – path to primordial lithium exists. Lithium has been observed in the ISM of the Galaxy (Hobbs 1984; White 1986) and, searched for in the ISM of the LMC (in front of SN87A; Baade et al. 1991). The interstellar data has assets and liabilities of its own which, however, are *different* from those of the stellar data. Among the liabilities is a large and uncertain ionization correction since LiI is observed but most ISM ^{7}Li is LiII. Another problem is the correction for lithium removed from the gas phase of the ISM (where it is observed) by grains and/or molecules (where it is unobserved). Steigman (1994a) has proposed avoiding these obstacles by comparing lithium to potassium (which shares the ionization/depletion problems with lithium) and evaluating the *relative* abundances (Li/K rather than Li/H). Comparing Galactic ($[\text{Fe/H}] \approx 0$) Li/K with the absence of Li and the presence of K in the LMC ($[\text{Fe/H}]_{LMC} \approx -0.3$), Steigman (1994a) has concluded that $(\text{Li/K})_{LMC} \lesssim 1/2(\text{Li/K})_{GAL}$. Since potassium has no primordial component, this bound can be used to derive an upper bound to primordial lithium (Steigman 1994a): $[\text{Li}]_P \lesssim 2.3 - 2.8$. Thus, although it appears that the Spite Plateau bound $[\text{Li}]_{BBN} \lesssim 2.3$ is supported, a higher value cannot be entirely excluded. Here, in the absence of evidence to the contrary, I will use the above bound $((\text{Li/H})_{BBN} \lesssim 2 \times 10^{-10})$ to constrain η,

$$^7\text{Li} : 1.6 \lesssim \eta_{10} \lesssim 4.0. \tag{10}$$

5   Consistency Among D, ^{3}He & ^{7}Li?

Before moving on to the keystone of BBN, helium-4, it is useful to pause at this point to consolidate the progress thus far. Solar system and interstellar observations of D and ^{3}He have been employed to set lower and upper bounds to primordial deuterium ($1.6 \lesssim 10^5 X_{2P} \lesssim 11$) which result in bounds on the nucleon abundance ($3.1 \lesssim \eta_{10} \lesssim 9.0$). PopII and ISM observations of lithium are consistent with an upper bound on primordial lithium which may be as small as $(\text{Li/H})_P \lesssim 2 \times 10^{-10}$ but, which could also be consistent with a larger value $(\text{Li/H})_P \lesssim 6 - 8 \times 10^{-10}$. Utilizing the more restrictive lithium bound, consistency among the BBN predicted abundances is achieved provided that η is restricted to a relatively narrow range,

$$D, \,^3\text{He}, \,^7\text{Li} : 3.1 \lesssim \eta_{10} \lesssim 4.0. \tag{11}$$

From (1) it follows that the present density in nucleons is similarly restricted,

$$0.045 \lesssim \Omega_N h_{50}^2 \lesssim 0.059 \tag{12}$$

which, for $40 \leq H_0 \leq 100 kms^{-1}Mpc^{-1}$, corresponds to, $0.011 \lesssim \Omega_N \lesssim 0.093$. The lower bound $\Omega_N \gtrsim 0.01$ exceeds the estimate of the mass associated with "luminous" matter, suggesting the presence of Baryonic Dark Matter, while the restrictive upper bound $\Omega_N \lesssim 0.09$ is strong evidence for the existence of Non-Baryonic Dark Matter.

6 Helium-4

The good news about ^{4}He is that it is ubiquitous and can be seen everywhere in the Universe. And, since its abundance is large, its value can be determined with high statistical accuracy. The bad news is that the path from observations to abundances to primordial helium is strewn with corrections which are accompanied by potentially large systematic uncertainties.

As stars burn, hydrogen is consumed producing ^{4}He which is returned to the galactic pool out of which subsequent generations of stars form. Thus, any observed abundances must be corrected for the ^{4}He enhancement from the debris of earlier generations of stars. To minimize this correction and its attendant uncertainties, the most valuable observational data is that from the low metallicity, extragalactic HII regions (e.g., Pagel et al. 1992). It is the emission lines from the recombination of ^{4}He$^+$ and ^{4}He^{++} (as well as H$^+$) which are observed from these regions. Since neutral helium (in the zone of ionized hydrogen) is unobserved, its correction – which carries with it systematic uncertainties – is minimized by restricting attention to the hottest, highest excitation regions (Pagel et al. 1992) where the correction may be negligible (or, even, negative in the sense that HII regions ionized by very hot – metal-poor – stars may have HeII zones larger in extent than the HII zones). Finally, to benefit from the high statistical accuracy of the observational data, corrections for collisional excitation, radiation trapping and destruction by dust, etc. must be considered.

The best (i.e., most coherent) data set of Pagel et al. (1992) has recently been supplemented (Skillman et al. 1993) by the addition of $\sim$ a dozen very low metallicity HII regions. Olive and Steigman (1994) have analyzed this data; there are some four dozen HII regions whose oxygen abundances extend down to $\sim 1/50$ solar and whose nitrogen abundances go down to $\sim 1\%$ of solar. For this data Olive and Steigman (1994) find that an extrapolation to zero metallicity yields,

$$Y_P = 0.232 \pm 0.003, \tag{13}$$

where the uncertainty is a 1σ *statistical* uncertainty. Thus, at 2σ, $Y_{BBN} \lesssim 0.238$. It is difficult to estimate the possible *systematic* uncertainty; Pagel (1993), WSSOK, and Olive & Steigman (1994) suggest ± 0.005 (i.e., $\sim 2\%$). If so, the upper bound may be relaxed to $Y_{BBN} \lesssim 0.243$ which, as will be seen shortly, may be crucial.

The BBN predicted ^{4}He mass fraction is known to high accuracy (as a function of η). For the standard case of three light neutrinos ($N_\nu = 3$) and a neutron lifetime in the range $\tau_n = 889 \pm 4(2\sigma)$ sec, the bounds from observation

$Y_{BBN} \leq 0.238(0.243)$ require $\eta_{10} \leq 2.5(3.9)$. Here, we have the first serious crisis confronting BBN! Unless systematic corrections increase the primordial abundance of helium inferred from the observational data, the upper bound on η from ^{4}He is exceeded by the lower bound on η from D (and ^{3}He). With, however, allowance for a possible $\sim 2\%$ uncertainty, consistency is maintained. Thus, for D, ^{3}He, ^{7}Li and $Y_{BBN} \leq 0.243$,

$$3.1 \leq \eta_{10} \leq 3.9. \tag{14}$$

Of course, the upper bound to η from ^{4}He will reflect the uncertainty in the obserational bound to Y_P. For $\eta_{10} \sim 4$, $\Delta Y_{BBN} \approx 0.012(\Delta\eta/\eta)$ so that an uncertainty of 0.003 in Y corresponds to a 25% uncertainty in $\eta(\Delta\eta_{10} \approx \pm 1)$.

The importance of ^{4}He is that the predicted primordial abundance is robust – relatively insensitive to η and, as a function of η, accurately calculated (to better than ± 0.001). And, being abundant, ^{4}He is observable throughout the Universe and, systematic uncertainties aside, the derived abundance is known to high statistical accuracy ($\lesssim \pm 0.003$). Thus, ^{4}He is the keystone to testing the consistency of BBN.

7 A Helium-4 Crisis?

Solar system data on D and ^{3}He, along with a "generic" model for galactic evolution (Steigman & Tosi 1994) leads to a *lower* bound to η ($\eta_{10} \gtrsim 3.1$) and, therefore, to a *lower* bound to the predicted BBN abundance of ^{4}He; for $N_\nu = 3$, $\tau_N \geq 885$ sec and $\eta_{10} \geq 3.1$, $Y_{BBN} \geq 0.241$. In contrast, accounting only for statistical uncertainties, $Y_P \leq 0.238$ (at 2σ; Olive & Steigman 1994). Thus, the issue of whether or not this is a crisis for BBN hinges on whether or not Y_P is known to three significant figures. Allowance for a possible, modest ($\sim 2\%$), systematic uncertainty of order 0.005 would transform this potential crisis to consistency.

8 A Deuterium Crisis?

Recently, two groups have independently reported the possible detection of extragalactic deuterium in the spectrum of a high z (redshift), low Z (metallicity) QSO absorption system (Songaila et al. 1994; Carswell et al. 1994). If, indeed, the absorption is due to deuterium, the inferred abundance is surprisingly high: $D/H \approx 19 - 25 \times 10^{-5}$. This high abundance – an order of magnitude larger than the pre- solar or ISM values – poses no problem for cosmology in the sense that for $(D/H)_{BBN} \sim 2\times 10^{-4}$, $\eta_{10} \sim 1.5$ and $Y_{BBN} \sim 0.23$ and $(^7\text{Li/H})_{BBN} \sim 2 \times 10^{-10}$, which are in excellent agreement with the observational data. If, indeed, $\eta_{10} \sim 1.5$, then $\Omega_N h_{50}^2 \sim 0.022$, reinforcing the argument for non-baryonic dark matter (for $H_0 \gtrsim 40 km s^{-1} Mpc^{-1}$, $\Omega_N \lesssim 0.034$)

But, such a high primordial abundance does pose a serious challenge to our understanding of the stellar and galactic evolution of helium-3. The issue is that if $\sim 90\%$ of primordial deuterium has been destroyed prior to the formation of the solar system, then the solar nebula abundance of ^{3}He should be much larger than observed (Steigman 1994b) since D burns to ^{3}He and some ^{3}He survives. Earlier, we have used the solar system data to infer a primordial bound $y_{2P} < 7.4 \times 10^{-5}$ (for $g_3 \geq 1/4$). A primordial abundance as large as $\sim 2 \times 10^{-4}$ would require much more efficient stellar destruction of ^{3}He ($g_3 \lesssim 0.09$).

It is, however, possible that the observed absorption feature is not due to high z, low Z deuterium at all but, rather, to a hydrogen interloper (Steigman 1994b). That is, the absorption may be from a very small cloud of neutral hydrogen whose velocity is shifted from that of the main absorber by just the "right" amount so that it mimics deuterium absorption. As Carswell et al. (1994) note, the probability for such an accidental coincidence is not negligible ($\sim 15\%$). This possibility can only be resolved statistically when there are other candidate D-absorbers. Data from Keck and the HST is eagerly anticipated.

9   The X-Ray Cluster Crisis?

This overview concludes with a glimpse of yet another potential crisis for BBN. Large clusters of galaxies are expected to provide a "fair sample" of the Universe in the sense that, up to factors not much different from unity, the baryon fraction in clusters should be the same as the universal baryon fraction.

$$f_B = \Omega_B/\Omega \approx (M_B/M_{TOT})_{Clusters} \qquad (15)$$

In (15) the baryon density parameter Ω_B is another name for what we have been calling the nucleon density parameter Ω_N and Ω is the ratio of the total density to the critical density. For x-ray clusters M_B is dominated by the mass in hot, x-ray emitting, intercluster gas ($M_B \approx M_{HG} + M_{GAL} \gtrsim M_{HG}$) so that,

$$\Omega \lesssim \Omega_{BBN}/f_{HG}, \qquad (16)$$

where $\Omega_{BBN} = 0.015\eta_{10}h_{50}^{-2}$ and f_{HG} is the hot gas fraction in x-ray clusters. Since M_{HG} and M_{TOT} scale differently with the distance to the cluster, f_{HG} depends on the choice of Hubble parameter (e.g., Steigman 1985): $f_{HG} = A_{50}h_{50}^{-3/2}$. Thus, (16) may be written as,

$$\Omega h_{50}^{1/2} \lesssim 0.6 \left(\frac{0.10}{A_{50}} \right) \left(\frac{\eta_{10}}{4.0} \right), \qquad (17)$$

where x-ray observations yield A_{50}.

The x-ray cluster crisis was perhaps first noted by White et al. (1993) for Coma where: $A_{50}(Coma) = 0.14 \pm 0.04$. For $A_{50} \gtrsim 0.10$ and $\eta_{10} \lesssim 4.0$, $\Omega = 1$ requires $H_0 \lesssim 18 kms^{-1}Mpc^{-1}$! Thus, either $\Omega < 1$ or the BBN upper bound to η is wrong. Further x-ray data, however, makes this latter choice less likely. White et al. (1994) find for Abell 478, $A_{50}(A478) = 0.28 \pm 0.01$, a result supported by White & Fabian's (1994) survey of 19 x-ray clusters which finds, at an Abell radius of $\sim 3 h_{50}^{-1} Mpc$, $A_{50} \approx 0.24$. For $A_{50} \approx 0.24$, $\Omega h_{50}^{1/2} \lesssim \eta_{10}/16$, strongly hinting at $\Omega < 1$. For $\Omega = 1$ and any sensible choice of H_0, η_{10} would have to be so large as to violate – separately – the observational bounds on D, ^{4}He, and ^{7}Li. The x-ray cluster crisis – if real – is a crisis for $\Omega = 1$ but, not for BBN.

BBN is alive and well and the healthy confrontation of theory with observation continues.

References

Baade, D., Cristiani, S., Lanz, T., Malaney, R.A., Sahu, K. C. & Vladilo, G. 1991, *A & A*, **251**, 253

Balachandran, S. 1994, Preprint (Submitted to *ApJ*)

Beckman, J. E., Rebolo, R. & Molaro, P. 1986, "Advances in Nuclear Astrophysics" (eds. E. Vangioni-Flam, J. Audouze, M. Cassé, J. P. Chièze & J. TranThanh Van; Editions Frontiéres) p. 29

Boesgaard, A. M. & Steigman, G. 1985, *Ann. Rev. Astron. Astrophys.* **23**, 319

Boesgaard, A. M. & King, J. R. 1993, *AJ* **106**, 2309

Carswell, R. F., Rauch, M., Weymann, R. J., Cooke, A. J. & Webb, J. K. 1994, *MNRAS* **268**, L1

Chaboyer, B., Deliyannis, C. P., Demarque, P., Pinsonneault, M. H. & Sarajedini, A. 1992, *ApJ*, **388**, 372

Charbonnel, C. & Vauclair, S. 1992, *A & A*, **265**, 55

Geiss, J. 1993 "Origin and Evolution of the Elements" (eds. N. Prantzos, E. Vangioni-Flam & M. Cassé; Cambridge Univ. Press) p. 89

Gilmore, G., Gustafsson, B., Edvardsson, B. & Nissen, P. E. 1992, *Nature*, **357**, 379

Grevesse, N. & Anders, E. 1989, in *ATP Conf. Proc. 183*, "Cosmic Abundances of Matter", (ed. C. J. Waddington; AIP), p.1

Hobbs, L. M. 1984, *ApJ* **286**, 252

Hobbs, L. M. & Thorburn, J. A. 1994, *ApJ* **428**, L25

Linsky, J. L., Brown, A., Gayley, K., Diplas, A., Savage, B. D., Ayres, T. R., Landsman, W., Shore, S. W. & Heap, S. 1993, *ApJ* **402**, 694

Olive, K. A. & Steigman, G. 1994, *ApJS* In Press

Pagel, B. E. J., Simonson, E. A., Terlevich, R. J. & Edmunds, M. G. 1992, *MNRAS* **255**, 325

Pagel, B. E. J. 1993, *Proc. Nat. Acad. Sci.* **90**, 4789

Pinsonneault, M. H., Deliyannis, C. P. & Demarque, P. 1992, *ApJS* **78**, 179

Skillman, E. D., Terlevich, R. J., Terlevich, E., Kennicutt, R. C. & Garnett, D. R. 1993, *Ann. N. Y. Acad. Sci.* **688**, 739

Smith, M. S., Kawano, L. H. & Malaney, R. A. 1993, *ApJS* **85**, 219

Songaila, A., Cowie, L. L., Hogan, C. & Rugers, M. 1994, *Nature* **368**, 599

Spite, M. & Spite, F. 1982a, *Nature* **297**, 483

Spite, F. & Spite, M. 1982b, *A & A*, **115**, 357

Steigman, G. 1985, "Theory and Observational Limits in Cosmology" (ed. W. R. Stoeger; Specola Vaticana), p. 149

Steigman, G. 1994a, "Cosmic Lithium: Going Up Or Coming Down?" (Preprint) OSU-TA-18/94

Steigman, G. 1994b, *MNRAS*, **269**, L53

Steigman, G. & Tosi, M. 1992, *ApJ* **401**, 150

Steigman, G. & Tosi, M. 1994, "Generic Evolution of Deuterium and Helium-3" (Preprint) OSU-TA-12/94

Steigman, G. & Walker, T. P. 1992, *ApJ* **385**, L13

Thomas, D., Hata, N., Scherrer, R., Steigman, G. & Walker, T. 1994, In Preparation

Thorburn, J. A. 1994, *ApJ* **421**, 318

Walker, T. P., Steigman, G., Schramm, D. N., Olive, K. A. & Kang, H.-S, 1991, *ApJ* **376**, 51 (WSSOK)

White, R. E. 1986, *ApJ* **307**, 777

White, D. A., Fabian, A. C., Allen, S. W., Edge, A. C., Crawford, C. S., Johnstone, R. M., Stewart, G. C., & Voges, W. 1994, *MNRAS* **269**, 598

White, D. A. & Fabian, A. 1994 Preprint (Submitted to MNRAS)

White, S. D. M., Navarro, J. F., Evrard, A. E. & Frenk, C. S. 1993, *Nature* **366**, 429

Yang, J., Turner, M. S., Steigman, G., Schramm, D. N. & Olive, K. A. 1984, *ApJ* **281**, 493 (YTSSO)

Light Nuclei
in the Quasi-Steady State Cosmological Model

F. Hoyle[1], G. Burbidge[2] and J.V. Narlikar[3]

[1] 102 Admirals Walk, Bournemouth BH2 5HF Dorset, England
[2] Center for Astrophysics and Space Sciences and Department of Physics University of California, San Diego, La Jolla, California 92093-0111, U.S.A.
[3] Inter-University Centre for Astronomy and Astrophysics, Post Bag 4, Ganeshkind, Pune 411 007 India

Abstract. A model for the decay of Planck particles is specified and the light elements synthesis resulting from it is described. The calculated values of $^4He/H$, $^7Li/H$, and $^{12}C/H$ are in close agreement with observations while those of D/H and $^3He/H$ are in agreement to within a factor of about 2. The model predicts a plateau under ^9Be but seemingly not under ^{11}B. The plateau under ^9Be corresponds to a freezing temperature $T_9 = 0.5$ whereas the calculated freezing temperature is $T_9 \simeq 0.62$.

1 Introduction

A series of papers (Hoyle, Burbidge and Narlikar, 1993, 1994a,b and 1995) have developed a cosmological model based on our belief that the gravitational theory must be scale invariant, along with the rest of physics. We have taken the view that the widely popular Big-Bang cosmology is logically flawed to an extent that we consider fatal. The gravitational theory on which Big-Bang cosmology is based, general relativity, assumes the world lines of particles considered classically to be unbounded. Then Big-Bang cosmology deduces the opposite, namely that world lines are bounded in the past at the Big-Bang. Our work referred to above is based on two requirements, one of resolving this contradiction and the other of constructing a scale invariant theory of gravitation.

In our theory particles at the past bounds of their world lines must be Planck particles, which subsequently decay into showers containing large numbers of familiar particles. The inverse process in which familiar particles come together, some 10^{19} of them into a region of dimension $\sim 10^{-33}\,cm$ can in principle lead to the termination of world lines in the future, but since this is a highly improbably configuration, it can be considered not to happen, thereby giving a one-sided time sense to the universe. Classical particles begin but they do not end.

A Planck particle is defined as one whose gravitational radius is comparable to its Compton wavelength, which requires in units with $c = \hbar = 1$, that the mass of the Planck particle is of the order of $G^{-1/2}$, G being the gravitational constant. When interactions other than gravitation are included Planck particles decay, ultimately into $\sim 10^{19}$ particles of familiar kinds. At least they do in our theory in which particles can be endowed at birth with properties defining the manner of this decay. Such particles expanding at a speed of order, but less than, c will be referred to as a Planck fireball.

The question to be addressed in this contribution is the synthesis of light elements in such fireballs.

2 The Model

Because the early stages in the development of a Planck fireball belong to the realm of unknown physics, it is necessary to begin with a specification of initial conditions. Fermions of familiar types are necessarily excluded by degeneracy conditions at early stages when the fireball dimension is only $\sim 10^{-33}\,cm$. Indeed, fermions of familiar types cannot appear until the interparticle spacing within the expanding fireball has increased to $\sim 10^{-13}\,cm$.

We take the view in specifying the model to be investigated that energy considerations discriminate against charmed, bottom and top quarks. We also take the view that degeneracy considerations, together with the need for electrical neutrality, prevent the strange quark from being discriminated against. When the up, down and strange quarks combine to baryons, equal numbers of N, P, Λ, $\Sigma^{\pm}$, Σ^0, Ξ^0, Ξ^- are thus formed, with only a negligible amount of Ω^-. Because of the long lifetimes, $\sim 10^{-10}$ seconds, of Λ, $\Sigma^{\pm}$, Ξ^0, Ξ^-, the strange quark survives the effective stages in the expansion of the fireballs, although Σ^0 goes to Λ plus a γ-ray at a stage proceeding the synthesis of the light elements. Finally, we consider that baryons containing the strange quark do not form stable nuclei. Ultimately they decay into N and P, but only after the particle density has fallen so far that the production of light elements has stopped. With N going on a much longer time scale (10 minutes) into P, six of the baryons of the octet go at last into hydrogen. Thus we see immediately that the fraction by mass of helium, Y, to emerge from Planck fireballs is given by

$$Y = 0.25(1 - y), \tag{1}$$

where $1 - y$ is the fraction of the original N and P to go to ^4He. Anticipating that y will be shown in the next section to be ~ 0.085, equation (1) gives $Y = 0.229$ somewhat lower than the value of ~ 0.237 obtained previously (Hoyle, 1992).

The numerical values used in the detailed calculations of later sections are given in the Table 1.

Table 1. Densities and Temperatures at $1 < r < 4$ in the expansion of a Planck Fireball

Here N is the number per cm^3 of each baryon type, the values in the table being such that N declines with increasing r as r^{-3}. The unit of r depends on a specification of the total number of baryons in the fireball. Thus for a total of 5.10^{18} the unit of r is $5.10^{-7}cm$. However, since this total is uncertain, because the Planck mass, usually given as $(3\hbar c/4\pi G)^{1/2}$, is theoretically uncertain to within factors such as 4π, we prefer to leave the unit of r unspecified – we shall not need it in the calculations. Suffice it that there will always be a unit for r such that N has the values in the table.

Taking the expansion of the fireball to occur at a uniform speed v, the time t of the expansion to radius r is proportional to r, $t \propto r$. In specifying the model we take the factor of proportionality here to be 10^{-16} seconds. With the unit of r chosen as $5.10^{-7}cm$ this requires $v = 5.10^9 cm\ s^{-1}$, a rather low speed. But for a Planck mass increased by 4π above $(3\hbar c/4\pi G)^{1/2}$ the expansion speed is raised by $(4\pi)^{1/3}$ to $1.16 \times 10^{10} cm\ s^{-1}$. Thus

$$t = 10^{-16}r \text{ seconds}, \tag{2}$$

thereby relating t to N and T_9 through the values in Table 1. The numerical coefficient of equation (2) can be regarded as a parameter of the theory, but it is not a parameter that can be varied by more than a small factor, unlike the parameter η in Big-Bang nucleosynthesis which could be varied by many orders of magnitude for all one knows from the theory.

The temperature values in Table 1 are calculated from a heating source which comes into play at $r = 1$, i.e. at $t = 10^{-16}s$. The heating source is from the decay of π^0 mesons with a mean life of $8.4 \times 10^{-17}s$. The temperature values in Table 1 correspond to a π^0 meson concentration of $2/3N\ cm^{-3}$, which is to say one π meson to each neutron and each proton, with π^0, $\pi^\pm$ in equal numbers.

The decay of a π^0 meson into two 75 Mev γ-rays does not immediately deposit energy into the temperature T_9 of the heavy particles. It does not even lead to more than a limited production of $e^\pm$ pairs, because at these densities this is prevented by electron degeneracy. Thus the energy of π^0 decay is at first stored in the form of relativistic particles, quanta and some $e^\pm$, the latter being adequate, however, to prevent the γ-rays from escaping out of the fireball.

As the fireball expands, confined relativistic particles lose energy proportional to $1/r$, the energy loss going to the heavy particles, for each type of which there is a conservation equation of the form $dQ = dE + PdV$, viz

$$-\alpha d(1/r) = 3/2kdT + 3kTdV/V, \tag{3}$$

an equation that integrates to give

$$T_9 = \frac{2}{3}\frac{\alpha}{k}\frac{r-1}{r^2}, \tag{4}$$

the constant of integration being chosen to given $T_9 = 0$ at $r = 1$. The constant α in (3) and (4) is easily determined from the energy yield of the π^0 mesons. Sharing the energy communicated to the heavy particles equally among all of them, leads to the values of T_9 in Table 1.

The energy is considered to have all gone to the heavy particles by the stage of the expansion when r reaches 4, after which T_9 declines as r^{-2}, i.e. adiabaticlly, the heavy particles being non-relativistic in their thermal motions. Thus for $r > 4$ we have

$$T_9 = 16.3. \left(\frac{4}{r}\right)^2 = \frac{260.8}{r^2}, \tag{5}$$

$$t = 10^{-16} r = \frac{1.62 \times 10^{-15}}{T_9^{1/2}} \text{seconds}, \tag{6}$$

while the particle densities decline as r^{-3}.

3   The Abundance of ^4He

It will be shown in this section that neutrons and protons are in statistical equilibrium with ^4He up to $r = 3$ in Table 1, but not for $r > 3$. Defining a parameter ζ by

$$\log \zeta = \log N - 34.07 - \frac{3}{2} \log T_9 \tag{7}$$

it was shown by Hoyle (1992) that the fraction y of neutrons and protons remaining free at temperature T_9 and particle density N for each nucleon type is given in statistical equilibrium by

$$\log \frac{1-y}{y^4} = 0.90 + 3 \log \zeta + \frac{142.6}{T_9}, \tag{8}$$

the values of T_9 and N in Table 1 at $r = 3$ giving $y = 0.085$, leading to the value $Y = 0.229$ given above. A similar calculation at $r = 2.5$ yields $y = 0.083$, much the same as at $r = 3$. For $r < 2.5$ the values of y fall away to ~ 0.06. Thus in moving to the right in the table the values of y increase towards $r = 3$, where the falling value of T_9 eventually freezes the equilibrium.

The condition for freezing is that the break-up of ^4He by $^4He(2N, T)T$, followed by the break-up of T and D into neutrons and protons should just be capable of supplying the densities of P and N, $n(P) = n(N) \simeq 5.10^{33} cm^{-3}$ for the range of r from 2.5 to 3 and $y \simeq 0.085$. The time available for this break-up of ^4He is that for r to increase from ~ 2.5 to ~ 3, i.e. 5.10^{-17} seconds. In this time the break-up of $n(A) \simeq 2.9 \times 10^{34} cm^{-3}$ using the reaction rates of Fowler, Coughlan and Zimmerman (1975) we verify that this is so, viz

$$\frac{1.67 \times 10^9}{T_9} \cdot \frac{3.28 \times 10^{-10}}{T_9^{2/3}} \exp -\frac{4.872}{T_9^{1/3}} \cdot \exp -\frac{131.51}{T_9}$$

$$(1 + 0.086 T_9^{1/3} - 0.455 T_9^{2/3} - 0.271 T_9 + 0.108 T_9^{4/3} + 0.225 T_9^{5/3})$$

$$\left(\frac{n(N)}{6.022 \times 10^{23}}\right)^2 n(A). 5 \times 10^{-17} \tag{9}$$

is required to be $\frac{1}{2} n(N) = 2.5 \times 10^{33} cm^{-3}$. Taking $T_9 \simeq 20$ for the range of r from 2.5 to 3, and putting $n(N) = 5.10^{33} cm^{-3}$, $n(A) = 2.9 \times 10^{34} cm^{-2}$, the value of (9) is $2.85 \times 10^{33} cm^{-3}$, adequately close to the required value of $2.5 \times 10^{33} cm^{-3}$.

This is already an astonishing result. That so complicated an expression as (9) should combine so exactly to produce such an outcome is not a consequence of the parametric choice of the model. The freedom of choice of the numerical coefficient in (2) is entirely dwarfed by the factors 10^{34}, 10^{33}, 10^9, 10^{-10}, 10^{-17} in (9), while even some variation in the parameter α in (4), as it affects the value of the factor $\exp{-131.51/T_9} \approx 2.5 \times 10^{-3}$, is also dwarfed by the much larger powers in (9). The most license that can be permitted to a critic would be to accept the above result as model-dependent to the extent that it already consumes essentially all the available degrees of freedom of the model, leaving all further results to be judged as effectively parameter independent.

4   The Abundances of D and ^3He

Because of space restrictions, we have omitted the analysis which leads to the values(given in the summary Table 2)

$$D/H =^3 He/H \simeq 5 \times 10^{-5}$$

5   The Abundance of ^7Li

Writing $n(P)$, $n(A)$ for the densities of protons and alpha particles we have

$$n(P) = 1.58 \times 10^{33} \left(\frac{T_9}{16.3}\right)^{3/2} cm^{-3}, n(A) = 8.5 \times 10^{33} \left(\frac{T_9}{16.3}\right)^{3/2} \quad (10)$$

The ratio of the abundance of ^7Li to ^8Be established in statistical equilibrium at temperature T_9 is given by

$$\log \frac{^7Li}{^8Be} = \frac{3}{2} \log \frac{7}{8} + \log 4 - \log n(P) + 34.07$$

$$+ \frac{3}{2} \log T_9 - \frac{5.04}{T_9} \times 17.35,$$

$$= 3.20 - \frac{87.44}{T_9}, \quad (11)$$

with

$$\log \frac{^8Be}{^4He} = \frac{3}{2} \log 2 + \log n(A) - 34.67 - \frac{3}{2} \log T_9 = -2.11 \quad (12)$$

also given by statistical equilibrium.

The abundance of ^7Li established at T_9 according to (23) will, however, be subject to attenuation as the temperature declines from T_9, according to an attenuation factor

$$A \int_0^{T_9} \exp{-\frac{30.443}{T_9}} dt, \quad (13)$$

with

$$dt = \frac{8.1 \times 10^{-16}}{T_9^{3/2}} dT_9 \tag{14}$$

as before and A a numerical coefficient obtained from the reaction rate for $^7Li(P, A)^4He$ given by FCZ, viz

$$A = 1.7 \times \frac{1.05 \times 10^{10}}{T^{3/2}} \cdot \frac{2.40 \times 10^{31}}{6.022 \times 10^{23}} T^{3/2} = 7.25 \times 10^{17} s^{-1} \tag{15}$$

The factor 1.7 here arises from an estimate of the combined effect of various terms adding to the rate of $^7Li(P, A)^4He$, the rest of A being the main term. Evaluating (13) leads to

$$\sim 19.3 T_9^{1/2} \exp -\frac{30.443}{T_9} \tag{15}$$

as the attenuation factor to be applied to the abundance of ^7Li given by (11). With $\log \frac{^4He}{H} = -1.08$ we thus have

$$\log \frac{^7Li}{H} = 3.20 - 2.11 - 1.08$$

$$-19.3 \times 0.4343 T_9^{1/2} \exp -\frac{30.443}{T_9} \tag{16}$$

$$= 0.01 - 8.38 T_9^{1/2} \exp -\frac{30.443}{T_9} \tag{17}$$

which has a maximum of -9.60 at $T_9 \simeq 12$. Thus the surviving lithium abundance is

$$\frac{^7Li}{H} \simeq 2.50 \times 10^{-10}, \tag{18}$$

a result in good agreement with the observational requirement, again calculates from highly complicated formulae, again without any model adjustment.

6 The Abundance of ^{11}B

A similar calculation for ^{11}B leads to $^{11}B/H \simeq 10^{-18}$, below the observational detection limit. This is significantly lower than the value calculated by Hoyle (1992) who used an attenuation factor that was not sufficient. From an observational point of view the model therefore predicts that there is effectively no 'plateau' under boron. Such boron as exists is required to come from cosmic-ray spallation on ^{12}C, ^{16}O.

7 The Abundance of ^9Be

As noted in Hoyle 1992, the nucleus of ^9Be is exceptionally fragile, leading to a particularly low freezing temperature. Statistical equilibrium at higher temperatures establishes

$$\log \frac{^9Be}{H} = \frac{3}{2} \log \frac{9}{8} + \log \frac{4}{3} - 0.15 + \log D/H$$

$$+ \log \frac{^8Be}{^4He} + \log \frac{^4He}{n(P)} - \frac{3.28}{T_9} \qquad (19)$$

with respect to the reaction $^9Be(P,D)2^4He$. Using $\log D/H = -4.30$ already calculated, $\log {^8Be}/^4He = -2.11$, $\log {^4He}/n(P) = 0.73$, gives $-5.63 - 3.28/T_9$ for the right hand side of (19). Because $^9Be(P,A)^6Li$ contributes equally with $^9Be(P,D)2^4He$ to the destruction of ^9Be, whereas at $T_9 \simeq 1$ it contributes essentially nothing to the production of ^9Be, the equilibrium concentration of ^9Be is lowered by a further factor 2, so that

$$\log \frac{^9Be}{H} = -5.93 - \frac{3.28}{T_9}. \qquad (20)$$

Freezing of the equilibrium condition at $T_9 = 0.50$ for ^9Be would thus give

$$\log \frac{^9Be}{H} = -12.5 \qquad (21)$$

in satisfactory agreement with the apparent observed plateau under ^9Be (A.M. Boesgaard, this conference).

The estimated freezing temperature according to the model can be obtained by requiring that the product of the expansion time scale, $1.62 \times 10^{-15}/T_9^{1/2}$ seconds at temperature T_9 and the sum of the reaction rates terms for $^9Be(P,D)2 {^4He}$ and of those for $^9Be(P,A)^6Li$ be unity, viz

$$2.\frac{1.03 \times 10^9}{T_9} \cdot \frac{2.40 \times 10^{31}}{6.033 \times 10^{23}} \cdot \frac{1.62 \times 10^{-15}}{T_9^{1/2}} T_9^{3/2} \cdot \exp -\frac{3.046}{T_9} = 1. \qquad (22)$$

The factor 2 on the left of this formula comes from the circumstance that at the values of T_9 in question the highly complicated non-resonant contribution given by FCZ about doubles the resonant reaction rates. Equation (22) determines a freezing temperature $T_9 = 0.623$, reasonably close to the required value of 0.5.

8 The Abundances of ^{12}C and ^{16}O

The reaction rate on ^9Be from $^9Be(A,N)^{12}C$ as given by FCZ is

$$\sim \frac{2.40 \times 10^8}{T_9^{3/2}} \frac{n(A)}{6.023 \times 10^{23}} \cdot \exp -\frac{12.732}{T_9}. \qquad (23)$$

Using (10) for $n(A)$ and putting $T_9 \simeq 10$, at which temperature most of the production of ^{12}C takes place, gives 1.44×10^{16} for (23). Multiplying by

Table 2. Summary of Results

$$^4He/H = Y \simeq 0.229$$
$$\frac{D}{H} \simeq \frac{^3He}{H} \simeq 5.10^{-5}$$
$$\frac{^7Li}{H} = 2.5 \times 10^{-10}$$
$$\frac{^{11}B}{H} \text{ very small}$$
$$\frac{^{12}C}{H} \simeq \frac{^{16}O}{H} \simeq 4.1 \times 10^{-6}$$

the time-scale $1.62 \times 10^{-15}/T_9^{1/2}$ then gives ~ 7.4, implying that an abundance $^9Be/H \simeq 5.5 \times 10^{-7}$ given by (20) is converted 7.4 times over to ^{12}C, leading to

$$\frac{^{12}C}{H} \simeq 4.1 \times 10^{-6} \tag{24}$$

The value of $^{16}O/H$ is of a similar order.

9. The External Medium

All of the above followed from just the N and P members of the baryon octet. The other six baryons are considered not to form stable nuclei. They decay in $\sim 10^{-10}$ seconds, by which time a Planck fireball has effectively expanded into its surroundings, which according to the QSSC model (Hoyle et al. 1993,1994a,b) is necessarily a strong gravitational field in which the decay products of Λ, $\Sigma^{\pm}$, Ξ^0 and Ξ^- may be expected rapidly to lose energy. The Ξ^0 baryon decays to Λ and π^0 in a mean life of $3.0 \times 10^{-10}s$, Σ^+ which decays in a mean life of $8.0 \times 10^{-11}s$, gives a π^0 meson in about a half of the cases, so that together with Λ, which decays in a mean life of $2.5 \times 10^{-10}s$, there is a late production of about $2.5\pi^0$ per baryon octet, yielding ~ 5 late γ-rays per octet, typically with energies ~ 100 Mev. It is these γ-rays and their products that are expected to be subjected to energy loss in strong gravitational fields.

The production of Planck particles near large masses of the order of galactic clusters occurs typically in an environmental density $\sim 10^{-16}g$ cm^{-3} at which density γ-rays of 100 Mev have path lengths of $\sim 10^{18}cm$, ample for considerable redshifting effects to occur, when quanta and particles in the $1 - 100$ kev range would arise. Although such particles and quanta are readily shielded against, it is an interesting speculation that pathways into the external universe may be briefly opened and that the mysterious γ-ray bursts arise in such situations.

10. Summary of Abundances and Conclusions

The calculations outlined here are more accurate than those described earlier (Hoyle, et al. 1993, Hoyle 1992). They lead to the abundances and results shown in the following table.

To obtain a ratio $\frac{^9Be}{H} \simeq 3.10^{-13}$ requires a freezing temperature $T_9 \simeq 0.5$ which is close but not equal to the calculated freezing temperature $T_9 \simeq 0.62$.

We conclude that a certain model of the decay of Planck particles leads to interesting values for the abundances of the light elements. The work is deductive, and in this sense the model used is not subject to negotiation, any more than the axioms on which a mathematical theorem is proved are subject to negotiation. Or any more than supporters of Big-Bang nucleosynthesis regard the choice of their parameter η as a matter of negotiation. Thus the only basis for judging the situation is to assess how good, or bad, are the results. Our assessment is the following:

(i) Our result for $^4He/H$ is very good.

(ii) The ratios D/H, $^3He/H$ too high by factor ~ 2. A more detailed calculation might well lower $^3He/H$ to its observational value. But at the expense of a further increase in D/H, necessitating an epicycle for the theory in which the observed $D/H \simeq 1.5 \times 10^{-5}$ is due to environmental effects.

(iii) The ratio $^7Li/H$ is very good.

(iv) The prediction of essentially no 'floor' under ^{11}B is subject to test. The 'floor' under ^9Be requires a freezing temperature $T_9 \simeq 0.50$, whereas the calculated freezing temperature was $T_9 \simeq 0.62$. Considering the very complicated expressions of FCZ, especially that involved in a cut-off procedure for non-resonant contributions, this correspondence is adequately close.

Finally we may ask how this situation for the synthesis of light elements from Planck particles compares with the situation in Big-Bang nucleosynthesis. In that case

(a) The classic choice $\eta = 3.10^{-10}$ for the baryon to photon ratio is good for $^3He/H$ but is too low for $^7Li/H$ and too high for Y and D/H.

(b) While reducing η brings Y and $^7Li/H$ into good agreement with observation the value of D/H becomes so large that the theory requires an astration epicycle to save itself.

(c) Raising η to $\sim 6.10^{-10}$ gives good results for D/H, $^3He/H$ and $^7Li/H$ but the resulting value $Y = 0.25$ is too high, and hardly savable by any epicycle or combination of epicycles.

(d) The theory predicts no plateau under ^9Be, which seems wrong. A recourse to inhomogeneous cosmological models would be to make the theory wildly epicyclic.

(e) Big-bang nucleosynthesis, but not the present model, predicts a present-day average baryon density in the universe much below the cosmological closure value, either forcing a change to a so-called open model (when galaxy formation is made difficult or impossible) or leading to the proposal that most of the material in the universe must be dark and non-baryonic. It is this argument that has led to the proposal that non-baryonic matter dominates the universe. None has so far been found and we find the argument far from overwhelming.

In view of these points (a) to (e) it is difficult to understand why supporters of Big-Bang cosmology claim that the synthesis of the light elements strongly supports their theory. Considering also the arbitrariness of the choice of η, one might rather say the reverse.

On several occasions in making our calculations above we have referred to our surprise at finding highly complex expressions for reaction rates, containing numerical factors involving high powers of 10, turning out to yield numbers agreeing with observation to within factors ~ 2. When one experiences complexities that simplify repeatedly into agreement with observations, the tendency is to make an inversion of logic. Instead of arguing from hypotheses to conclusions, the temptation is to argue in the contrary direction, from the conclusions to the hypotheses, i.e. to argue that because the conclusions are correct, or nearly so, the hypotheses must be true. Undeniably, science has made its most progressive steps from inversions of this kind. But also undeniably, science goes most wrong from them as well. Science goes wrong, because invalid inversions lead to a theological style of thinking. A theological style of thinking is one in which we begin from an implicit belief in the correctness of some hypothesis that has not been explicitly proved by observation or experiment, as many people believe today in the Big-Bang. When departures in deductive logic then appear, as in (a) to (e) above, epicycles are invented and the epicycles then become 'true' in the eyes of the believers. Escape from theological thinking is relatively easy when only a score or so of people are effectively involved, as it was for physics in the 1920s. But when thousands are involved as in modern science, or millions as in medieval theology, escape becomes a lengthier and more fraught procedure. The historical caution against this syndrome is usually attributed to William of Ockham, who said one should not invent hypotheses to 'save' appearances, which meant that one should not invent epicycles to bolster one's belief, a perceptive statement that is often misquoted and misunderstood.

References

Fowler, W.A., Caughlan, G. and Zimmerman, B.A. (1975): *ARA&A*, **13**, 69
Hoyle, F. (1992): *Astrophys. Sp. Sci.*, **198**, 177
Hoyle, F., Burbidge, G. and Narlikar, J.V. (1993): *Ap.J.*, **410**, 437
Hoyle, F., Burbidge, G. and Narlikar, J.V. (1994): *M.N.R.A.S.*, **267**, 1007
Hoyle, F., Burbidge, G. and Narlikar, J.V. (1994): *A. & A.*, in press.
Hoyle, F., Burbidge, G. and Narlikar, J.V. (1994): *Proc. of Roy. Soc. A.* in press.

Primordial Heavy Element Production

T. Rauscher[1], F.-K. Thielemann[2]

[1] Institut für Kernchemie, Universität Mainz, D-55099 Mainz, Germany
[2] Institut für theoretische Physik, Universität Basel, CH-4046 Basel, Switzerland

Introduction

A number of possible mechanisms have been suggested to generate density inhomogeneities in the early Universe which could survive until the onset of primordial nucleosynthesis (Malaney and Mathews 1993). In this work we are not concerned with how the inhomogeneities were generated but we want to focus on the effect of such inhomogeneities on primordial nucleosynthesis. One of the proposed signatures of inhomogeneity, the synthesis of very heavy elements by neutron capture, was analyzed for varying baryon to photon ratios η and length scales L. A detailed discussion is published in (Rauscher et al. 1994b). Preliminary results can be found in (Thielemann et al. 1991; Rauscher et al. 1994a).

Method

After weak decoupling the vastly different mean free paths of protons and neutrons create a very proton rich environment in the initially high density regions, whereas the low density regions are almost entirely filled with diffused neutrons. Since the aim of the present investigation was to explore the production of heavy elements we considered only the neutron rich low density zones. High density, proton rich, environments might produce some intermediate elements via the triple-alpha-reaction but will in no case be able to produce heavy elements beyond iron. However, we included the effects of the diffusion of neutrons into the proton rich zones. Using a similar approach as introduced in (Applegate 1988; Applegate et al. 1988), the neutron diffusive loss rate κ is given by

$$\kappa = \frac{4.2 \times 10^4}{(d/a)_{\mathrm{cm\,MeV}}} T_9^{5/4} (1 + 0.716 T_9)^{1/2}\,\mathrm{s}^{-1} \tag{1}$$

in the temperature range $0.2 < T_9 < 1$. We want to emphasize that this analytical treatment is comparable in accuracy to numerical methods using high resolution grids. Thus, the only open parameter in the neutron loss due to diffusion is the comoving length scale of inhomogeneities (d/a). Small separation lengths between high density zones make the neutron leakage out of the small low density zones most effective. Large separation lengths make it negligible. (For a detailed derivation of (1), see also Rauscher et al. 1994b).

Our reaction network consists of two parts, one part for light and intermediate nuclei ($Z \leq 36$), being a general nuclear network of 655 nuclei. The second part is an r-process code (including fission) extending up to $Z = 114$ and containing all (6033) nuclei from the line of stability to the neutron-drip line (see also Cowan et al. 1983). These two networks were coupled together such that they both ran simultaneously at each time step, and the number of neutrons produced and captured was transmitted back and forth between them. (For details of the included rates and new rate determinations see Rauscher et al. 1994b).

Results and Discussion

The most favorable condition for heavy element formation is an initial neutron abundance of $X_n = 1$ (i.e. only neutrons) in the low density region, leading to a density ratio $\rho_{\text{low}}/\rho_b = 1/8$ (Rauscher et al. 1994b). This leaves as open parameters the baryon to photon ratio $\eta = n_b/n_\gamma = 10^{-10}\eta_{10}$ and the comoving length scale (d/a). Four sets of calculations have been performed, employing η_{10} values of 416, 104, 52, and 10.4. Using the relation (Börner 1988)

$$\Omega_b h_{50}^2 = 1.54 \times 10^{-2}(T_{\gamma o}/2.78\text{K})^3 \eta_{10} \quad , \tag{2}$$

with the present temperature of the microwave background $T_{\gamma o}$ and the Hubble constant $H_o = h_{50} \times 50\,\text{km}\,\text{s}^{-1}\,\text{Mpc}^{-1}$, this corresponds to possible choices of (h_{50}, Ω_b) being (2.5,1), (1.3,1), (1,0.8), and (1,0.16). The range covered in η_{10} extends from roughly a factor of 2.2 below the lower limit to a factor of 13 above the upper limit for η in the standard big bang. For each of the η-values we considered four different cases of d/a: (0) ∞ (negligible neutron back diffusion), (1) $10^{7.5}$ cm MeV, (2) $10^{6.5}$ cm MeV, and (3) $10^{5.5}$ cm MeV. (This corresponds to distances between nucleation sites of ∞, 2700, 270, and 27 m, respectively, at the time of the quark-hadron phase transition).

An exponential increase in r-process abundances with increasing η was found. This is due to "fission cycling", whereby each of the fission fragments can form a fissionable nucleus again by neutron captures. This is of particular importance in environments with a long duration of high neutron densities. fission cycling the production of heavy nuclei is not limited to the r-process flow coming from light nuclei but requires only a small amount of fissionable nuclei to be produced initially. The total mass fraction of heavy nuclei is doubled with each fission cycle and can thus be written as $X_r = 2^n X_{\text{seed}}$, with X_{seed} denoting the initial mass fraction of heavy nuclei. The cycle number n is decreasing with decreasing neutron number density n_n (and increasing temperature T) because the reaction flux experiences longer half-lives when the r-process path is moving closer to stability.

Since the formation of heavy elements beyond Fe and Kr is a very sensitive measure of η it can be used to provide an independent upper limit for the product $\Omega_b H_0^2$. Figure 1 shows observational (upper) limit on possible primordial heavy element abundances (Cowan et al. 1991; Beers et al. 1992; Mathews et al. 1992) compared to our results. Also shown are the limits for $\Omega_b H_0^2$ from comparison of observed (Meyer et al. 1991; Kurki-Suonio et al. 1990; Ryan et al. 1992; Duncan et al. 1992) and calculated light element abundances. The tightest constraints are given by the light elements including Li, Be, and B (however, see recent doubts on the primordial ^{7}Li abundance in (Deliyannis et al. 1993)) for which the conditions cannot differ much from the standard big bang.

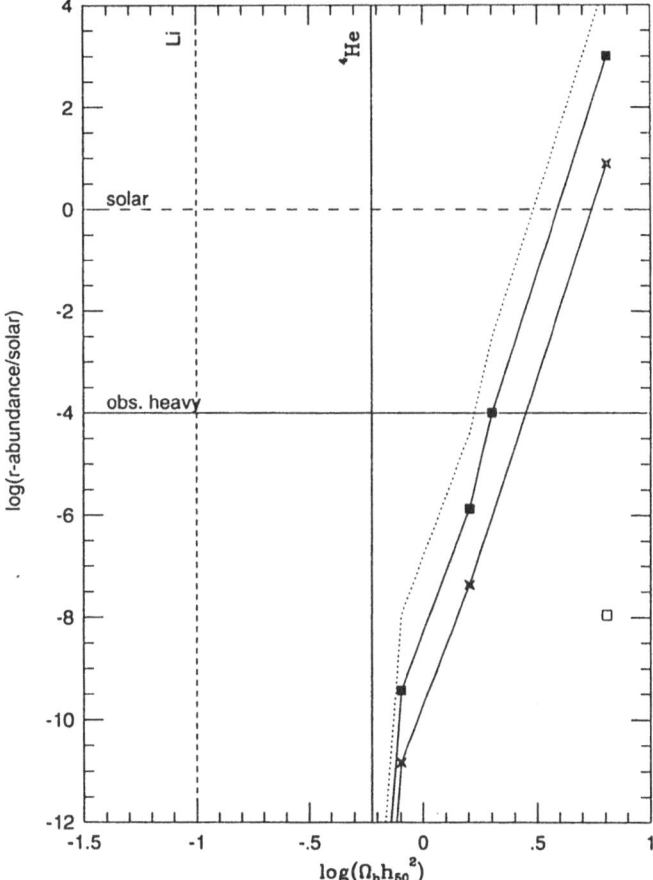

Fig. 1. Limits on $\Omega_b h_{50}^2$ from light and heavy element abundances. Shown are the results for cases 0 (full sq.), 1 (crosses), 2 (open sq.), and case 0 with enhanced rates (dotted). (The lines are drawn to guide the eye). The limits resulting from the calculated values for the *light* elements are given by the vertical lines. (See text)

To study the influence of uncertainties in the reaction rates some test calculations were performed with a variation of the ^{8}Li$(\alpha,n)^{11}$B rate. Recent experiments (Mao et al. 1994) seem to suggest that the rate used in our calculations (Rauscher et al. 1992) has to be increased by a factor of 3. Since the Li-rate is only affecting the seed abundances X_{seed}, only a linear dependence of the resulting r-process abundances ($X_r = 2^n X_{seed}$) was found. The same is true for the

$^{18}O(n,\gamma)^{19}O$ rate which was changed by a factor of 10 in a recent investigation (Beer et al. 1994). The total change in heavy element abundances by a factor of 30 is also shown in Fig. 1. However, this underlines that not all reactions of importance are fully explored, yet, and future changes can be expected.

Provided that density fluctuations exist with large length scales compared to the neutron diffusion length, the limits for η_{10} or $\Omega_b h_{50}^2$ change to 104 and 1.6, respectively, at which heavy element abundances are produced in inhomogeneous big bang models at a level comparable to the ones seen at lowest observable metallicities. This reduces the difference between the constraints from light and heavy elements, although the light element constraint is still tighter.

Acknowledgement: TR is an Alexander von Humboldt fellow.

References

Applegate, J.H. (1988): Phys. Rep. **163** 141

Applegate, J.H., Hogan, C.J., Scherrer, R.J. (1988): Ap. J. **329** 572

Beer, H., Käppeler, F., Wiescher, M. (1994): in *Capture Gamma-Ray Spectroscopy*, ed. by J. Kern (IOP, Bristol), p. 756

Beers, T., Preston, G.W., Shectman, S.A. (1992): Astron. J. **103** 1987

Börner, G. (1988): *The Early Universe* (Springer, New York)

Cowan, J.J., Cameron, A.G.W., Truran, J.W. (1983): Ap. J. **265** 429

Cowan, J.J., Thielemann, F.-K., Truran, J.W. (1991): Phys. Rep. **208** 267

Deliyannis, C.P., Pinsonneault, M.H., Duncan, D.K. (1993): Ap. J. **414** 740

Duncan, D.K., Lambert, D.L., Lemke, D. (1992): Ap. J. **401** 584

Kurki-Suonio, H., Matzner, R.A., Olive, K.A., Schramm, D.N. (1990): Ap. J. **353** 406

Malaney, R.A., Mathews, G.J. (1993): Phys. Rep. **229** 145

Mao, Z.Q., Vogelaar, R.B., Champagne, A.E., Blackmon, J.C., Das, R.K., Hahn, K.I., Yuan, J. (1994): Nucl. Phys. **A567** 125

Mathews, G.J., Bazan, G., Cowan, J.J. (1992): Ap. J. **391** 719

Meyer, B.S., Alcock, C.R., Mathews, G.J., Fuller, G.M. (1991): Phys. Rev. D **43** 1079

Rauscher, T., Applegate, J.H., Cowan, J.J., Thielemann, F.-K., Wiescher, M. (1994a): in Proc. Europ. Workshop on Heavy Element Nucleosynthesis, ed. by E. Somorjai, Z. Fülöp (Inst. Nucl. Res., Debrecen), p. 121

Rauscher, T., Applegate, J.H., Cowan, J.J., Thielemann, F.-K., Wiescher, M. (1994b): Ap. J. **429** 499

Rauscher, T., Grün, K., Krauss, H., Oberhummer, H., Kwasniewicz, E. (1992): Phys. Rev. C **45** 1996

Ryan, S.G., Norris, J.E., Bessel, M.S., Deliyannis, C.P. (1992): Ap. J. **388** 184

Thielemann, F.-K., Applegate, J.H., Cowan, J.J., Wiescher, M. (1991): in *Nuclei in the Cosmos*, ed. by H. Oberhummer (Springer, Heidelberg), p. 147

Scalar-Tensor Gravity Theories and Baryonic Density

Jean-Michel Alimi[1], Arturo Serna[1]

[1] Laboratoire d'Astrophysique Extragalactique et de Cosmologie, Observatoire de Paris-Meudon, 92195 Meudon Cedex, FRANCE

Abstract. We briefly present some results concerning cosmological implications of scalar-tensor gravity theories. We show that scalar-tensor gravity theories can be a way to relax the primordial nucleosynthesis bound to the baryon density

1 Introduction

It is known that in the standard picture of Big Bang nucleosynthesis (Contributions of J. Audouze and G. Steigman in this volume), yields of light elements only agree with their inferred primordial abundances if the baryonic density parameter, Ω_b is smaller than 0.1 or may be less with recent measures of deuterium abundance (Songailia *et al.* 1994). In the opposite, observational estimations of the total density parameter suggest Ω_0 larger than 0.1 (Peebles 1993).

The difference between the total density parameter and the baryonic density parameter can be due to the presence of non-baryonic dark matter. This hypothesis has been extensively studied (Peebles 1993). However until now, we have no observational evidence for the existence of such a non-baryonic dark matter.

Alternate scenarios which explicitly modify nuclear processes by introducing a decaying particle, or by considering inhomogeneities in baryon densities, or more exotic procedures have also been proposed. However it has been shown that when these scenarios are made to fit the observed abundances accurately, the resulting conclusions on the baryonic density relative to the critical density remains approximatively the same as in the standard homogeneous case (Schramm 1991).

We propose a different approach which considers a scalar-tensor gravity theory.

2 Scalar-Tensor Gravity Theories

The most natural alternatives to General Relativity are scalar-tensor theories, which contain the metric tensor $g_{\mu\nu}$ and an additional scalar field Φ. This class of theories arise currently in primordial cosmology (inflation), and particle physics (Kaluza-Klein, supersymmetric, low-energy limit of superstrings, theories).

The most general action describing a scalar-tensor theory of gravitation is

$$S = \frac{1+}{16\pi} \int (\phi\mathcal{R} - \frac{\omega(\phi)}{\phi}\phi_{,\mu}\phi^{,\mu})\sqrt{g}d^4x + S_M, \tag{1}$$

where $\mathcal{R}$ is the curvature scalar of the metric $g_{\mu\nu}$, $g \equiv det(g_{\mu\nu})$, ϕ is the scalar field, and $\omega(\phi)$ is an arbitrary coupling function determining the relative importance of the scalar field. The action (1) has been expressed in terms of the metric tensor $g_{\mu\nu}$, called the "Jordan frame" (Holman et al. 1990, 1991) to which matter is universally coupled. In this frame, $g_{\mu\nu}$ is measured by using non-gravitational rods and clocks.

In order to build up cosmological models, we consider an homogeneous and isotropic universe. The line-element has then a Robertson-Walker form and the energy-momentum tensor corresponds to a perfect fluid. The field equations then are

$$\frac{8\pi}{3\phi}\rho = \frac{c^2k}{R_0^2a^2} + \frac{\dot{a}^2}{a^2} - \frac{\omega}{6}\frac{\dot{\phi}^2}{\phi^2} + \frac{\dot{a}\dot{\phi}}{a\phi} \tag{2a}$$

$$\ddot{\phi} + 3\frac{\dot{a}}{a}\dot{\phi} = \frac{1}{(3+2\omega)}[8\pi(\rho - 3P/c^2) - \omega'\dot{\phi}^2] \tag{2b}$$

where $k = 0, \pm1$, ρ and P are the energy-mass density and pressure, respectively, $a \equiv R/R_0$ is the dimensionless scale factor and dots mean time derivatives.

It is convenient to consider the speed-up factor, $\xi \equiv H/H^{FRW}$, where $H \equiv \dot{a}/a$ is the Hubble parameter, while H^{FRW} is that predicted by GR at the same temperature. As we are only interested on viable scalar-tensor cosmological models, that is, those which are compatible with present solar-system observations but where the early evolution of the Universe is distinguishable from that obtained in GR, ξ is normalized so that $\xi \to 1$ today.

Analytical asymptotic solutions of all possible scalar-tensor cosmological models converging to GR have been obtained and discussed (Serna & Alimi 1994a). However, the evolution from the asymptotic behaviors until convergency to GR, have been numerically found by specifying $\omega(\Phi)$ as the following Taylor series.

$$\frac{1}{|\omega - b|^{1/\epsilon}} = \sum_{i=0}^{\infty} a_i |\Phi - 1|^i \tag{3}$$

the first order of which is

$$\frac{1}{|\omega - b|^{1/\epsilon}} = a_1 |\Phi - 1| \tag{4}$$

where $1/2 < \epsilon < 2$ (Serna & Alimi 1994a) and we have taken $a_0 = 0$ in order to have convergency to GR. Eq. (4) is a fist-order approximation to any function $\omega(\Phi)$ provided that Φ is not very different from unity. Moreover it gives an exact representation for most of the particular scalar-tensor theories proposed up to date and it contains all possible asymptotic behaviors (Serna & Alimi 1994a).

Our main results are:

a. The speed-up factor is in all generality, variable in time. If it is monotonic in time, it can decrease or increase, and the corresponding cosmological models can be independently singular or not singular.

b. It is also possible to construct a class of Scalar-Tensor cosmological models with a non-monotonic in time speed-up factor. In Figure 1, we present an example of such a Scalar-Tensor cosmological model which is defined by the values parameters $a_1 = 1$ and $b = -1$. The conditions on the parameters of our representation of the coupling function $\omega(\Phi)$ to get such a remarkable time dependance of the speed-up factor are presented in detail elsewhere (Serna & Alimi 1994a).

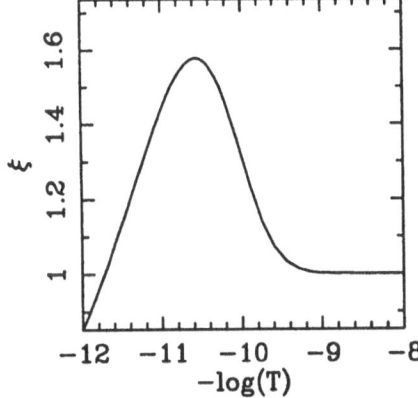
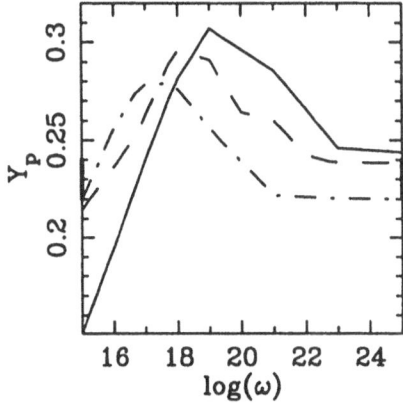

Fig. 1. Speed-up factor evolution for the Scalar-Tensor model defined by $a_1 = 1$, $b = -1$.

Fig. 2. Primordial He abundance for the same model, as a function of ω_0.

3   Primordial Nucleosynthesis and Scalar-Tensor Gravity Theories

Light element production in scalar-tensor theories has been computed (Serna & Alimi 1994b) by using updated reaction rates (Serna & Alimi 1994b). In order to avoid an excessive number of free parameters, we have taken the average values of the present photon temperature $T_{\gamma 0} = 2.73 \pm 0.02$ (2σ) and of the neutron mean-life, $\tau_n = 889 \pm 4$ s. The present Hubble parameter has been taken to be 50 Km s^{-1}Mpc^{-1} and it has been assumed that the number of light neutrino families, N_ν, is three, in accordance with the LEP and SLC results. The yields of light elements have been then calculated as a function of the baryon-to-photon ratio in units of 10^{-10}, η_{10}, the present value of the coupling function, ω_0, and the parameters a_1, b, and ϵ characterizing the form of $\omega(\Phi)$.

In order to reproduce the right primordial abundances, all scalar-tensor cosmological models with a monotonic time evolution of the speed-up factor cannot differ from GR by a factor larger than 2% during the primordial nucleosynthesis. ω_0 in all these models must be larger than 10^{21}. Consequently in all these Scalar-Tensor models, the constraint on the baryon density parameter is not modified with respect to the prediction of GR. It is important to notice that this conclusion does not depend on the form of the coupling function that we have chosen, because for all Scalar-Tensor theories with a monotonic time evolution of the speed-up factor which correctly reproduces the primordial light elements, the Φ field is never very different to 1.

For Scalar-Tensor theories which admit a non-monotonic time evolution speed-up factor, conclusions are completely different. We have now two ranges of ω_0 values which allow to reproduce the right primordial abundance (see Figure 2 for Helium abundance). For very large values of ω_0 (larger than 10^{21}), theories are not significantly different from GR during primordial nucleosynthesis and, as before, they imply a constraint on Ω_b similar to that imposed by GR. Nevertheless, right primordial abundances can be also obtained for intermediate values of ω_0. In this last case, theories are very different from GR during primordial nucleosynthesis ($0.7 < \xi < 1.6$ in previous example (Figure 1)), while the primordial light elements productions remains right (Helium, Deuterium and Lithium primordial abundances). Consequently, the allowed range for the η parameter is now larger than in GR. We plot on Figure 3 and Figure 4 the theoretical light elements production as a function of η parameter, for three different given ω_0.

4 Conclusion and Perspectives

We have succeeded in constructing a class of scalar-tensor models which cor-rectlly predict primordial light elements abundances and which are effectively different from GR during primordial nucleosynthesis. This class of scalar-tensor models allows a much higher baryon density parameter than in GR. In the ex-ample presented in this paper, we get $\Omega_b h^{-2} \sim 4 * \Omega_b^{GR} h^{-2}$. It is possible to chose the parameter a_1 and b in order to allow for higher baryonic densities (Alimi & Serna 1994). In such models, the inflation of the primordial fluctuation spectrum, and large-scale structure formation are open questions. Finally, it is important to notice that it is always possible to chose a combination of param-eters a_1 and b in order to get a purely baryonic universe, even with the recent measure of primordial abundance of deuterium (Songailia *et al.* 1994).

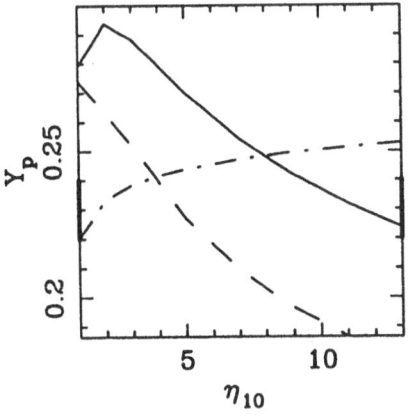

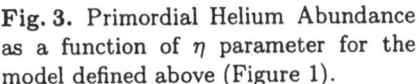

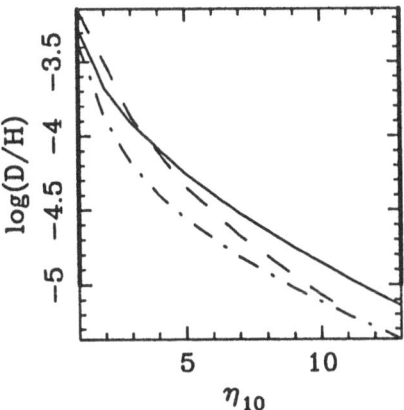

Fig. 3. Primordial Helium Abundance as a function of η parameter for the model defined above (Figure 1).

Fig. 4. Primordial Deuterium Abundance.

References

J.-M. Alimi & A. Serna, *submitted to Nature* (1994).

R. Holman, E.W. Kolb, & Y. Wang, Phys. Rev. Lett., **65**, 17 (1990). R. Holman, E.W. Kolb, S. L. Vadas & Y. Wang, Phys. Rev. D **43**, 3833 (1991); Phys. Lett. B **269**, 252 (1991).

P.J.E. Peebles, Principles of Physical Cosmology, *Princeton University Press* (1993).

D.N. Schramm, Physica Scripta **T36**, 22 (1991).

A. Serna, & J.-M. Alimi, *submitted to Phys. Rev. D* (1994a).

A. Serna, & J.-M. Alimi, *submitted to Phys. Rev. D* (1994b).

A. Songailia, L.L. Cowie, C.J. Hogan, & M. Rugers Nature, **368**, 599 (1994).

On the Destruction of Primordial Deuterium

Keith A. Olive[1]

[1] School of Physics and Astronomy, University of Minnesota Minneapolis, MN, 55455, USA

Abstract. Agreement between the predictions of standard big bang nucleosynthesis and the observed abundances of the light elements is achieved for a small range in values of the baryon to photon ratio, η near 3×10^{-10}. At that value of η, the primordial deuterium abundance is about 7.5×10^{-5}. ISM measurements of D indicate a present abundance 5 times smaller. Chemical evolution models are explored to achieve this degree of D destruction. Consistency with the ^{3}He production in low mass stars and with the abundance of cosmo-chronometers is also discussed.

Overall, the standard model of big bang nucleosynthesis is regarded to be a consistent theory on the abundances of the light elements D, ^{3}He, ^{4}He, and ^{7}Li (Walker et al., 1991). What's more, consistency is achieved for a narrow range of what is now perhaps the single remaining parameter of the theory, namely the baryon-to-photon ratio $\eta = n_B/n_\gamma \simeq 3 \times 10^{-10}$. At this value of η, the primordial mass fraction of ^{4}He is $Y_P \simeq 0.24$, and the abundances by number with respect to H, for D, ^{3}He and ^{7}Li are: D/H $\simeq 7.5 \times 10^{-5}$, ^{3}He/H $\simeq 1.5 \times 10^{-5}$ and ^{7}Li/H $\simeq 10^{-10}$. This ^{4}He abundance is consistent (though on the high side) with observations (see eg. Olive & Steigman, 1994) as is the ^{7}Li abundance (Spite & Spite, 1993). It is difficult to compare the primordial deuterium and ^{3}He abundances with the observations. Despite the fact that all observed deuterium is primordial, deuterium is destroyed in stars. A comparison between the predictions of the standard model and observed solar and interstellar values of deuterium must be made in conjunction with models of galactic chemical evolution. The problem concerning ^{3}He is even more difficult. Not only is primordial ^{3}He destroyed in stars but it is very likely that low mass stars are net producers of ^{3}He. Thus the comparison between theory and observations is complicated not only by our lack of understanding regarding chemical evolution but also by the uncertainties in production of ^{3}He in stars.

Given the above primordial D abundance, it appears that D/H has decreased over the age of the galaxy. The (pre)-solar system abundance of deuterium is D/H $\approx (2.6 \pm 1.0) \times 10^{-5}$ (Geiss, 1993), corresponding to an age $t \sim 9$ Gyr, while the present ($t \sim 14$ Gyr) ISM abundance of D/H is 1.65×10^{-5} (Linsky et al., 1992). Thus it would appear that significant destruction of deuterium is necessary. Note that there is a reported detection of D in a high redshift, low metallicity quasar absorption system (Carswell et al., 1994; Songaila et al., 1994) with an abundance which may be the primordial one. Due to the still some what preliminary status of this observation and the fact that it can also be interpreted as a H detection in which the absorber is displaced in velocity by 80 km s^{-1} with respect to the quasar (Steigman, 1994; Linsky, 1994), we do not fix the primordial abundance with that value. In any case, such a high value for the D abundance would require an even greater degree of D destruction over the age of the galaxy.

To obtain an overall destruction factor of ~ 5 for deuterium, Vangioni-Flam, Olive & Prantzos (1994), considered a number of simple chemical evolution models with various star formation rates (SFRs) and initial mass functions (IMFs).

Table 1. Simple chemical evolution models

	IMF	SFR
I_a	$\propto m^{-2.7}$	$\propto M_{\mathrm{Gas}}$
I_b	Tinsley	$\propto M_{\mathrm{Gas}}$
II_a	Scalo	$\propto e^{-t/\tau}$
II_b	$\propto m^{-2.7}$	$\propto e^{-t/\tau}$

The evolution of D is a straightforward calculation. It is assumed that D is totally destroyed in stars. The instantaneous recycling approximation can give us a good indication of which models will destroy D efficiently (though this approximation was not used in the calculations). The rate of change for the mass in gas, $\sigma = M_{\mathrm{Gas}}/M_{\mathrm{tot}}$ is given by

$$\frac{d\sigma}{dt} = -(1 - R)\psi(t) \tag{1}$$

where R is the return fraction

$$R = \int (m - m_{\mathrm{rem}})\phi(m)dm \tag{2}$$

for an IMF, ϕ, and a SFR, ψ. m_{rem} is the mass of a remnant. (See Vangioni-Flam et al. (1994) for details.) The evolution of the D abundance, X_D, is described by

$$\frac{dX_D \sigma}{dt} = -X_D \psi(t) \tag{3}$$

giving a final D abundance relative to the primordial D, D_p

$$\frac{D}{D_p} = \sigma^{R/(1-R)} \tag{4}$$

With a present-day gas mass fraction $\sigma = 0.1$, one sees that $R \sim 0.2$ gives $D/D_p \sim 1/2$ whereas models with a return fraction $R \sim 0.5$ will give $D/D_p \sim 1/10$. Clearly, the degree to which deuterium is destroyed is very sensitive to the chemical evolution model.

In Figs. 1 and 2, the evolution of D/H is shown for the models considered. As one can see, a total deuterium destruction factor of 5 is certainly achievable in these simple models.

The destruction of deuterium, however, is intimately connected to the question of the ^{3}He abundance. First of all, during the pre-main-sequence, essentially all of the primordial D is converted into ^{3}He. Some of the pre-main-sequence produced and primordial ^{3}He will survive in stars and may even be produced (Iben,

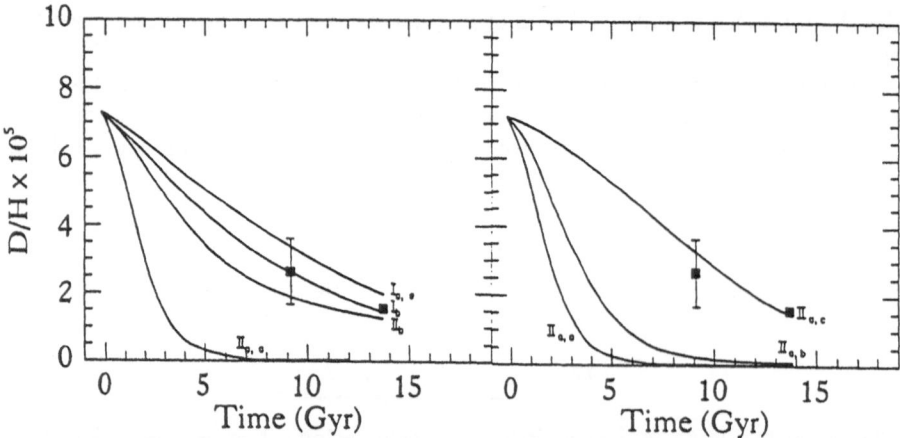

Fig. 1. The evolution of the D/H ratio as a function of time for the different models

Fig. 2. As in Fig. 1 for model IIa with different time constants $\tau = 3$ (a); $\tau = 5$ (b); $\tau = 10$ (c)

1967; Rood, 1972; Iben & Truran, 1978). If $g_3 = \frac{(^3He/H)_f}{((D+^3He)/H)_i}$ is the fraction of ^{3}He that survives stellar processing, then the ^{3}He abundance at a time t is at least

$$\left(\frac{^3He}{H}\right)_t \geq g_3 \left(\frac{D + ^3He}{H}\right)_p - g_3 \left(\frac{D}{H}\right)_t \tag{5}$$

The inequality comes about by neglecting any net production of ^{3}He. Of course, this equation can be rewritten as an upper limit on $(D + ^3He)/H$ in terms of the observed pre-solar abundances ($t = \odot$) and g_3 as was done in Yang et al. (1984) and yields a lower bound to η. Neglecting ^{3}He production, and assuming only that $g_3 > 0.25$ a limit $\eta > 2.8 \times 10^{-10}$ is derived. The evolution of $(D + ^3He)/H$ for the models shown in Figs. 1 and 2, is shown in Figs 3 and 4.

Finally, I address the question regarding the consistency of deuterium destruction and the nuclear cosmochronometers ^{232}Th/^{238}U and ^{235}U/^{238}U (Scully & Olive, 1994). In addition to choosing a SFR and an IMF, to obtain the correct ratio for the chronometers corresponding to a reasonable age to the Galaxy requires some infall of gas onto the Galaxy (assumed primordial) and a burst of star formation providing an initial enrichment (Cowan, Thielemann, & Truran, 1987). In general the abundance ratio of the chronometers will depend on the lifetime of the radioactive isotopes, their stellar production factors, the infall rate rate, and the degree of initial enrichment. By choosing the latter two, fitting the abundance ratios to their observed values leads to an age for the Galaxy. Indeed, many such models were found in Scully & Olive (1994) which gave consistent ages ranging from 9.8 to 21.6 Gyrs. Best fit solutions however, tended to result in a somewhat low present day gas mass fraction (5%).

Thus, it is clear that models of galactic chemical evolution can be found which destroy D by a significant factor. Cosmochronolgy imposes restrictions on these models but many solutions can still be found. The main problem will

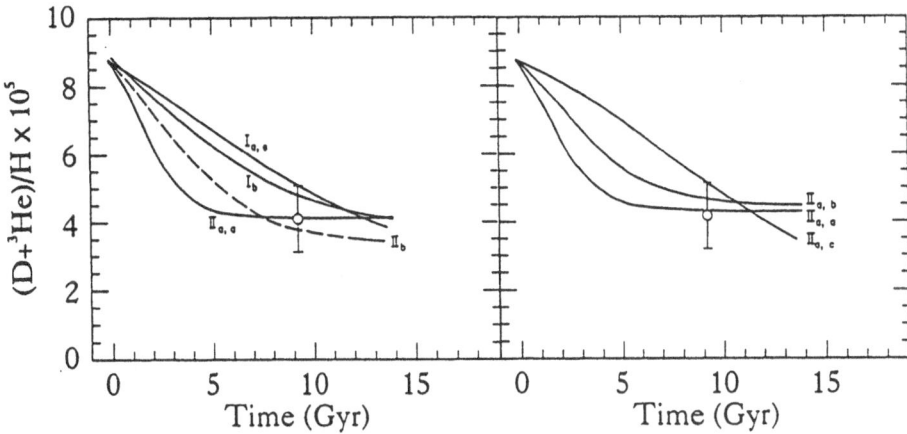

Fig. 3. As in Fig. 1, for the evolution of (D + ^{3}He)/H

Fig. 4. As in Fig. 2, for the evolution of (D + ^{3}He)/H

reside with the ^{3}He abundance when production factors are taken into account (Galli, 1994; Tosi, 1994; Olive et al. 1994).

References

Carswell, R.F., Rauch, M., Weymann, R.J., Cooke, A.J., & Webb J.K. 1994, MNRAS, 268, L1

Cowan, J.J., Thielemann, F.-K., & Truran, J.W. 1987, ApJ, 323, 543

F. Galli 1994, these proceedings

Geiss, J. 1993, in *Origin and Evolution of the Elements* eds. N. Prantzos, E. Vangioni-Flam, & M. Cassé (Cambridge:Cambridge University Press), p. 89

Iben, I. 1967, ApJ, 147, 624

Iben, I., & Truran, J.W. 1978, ApJ, 220, 980

Linsky, J., et al. 1992, ApJ, 402, 694

Linsky, J., 1994, these proceedings

Olive, K.A., Rood, R.T., Schramm, D.N., Truran, J.W., & Vangioni-Flam, E. 1994, Univeristy of Minnesota preprint UMN-TH-1305/94

Olive, K.A., & Steigman G. 1994, ApJ Supp, in press

Rood, R.T. 1972, ApJ, 177,681

Scully, S., & Olive, K.A. 1994, Univeristy of Minnesota preprint UMN-TH-1260/94

Songaila, A., Cowie, L.L., Hogan, C., & Rugers, M. 1994 Nature, 368, 599

Spite, F., & Spite, M. 1993, in *Origin and Evolution of the Elements* eds. N. Prantzos, E. Vangioni-Flam, & M. Cassé (Cambridge:Cambridge University Press), p.201

Steigman, G.1994, Ohio State University preprint OSU-TA-7/94

Tosi, M. 1994, these proceedings

Vangioni-Flam, E., Olive, K.A., & Prantzos, N. 1994, ApJ, 427, 618

Walker, T.P., Steigman, G., Schramm, D.N., Olive, K., & Kang, H. 1991, ApJ, 376, 51

Yang, Y., Turner, M.S., Steigman, G., Schramm, D.N., & Olive, K. 1984, ApJ, 281, 493

Primordial Deuterium, Dark Matter and Chemical Evolution of the Galaxy

M. Cassé[1] and E. Vangioni–Flam[2]

[1] CE–Saclay, DSM/DAPNIA/Service d'Astrophysique, F91191 Gif sur Yvette cedex
[2] Institut d'Astrophysique, 98 bis boulevard Arago, F75014 Paris

Abstract. We explore the cosmological and astrophysical consequences of the high deuterium over hydrogen ratio observed recently on the line of sight of a distant quasar.

1 Introduction

The written story of deuterium has been presented at length in this meeting by J. Audouze, G. Steigman and K. Olive, thus we will limit ourselves to explore the consequences of the first measurement of the deuterium/hydrogen ratio in a high redshift very metal poor absorbing cloud complex (Songalia et al. 1994, Carswell et al. 1994), well aware of the preliminary character of the result since a possible corruption of the signal by an errant hydrogen cloud of misfortunate velocity (80 km s^{-1} with respect to the main cloud) cannot be discarded. Taken at face value the high primordial D abundance observed has three important cosmological and astrophysical consequences: 1. on the coproduction of ^{3}He, ^{4}He and ^{7}Li in the big-bang; 2. on the baryon density of the universe through the baryon/photon ratio η and consequently on the problem of dark matter in halos and clusters of galaxies; 3. on the evolutionary models of the Galaxy.

2 Light Elements Coproduced with D

Fixing the primordial deuterium over hydrogen ratio$(D/H)_p$ to the recently observed value (1.9 to $2.5\,10^{-4}$), the primordial abundance of other light isotopes synthesized in the big-bang ensues directly from precise nucleosynthesis calculations (Smith et al.,1993; Olive and Steigman,1994). One gets $Y = 0.228$ to 0.236 and ^{7}Li/H $= 1$ to $3.5\,10^{-10}$, including the uncertainties on the relevant reactions rates and the possible effects of inhomogeneities related to the quark–hadron phase transition (Reeves and Terasawa,1994). These values are entirely compatible with the observed ones. All light elements are in line at $\eta_{10} = 1.3$ to 2. Note that the derived primordial ^{3}He/H ratio $(2\,10^{-5})$ is close to the presolar one (Geiss 1993). This near equality will cause a serious problem, as discussed in the following section.

3 Dark Matter

The baryon density of the universe expressed in units of the closure density is related to η by $\Omega_b = 0.015\,\eta\,h_{50}^{-2}$. The main uncertainty on Ω_b is brought by the Hubble parameter H which is still dichotomic ($H = (50 \pm 10)$ or $(80 \pm 10)\,\mathrm{km\ s^{-1}\ Mpc^{-1}}$ according to different authors, see for a review, Fukugita et al. 1994). It is instructive to compare it with the density of luminous matter (including the hot X–ray emitting gas in clusters of galaxies), Ω_L, the total density of matter (including dark halos) derived from dynamical studies of galaxies, Ω_h and, at larger scale, the total density of the material included in groups and clusters of galaxies (luminous + dark matter), Ω_c. As shown in Vangioni-Flam and Casse 1994, fig. 1, no room is left for baryonic extended halos around elliptical galaxies if H is about $100\,\mathrm{km\ s^{-1}\ Mpc^{-1}}$. On the other hand, non baryonic matter is definitly necessary to explain the matter density of clusters of galaxies, irrespective to the value of H.

4 Chemical Evolution of the Galaxy

Up to now only local and present values of D/H were available through observation of the nearby interstellar medium (Vidal-Madjar 1991, Ferlet 1992, Linsky et al. 1993, Lemoine et al. 1994), and it was necessary to use a galactic evolutionary model to derive the primordial ratio. The result is model dependent through the choise of the Initial Mass Function, of the Star Formation Rate and Infall Rate. According to different authors, the D destruction factor ranged from 2 (Audouze and Tinsley 1976, Steigman and Tosi 1993) to about 10, at maximum (Vangioni–Flam and Audouze 1988, Vangioni–Flam et al. 1994). The great interest of the new measurement, if confirmed, is to provide the pristine D abundance and thus by comparison with the present value ($1.65\ 10^{-5}$, Linsky et al. 1993, Lemoine et al. 1994) the destruction factor ($D/D_p = 12$ to 17). This high value, if confirmed, would not only necessitate a drastic revision of the galactic evolutionary models but also of stellar models, to avoid strong ^{3}He overproduction. The challenge is now to destroy thoroughly D without overproducing both ^{3}He and metals. Since D is processed into ^{3}He in pre-main sequence stars the only way out is to assume that ^{3}He is efficiently destroyed even in stars between 1 and $3\,\mathrm{M_\odot}$, contrary to conventional wisdom. A net production of ^{3}He (Iben and Truran 1978) should be absolutely avoided. Note that the observation of large ^{3}He/H ratios in low mass planetary nebulae (Wilson, this meeting), if representative, would pleade against this drastic solution. On the other hand the hint for a galactic gradient opposite to that of oxygen (Balser et al. 1994) would indicate that ^{3}He is more depleted in internal region of the galaxy where matter has been more processed. We urge stellar modelers to reasess the question. Galactic evolutionary models have to be revised to deplete D by factors larger than 10 without overproducing oxygen and metals. A limited class of models including an early bimodal star formation rate and a galactic wind driven by supernovae and/or cosmic rays achieves this goal (Vangioni-Flam and Cassé 1994). A specific example is shown in Figure 1 based on the following parametrization :

massive mode $\phi(M)\,\mathrm{d}M \propto M^{-3}$ for $1.5 \leq M \leq 100\,\mathrm{M_\odot}$; SFR $\propto \exp(-t/1\mathrm{Gyr})$ normal mode $\phi(M)\,\mathrm{d}M \propto M^{-3}$ for $0.4 \leq M \leq 100\,\mathrm{M_\odot}$; SFR $\propto \sigma$, where σ denotes the gas mass fraction.

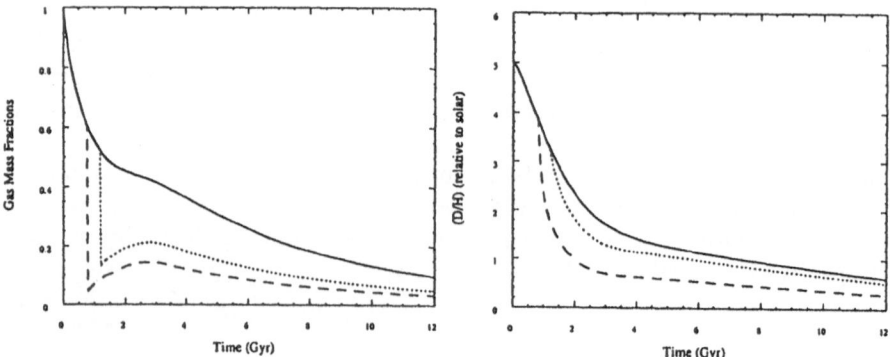

Fig. 1. D/H (left) and gas mass fraction σ (right) versus time. Full line : no galactic wind. Dotted line : galactic wind beginning at 1.2 Gyr. Dashed line : galactic wind beginning at 0.8 Gyr. The D-destruction are respectively 8.5, 10 and 18 and the corresponding metallicity is 1.5, 1.1 and 0.9 relative to solar value. The ejected masses are 0, 0.4 and 0.55 times the initial galactic mass.

The results are sensitive to the time at which the wind is established and its intensity (or equivalently the mass lost by the galaxy). The behaviour of the gas mass fraction and that of deuterium which are strongly correlated are shown in Figure 1. Qualitatively, the first generations of massive stars destroy deuterium efficiently and reject rapidly in the ISM the D-free, Z-rich material, then the wind blows, depressing the ISM. After the wind, the star formation is reduced due to the fall of σ and the production of metals is moderated. Globally one arrives, after 12 Gyr of galactic evolution, to a destruction factor of about 15 together with a near-solar metallicity. Finally, the D-free ejecta of stars of lower mass dilutes efficiently the deuterium present in the ISM. Details of this scenario are discussed in Vangioni–Flam and Cassé (1994). This tentative scenario needs to be confirmed on dynamical grounds. Note that a similar process has been invoked by M. Peimbert at this conference to alleviate the $\Delta Y/\Delta Z$ problem. It is worth mentionning that infall of primordial matter would lead to a small destruction factor contrary to the goal persued.

5 Provisional Conclusions

1. The abundance of D, ^{4}He and ^{7}Li converge with theoritical big bang nucleosynthesis calculations, at $\eta = 1.3 to 2..$.
2. Ω_b is low $(0.025 \pm 0.005)\, h_{50}^{-2}$. No room is left for baryonic extended dark halos around elliptical galaxies if H is close to $100\,\mathrm{km\ s^{-1}\,Mpc^{-1}}$. Most of the matter in clusters of galaxies is non-baryonic.
3. ^{3}He should be thoroughly transformed into ^{4}He and not produced by stars between 1 to $3\,\mathrm{M_\odot}$. A revision of stellar evolution from main sequence to planetary nebulae would be necessary in that case.
4. If the high primordial D is confirmed, galactic evolutionary models will be forced to evolve themselves. As a first step towards this direction we suggest that a destruction of D by a factor of 15 and even more would result from an early formation of massive stars generating a galactic wind.

Acknowlegements : We thank D. Elbaz and R. Lehoucq for permanent help. This work was supported by PICS No114 "Origin and Evolution of the light elements," CNRS.

References

Audouze J., Tinsley B., 1976, ARA&A, 14, 43

Balser D. S., Bania T. M., Brockway C. J., Rood R. T., & Wilson T. L., 1994, preprint

Carswell R.F., Rauch M., Weymann R.J., Cooke A.J., & Webb J.K., 1994, MNRAS, 268, L1.

Ferlet R., 1992,in IAU Symposium 150, 85

Fukugita M., Hogan C.J. & Peebles P.J.E., 1993, Nature, 366, 309

Geiss J., 1993,in Origin and Evolution of the Elements, ed. N. Prantzos, E. Vangioni-Flam & M. Cassé (Cambridge University Press), 89

Iben I., Truran J.W., 1978,ApJ, 220, 980

Lemoine M., Vidal-Madjar A., Ferlet R., Bertin P., Gry C., & Lallement R., 1994, submitted to A&A

Linsky J.L., Brown A., Gayley K., Diplas A., Savage B.D., Ayres T.R., Landsman W., Shore S.W. & Heap S.R., 1993, ApJ, 402, 694

Olive K., Steigman G., 1994, preprint

Reeves H. & Terasawa K., 1994, preprint

Smith M.S., Kawano L.H. & Malaney R.A., 1993, ApJS, 85, 219

Songaila A., Cowie L.L., Hogan C. & Rugers M., 1994, Nature, 368, 599.

Steigman G. & Tosi M. 1992, ApJ, 401, 150

Vangioni–Flam E. & Audouze J., 1988, A&A, 193, 81

Vangioni–Flam E., Olive K A., & Prantzos N., 1994, ApJ, 427, 618

Vangioni–Flam E. and Cassé M., 1994, ApJ, accepted

Vidal–Madjar A., 1991, Adv. Space Res., 11, 97

Primordial Abundances of Light Elements in Quasi-Homogeneous Models

V.Chuvenkov, V.Alyshaev, A.Glukhov

Department of Space Physics, Rostov University, 5 Zorge st., 344104 Rostov-on-Don, Russia

Abstract. Process of primordial nucleosynthesis is considered and yields of light element $(D, He^3, He^4, Li^7, Be^9, B^{11})$ are calculated in the approximation of inhomogeneous model with small fluctuations of baryon density and neutron-proton ratio due to quark-hardron phase transition. It is shown that in the simplified 2-zone model final abundances of the light elements correspond to recent observations if average baryon density parameter $\Omega_b \sim 0.2$ and value of initial fluctuations $\frac{\delta\rho}{\rho} \sim 0.4$. This result is in good agreement with baryonic nature of virial dark matter.

1 Introduction

Numerous calculations of primordial nucleosynthesis have shown that the best agreement between theoretical and observed abundances of light elements is in in the standard model of Big Bang nucleosynthesis (SBBN) with a range of baryon density parameter $0.01 \leq \Omega_b \leq 0.09$ (Smith, Kawano and Malaney 1993). This limitation is inconsistent with conclusions of inflationary scenarios ($\Omega_{tot} = 1$) and even with estimations of Ω, following from dynamical properties of galaxy clusters ($\Omega_{vir} \simeq 0.2 \div 0.3$). So, the conclusion follows, that main share of matter in the Universe (at least $10\% \div 20\%$ according to determination of Ω_{vir}) must exist in the form of non-baryonic particles, that contradicts recent conclusions of large scale structure theories. As the main point of this problem is concerned with SBBN, either the existense of significant mass of non-baryonic dark matter in galaxies should be assumed, or the theory of SBBN should be modified to obtain realistic yields of light elements if, at least, $\Omega_b \simeq \Omega_{vir}$. Recently, the main way of this modification is concerned with theories including the quark-hardron phase transition that leads to an inhomogeneous distribution of baryon density and of neutron and proton concentrations (Applegate, Hogan and Sherrer 1987). We shall concentrate attention on one of these models with $\Omega_b \simeq \Omega_{vir}$ and small level of initial fluctuations with partial neutron diffusion.

2 Quasi-Homogeneous 2-Zone Model

Assume that at the temperature T_c the quark-hardron phase transition occurs and the lumps of nucleon matter arise in the quark-gluon plasma. The size of the lump l and mean distance between the lumps L can be presented by the following equations (Kajino 1990):

$$l(T) = 3 \cdot 10^{10} \alpha T_c^{-1} T^{-1} \quad [\text{cm}], \tag{1}$$

$$L(T) \simeq 3 \cdot 10^6 \sigma^{3/2} T_c^{-13/2} T^{-1} \quad [\text{cm}], \tag{2}$$

where $\alpha \leq 1$ is ratio of initial lump size to horizon size, σ is surface tension of bubble of baryonic matter (in MeV^3). Temperatures T_c and T are presented in MeV.

After the phase transition the neutrons diffuse from baryonic lumps. At the temperature of freeze up $T_f \simeq 0.5$ MeV, the neutron diffusion coeffitient can be estimated as

$$D_n = \left(D_{np}^{-1} + D_{ne}^{-1}\right)^{-1}, \tag{3}$$

where

$$D_{np} \simeq 8 \cdot 10^8 T^{5/2} \left(1 - X_{nl}(T_f)\right)^{-1} \Omega_{bl}(T_f) \quad [\text{cm}^2 \text{s}^{-1}], \tag{4}$$

$$D_{ne} \simeq 3.9 \cdot 10^9 \exp(\xi^{-1}) \xi^{-1} f^{-1}(\xi) \quad [\text{cm}^2 \text{s}^{-1}], \tag{5}$$

X_{nl} is mass concentration of neutrons in lumps, Ω_{bl} is Ω_b in lumps, $\xi^{-1} = 0.511/T$, $f(\xi) = 1 + 3\xi + 3\xi^2$ (Terasawa and Sato 1989).

Shape of neutron distribution in space after freeze up is determined by a ratio of neutron diffusion length to mean distance between lumps:

$$N_n = (6D_n)^{1/2} L^{-1} = 8 \cdot 10^{-7} D_n^{1/2} \sigma^{-3/2} T_c^{13/2} \tag{6}$$

It is seen, that at $T = T_f$ the 3 different cases follow from value of N_n:

a) $N_n \gg 1$, that results in homogeneous neutron distribution in space;

b) $N_n \ll 1$, that is concerned with concervation of initial distribution;

c) $N_n \lesssim 1$, that leads to partial neutron diffusion from lumps.

In this work we shall consider case (c). Since the proton diffusion coefficient D_p less than D_n about 10^6 times, before the beginning of nucleosynthesis the space is occupied by two zones: i) protons and neutrons in the lumps; ii) neutrons diffused from lumps and distributed in space between lumps. As a first order approximation, neutron distribution may be considered as homogeneous.

Introducing the parameters $\mu = (\Omega_{bl} - \Omega_{bv})/\Omega_{bv}$ and $f_l = V_l/V$, where index l cooresponds to lumps, v - to voids between lumps, the average final abundance of i-th element can be found as:

$$\overline{X_i} = \frac{f_l(\mu + 1)X_{il} + (1 - f_l)X_{iv}}{\mu f_l + 1} \tag{7}$$

It is seen that the best agreement with observations if $\Omega_b = 0.2$ is in the model version with the following values of initial parameters: $\mu \simeq 8 \div 10$ and $f_l \simeq 0.5$. The large final abundances of Li, Be and B is due to reason that we have not take into account late-time neutron diffusion (Sato and Terasawa 1991). The main conclusion of the work is following: the value of $\Omega_b = 0.2$ is the upper limit for realistic yields of light elements in inhomogeneous models with quark-hardron phase transition. However, for confirmation of this preliminary conclusion and refinement of Ω_b the consideration of model with accounting for the shape of neutron distribution and late-time neutron diffusion is needed.

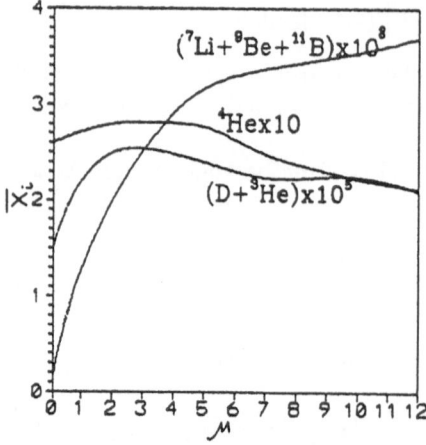

Fig. 1. Average element yields vs μ, if $\overline{\Omega_b} = 0.2$ and $f_l = 0.5$

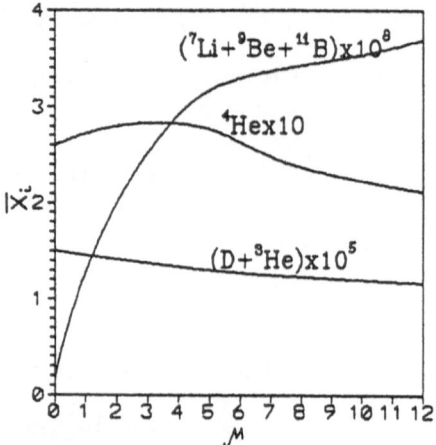

Fig. 2. Same as in Fig. 1, if $f_l = 0.7$

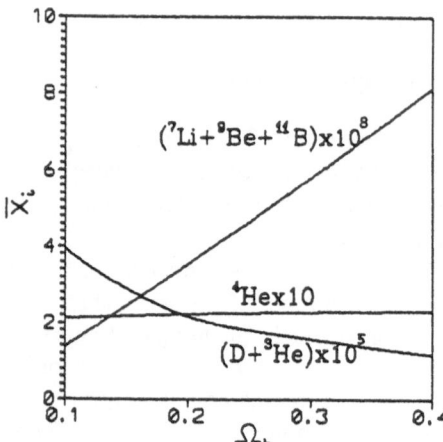

Fig. 3. Average element yields vs $\overline{\Omega_b}$, if $\mu = 10$ and $f_l = 0.5$

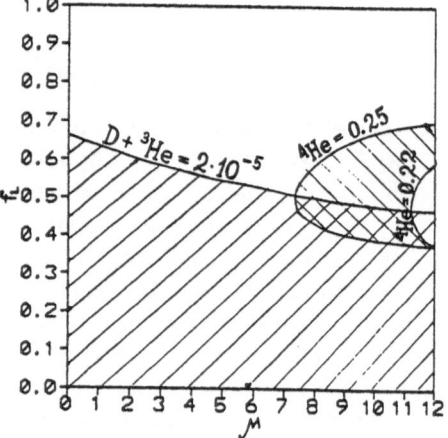

Fig. 4. Plane of parameters f_l and μ with bound of element production if $\overline{\Omega_b} = 0.2$

References

Smith, M.S., Kawano, L.H., Malaney, R.A. (1993): Astrophys. J. Suppl. **85** 219

Applegate, J.H., Hogan, C.J., Sherrer, R.J. (1987): Phys. Rev. D **35** 1151

Kajino, T. (1990): Gansikaku Kankju **35** 19

Terasawa, N., Sato, K. (1989): Phys. Rev. D **39** 2893

Sato, K., Terasawa, N. (1991): Physica Scripta **T36** 60

Primordial Nucleosynthesis and Light Element Abundances

David N. Schramm

The University of Chicago, 5640 S. Ellis Avenue, Chicago,IL 60637
and
NASA/Fermilab Astrophysics Center, Fermi National Accelerator Laboratory,
Box 500, Batavia, IL 60510-0500

Abstract. Primordial nucleosynthesis has established itself as one of the three pillars of Big Bang cosmology. This symposium summarized the current observational and theoretical situation with regard to Big Bang Nucleosynthesis. In particular, it is shown that the Pop II ^{6}Li results strongly support the argument that the Spite Plateau lithium is a good estimate of the primordial value. The ^{6}Li is consistent with the Be and B found in Pop II stars, assuming those elements are cosmic ray produced. The HST ^{2}D value tightens the ^{2}D arguments. The recent reports of extragalactic D/H at redshift $Z \sim 3$ are discussed. The new low metalicity ^{4}He determinations appear to raise slightly the best primordial ^{4}He number and thus make a better fit and avoid a potential problem. The new ^{3}He measurements in planetary nebulae strongly support the basic concept that ^{3}He is produced in low mass stars as well as the Big Bang. However, the variation of ^{3}He/H inversely with H-II region mass argues that destruction in massive stars is probably occurring. In general, it was shown that abundance determinations are dominated by systematic rather than statistical errors, and work on bounding systematics is crucial. The quark-hadron inspired inhomogeneous calculations now unanimously agree that only relatively small variations in Ω_B are possible vis-à-vis the homogeneous model; hence, the robustness of $\Omega_b \sim 0.05$ is now apparent. This argues that the bulk of the baryons in the universe are not producing visible light. A comparison with the ROSAT cluster data is also shown to be consistent with the standard BBN model. $\Omega_b \sim 1$ seems to be definitely excluded, so, if $\Omega = 1$, as some recent observations may hint, then non-baryonic dark matter is required. The implications of the recently reported halo microlensing events are discussed. If real, they indicate the location of the baryons in compact halo objects.

1 Introduction

The study of the light element abundances has undergone a recent burst of activity on many fronts. News results on each of the cosmologically significant abundances have sparked renewed interest and new studies. The bottom line remains: primordial nucleosynthesis has joined the Hubble expansion and the microwave background radiation as one of the three pillars of Big Bang cosmology. Of the three, Big Bang Nucleosynthesis, BBN, probes the universe to far earlier times (~ 1 sec) than the other two and led to the interplay of cosmology with nuclear and particle physics. Furthermore, since the Hubble expansion is also part of alternative cosmologies such as the steady state, it is BBN and the microwave background that really drive us to the conclusion that the early universe was hot and dense.

Recent heroic observations of ^{6}Li, Be and B, as well as ^{2}D, ^{3}He and new ^{4}He determinations, have all gone in the direction of strengthening the basic picture of cosmological nucleosynthesis. In particular, theoretical calculations of cosmic ray production of ^{6}Li, Be and B have fit the observations remarkably well, thus preventing these measurements from disturbing the standard scenario (Olive et al. 1990). In this regard, the recent reports of D/H in quasar absorption systems at redshift $Z \sim 3$ are particularly interesting (Songaila et al. 1994; Carswell et al 1994) since BBN requires that fragile deuterium be found in primitive material. However, the possible variation of D/H in different lines of sight at $Z \sim 3$ argues that perhaps hydrogen cloud interlopers may be immitating deuterium, at least in lines of sight having higher apparent D/H. Such an interpretation avoids a potential conflict with ^{3}He arguments. Furthermore, recent theoretical calculations have confirmed that quark-hadron inspired inhomogenous Big Bang Nucleosyntheis does not significantly alter the basic conclusions of standard BBN. We will also briefly discuss the possible impact on BBN of the recent ROSAT cluster results and the recent halo microlensing results. This summary will attempt to put it all together within an historical framework. The bottom line that emerges is how dramatically robust BBN is. Alternative models such as that presented by Hoyle at the Elba meeting are unable to first observational data without invoking large numbers of ad hoc parameters and/or new and otherwise unmotivated physics.

Let us now briefly review the history, with special emphasis on the remarkable agreement of the observed light element abundances with the calculations. This agreement works only if the baryon density is well below the cosmological critical value. This summary draws on the reviews of Walker et al (1991), Schramm (1991), Schramm et al. (1992), Schramm (1993) and Schramm (1994) as well as the other talks at Elba.

2   History and Highlights of Big Bang Nucleosynthesis

It should be noted that there is a symbiotic connection between BBN and the 3K background dating back to Gamow and his associates Alpher and Herman. The initial BBN calculations of Gamow's group (Alpher et al. 1948) assumed pure neutrons as an initial condition and thus were not particularly accurate, but their inaccuracies had little effect on the group's predictions for a background radiation.

Once Hayashi (1950) recognized the role of neutron-proton equilibration, the framework for BBN calculations themselves has not varied significantly. The work of Alpher, Follin and Herman (1953) and Taylor and Hoyle (1964), preceeding the discovery of the 3K background, and Peebles (1966) and Wagoner, Fowler and Hoyle (1967), immediately following the discovery, and the more recent work of our group of collaborators (Olive et al 1990; Walker et al 1991; Schramm and Wagoner 1977; Boesgaard and Steigman 1985; Yang et al 1984;

Kawano et al. 1988) all do essentially the same basic calculation, the results of which are shown in Fig. 1. As far as the calculation itself goes, solving the reaction network is relatively simple by the standards of explosive nucleosynthesis calculations in supernovae, with the changes over the last 25 years being mainly in terms of more recent nuclear reaction rates as input, not as any great calculational insight, although the current Kawano code (Kawano et al. 1988) is somewhat streamlined relative to the earlier Wagoner code (Wagoner et al. 1967). In fact, the early Wagoner code is, in some sense, a special adaptation of the larger nuclear network calculation developed by Truran (1965; Truran et al 1966) for work on explosive nucleosyntheis in supernovae. With the possible exception of Li yields, the reaction rate changes over the past 25 years have not had any major affect (see Yang et al. (1984) and Krauss and his collaborators (Krauss and Romanelli 1990; Kernan and Krauss 1994) or Copi, Schramm and Turner (1994) for discussion of uncertainties). The one key improved input is a better neutron lifetime determination (Mampe et al. 1989).

With the exception of the effects of elementary particle assumptions to which we will return, the real excitement for BBN over the last 25 years has not really been in redoing the basic calculation. Instead, the true action is focused on understanding the evolution of the light element abundances and using that information to make powerful conclusions. In the 1960's, the main focus was on ^{4}He which is very insensitive to the baryon density. The agreement between BBN predictions and observations helped support the basic Big Bang model but gave no significant information, at that time, with regard to density. In fact, in the mid-1960's, the other light isotopes (which are, in principle, capable of giving density information) were generally assumed to have been made during the T-Tauri phase of stellar evolution (Fowler et al 1962), and so, were not then taken to have cosmological significance. It was during the 1970's that BBN fully developed as a tool for probing the Universe. This possibility was in part stimulated by Ryter et al. (1970) who showed that the T-Tauri mechanism for light element synthesis failed. Furthermore, ^{2}D abundance determinations improved significantly with solar wind measurements (Geiss and Reeves 1971; Black 1971) and the interstellar work from the Copernicus satellite (Rogerson and York 1973). (Recent HST observations reported at this meeting by Linsky have compressed the ^{2}D error bars considerably.) Reeves, Audouze, Fowler and Schramm (1973) argued for cosmological ^{2}D and were able to place a constraint on the baryon density excluding a universe closed with baryons. Subsequently, the ^{2}D arguments were cemented when Epstein, Lattimer and Schramm (1976) proved that no realistic astrophysical process other than the Big Bang could produce significant ^{2}D. It was also interesting that the baryon density implied by BBN was in good agreement with the density implied by the dark galactic halos (Gott et al 1974).

By the late 1970's, a complimentary argument to ^{2}D had also developed using ^{3}He. In particular, it was argued (Rood et al 1976) that, unlike ^{2}D, ^{3}He was made in stars; thus, its abundance would increase with time. Since ^{3}He like ^{2}D monotonically decreased with cosmological baryon density, this argument

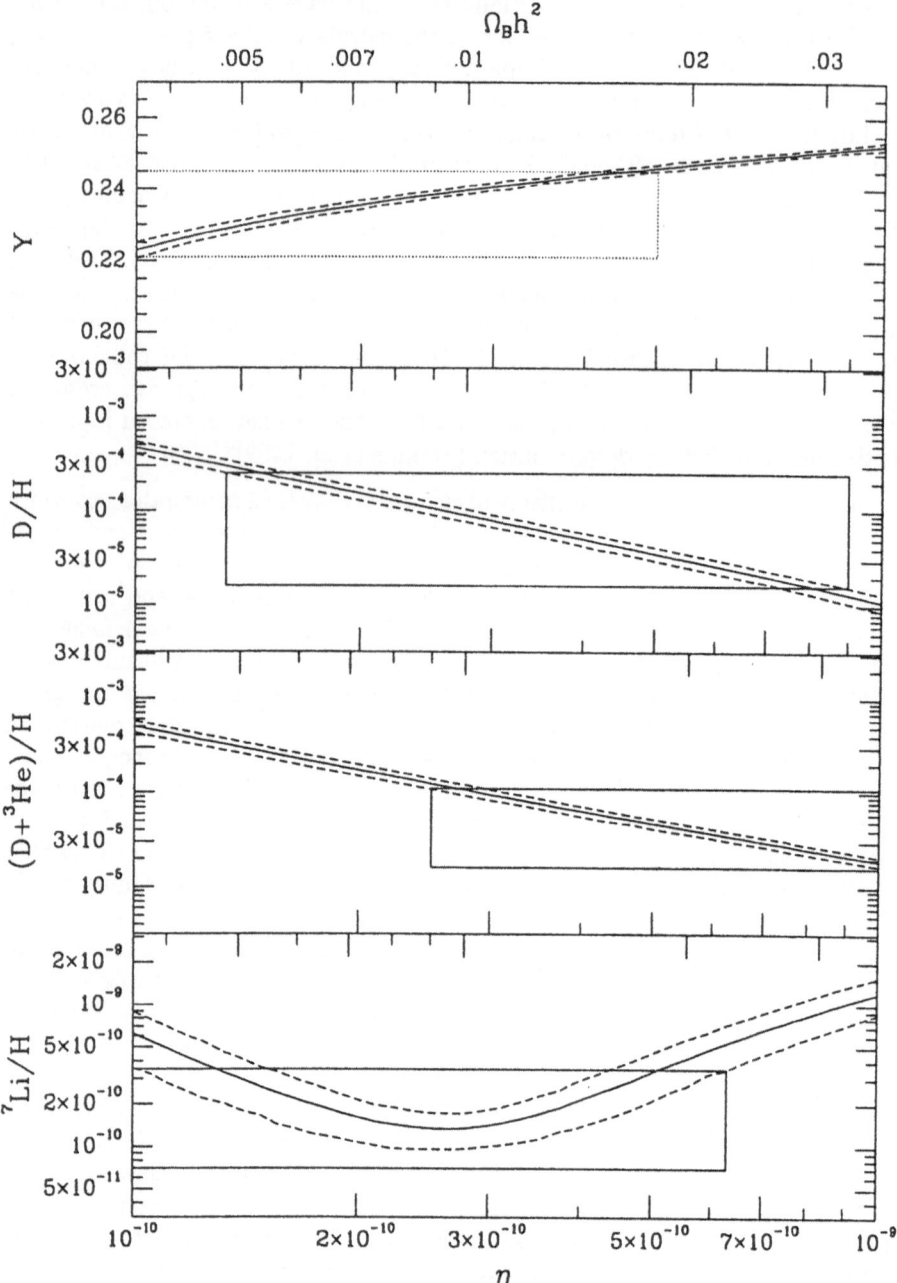

Fig. 1. Big Bang Nucleosynthesis abundance yields versus baryon density (Ω_b) and $\eta \equiv \frac{n_b}{n_\gamma}$ for a homogeneous universe. ($h \equiv H_0/100$ km/sec/Mpc; thus, the concordant region of $\Omega_b h^2 \sim 0.013$ corresponds to $\Omega_b \sim 0.05$ for $H_0 = 50$ km/sec/Mpc.) Figure is from Copi, Schramm and Turner (1994). Note concordance region is slightly larger than Walker et al. (1991) due primarily to inclusion of possible systematic errors on Li/H.

could be used to place a lower limit on the baryon density (Yang et al 1979) using ^{3}He measurements from solar wind (Geiss and Reeves 1971; Black 1971) or interstellar determinations Wilson et al 1983). Since the bulk of the ^{2}D was converted in stars to ^{3}He, the constraint was shown to be quite restrictive (Yang et al. 1984). Rood, Bania and Wilson (1992) showed in print and at Elba that ^{3}He was indeed enriched in planetary nebulae; hence, the argument that ^{3}He increases with time is now strengthened. However, there is nonetheless the worry that interstellar ^{3}He measurements (Balser et al. 1994) do vary with location more than one might expect for an isotope produced primarily in low mass stars. This latter point was the subject of much discussion at Elba.

It was interesting that the lower boundary from ^{3}He and the upper boundary from ^{2}D yielded the requirement that ^{7}Li be near its minimum of ^{7}Li/H $\sim$ 10^{-10}, which was verified by the Pop II Li measurements of Spite and Spite (1982; Rebolo et al. 1988; Hobbs and Pilachowski 1988), hence, yielding the situation emphasized by Yang et al. (1984) that the light element abundances are consistent over nine orders of magnitude with BBN, but only if the cosmological baryon density, Ω_b, is constrained to be around 5% of the critical value (for $H_0 \simeq$ 50 km/sec/Mpc).

The Li plateau argument was further strengthened with the observation of ^{6}Li in a Pop II star by Smith, Lambert and Nissen (1982). Since ^{6}Li is much more fragile than ^{7}Li, and yet it survived, no significant nuclear depletion of ^{7}Li is possible (Olive and Schramm 1992; Steigman et al. 1993). This observation of ^{6}Li has now been verified by Hobbs and Thorburn (1994), and a detection in a second Pop II star has been reported.

The other development of the 70's for BBN was the explicit calculation of Steigman, Schramm and Gunn (1977), showing that the number of neutrino generations, N_ν, had to be small to avoid overproduction of ^{4}He. (Earlier work (Taylor and Hoyle 1964; Schwarzman 1969; Peebles 1971) had commented about a dependence on the energy density of exotic particles but had not done an explicit calculation probing N_ν.) This will subsequently be referred to as the SSG limit. To put this in perspective, one should remember that the mid-1970's also saw the discovery of charm, bottom and tau, so that it almost seemed as if each new detector produced new particle discoveries, and yet, cosmology was arguing against this "conventional" wisdom. Over the years, the SSG limit on N_ν improved with ^{4}He abundance measurements, neutron lifetime measurements, and with limits on the lower bound to the baryon density, hovering at $N_\nu \lesssim 4$ for most of the 1980's and dropping to slightly lower than 4 just before LEP and SLC turned on (Olive et al 1990; Walker et al. 1991; Schramm and Kawano 1989; Patel 1990). This was verified by the LEP results (ALEPH et al. 1993) where now the overall average is $N_\nu = 2.99 \pm 0.03$.

An exciting new observation has been reported by Songaila et al. (1994) and by Carswell et al. (1994) of a possible D/H measurement in a QSO absorption system at $Z = 3.3$. Such a detection would be an extremely important confirmation of the basic BBN argument that deuterium is primordial. However, as Steigman (1994) points out, there is a fair probability that this observation is merely an interloping H cloud with a velocity of 82 km/sec relative to the main cloud and thus mimicing D. Therefore, the implied D/H of $\sim 2.5 \times 10^{-4}$ should be viewed as an upper limit at the present time. It should be noted that such a high value would be hard to fit with relatively low ^{3}He values (few $\times 10^{-5}$) observed in the ISM today unless something is seriously wrong with our understanding of ^{3}He evolution (the D primarily burns to ^{3}He in stars) or the ^{3}He observations are in error. One of these possibilities is not categorically excluded since ^{3}He evolution models, as noted at Elba by Matteuchi and by Tosi, do have trouble simultaneously fitting the solar wind measurements and the recent ISM measurements [see also Vangioni-Flam, Olive and Prantzos (1994) and Olive et al. (1994)]. If an extragalactic D/H determination can eventually be confirmed in several directions, it would provide the firmest determination of the baryon density and may enable a collpase of the present range for ρ_b to an even narrower band. Unfortunately, unpublished reports at Elba of another observation by Carswell et al. of a different QSO system at $Z \sim 3$ appear to set an upper limit on D/H of 6×10^{-5}, which would imply that the higher value was merely a hydrogen interloper.

Table 1 summarizes how attitudes about the various light elements have changed in the past 25 years.

Table 1. Twenty-five Years for the Light Elements

	25 YEARS AGO		PRESENT	
Isotope	Best Observations	Presumed Origin	Best Observations	Presumed Origin
^{2}D	Sea Water	T-Tauri Stars	ISM (HST)	BBN
^{3}He	Solar Flares	Low Mass Stars	Galactic H-II regions planetary nebulae	BBN plus low mass stars
^{4}He	Indirect	BBN	Extragalactic H-II regions	BBN
^{7}Li	Pop I stars	T-Tauri stars	Pop II stars	BBN (Pop I has additional sources
^{6}Li	Meteorites	T-Tauri stars	Pop II stars	Cosmic Ray Spallation
Be	Pop I stars	T-Tauri stars	Pop II stars	Cosmic Ray Spallation
B	Meteorites	T-Tauri stars	Pop II stars (HST)	Cosmic Ray Spallation (possible additional source for ^{11}B)

The power of homogeneous BBN comes from the fact that essentially all of the physics input is well determined in the terrestrial laboratory. The appropriate temperature regimes, 0.1 to 1 MeV, are well explored in nuclear physics labs. Thus, what nuclei do under such conditions is not a matter of guesswork, but is precisely known. In fact, it is known for these temperatures far better than it is for the centers of stars like our sun. The center of the sun is only a little over 1 keV, thus, below the energy where nuclear reaction rates yield significant results in laboratory experiments, and only the long times and higher densities available in stars enable anything to take place.

To calculate what happens in the Big Bang, all one has to do is follow what a gas of baryons with density ρ_b does as the universe expands and cools. As far as nuclear reactions are concerned, the only relevant region is from a little above 1 MeV ($\sim 10^{10}$K) down to a little below 100 keV ($\sim 10^9$ K). At higher temperatures, no complex nuclei other than free single neutrons and protons can exist, and the ratio of neutrons to protons, n/p, is just determined by $n/p = e^{-Q/T}$, where $Q = (m_n - m_p)c^2 \sim 1.3$ MeV. Equilibrium applies because the weak interaction rates are much faster than the expansion of the universe at temperatures much above 10^{10}K. At temperatures much below 10^9K, the electrostatic repulsion of nuclei prevents nuclear reactions from proceeding as fast as the cosmological expansion separates the particles.

After the weak interaction drops out of equilibrium, a little above 10^{10}K, the ratio of neutrons to protons changes more slowly due to free neutrons decaying to protons, and similar transformations of neutrons to protons via interactions with the ambient leptons. By the time the universe reaches 10^9K (0.1 MeV), the ratio is slightly below 1/7. For temperatures above 10^9K, no significant abundance of complex nuclei can exist due to the continued existence of gammas with energies greater than MeV. Note that the high photon to baryon ratio in the universe ($\sim 10^{10}$) enables asignificant population in the MeV high energy Boltzman tail until $T \lesssim 0.1$MeV.

Once the temperature drops to about 10^9K, sufficient abundances of nuclei can exist in statistical equilibrium through reactions such as $n + p \leftrightarrow {}^2D + \gamma$ and ${}^2D + p \leftrightarrow {}^3He + \gamma$ and ${}^2D + n \leftrightarrow {}^3H + \gamma$, which in turn react to yield ${}^4He$. Since ${}^4He$ is the most tightly bound nucleus in the region, the flow of reactions converts almost all the neutrons that exist at 10^9K into ${}^4He$. (See Fig. 2 for the full reaction rate network used by Copi, Schramm and Turner (1994). The flow essentially stops there because there are no stable nuclei at either mass-5 or mass-8. Since the baryon density at Big Bang Nucleosynthesis is relatively low (about 1% the density of terrestrial air) and the time-scale short ($t \lesssim 10^2$ sec), only reactions involving two-particle collisions occur. It can be seen that combining the most abundant nuclei, protons and ${}^4He$ via two body interactions always leads to unstable mass-5. Even when one combines ${}^4He$ with rarer nuclei like ${}^3H$ or ${}^3He$, we still get only to mass-7, which, when hit by a proton, the most abundant nucleus around, yields mass-8. (As we will discuss, a loophole around the mass-8 gap can be found if $n/p > 1$, so that excess neutrons exist, but for the standard case $n/p < 1$). Eventually, ${}^3H$ decays to ${}^3He$, and any mass-7 made decays to ${}^7Li$. Thus, Big Bang Nucleosynthesis makes ${}^4He$ with traces of ${}^2D$, ${}^3He$, and ${}^7Li$. (Also, all the protons left over that did not capture neutrons remain as hydrogen.) For standard homogeneous BBN, all other chemical elements are made later in stars and in related processes (such as cosmic ray spallation). (Stars jump the mass-5 and -8 instability by having gravity compress the matter to sufficient densities and have much longer times available so that three-body collisions, $3{}^4He \rightarrow {}^{12}C$, can occur.)

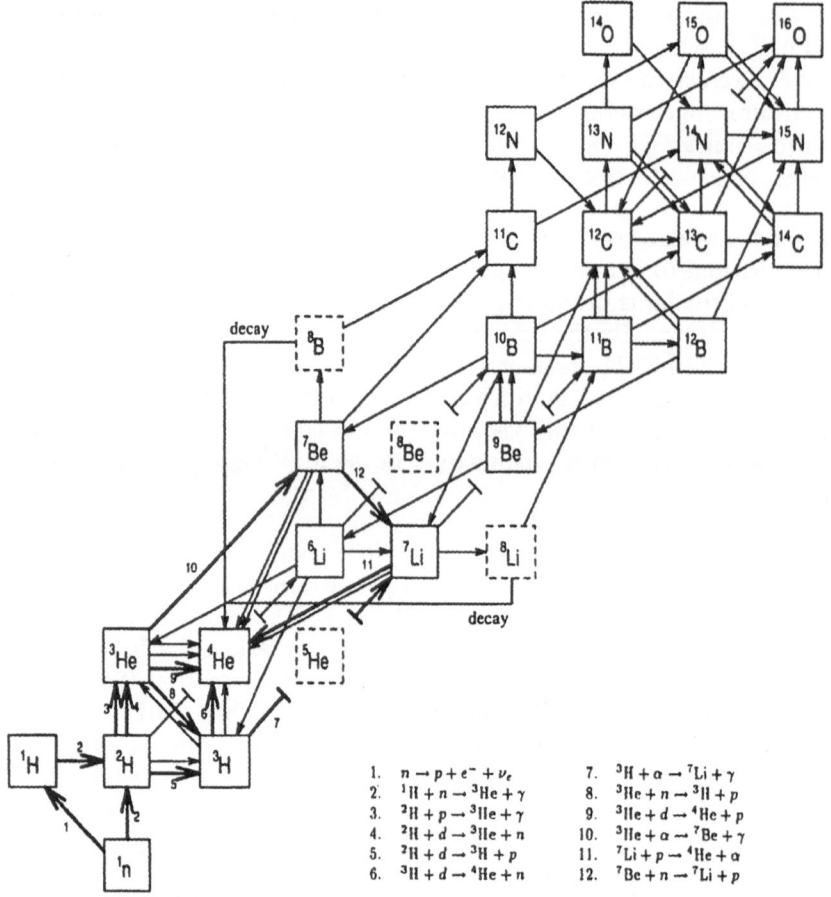

Fig. 2. Full reaction network used by Copi, Schramm and Turner (1994). Note that only lower half of network is all that matters for standard BBN.

3   Systematic Errors

Perhaps the most pervasive theme of the Elba meeting was the focus on potential systematic errors in light element abundance determinations. Comparisons between stellar nucleosynthesis theory and observed abundances usually need only relative abundances of heavy elements so that many systematic errors cancel. Furthermore, most comparisons are content with agreement at the factor of a few level. This, in part, has resulted in astronomical spectroscopists not explicitly showing their full range of systematic error with their results. Comments, such as "model atmosphere used," are in the text, but explicit separate error ranges for different model atmosphere assumptions are rarely provided. Unfortunately for BBN, the absolute abundances relative to H are critical, and model atmosphere effects do not "cancel." This latter point was brought into great focus by the factor of almost two differences in the value of Li/H on the Spite plateau when different atmosphere assumuptions were made. While the plateau remains robust, its absolute value is more uncertain than one might suspect from reading any single lithium observation paper. Furthermore, the possibility of stellar lithium depletion at a level of a factor of two cannot be ruled out by the ^{6}Li observations, although higher values are unlikely. (The Pop I depletion model of Faulkner presented at Elba suggested the lack of any need for forcing large depletions on Pop II stars.) Note that most systematics do cancel for ^{6}Li/^{7}Li.

Perhaps the least susceptible to systematics is the D/H value from HST in the ISM. But this only gives a lower limit to the primordial D/N, since D is destroyed in stars. Furthermore, one also has to worry about H interlopers. The helium-3 variations in galactic H-II regions show that ^{3}He destruction in massive stars is probably non-negligible. Thus, it is probably inappropriate to use the H-II region data (it would be like the drunk looking for his keys under the lampost even though he dropped them in a dark alley). However, the observation of ^{3}He/H $\sim 10^{-3}$ in planetary nebulae is in good agreement with low mass stellar evolution theory. Therefore, using the solar system $\frac{^3He+D}{H}$ value as determined in the solar wind and in meteorites may be the best value to use. Potential Helium-4 systematics were discussed in detail by Pagel (1994) at Elba and by Skillman (1994) at Snowmass. Pagel showed that a systematic shift for Y to the high side, as large as 0.01, was allowed and to the low side, 0.005.

Given that systematics are so important for light element abundances, observers should be encouraged to quote explicitly the possible range of systematics as an additional error on top of and separate from the traditional statisical error. This is the practice in particle physics, and since BBN is perhaps approaching a similar level of fundamental impact, we should be similarly cautious in how we quote results. It should also be noted that systematic errors are frequently not gaussian distributed. Thus, it is quite inappropriate to add them in quadrature with the statistical errors. It is for this reason that complex maximum likelihood calculations of BBN are also quite inappropriate and misleading (at least when only gaussian distributions are presumed).

Table 2 summarizes my interpretation of the current light element abundances with my estimate of their uncertainties. The numbers are drawn from discussions at Elba and Snowmass and from Copi, Schramm and Turner's (1994) summary. The only significant difference with Walker et al.(1991) is in the opening of the lithium range due to systematic errors that were not explicitly presented in the observational papers at that time.

Table 2. Current Estimates of Primordial Light Element Abundances

(First ± is statistical, second is systematic. Limits are extreme of systematic + 2σ statistical.)

Isotope	Primordial Value	Note	Reference
^{4}He	$Y_o = 0.232 \pm 0.03^{+0.01}_{-0.005}$	Ex-H-II region	Olive & Steigman (1994)
^{2}D + ^{3}H	$\left.\frac{D+^3H}{H}\right\|_o \leq 1.1 \times 10^{-4}$	Solar wind with Max-correction for possible ^{3}He depletion	Pagel (1990) Skillman (1994) Geiss (1993)
^{2}D	$\left.D/H\right\|_o > 1.4 \times 10^{-5}$	ISM (HST)	Linsky et al. (1933)
^{6}Li	$\left.\frac{^7Li}{H}\right\|_o = 1.4 \pm 0.3^{+1.8}_{-0.4} \times 10^{-10}$	Systematic includes possible depletion in stars and spallogenic addition	Spite & Spite (1982) Thorburn (1994)

The limits shown above (or any others for that matter) can be plugged into the following fitting formula to obtain a limit on the number of equivalent neutrino species allowed by BBN:

$$N_\nu \lesssim 3.4 + 0.08 \left[\frac{\left.Y_0\right|_{MAX} - 0.245}{0.001} \right] + \left[\frac{\left.\frac{D+^3H_t}{H}\right|_0 - 1.1 \times 10^{-4}}{10^{-4}} \right]$$

4 Inhomogeneous Big Bang Nucleosynthesis

As noted above, BBN yields all agree with observations using only one freely adjustable parameter, $\eta = n_b/n_\gamma$, or equivalently, the baryon density, ρ_b (since n_γ is directly related to the temperature of the microwave background.) Thus, BBN can make strong statements regarding ρ_b if the observed light element abundances cannot be fit with any alternative theory. The most significant alternative that has been discussed involves quark-hadron transition inspired inhomogeneities (Scherrer et al. 1987; Alcock et al. 1987; Fowler and Malaney 1988). While inhomogeneity models had been looked at previously (Yang et al. 1984) and were found to make little difference, the quark-hadron inspired models had the added ingredient of variations in n/p ratios. Cosmologists are well aware that current trends in lattice gauge calculations imply that the transition is probably second order or not a phase transition at all. Nevertheless, it has been important to explore the maximal cosmolgical impact that can occur. This maximal impact requires a first-order phase transition.

The initial claim by Applegate et al. (1988), followed by a similar argument from Alcock et al. (1987), that $\Omega_b \sim 1$ might be possible, created tremendous interest. Their argument was that if the quark-hadron transition was a first-order phase transition, then it was possible that large inhomogeneities could develop at T $\gtrsim$ 100 MeV. The preferential diffusion of neutrons versus protons out of the high density regions could lead to Big Bang Nucleosynthesis occurring under conditions with both density inhomogeneities and variable neutron/proton ratios. In the first round of calculations, it was claimed that such conditions might allow $\Omega_b \sim 1$, while fitting the observed primordial abundances of ^{4}He, ^{2}D, ^{3}He with an overproduction of ^{7}Li. Since ^{7}Li is the most recent of the cosmological abundance constraints and has a different observed abundance in Pop I stars versus the traditionally more primitive Pop II stars (Spite and Spite 1982; Rebolo et al. 19088; Hobbs and Pilachowski 1988), some argued that perhaps some special depletion process might be going on to reduce the excess ^{7}Li. Today we know from the ^{6}Li observation (Smith et al. 1982) that such depletion is unlikely but, at the time, this option was more viable.

At first it appeared that if the lithium constraint could be surmounted, then the density constraints of standard Big Bang Nucleosynthesis might disintegrate. To further stimulate the flow through this new loophole, Fowler and Malaney (1988) showed that, in addition to looking at the diffusion of neutrons out of high density regions, one must also look at the subsequent effect of excess neutrons diffusing back into the high density regions as the nucleosynthesis goes to completion in the low density regions. (The initial calculations treated the two regions separately.) Fowler and Malaney (1988) argued that for certain phase transition parameter values (e.g. nucleation site separations ~ 10m at the time of the transition), this back diffusion could destroy much of the excess lithium.

However, Kurki-Suonio, Matzner, Olive and Schramm (1990), the Tokyo group (Sato and Tarasawa 1991), and the Livermore group (Meyer et al 1991) have eventually argued that in their detailed diffusion models, the back diffusion not only affects ^{7}Li, but also the other light nuclei as well. They find that for $\Omega_b \sim 1$, ^{4}He is also overproduced (although it does go to a minimum for similar parameter values as does the lithium). Added to this has been the detailed work of Jedamzik and Fuller (1994) on lithium in inhomogeneous models showing that, despite its sensitivity to assumptions, it nonetheless supports the conclusion that inhomogeneous models yield essentially the same constraints on Ω_b as homogenous models do. One can understand why these models might tend to overproduce ^{4}He and ^{7}Li by remembering that in standard homogeneous Big Bang Nucleosynthesis, high baryon densities lead to excesses in these nuclei. As back diffusion evens out the effects of the initial fluctuation, the averaged result should approach the homogeneous value. Furthermore, it can be argued that any narrow range of parameters, such as those which yield relatively low lithium and helium, are unrealistic since in most realistic phase transitions there are distributions of parameter values (distribution of nucleation sites, separations, density fluctuations, etc.). Therefore, narrow minima are washed out which would bring the ^{7}Li and ^{4}He values back up to their excessive levels for all parameter values with $\Omega \sim 1$. Furthermore, Freese and Adams (1990) and Link (1992) have argued that the boundary between the two phases may be fractal-like rather than smooth. The large surface area of a fractal-like boundary would allow more interaction between the regions and minimize exotic effects.

Fig. 3 shows the updated results of Kurki-Suonio et al. (1990) for varying spacing l with the constraints from the different light element abundances. This can be compared to the current state-of-the-art work of Thomas et al. (1994) where essentially the same conclusions are reached. Notice that the Li, ^{2}D, and even the ^{4}He constraint do not allow $\Omega_b \sim 1$. Note also that with the Pop II ^{7}Li constraint, the results for Ω_b are quite similar to the standard model. Only by relaxing the ^{7}Li constraint and by using the ^{4}He and/or ^{2}D to bound the baryon density can even factors of ~ 2 variations in Ω_b be allowed. Thus, even an optimally tuned first order quark-hadron transition is not able to alter the basic conclusions of homogeneous BBN regarding Ω_b. (It also cannot significantly change the N_ν argument.) In fact, the main role that a quark-hadron option has played for BBN is to show how robust the standard model results are.

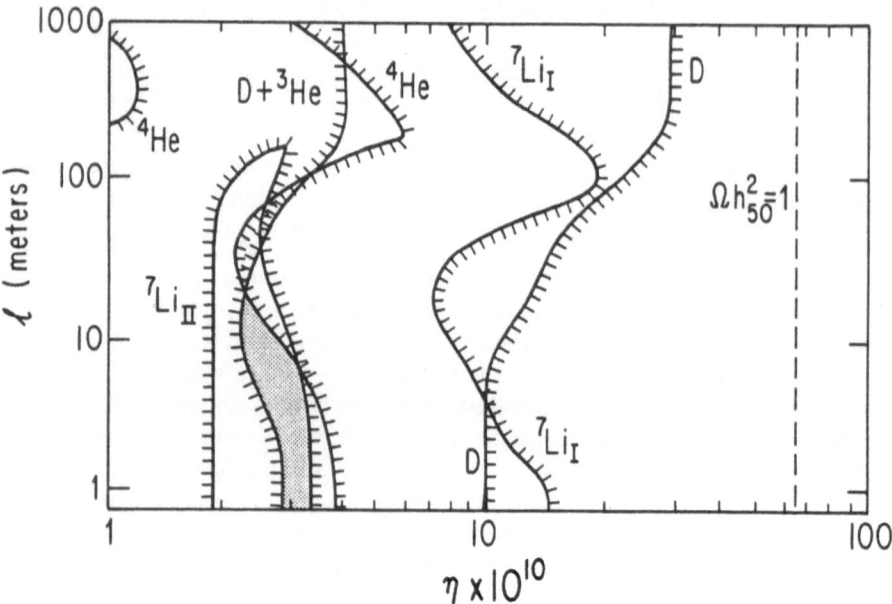

"QUARK-HADRON" NUCLEOSYNTHESIS

Fig. 3. The updated Kurki-Suonio, Matzner, Olive and Schramm (1990) results showing that even allowing for a 1st order quark-hadron transition with variable spacing l
between nucleation sites, the light element abundances constrain the baryon to photon
ratio, η, and thus Ω_b to essentially the same values as those obtained in the homogeneous case with only slightly large Ω_b's possible with $l \sim 100$. Note, even if the
Pop II lithium constraint is dropped, the ^{4}He constraint excludes values of η above
$\sim 5 \times 10^{-10}$, still far below $\Omega = 1$.

5 Boron, Beryllium and the Spallation Process

While quark-hadron inspired variations have not been able to alter the basic
conclusions of BBN, an important question remains, namely, is there an observable signature that could differentiate quark-hadron inspired variations from the
homogeneous model? On the theoretical side, this point was initially debatable.
Several authors (Boyd and Kajino 1990; Applegate et al. 1988) have argued that
because of the high n/p region in the inhomogeneous models, leakage beyond the
mass 5 and 8 instability gaps can occur and traces of ^{9}Be, ^{10}B, ^{11}B and maybe
even r-process elements can be produced. Thus, detection of nuclei beyond ^{7}Li
in primitive objects may be a signature. However, Sato and Tarasawa (1991) and
Thomas et al. (1994) have argued that such leakage is negligible.

Let us now comment on some recent observations of Be and B in primitive
Pop II stars. In particular, there has been much recent attention given to reports
(Gilmore et al. 1991; Ryan et al. 1990; Ryan et al. 1992) of beryllium lines being

observed in extreme Pop II stars. Boesgaard reported on these at Elba. For one very metal poor Pop II star, HD 140283, boron was also observed (Duncan et al. 1992). These observations showed Be/H and B/H ratios that scaled with the metalicity of the observed stars. Thus, contrary to some initial claims, the Be/H and B/H observations (Kiselman 1994) do not require cosmological origin, only a scaling with oxygen of the same process that produced Be and B in the Pop I stars. Recent arguments discussed at Elba on non-LTE effects in the Be and B determinations do not in any way alter these conclusions, although they may imply a higher B/Be. [See discussion of Fields, Olive and Schramm (1994).]

The presumed process that produced Be and B in Pop I stars (as well as the ^{6}Li), is thought to be cosmic ray spallation (Reeves et al. 1970). For Be and B, such spallation comes from the breakup of heavy nuclei such as CNO and Ne, Mg, Si, S, Ca and Fe by protons and alphas. As noted by Epstein, Arnett and Schramm (1974; 1976) for lithium one must also include alpha plus alpha fusion processes as well. This latter point was well noted by Steigman and Walker (1992) and by Walker et al. (1993) who emphasized that Be and B spallation production on Pop II abundances would imply a significant enhancement of lithium from alpha-alpha relative to the reduced production of Be and B from depleted heavy nuclei. While the ^{6}Li so produced would be destroyed at the base of the convective zones in most stars observed (Brown and Schramm 1988; Pinsonneault et al. 1992) it did manage to survive in at least one extremely high temperature (thin convection zone) star (Smith et al. 1982; Hobbs and Thorburn 1994).

Perhaps most critical to any spallation origin is the resultant B/Be ratio. It is also known, from actual measurements, that the cosmic rays themselves (Simpson 1983) show B/Be $\sim$ 14 (and B and Be are pure spallation products in the cosmic rays) with a carbon to oxygen ratio exceeding unity (Pop I has C/O $\lesssim$ 0.5). Since spallation off carbon favors B relative to Be (mass 11 requires only a single nucleon ejected from mass 12), whereas oxygen being farther from either shows less favoritism, the cosmic ray observations are actually an upper limit on what B/Be ratio one might expect in Pop II cosmic rays. [See also Fields, Olive and Schramm (1994).] However, of more concern here is the lower limit on B/Be achievable by a spallation process (Walker et al. 1993). The resultant limit is B/Be $\simeq$ 7 which means that cosmic ray spallation can yield $7 \lesssim B/Be \lesssim 14$. In other words, yields can still not get a B/Be ratio below 7. Furthermore, non-spallation mechanisms such as ν-process (Woosley et al. 1990) in supernovae [see also Dearborn et al. (1989)] or a low energy cosmic ray spike (Meneguzzi et al. 1971; Fields et al 1994) can enhance B/Be, but no known process can produce Be without making B. Thus, the lower bound on B/Be is firm, whereas values above 14 would only be indicative of a B enhancing addition.

Figures 4A and 4B show the trace element yields in a standard homogeneous BBN calculation with Figure 4A showing ^{2}H, ^{3}He, ^{6}Li and ^{7}Li, and Figure 4B showing the ^{9}Be, ^{10}B and ^{11}B yields. This work is part of an extensive study of $A \geq 6$ BBN by Thomas et al. (1993) using a more extensive reaction network than previously used (see Figure 2). Note, in particular, that ^{9}Be/H yields are always less than 10^{-14} regardless of η. (Also note that for the standard model, B/Be $\gg 10$ unless $\eta \lesssim 3 \times 10^{-10}$.) Also note that for $\eta \sim 3 \times 10^{-10}$ where agreement with the traditional BBN light element abundances occurs, Be/H and B/H $\sim 10^{-18}$. In other words, homogeneous BBN cannot yield Be/H consistent with the Pop II stellar observations. As Thomas et al. (1994) noted, even inhomogeneities do not significantly alter these conclusions. Thus, at first glance an observed plateau in Be/H and B/H versus low metalicity might be a problem for BBN. However, Mathews (1994) has argued that galactic evolution can produce such a partial plateau for extremely low metalicities with no cosmological origin for Be and B.

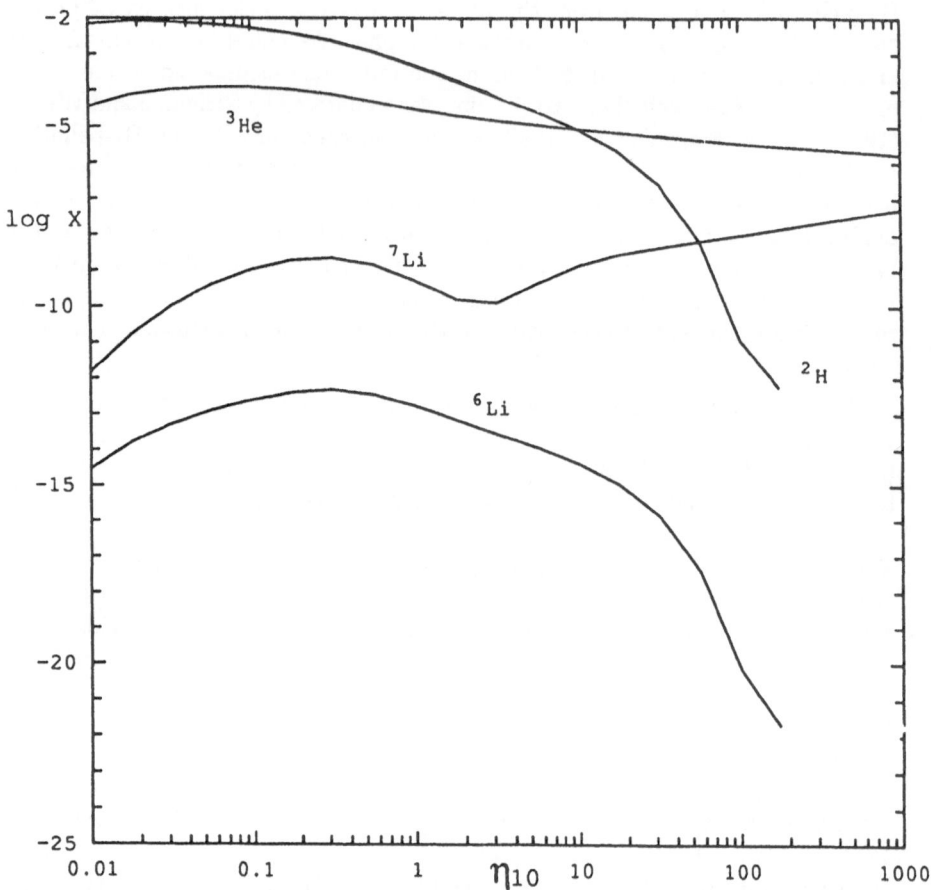

Fig. 4A. The standard homogeneous BBN yields showing ^{2}H, ^{3}He, ^{6}Li and ^{7}Li for 6 orders of magnitude in n_b/n_γ. Note that ^{6}Li is always negligible relative to ^{7}Li.

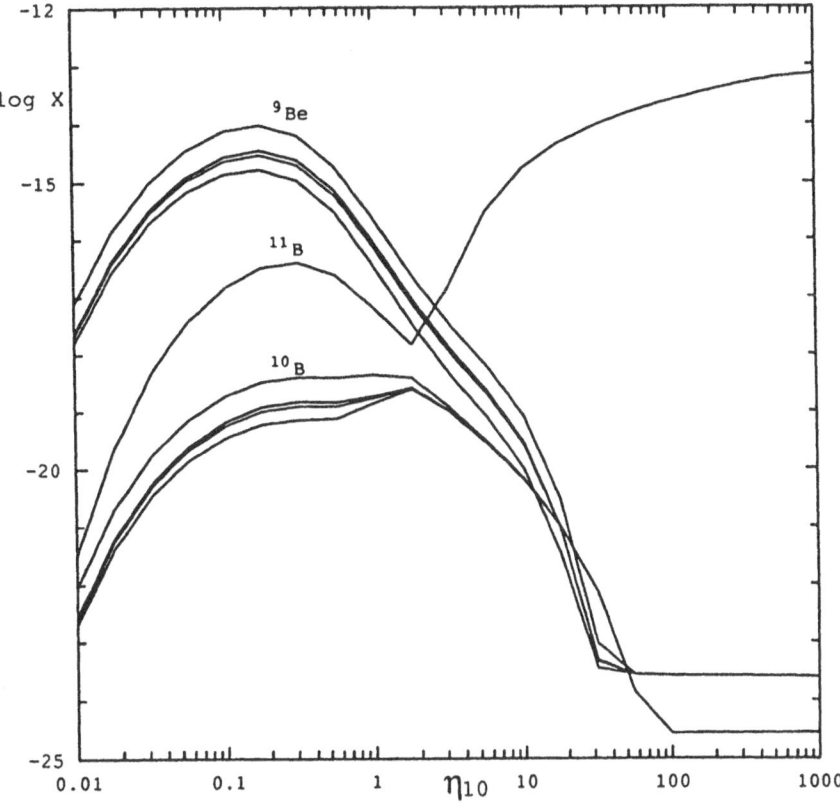

Fig. 4B. The standard homogeneous BBN yields for ^{9}Be, ^{10}B and ^{11}B. The various curves for ^{9}Be and ^{10}B represent different cross section assumptions. The ^{11}B yield is double humped due to production both directly as ^{11}B and also as ^{11}C which beta decays to ^{11}B. The appropriate region of this graph which fits $A \leq 7$ abundances is near $\eta_{10} \sim 3$, so B/H $\sim$ Be/H $\sim 10^{-18}$.

6 Limits on Ω_b and Dark Matter Requirements

The success and robustness of BBN in the face of the Be and B results as well as the quark-hadron variations gives renewed confidence to the limits on the baryon density constraints. Let us convert this density regime into units of the critical cosmological density for the allowed range of Hubble expansion rates. For the Big Bang nucleosynthesis constraints of Table 2, the dimensionless baryon density, Ω_b, that fraction of the critical density that is in baryons, is less than 0.14 and greater than 0.02 for $0.04 \lesssim h_0 \lesssim 0.7$, where h_0 is the Hubble constant in units of 100km/sec/Mpc. The lower bound on h_0 comes from direct observational limits and the upper bound from age of the universe constraints (Freese and Schramm 1984). The constraint on Ω_b still means that the universe *cannot be closed with baryonic matter.* [This point was made over twenty years ago (Reeves et al. 1970) and has proven to be remarkably strong.] If the universe is truly at its critical density, then nonbaryonic matter is required. This argument has led to one of the major areas of research at the particle-cosmology interface, namely, the search for non-baryonic dark matter.

Another important conclusion regarding the allowed range in baryon density is that it is in very good agreement with the density implied from the dynamics of galaxies, *including their dark halos*. An early version of this argument, using only deuterium, was described over twenty years ago (Gott et al. 1974). As time has gone on, the argument has strengthened, and the fact remains that galaxy dynamics and nucleosynthesis agree at about 5% of the critical density. The recent MACHO and EROS (Alcock et al. 1993; Aubourg et al. 1993) reports of halo microlensing may well indicate that at least some of the dark baryons are in the form of brown dwarfs in the halo. Note also that if the universe is indeed at its critical density, as many of us believe, it requires most matter not to be associated with galaxies and their halos, as well as to be nonbaryonic. Let us put the nucleosynthetic arguments in context.

The arguments requiring some sort of dark matter fall into separate and quite distinct areas. These arguments are summarized in Fig. 5. First are the arguments using Newtonian mechanics applied to various astronomical systems that show that there is more matter present than the amount that is shining. It should be noted that these arguments reliably demonstrate that galactic halos seem to have a mass ~ 10 times the visible mass.

Note however that Big Bang nucleosynthesis requires that the bulk of the baryons in the universe be dark since $\Omega_{vis} \ll \Omega_b$. Thus, the dark halos could in principle be baryonic (Gott et al 1974; Alcock et al. 1993; Aubourg et al. 1993). Recently, arguments (Fisher 1992) on very large scales (bigger than clusters of galaxies) hint that Ω on those scales is indeed greater than Ω_b, thus forcing us to need non-baryonic matter.

An Ω of unity is, of course, preferred on theoretical grounds since that is the only long-lived natural value for Ω, and inflation (Guth 1981; Linde 1990) or something like it provided the early universe with the mechanism to achieve that value and thereby solve the flatness and smoothness problems. Note that our need for exotica is not dependent on the existence of dark galatic halos. This point is frequently forgotten, not only by some members of the popular press but occasionally by active cosmologists.

Non-baryonic matter can be divided following Bond and Szalay (1992) into two major categories for cosmological purposes: hot dark matter (HDM) and cold dark matter (CDM). Hot dark matter is matter that is relativistic until just before the epoch of galaxy formation, the best example being low mass neutrinos with $m_\nu \sim 20eV$. (Remember $\Omega_\nu \sim m_\nu(eV)/100h_0^2$). Cold dark matter is matter that is moving slowly at the epoch of galaxy formation. Because it is moving slowly, it can clump on very small scales, whereas HDM tends to have more difficulty in being confined on small scales. Examples of CDM could be massive neutrino-like particles with masses, M_x, greater than several GeV or the lightest super-symmetric particle which is presumed to be stable and might also have masses of several GeV. Following Michael Turner (Turner et al. 1993), all such weakly interacting massive particles are called "WIMPS." Axions, while very

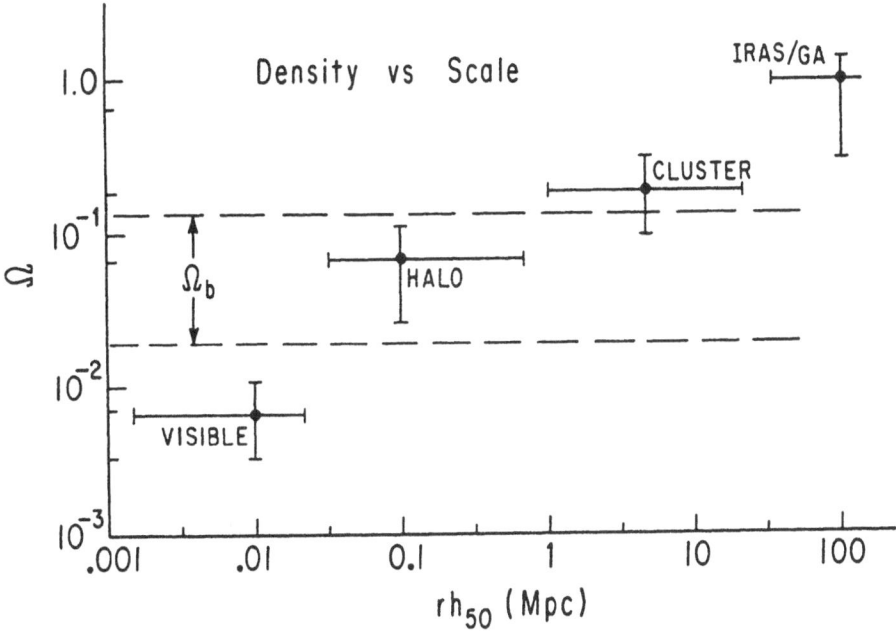

Fig. 5. Implied densitites versus the scale of the measurements.

light, would also be moving very slowly and thus would clump on small scales. Or, one could also go to non-elementary particle candidates, such as planetary mass blackholes or quark nuggets of strange quark matter, possibly produced at the quark-hadron transition (Crawford and Schramm 1982; Alcock and Olinto 1988). Another possibility would be any sort of massive toplogical remnant left over from some early phase transition. Note that CDM would clump in halos, thus requiring the dark baryonic matter to be out between galaxies, whereas HDM would allow baryonic halos. The MACHO and EROS (Alcock et al. 1993; Aubourg et al. 1993) events may favor HDM depending on what fraction of the halos can eventually be explained by the MACHO/EROS events.

While the exact nature of the seeds to stimulate the clumping has yet to be fully determined, COBE's discovery of large scale fluctuations in the microwave background probably shows that some sort of seeds did indeed exist and that the basic scenario is probably on the right track.

Some baryonic dark matter must exist since we know that the lower bound from Big Bang Nucleosynthesis is greater than the upper limits on the amount of visible matter in the universe. However, unless the microlensing turns out to make up the bulk of the halo, we do not know what form this baryonic dark matter is in. It could be either in condensed objects in the halo, as MACHO and

EROS may imply, such as brown dwarfs and jupiters (objects with $\lesssim 0.08 M_\odot$ so they are not luminous stars), or in black holes (which at the time of nucleosynthesis would have been baryons). Or, if the baryonic dark matter is not in the halo, it could be in hot intergalactic gas, hot enough not to show absorption lines in the Gunn-Peterson test, but not so hot as to be seen in the x-rays. The exciting report by Jakobsen et al. (1994) at Elba of a Gunn-Peterson effect observed with HST for He-II at high Z showed that at least some hot IGM exists (and verifies that He seems primordial).

Another possible hiding place for the dark baryons would be failed galaxies, large clumps of baryons that condense gravitationally but did not produce stars. Such clumps are predicted in galaxy formation scenarios that include large amounts of biasing where only some fraction of the clumps shine. Evidence for some hot gas is found in clusters of galaxies from ROSAT. In particular, Mushotzky (1993) and White et al. (1993) have discussed how certain observed rich clusters have $M_{HOTGAS}/M_{TOTAL} \sim 1/4$, which, with $\Omega_{CLUSTER} \sim 0.2$ and $\Omega_b \sim 0.05$, would imply no conflict with BBN, but that clusters are not fair samples of the baryon to non-baryon ratio in the universe. Apparent variation (Mushotzky 1993)in M_{HOT}/M_{TOTAL} for small groups relative to clusters seems to support the point of view that megaparsec scales are not always fair samples. If true, this is difficult to reconcile with cold dark matter models but can be fit with topological defects and HDM or with mixed models. (However, present uncertainties do not confidently exclude eventual agreement all the way around which would prove supportive for CDM.)

Hegyi and Olive (1986) have argued that dark baryonic halos are unlikely. However, they do allow for the loopholes mentioned above of low mass objects or of massive black holes. It is worth noting that these loopholes are not that unlikely. If we look at the initial mass function for stars forming with Pop I composition, we know that the mass function falls off roughly like a power law for standard size stars as was shown by Salpeter. Or, even if we apply the Miller-Scalo mass function, the fall off is only a little steeper. In both cases there is some sort of lower cut-off near $0.1 M_\odot$. However, we do not know the origin of this mass function and its shape. No star formation model based on fundamental physics predicts it.

We do believe that whatever is the origin of this mass function, it is probably related to the metalicity of the materials, since metalicity affects cooling rates, etc. It is not unreasonable to expect the initial mass function that was present in the primordial material which had no heavy elements (only the products of big bang nucleosynthesis) would be peaked either much higher than the present mass function or much lower—higher if the lower cooling from low metals resulted in larger clumps, or lower if some sort of rapid cooling processes ("cooling flows") were set up during the initial star formation epoch, as seems to be the case in some primitive galaxies. In either case, moving either higher or lower produces the bulk of the stellar population in either brown dwarfs and jupiters or in massive black holes. Thus, the most likely scenarios are that a first generation of condensed objects would be in a form of dark baryonic matter that could make up the halos, and could explain why there is an interesting coincidence between the implied mass in halos and the implied amount of baryonic material. However, it should also be remembered that a consequence of this scenario is to have the condensation of the objects occur prior to the formation of the disk. If the first large objects to form are less than galactic mass, as many scenarios imply, then mergers are necessary for eventual galaxy size objects. Mergers stimulate star formation while putting early objects into halos rather than disks. Mathews and Schramm (1993) have recently developed a galactic evolution model which does just that and gives a reasonable scenario for chemical evolution. Thus, while making halos out of exotic material may be more exciting, it is certainly not impossible for the halos to be in the form of dark baryons. The microlensing reported by MACHO and EROS is not a problem but may tell us something about early star formation.

7 Conclusion

Primordial nucleosynthesis is more robust than ever and has become a pillar of the Big Bang. Recent action from new observations has strengthened and enriched the picture. However, one must be more explicit and careful in how systematic errors are presented.

8 Acknowledgments

I would like to thank my recent collaborators, David Dearborn, Brian Fields, Dave Thomas, Gary Steigman, Brad Meyer, Craig Copi, Keith Olive, Frank Timmes, Michael Turner, Rocky Kolb, Grant Mathews, Bob Rood, Jim Truran and Terry Walker for many useful discussions. I would further like to thank Poul Nissen, Jeff Linsky, Julie Thorburn, Doug Duncan, Lew Hobbs, Evan Skillman, Bernard Pagel and Don York for valuable discussion regarding the astronomical observations. This work was supported in part by NSF Grant AST 90-22629, by NASA Grant NAGW-1321 and by DoE Grant DE-FG02-91-ER40606 at the University of Chicago, and by the DoE and NASA Grant NAGW-2381 at the NASA/Fermilab Astrophysics Center. This manuscript was prepared at the Aspen Center for Physics.

References

Alcock, C. et al. (1993) Nature 365 621
Alcock, C., Fuller, G., Mathews, G. (1987) Astrophys. J. 320 439
Alcock, C., Olinto, A. (1988) Ann. Rev. Nuc. Part. Phys. 38 161
ALEPH, L3, OPAL, DELPHI results (1993) 1993 Lepton-Photon meeting at Ithaca, NY
Alpher, R.A., Bethe, H., Gamow, G. (1948) Phys. Rev. 73 803
Alpher, R.A., Follin, J.W., Herman, R.C. (1953) Phys. Rev. 92 1347

Applegate, J., Hogan, C., Scherrer, R. (1988) Astrophys. J. **329** 572

Aubourg, E. et al. (1993) Nature **365** 623

Balser, D.S. et al., (1994) Astrophys. J., sumbitted

Black, D. (1971) Nature **234** 148

Boesgaard, A., Steigman, G. (1985) Ann. Rev. of Astron. and Astrophys. **23** 319

Bond, R., Szalay, A. (1992) in Proc. Texas Relativistic Astrophysical Symposium, Austin, Texas.

Boyd, R., Kajino, T. (1990) Astrophys. J. **359** 267

Brown, L., Schramm, D.N. (1988) Astrophys. J. Lett. **329** L103

Carswell, R.F. et al. (1994) MNRAS, in press

Copi, C., Schramm, D.N., Turner, M. (1994) FERMILAB-pub-94/174-A and Science, submitted

Crawford, M., Schramm, D.N. (1982) Nature **298** 538

Dearborn, D., Schramm, D.N., Steigman, G., Truran, J.W. (1989) Astrophys. J. **347** 455

Duncan, D., Lambert, D. Lemke, D. (1992) Astrophys. J. **401** 584

Epstein, R., Arnett, W.D., Schramm, D.N. (1974) Astrophys. J. **190** L13

Epstein, R., Arnett, W.D., Schramm, D.N. (1976) Astrophys. J. Supp. **31** 111

Epstein, R., Lattimer, J., Schramm, D.N. (1976) Nature **263** 198

Fields, B., Olive, K., Schramm, D.N. (1994) Astron. and Astrophys., in press

Fisher, C. (1992) Ph.D. Thesis, University of California at Berkeley, and references therein

Fowler, W.A., Greenstein, J., Hoyle, F. (1962) Geophys. J.R.A.S. **6** 6

Fowler, W.A., Malaney, R. (1988) Astrophys. J. **333** 14

Freese, K., Adams, F. (1990) Phys. Rev. D **41** 2449

Freese, K., Schramm, D.N. (1984) Nucl. Phys. B **233** 167

G eiss, J. (1993) in Origin and Evolution of the Elements, ed. by N. Prantzos et al. (Cambridge Univ. Press, Cambridge) p. 89.

Geiss, J., Reeves, H. (1971) Astron. and Astrophys. **18** 126

Gilmore, G., Edvardsson, B., Nissen, P. (1991) Astrophys. J. **378** 17

Gott, J.R.,III, Gunn, J., Schramm, D.N., Tinsley, B.M. (1974) Astrophys. J. **194** 543

Guth, A. (1981) Phys. Rev. D **23** 347

Hayashi, C. (1950) Prog. Theor. Phys. **5** 224

Hegyi, D., Olive, K. (1986) Astrophys. J. **303** 56

Hobbs, L., Pilachowski, C. (1988) Astrophys. J. Lett. **326** L23

Hobbs, L., Thorburn, J. (1994) Astrophys. J. Lett. **428** L25

Jakobsen, P., Boksenberg, A., Deharveng, J.M., Greenfield, P., Jedrzewski, R. Paresce, F. (1994) Nature, in press

Jedamzik, K., Fuller, G., Mathews, G., Kajino, T. (1994) Astrophys. J. **422** 423

Kawano, L., Schramm, D.N., Steigman, G. (1988) Astrophys. J. **327** 750

Kernan, P., Krauss, L. (1994) Phys. Rev. Lett. **72** 3309

Kiselman, D. (1994) Astron. and Astrophys., in press (1994).

Krauss, L., Romanelli, P. (1990) Astrophys. J.**358** 47

Kurki-Suonio, H., Matzner, R., Olive, K., Schramm, D.N. (1990) Astrophys. J. **353** 406

Linde, A. (1990) Particle Physics and Inflationary Cosmology (Harwood Academic Publishers, N.Y.)

Link, B. (1992) Phys. Rev. Lett. **68** 2425

Linsky, J. et al. (1993) Astrophys. J. **402** 694

Mampe, W., Ageron, P., Bates, C., Pendlebury, J.M., Steyerl, A. (1989) Phys. Rev. Lett. **63A** 593

Mathews, G. (1994) in Proc. Snowmass Workshop on Particle Astrophysics in the Next Millenium, in press

Mathews, G., Schramm, D.N. (1993) Astrophys. J.**404** 468

Meneguzzi, M., Audouze, J., Reeves, H. (1971) Astron. and Astrophys. **15** 337

Meyer, B., Alcock, C., Mathews, G., Fuller, G. (1991) Phys. Rev. D **43** 1079

Mushotsky, R. (1993) in Texas/PASCOS 92: Relativistic Astrophysics and Particle Cosmology, ed. by C. W. Akerlof and M. A. Srednicki, Annals of the N.Y. Academy of Sciences **688** 184

Olive, K., Schramm, D.N. (1992) Nature **360** 434

Olive, K., Schramm, D.N., Steigman, G., Turner, M., Yang, J. (1981) Astrophys. J. **246** 557

Olive, K., Schramm, D.N., Steigman, G., Walker, T. (1990) Phys. Lett. B **236** 454

Olive, K. Schramm, D.N., Truran, J.W., Rood, R., Vangioni-Flam, E. (1994) U. Minnesota preprint UMN-TH-1305/94 and Astrophys. J., sumitted

Olive, K. and Steigman, G. (1994) Minnesota preprint UMN-TH-1230/94

Pagel, B. (1990) in Proc. of 1989 Rencontres de Moriond

Pagel, B. (1994) in Proc. of the ESO/EIPC Workshop on The Light Element Abundances, Elba, May, 1994, ed. by P. Crane (Springer-Verlag, Berlin), in press.

Peebles, P.J.E. (1966) Phys. Rev. Lett. **16** 410

Peebles, P.J.E. (1971) Physical Cosmology (Princeton University Press, Princeton)

Pinsonneault, M., Deliyannis, C., Demarque, P. (1992) Astrophys. J. Supplement **78** 179

Rebolo, R., Molaro, P., Beckman, J. (1988) Astron. and Astrophys. **192** 192

Reeves, H., Audouze, J., Fowler, W.A., Schramm, D.N. (1973) Astrophys. J. **179** 909

Reeves, H., Fowler, W.A., Hoyle, F. (1970) Nature **226** 727

Rogerson, J., York, D. (1973) Astrophys. J.**186** L95

Rood, R.T., Bania, T., Wilson, J. (1992) Nature **355** 618

Rood, R.T., Steigman, G., Tinsley, B.M. (1976) Astrophys. J. Lett. **207** L57

Ryan, S., Bessel, M., Sutherland, R., Norris, J. (1990) Astrophys. J. **348** L57

Ryan, S., Norris, J., Bessell, M., Deliyannis, C.O. (1992) Astrophys. J. **388** 184

Ryter, C., Reeves, H., Gradstajn, E., Audouze, J., Astron. and Astrophys. **8** (1970) 389.

Sato, K. Tarasawa, N. (1991) in The Birth and Early Evolution of Our Universe: Proceedings of Nobel Symposium 79, ed. by J. S. Nilsson, B. Gustafson, B.S. Skagerstam (World Scientific, Singapore) p. 60.

Schramm, D.N. (1991) Physica Scripta **T36** 22

Schramm, D.N. (1993) in Texas/PASCOS 92: Relativistic Astrophysics and Particle Cosmology, ed. by C.W. Akerlof and M.A. Srednicki (Annals of the N.Y. Academy of Sciences) Vol.688, p. 776.

Schramm, D.N. (1994) in Proc. of Yamada Conf. XXXVII - Evolution of the Universe and Its Observational Quest, Tokyo, June 1993 (Universal Academic Press, Tokyo), in press

Schramm, D.N., Fields, B., Thomas, D. (1992) Nucl. Phys. A **544** 267c

Schramm, D.N., Kawano, L. (1989) Nuc. Inst. and Methods A **284** 84

Schramm, D.N., Wagoner, R.V. (1977) Ann. Rev. of Nuc. Sci. **27** 37

Scherrer, R., Applegate, J., Hogan, C., (1987) Phys. Rev. D **35** 1151

Schvartzman, V.F. (1969) JETP Letters **9** 184

Simpson, J. (1983) Ann. Rev. Nuc. and Part. Sci. **33** 323

Skillman, E. (1994) Talk at particle and Nuclear Astrophysics and Cosmology in the Next Millennium, Snowmass, Colorado, July 1994.

Smith, V.V., Lambert, D.L., Nissen, P.E. (1992) Astrophys. J. **408** 262

Songaila, A., Cowie, L.L., Hogan, C.J., Rugers, M. (1994) Nature **368** 599

Spite, F., Spite, M. (1982) Astron. and Astrophys. **115** 357

Steigman, G. (1994) Nature, submitted

Steigman, G., Fields, B., Olive, K., Schramm, D.N., Walker, T. (1993) Astrophys. J. Lett. **415** L35

Steigman, G., Schramm, D.N., Gunn, J. (1977) Phys. Lett. B **66** 202

Steigman, G., Walker, T. (1992) Astrophys. J. Lett. **385**, L13

Taylor, R., Hoyle, F. (1964) Nature **203** 1108

Thomas, D., Schramm, D.N., Olive, K., Fields, B. (1993) Astrophys. J. **406** 569

Thomas, D., Schramm, D.N., Olive, K., Mathews, G., Meyer, B., Fields, B. (1994) Astrophys. J. **430** 291

Thorburn, J., (1994) Astrophys. J. **421** 318

Truran, J. (1965) Doctoral Thesis, Yale University; and Truran, J.W., Cameron, A.G.W., Gilbert, A. (1966) Can. Journal of Physics **44** 563

Turner, M., Wilczek, F., Zee, A., (1993) Phys. Lett. B **125** 35; **125** 519

Vangioni-Flam, E., Olive, K., Prantzos, N. (1994) Astrophys. J. **427** 618

Wagoner, R., Fowler, W.A., Hoyle, F. (1967) Astrophys. J. **148** 3

Walker, T., Steigman, G., Schramm, D.N., Olive, K., B. Fields, (1993) Astrophys. J. **413** 562

Walker, T., Steigman, G., Schramm, D.N., Olive, K., Kang, H.S. (1991) Astrophys. J. **376** 51

White, S.D.M., Navarro, J.F., Evrard, A.E., Frenck, C. (1993) Nature **366** 261

Wilson, R., Rood, R.T., Bania, T. (1983) in Proc. of the ESO Workshop on Primordial Helium, ed. by P. Shaver and D. Knuth, (European Southern Observatory, Garching)

Woosley, S.E. Hartmann, D.H., Hoffman, R.D., Haxton, W. (1990) Astrophys. J. **356** 272

Yang, J., Schramm, D.N., Steigman, G., Rood, R.T. (1979) Astrophys. J. **227** 697

Yang, J., Turner, M., Steigman, G., Schramm, D.N., Olive, K. (1984) Astrophys. J. **281** 493

Part II

Light Elements at High Redshift

Helium at High Redshift

Peter Jakobsen

Astrophysics Division, Space Science Department of ESA,
ESTEC, 2200 AG Noordwijk, The Netherlands

Abstract. The intergalactic clouds giving rise to the Lyman forest absorption seen in the spectra of high redshift quasars represent the best candidate examples of possibly primordial baryonic matter presently known. A longstanding goal of ultraviolet space astronomy has therefore been to gauge the helium content of the Lyman forest clouds and the ambient intergalactic medium clouds through observations of matching absorption in the resonance lines of neutral helium at λ584Å and once ionized helium at λ304Å. The theoretical background to this problem and the practical constraints in carrying out such observations are outlined, and the results obtained so far with the refurbished *Hubble Space Telescope* are discussed.

1. Introduction and Background

Since their discovery more than twenty years ago, the various types of absorption lines seen in quasar spectra have been a topic of intense astrophysical interest. Detailed ground-based studies of the redshifted ultraviolet Lyα absorption lines of neutral hydrogen and resonance lines of ionized species of the most abundant elements (CIV, NV, SiIV, MgII, etc), have provided considerable insight into the presence and physical state of gaseous matter in the early universe. The launch of the *Hubble Space Telescope*, with its ultraviolet spectroscopic capabilities, has not only enabled the more nearby universe at lower redshift to be probed using these same techniques, but also made the absorption lines of many previously unobserable key species accessible at higher redshift.

Most notable among the latter are the HeI λ584 and HeII λ304 resonance lines of neutral and singly ionized helium. The first glimpses of this cosmologically important element in quasar absorption systems are now being provided by *HST*. The remainder of this paper is concerned with briefly outlining the theoretical background to these results and what might be learned from them.

1.1 Quasar Absorption Lines in Brief

Roughly speaking, the various types of intervening absorption line systems seen in quasar spectra can be classified into two main classes: the "Lyman forest" systems and the "Metal line" or "Lyman limit" systems.

The ubiquitous Lyman forest systems are seen in the spectra of all quasars at all redshifts. These systems have modest neutral hydrogen column densities in the range $N_{HI} \simeq 10^{13} - 10^{17} \mathrm{cm}^{-2}$. The forest systems show little or no evidence for heavy elements and are therefore believed to be due to intergalactic clouds of possibly primordial material (Sargent et al. 1980)

In contrast, the scarcer, but denser, "Lyman limit" systems having HI column densities $N_{HI} \simeq 10^{17} - 10^{22}\text{cm}^{-2}$. Lyman limit systems always show matching absorption from heavy elements, and are therefore thought to be associated with the gaseous halos of galaxies.

The Lyman forest systems are extremely numerous and evolve rapidly with redshift. Their line-of-sight density evolution is usually parameterized in the form (Sargent et al. 1980, Murdoch et al. 1986, Hunstead 1988)

$$E[\frac{dn}{dz}] = N(z) = A(1 + z)^{\gamma} \tag{1}$$

with $A \simeq 10$ and $\gamma \simeq 2 - 3$. In comparison, the Lyman limit absorbers are roughly ten times scarcer ($A \simeq 1$) and evolve less rapidly ($\gamma \simeq 1$) than Lyman forest systems (Tytler 1982; Bechtold et al. 1984; Lanzetta 1988; Sargent, Steidel & Boksenberg 1989; Bahcall et al. 1993). The column density spectrum of both types of absorber can be approximated by a power law, $dP/dN_H \propto N_H^{-s}$, with slope $s \simeq 1.2 - 1.6$ (Tytler 1988, Sargent et al. 1989).

The measured occurrence and detailed statistics of the various types of HI containing quasar absorption systems are important for assessing the important question of the overall transparency of the UV universe out to high redshift (Section 1.3).

Studies of absorption in quasars are, however, not only concerned with discrete clouds of gas. Although extremely abundant – especially at the highest redshifts – the Lyman forest clouds still only fill a negligible fraction of the volume of intergalactic space. Since the process of galaxy formation was unlikely to have been so efficient as to leave the space between the galaxies and the forest clouds completely devoid of matter, it has long been suspected that a gaseous intergalactic medium (IGM) of some form must exist. One of the most sensitive probes of the diffuse IGM is the so-called Gunn-Peterson test, which places an extremely stringent constraint on the density of smoothly distributed intergalactic neutral hydrogen through the redshift-smeared Lyα trough that such gas would give rise to shortward of the emitted Lyα line in quasar spectra (Gunn & Peterson 1965). The fact that no such smooth Lyα absorption beyond the line blanketing of the Lyman forest has ever been detected (Steidel & Sargent 1987, Schneider, Schmidt & Gunn 1991, Webb et al. 1992, Meiksin & Madau 1993), implies that the IGM – if it exists – must be highly ionized.

1.2 The Ionization State of Intergalactic Helium

The gas of the Lyman forest clouds – and by implication the diffuse IGM – is believed to be kept highly photoionized by a metagalactic background flux of ionizing radiation. That the forest clouds are photoionized is suggested by the large $b \simeq 30$ km s^{-1} velocity width of the Lyman forest lines, which corresponds to gas temperature of $T \simeq 5 \times 10^4$K (Carswell et al. 1987, 1991). Another important piece of evidence is the so-called "inverse" or "proximity" effect (Carswell et al. 1982; Murdoch et al. 1986). As the emission redshift is approached, the Lyman forest systems in a given quasar tend to show a gradual under-density of absorbers with respect to the global density given by equation (1). This effect is interpreted as being caused by the radiation field of the background quasar enhancing the total ionizing flux above the baseline metagalactic level. Since the quasar flux can be estimated from the magnitude and spectrum of the background quasar, the background metagalactic ionizing background intensity can be estimated from the measured contrast of the proximity effect. Through this technique (Bajtlik, Duncan & Ostriker 1988; Lu, Wolfe & Turnshek 1991, Bechtold 1994) one infers that the intergalactic ionizing background in the redshift range $1.7 \lesssim z \lesssim 3.8$ is consistent with an $I_\nu \propto \nu^{-\alpha}$, $\alpha \simeq 0.6$ power law with intensity at the Lyman limit of order $I_{\nu_H} \approx 10^{-21}$ ergs s^{-1} cm^{-2} sr^{-1} Hz^{-1}.

The origin of this ionizing flux is a topic of some debate (Bechtold et al. 1987; Miralda-Escudé & Ostriker 1990,1992; Madau 1992). Obvious candidate sources are the integrated flux from quasars and radiation from primeval galaxies.

In quantitative terms, the absolute intensity of the metagalactic ionizing background inferred from the proximity effect, combined with the typical sizes of $D \approx 10$ kpc and corresponding neutral hydrogen densities of $n_{HI} \approx 10^{-8}$ cm^{-3} for the Lyman forest clouds derived from the study of correlated absorption in close quasar pairs, implies that the Lyman forest clouds possess a very high ionization parameter; $U \simeq \frac{4\pi}{hc\alpha}(\frac{I^o_{\nu_{HI}}}{n_H}) \approx 10^{-1}$; (e.g. Sargent et al. 1980, Carswell 1988). In this limit where the hydrogen is essentially completely ionized, standard photoionization theory (Osterbrock 1989) leads to the following simple expression for the relative degree of ionization of hydrogen

$$\left[\frac{HI}{H}\right] \simeq \frac{1}{U}\left(\frac{\alpha_{HI}}{c\sigma^o_{HI}}\right)\left[\frac{\alpha+3}{\alpha}\right] \tag{2}$$

where α is the spectral index of the ionizing flux and α_{HII} is the hydrogen recombination coefficient and σ^o_{HI} the HI photoionization cross section at threshold. For the anticipated values of $U \sim 10^{-1}$ and $\alpha \sim 1$ this equation predicts [HI/H] $\sim 10^{-4}$. The neutral hydrogen giving rise to the observed Lyα absorption is therefore believed to represent only a very minor fraction of the total mass present in the Lyman forest clouds.

One way of testing the photoionization picture outlined above, is to extend the quasar absorption observations to also cover the resonance lines of neutral and singly ionized helium. If the Lyman forest clouds are primordial in nature, then according to standard hot Big Bang nucleosynthesis theory they must also contain helium at an abundance relative to hydrogen of [He/H] $\simeq 8\%$ by number (Walker et al. 1991). The expressions governing the photoionization of this helium are

$$\left[\frac{HeII}{He}\right] \simeq \frac{1}{U}\left(\frac{\alpha_{HeII}}{c\sigma^o_{HeII}}\right)\left[\frac{\alpha+3}{\alpha}\right]\left(\frac{\nu_{HeII}}{\nu_{HI}}\right)^\alpha \tag{3}$$

$$\left[\frac{HeI}{He}\right] \simeq \left[\frac{HeII}{He}\right]\frac{1}{U}\left(\frac{\alpha_{HeI}}{c\sigma^o_{HeI}}\right)\left[\frac{\alpha+3}{\alpha}\right]\left(\frac{\nu_{HeI}}{\nu_{HI}}\right)^\alpha \tag{4}$$

where α_{HeI} and α_{HeII} are the HeI and HeII recombination coefficient and σ_{HeI}^o and σ_{HeII}^o the HeI and HeII photoionization cross sections at threshold at $\lambda = 504$ Å and $\lambda = 228$ Å, respectively. For the canonical values of $U \sim 10^{-1}$ and $\alpha \sim 1$ these equations predict that the intergalactic helium should also be highly ionized at a level corresponding to [HeI/He] $\sim 10^{-6}$ and [HeII/He] $\sim 10^{-2}$.

The anticipated strength of the observable absorption in HeI and HeII can be estimated by scaling to the absorption in Lyα. The ratio of the density of neutral helium to neutral hydrogen in an optically thin forest cloud is given by

$$\left[\frac{\text{HeI}}{\text{HI}}\right] \simeq \left[\frac{\text{He}}{\text{H}}\right]\left[\frac{\text{HeII}}{\text{He}}\right](\frac{\alpha_{\text{HeII}}}{\alpha_{\text{HII}}})(\frac{\sigma_{\text{HI}}^o}{\sigma_{\text{HeI}}^o})\left(\frac{\nu_{\text{HeI}}}{\nu_{\text{HI}}}\right)^\alpha \approx 0.1 \times 1.8^\alpha \left[\frac{\text{HeII}}{\text{He}}\right] \quad (5)$$

where $(\alpha_{\text{HeII}}/\alpha_{\text{HII}}) \simeq 1.2$, $(\sigma_{\text{HI}}^o/\sigma_{\text{HeI}}^o) \simeq 1$, and [He/H] $\simeq 0.08$ is the standard Big Bang helium to hydrogen ratio by number.

Equation (5) predicts that the abundance of neutral helium in the Lyα forest clouds should be far lower than that of neutral hydrogen; i.e. [HeI/HI] $\sim 10^{-3}$. Absorption in the HeI $\lambda584$ forest lines is therefore expected to be universally weak compared to Lyα.

In contrast, absorption due to singly ionized helium is predicted to be universally strong. The predicted density ratio of HeII to HI is

$$\left[\frac{\text{HeII}}{\text{HI}}\right] \simeq \left[\frac{\text{He}}{\text{H}}\right](\frac{\alpha_{\text{HeIII}}}{\alpha_{\text{HII}}})(\frac{\sigma_{\text{HI}}^o}{\sigma_{\text{HeII}}^o})\left(\frac{\nu_{\text{HeII}}}{\nu_{\text{HI}}}\right)^\alpha \approx 2.4 \times 4^\alpha \quad (6)$$

where in this case $(\alpha_{\text{HeIII}}/\alpha_{\text{HII}}) \simeq 6$ and $(\sigma_{\text{HI}}^o/\sigma_{\text{HeII}}^o) = 4$. It follows that for an ionizing background spectrum with slope in the range, $\alpha \simeq 0.5 - 2$, the column density of once ionized helium through a given Lyman forest cloud is predicted to exceed that of neutral hydrogen by a factor [HeII/HI] $\approx 5-40$. Softer photoionizing spectra lead to larger predicted [HeII/HI] ratios. For instance, if the ionizing background flux does not extend down to the HeII photoionization threshold at $\lambda228$, and is therefore only capable of ionizing helium once, then the relative abundance of HeII can reach values as high as [HeII/HI] $\simeq$ [He/H]/[HI/H] $\approx 10^3$.

Since the predicted values of [HeI/HI] and [HeII/HI] given by equations (5) and (6) only depend on the shape of the ionizing spectrum, the above considerations apply equally well to the case of a photoionized intergalactic medium. In particular, it has long been appreciated that a diffuse IGM could reveal its presence through redshift-smeared Gunn-Peterson absorption in the HeII $\lambda304$ line, and still display negligible Gunn-Peterson absorption in HI and HeI.

1.3 The Intervening Hydrogen Opacity Problem

In practice, observations of HeI $\lambda584$ and HeII $\lambda304$ absorption lines in quasar spectra are bedeviled by several factors. For one, since *HST* only transmits UV light at wavelengths $\lambda \gtrsim 1200$ Å, these transitions are only accessible with *HST* in the case of the most remote – and therefore faint – quasars at redshifts $z \gtrsim 1$ and $z \gtrsim 3$, respectively. Moreover, both transitions lie at wavelengths below the Lyman limit at $\lambda = 912$ Å, and are therefore subject to strong photoelectric absorption in the numerous Lyman forest and Lyman limit systems intercepted by the line of sight out to such large redshift. This opacity is especially severe at redshifts $z > 3$ where the pathlength and the number of intercepted absorbers is large.

The character and statistics of this cumulative absorption process have been described by Møller and Jakobsen (1990), who showed that the observed statistics of quasar absorption systems imply that all $z > 3$ quasars are severely affected by this absorption, to the point of all but a very small minority being completely undetectable at the wavelength of the redshifted HeII $\lambda304$ line.

This prognosis is been borne out observationally by *HST*. In spite of several systematic *HST* programs aimed at overcoming the hurdle of the Lyman continuum opacity by searching for high redshift quasars with unusually clear lines of sight (cf. Jakobsen et al. 1993), only six quasars have so far been detected shortward of the emitted HeI $\lambda584$ line (Tripp et al. 1990; Reimers et al. 1990; Beaver et al. 1991, 1992; Jakobsen et al. 1993). Of these, only one is known to display a detectable flux down to the rest wavelength of HeII $\lambda304$ (Jakobsen et al. 1994). Helium absorption has so far been reported in only two of these cases.

2. Detections of Helium Absorption

Although rather different in nature, the two successful detections of helium absorption in quasars obtained so far nonetheless provide clear confirmation of the theoretical expectations concerning the ionization state of intergalactic helium outlined in Section 1.2 above – namely that absorption in the HeI $\lambda584$ line of neutral helium should be very weak with respect to that of Lyα, whereas absorption in the HeII $\lambda304$ line of singly ionized helium should be strong.

The downside of this confirmation of the high ionization, however, is that since the observable species of HeI and HeII represent only a small fraction of the total helium present, the large and highly uncertain associated ionization corrections stand in the way of enabling the helium abundance of the absorbing gas to be measured at any accuracy of interest to nucleosynthesis theory. The existing detections of helium in quasar absorbers therefore provide only a qualitative – but nevertheless important – consistency check of standard Big Bang nucleosynthesis theory.

2.1 HeI in HS1700+6414

Absorption lines due to neutral helium have so far been detected in only one object, the high redshift quasar HS1700+6416 (Reimers et al. 1992). This quasar was discovered by Reimers et al. (1989) and is highly remarkable in several regards. For one, at $V \simeq 16$ it is exceptionally bright for its redshift of $z = 2.72$. Moreover, the line of sight to this quasar happens to be unusually devoid of massive Lyman limit absorption systems. Because of this, HS1700+6416 is also exceptionally bright in the ultraviolet.[1]

The UV continuum of HS1700+6416 first detected by Reimers et al. (1989) using *IUE*, and later re-observed at much higher resolution and signal-to-noise ratio using the FOS on board *HST* (Reimers et al. 1992) Unfortunately, at $z = 2.72$ the redshift of HS1700+6416 is too small to bring the HeII $\lambda304$ line of ionized helium within range of *HST*. However, the HeI $\lambda584$ line is accessible at redshifts $1 \lesssim z \lesssim 2.7$.

A detailed analysis of the HeI absorption in HS1700+6416 has been carried out by Reimers & Vogel (1993). In agreement with the theoretical expectations of Section 1.2, HeI $\lambda584$ lines corresponding to the Lyα forest are too weak to be detected in the FOS spectrum of HS1700+6416. Also – as is the case for all UV bright quasars observed so far – there is no evidence of a HeI $\lambda584$ Gunn-Peterson trough. Weak HeI lines are, however, detected in the case of four denser ($N_{HI} \sim 10^{17} \mathrm{cm}^{-2}$) Lyman limit systems seen in the spectrum. The four systems have redshifts in the range $z \simeq 1.85 - 2.43$, and all display absorption due to heavy elements.

Reimers & Vogel (1993) have attempted to model the ionization of these systems by matching photoionization calculations to the observed ionization ratios of the detected heavy elements. According to their analysis, the values of [HI/HeI]$\simeq 1 - 5 \ 10^{-2}$ inferred for the absorbers – although indeed low as expected for a photoionized gas – are still a factor of ~ 5 higher than predicted for the model that best fits the ionization of the heavy elements, assuming a cosmological helium abundance.

As discussed by Reimers & Vogel (1993) the discrepancy is not likely to imply a high helium abundance, but almost certainly reflects the huge uncertainties in understanding the ionization processes at work in quasar absorption systems. The emphasizes the point that neutral helium is a residual minority ionization species in quasar absorbers, and therefore not an accurate diagnostic of the total helium content.

Although the four systems in HS1700+6416 in which HeI absorption is seen are metal line systems, and not potentially primordial Lyman forest systems, it is still reassuring that these clouds of mildly stellar-processed material do contain helium, and that the neutral helium lines detected are indeed weak as anticipated.

---

[1] Picard & Jakobsen (1993) have estimated that this fortuitous combination of a very luminous quasar lying along an unusually clear line of sight is sufficiently scarce that the whole sky should contain only between ~ 1 and ~ 20 quasars at redshifts $z > 2$ comparable in brightness to HS1700+6416 in the UV.

2.2   HeII in Q0302−003

Absorption due to singly ionized HeII has very recently been detected by the refurbished *HST* in the $z = 3.286$ quasar Q0302−003. This object was identified as having an exceptionally clear line of sight during a dedicated survey of 25 selected $z > 3$ quasars observed with the aberrated *HST* using the far-UV prism of the Faint Object Camera (Jakobsen et al. 1993). It was then re-observed with the COSTAR-corrected FOC.

The low-resolution ($\Delta\lambda \simeq 20$ Å) FOC prism spectrum of Q0302−003 obtained by Jakobsen et al. (1994, Fig. 1) reveals a strong absorption trough shortward of emitted HeII $\lambda 304$, which is almost certainly due to intervening redshift-smeared HeII $\lambda 304$ line absorption in the gas giving rise to the Lyman forest clouds and in the intergalactic medium (IGM).

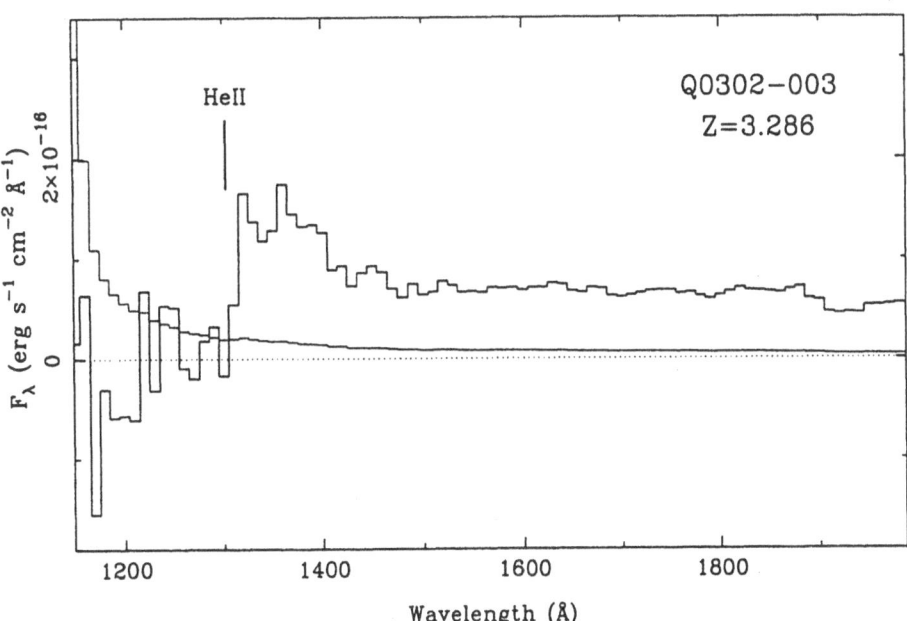

Fig. 1. FOC far-ultraviolet prism spectrum of the $z = 2.786$ quasar Q0302−003. The thin solid line gives the 1σ uncertainty per $\Delta\lambda = 10$ Å wavelength bin. The position of the HeII $\lambda 304$ line in the quasar restframe is marked.

In agreement with the theoretical predictions of Section 1.2, the HeII absorption detected is far stronger than the matching absorption in HI and HeI. In fact, the FOC prism spectrum of of Q0302−003 is seemingly completely absorbed shortward of the redshifted HeII line, with no flux detected below the edge at the sensitivity limit of the 3 hour FOC integration. This by itself proves that the intergalactic helium and hydrogen is indeed highly ionized.

The overriding open issue surrounding this first detection of intergalactic HeII absorption is whether or not the observations also mark the discovery of the long sought after IGM. The FOC prism spectrum of Q0302−003 is of too low spectral resolution to distinguish directly between the absorption from unresolved HeII $\lambda 304$ line-blanketing from the Lyman forest clouds and true HeII Gunn-Peterson absorption from a smooth intergalactic medium.

The present data only suffice to place a lower limit for the total effective HeII optical depth across the edge of $\tau_{\text{HeII}} > 1.7$ (90% confidence). As discussed in detail by Jakobsen et al. (1994), this quantitative limit on the total HeII absorption is still far larger than the corresponding wavelength-averaged Lyman forest optical depth of $\tau_{\text{HI}} \simeq 0.35$ measured shortward of Lyα at $z \simeq 3.2$, and is therefore not readily explained by HeII line blanketing in the known population of Lyman forest clouds alone. This suggests that there may be a significant contribution from true Gunn-Peterson absorption in the IGM. On the other hand, given the uncertainties in the observed occurrence and column density distribution of the forest systems, values of τ_{HeII} exceeding unity could in principle be reached provided that the gas in the Lyman forest clouds possess an extremely high HeII to HI ratio: [HeII/HI]$> 10^3$ − corresponding to a situation where the forest clouds are highly ionized in hydrogen ([HI/H]$< 10^{-4}$), but only singly ionized in helium ([HeII/He]~ 1). In other words, although unlikely, it cannot be completely ruled out that the HeII absorption detected in Q0302−003 is dominated by unresolved HeII forest line blanketing. Unfortunately, given the extreme faintness of Q0302−003, the goal of obtaining a higher resolution far-UV spectrum of the redshifted HeII region of this object is not technically feasible with the present set of *HST* grating spectrographs, but will have to await the installation of STIS later this decade.

The lack of knowledge of the detailed ionization conditions in the plasma giving rise to the HeII absorption in Q0302−003 obviously also affects the quantitative estimates of the total amount of baryonic matter detected. As discussed by Jakobsen et al. (1994), the detected intergalactic plasma is in all likelihood very tenuous ($\Omega \sim 10^{-5} - 10^{-2}$) and not likely to contain more mass than that already known to exist in galaxies. Consequently, this new source of intergalactic "dark baryonic matter" is not likely to conflict with constraints on the baryonic content of the universe stemming from considerations of nucleosynthesis theory, nor is it likely to contribute significantly toward solving the missing mass problem of astrophysics.

3. Conclusions

The first glimpses of intergalactic helium absorption in quasars provided by *HST* have proven that the Lyman forest clouds and the IGM; i.e. either of the two most paradigmatic examples of primordial matter known, do contain helium at an epoch corresponding to a lookback time of 89% of the present age of the universe. This in itself provides a dramatic − albeit qualitative − validation of Big Bang nucleosynthesis theory.

The observations also confirm that that the Lyman forest clouds and the IGM are indeed highly ionized and that rather harsh conditions existed in intergalactic space at high redshift. Unfortunately, this high level of ionization also means that the observable species of HI, HeI and HeII are mere trace constituents of the intergalactic gas, and that most of the mass is in the form of unobservable HII and HeIII . Consequently, the prospects for determining an accurate value for the primordial helium abundance of the intergalactic plasma through quasar absorption techniques are rather poor due to the huge uncertainties associated with the large ionization corrections.

Nonetheless, it is clear that with the first detection of neutral helium absorption in HS1700+6416 and the likely discovery of the intergalactic medium through the detection of strong singly ionized helium absorption in Q0302−003, a new and important tool for probing the early universe has become available.

References

Bahcall, J. N. et al. 1993 ApJS 87,1

Bajtlik, S., Duncan, R. C., & Ostriker, J. P. 1988 ApJ 327,570

Beaver, E. A, Burbidge, E. M., Cohen, R. D., Junkkarinen, V. T., Lyons, R. W., Rosenblatt, E. I., Hartig, G. F., Margon, B., & Davidsen, A. F. 1991 ApJ 377, L9

Beaver, E. A, Burbidge, E. M., Cohen, R., Junkkarinen, V., Lyons, R., & Rosenblatt, E. 1992, in Science with the Hubble Space Telescope, ed. P. Benvenuti & E. Schreier (Garching: European Southern Observatory), p.53

Bechtold, J. 1994 ApJS 91, 1

Bechtold, J., Green, R. F., Weymann, R. J., Schmidt, M., Estabrook, F. B., Sherman, R. D., Wahlquist, H. D., & Heckman, T. M. 1984 ApJ 281, 76

Bechtold, J., Weymann, R. J., Lin, Z., & Malkan, M. A. 1987 ApJ 315, 180

Carswell, R. F., 1988, in QSO Absorption Lines, Probing the Universe, ed. J. C. Blades, D. Turnshek & C. A. Norman (Cambridge: Cambridge University Press), p91

Carswell, R. F., Lanzetta, K. M., Parnell, H. C & Webb, J. K. 1991 ApJ 371, 36

Carswell, R. F., Webb, J. K., Baldwin, J. A. & Atwood, B. 1987 ApJ 319, 709

Carswell, R. F., Whelan, J. A. J., Smith, M. G., Boksenberg, A., & Tytler, D. 1982 MNRAS 198,91

Gunn, J. E., & Peterson, B. A. 1965 ApJ 142, 1633

Hunstead, R. W. 1988, in QSO Absorption Lines, Probing the Universe, ed. J. C. Blades, D. Turnshek & C. A. Norman (Cambridge: Cambridge University Press), p71

Jakobsen, P. et al. 1993 ApJ 417, 528

Jakobsen, P., Boksenberg, A., Deharveng, J. M., Greenfield, P., Jedrzejewski, R., & Paresce, F. 1994 Nature 370,35

Jenkins, E. B. & Ostriker, J. P. 1991 ApJ 376, 33

Lanzetta, K. M. 1988 ApJ 332, 96

Lu, L., Wolfe, A. M., & Turnshek, D. A. 1991 ApJ 367, 19

Madua, P. 1992 ApJ 389, L1

Meiksin, A. & Madau, P. 1993 ApJ 412, 34

Miralda-Escudé, J., & Ostriker, J. P. 1990 ApJ 350, 1

Miralda-Escudé, J., & Ostriker, J. P. 1992 ApJ 392, 15

Miralda-Escudé, J. 1993 MNRAS 262, 273

Murdoch, H. S., Hunstead, R. W., Pettini, M., & Blades, J. C. 1986 ApJ, 309

Møller, P., & Jakobsen, P. 1990 A&A 228, 299

Osterbrock, D. E. 1989, Astrophysics of Gaseous Nebulae and Active Galactic Nuclei (Mill Valley: University Science Books), p23

Picard, A. & Jakobsen, P. 1993 A&A 276, 331

Reimers, D. & Vogel, S. 1993 A&A 276, L13

Reimers, D., Vogel, S., Hagen, H. J., Engels, D., Groote, D., Wamsteker, W., Clavel, J., & Rosa, M. R. 1992 Nature 360, 561

Reimers, D., Clavel, J., Groote, D., Engels, D., Hagen, H. J., Naylor, T., Wamsteker, W., & Hopp, U. 1989 A&A 218, 71

Sargent, W. L. W., 1988, in QSO Absorption Lines, Probing the Universe, ed. J. C. Blades, D. Turnshek & C. A. Norman (Cambridge: Cambridge University Press), p1

Sargent, W. L. W., Steidel, C. C., & Boksenberg, A. 1989 ApJS 69, 703

Sargent, W. L. W., Young, P. J., Boksenberg, A., & Tytler, D. 1980 ApJS 42, 41

Schneider, D. P., Schmidt, M. & Gunn, J. E. 1991 AJ 101, 2004

Steidel, C. C., & Sargent, W. L. W. 1987 ApJ 318, L11

Tripp, T. M., Green, R. F., & Bechtold, J 1990 ApJ 364, L29

Tytler, D. 1982 Nature 298, 427

Walker, T. P., Steigman, G. Schramm, D. N., Olive, K. A. & Kang, H.-S. 1991 ApJ 376, 51

Webb, J. K., Barcons, X., Carswell, R. F. & Parnell, H. C. 1992 MNRAS 255, 319

Lithium at High Redshifts

Jorge Sánchez Almeida and Rafael Rebolo

Instituto de Astrofísica de Canarias, 38200 La Laguna, Tenerife, Spain

Abstract. We have studied the formation of lithium spectral lines by the cosmic plasma after the hydrogen recombination epoch. Specifically, we have considered: i) the formation of Li absorption lines and other features in the spectrum of high redshift quasars ($z = 2$ - 4) as a result of the residual primordial Li in the intergalactic medium and matter concentrations along the line of sight (intervening galaxies, protogalaxies, etc.); ii) the formation of Li lines during the recombination of Li I (at $z \sim 500$) which would give rise to spectral features in the submillimetric spectrum of the Cosmic Microwave Background. We briefly discuss the present capabilities to observe the expected signatures and the implications of their detection on the determination of the primordial Li abundance and cosmological parameters.

1 Introduction

The primordial abundance of ^{7}Li plays a critical role to test the consistency of Standard Big Bang Nucleosynthesis (SBBN). Observations of a "Li-plateau" in warm halo dwarfs suggested the pregalactic abundance of Li was Li/H$\sim 10^{-10}$ (Spite and Spite 1982, Rebolo et al. 1988, Thorburn 1994). However, Li is a fragile element easily destroyed in stellar interiors at relatively low temperatures. Several theoretical studies have suggested that the amount of Li presently observed in halo stars may have been depleted from an initially higher value Li/H$\sim 10^{-9}$ (see e.g Vauclair 1988, Deliyannis et al. 1990). Although large observational and theoretical effort has been dedicated to elucidate the effects of depletion, no firm conclusion has been reached and the precise value of the Li abundance produced in the Big Bang remains an open question. It is clear that a measurement of the Li abundance in the intergalactic medium at high redshift would be very valuable to solve this issue. We therefore decided to investigate astrophysical contexts where Li atoms could produce observable features in the Universe at high redshift. Here, we only present results for the Li I resonance doublet at λ 6708 Å. The resonance line of Li II at λ 199 Å might also produce interesting features.

2 Absorption Lines and Spectral Features in Spectra of Quasars

Absorption lines of various species have been observed in the spectra of quasars, in particular, very recently it has been claimed the detection of deuterium, a key SBBN product (Songalia et al. 1994). These spectral features are produced by clumps of matter along the-line-of-sight (proto-galaxies?). Would it be possible to detect Li I 6708 Å?

Using the radiative transfer equation in a expanding universe (e.g. de Bernardis et al. 1993), one can compute the equivalent width W of such an absorption line,

$$W/\lambda_0 = n_H A_{Li\ I}/2.2 \times 10^{16} \text{cm}^{-2}, \tag{1}$$

where λ_0, n_H (in cm^{-2}) and $A_{Li\ I}$ stand, respectively, for the observed wavelength of the spectral feature, the H column density, and the abundance of neutral Li relative to H. Assuming $A_{Li\ I} = 10^{-9}$, the spectral lines might reach sizeable equivalent widths: 30-40 Å (see Fig. 2 for other abundances, column densities, and redshifts of the background sources). We note that it might be possible with future infrared spectroscopy facilities in 10m-class telescopes to search for these features in low-ionization, high density, high redshift absorption systems. A resolving power $\lambda/\Delta\lambda$ ~5000 and S/N~10 in the near infrared continuum spectrum of quasars seems adequate. Look at the curves in Fig. 1 to notice the ability of this test to discriminate between high and low primordial Li abundance.

There is also another possibility for observing Li I lines in the spectra of distant sources. In case that the matter is not concentrated in clumps, but distributed along the line-of-sight, absorption features will be found at all redshifts, between zero and that of the source. If the distribution follow that of the standard cosmological models, the absorption will increase towards the red (following the increase of density of the universe). The absorption drops to zero at the exact redshift of the source, as the Li I atoms able to remove redder photons stand behind the source. Consequently, the absorption gives rise to a step-like feature in the spectra, which might be detectable. It is possible to show that the decrement in the spectra is given by

$$\Delta I_\nu/I = -3.5 \times 10^6 h \Omega_b A_{Li\ I}[(1+z_*)^2/(1+\Omega z_*)^{1/2}], \tag{2.a}$$

with

$$z_* = \nu_*/\nu_0 - 1. \tag{2.b}$$

The symbols ν_* and ν_0 represent, respectively, the laboratory frequency of the transition ($\sim 4.47 \times 10^{14}$ Hz) and the observed frequency ($= c/\lambda_0$). Empirical determination of these effects in the spectrum of distant quasars will require again, infrared spectroscopic facilities at 10m-class telescopes, lower resolving power than mentioned above (i.e. ~ 100) but a S/N significantly higher ($\sim$500). These features should be present in the spectrum of any quasar if Li is of cosmological origin and if it is mainly in a neutral state in the intergalactic medium.

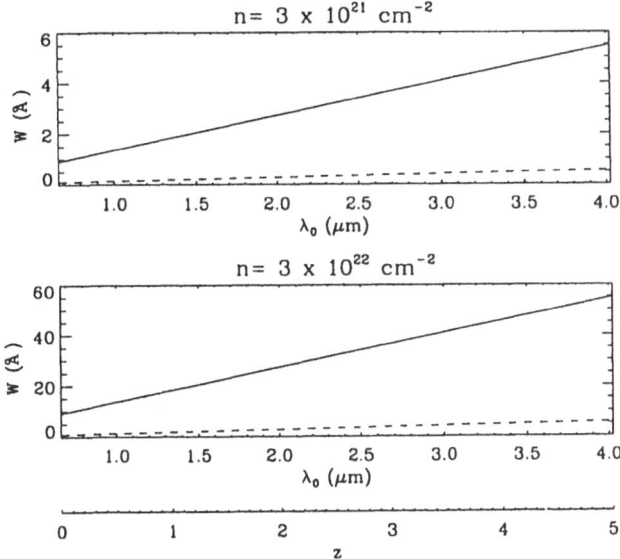

Fig. 1. Equivalent widths W of lines of Li I 6708Å produced by clumps of matter between a distant source and us. The labels on top of the figures quote the total column densities of H used in the simulations. λ_0 stands for the observed wavelength of the line. The z-scale corresponds to the redshift of the background clump. The relative abundance of Li I is 10^{-10}, the dashed lines, and 10^{-9}, the solid lines.

Again, the different behaviour for high and low values of the primordial Li abundances can be a valuable test for its cosmological production.

3 Lithium at Larger Redshifts

The radiative transfer equation in a expanding universe provides a way to determine the range of redshifts z where the photons of the CMB originate. The integrand of the formal solution is related to the probability that a detected photon comes from a given z (see, e.g., Jones & Wyse 1985). We have computed this so-called *contribution function* when, in addition to the scattering by free electrons, the opacity due to Li I 6708Å has been taken into account. The probability that a photon of the Universe is last-scattered by Li I is shown in Fig. 2. It turns out that most of the photons of the CMB with $\nu_0 > 25 \text{cm}^{-1}$ were last-scattered by Li at $z < 500$! (the exact values are model-dependent, though).

The fact the Li I opacity is larger than 1 at certain wavelengths, might have caused Li I lines at extremely large redshifts (~ 500). The tail of blue photons produced during the H recombination (Peebles 1968) might have been re-processed by Li. Spectral distortions would be produced by some sort of photon degradation mechanism, as that proposed by Lyubarsky & Sunyaev (1983). The magnitude of such a distortion has not been computed yet (the work is still in progress). However, even if the lines were too small to be detected (because

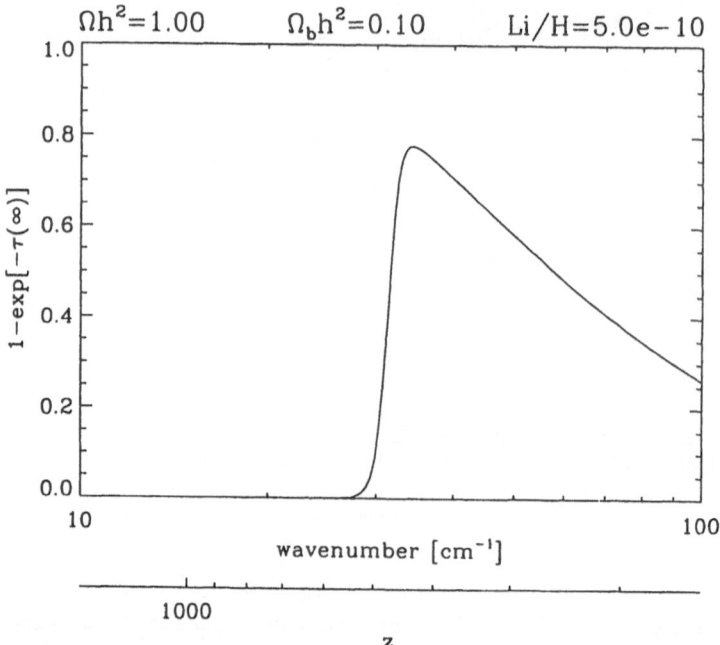

Fig. 2. Probability that a CMB photon was last-scattered by Li I 6708Å, versus observed wavenumber. The lithium abundance, as well as other cosmological parameters of the simulation, can be directly read in the figure.

of galactic emission; see Wright et al. 1991), an observational upper limit can be used to constrain the spectrum of the CMB at high frequencies. As far as we know, there is no other way to do it.

References

de Bernardis, P., Dubrovich, V., Encrenaz, P. et al., 1993, A&A 269, 1
Deliyannis, C. Demarque, P, Kawaler S.D 1990 ApJS 73, 21
Lyubarsky Y. E., Sunyaev, R. A., 1983, A&A 123, 171
Jones, B. J. T., Wyse, R. F. G., 1985, A&A 149, 144
Peebles, P. J. E., 1968, ApJ 153, 1
Rebolo, R., Molaro, P., Beckman, J.E. 1988, A&A 192, 192
Songalia, A., Cowie, L.L, Hogan, C.J., Rugers, M. 1994, Nature, 368, 599
Spite, F. Spite, M. 1982, AA 115, 357
Thorburn, J.A. 1994, ApJ 421, 318
Vauclair, S. 1988 ApJ 335, 971
Wright, E. L., Mather, J. C., Bennett et al., 1991, ApJ 381, 200

On the Detectability of Primordial Deuterium in QSO Absorption Systems

V.K. Khersonsky, F.H. Briggs, D.A. Turnshek

Department of Physics and Astronomy, University of Pittsburgh

Abstract. A detailed examination of the detectability of the Lyman series of deuterium in QSO absorption-line systems shows that observations of the high-order Lyman lines in clouds of very high column density, low velocity dispersion (such as "damped Lyα systems"), and low heavy element abundance might produce efficient and precise measures of the deuterium to hydrogen ratio at early epochs in the evolution of galaxies.

1 Introduction

Measurements of the deuterium cosmic abundance provide critical tests for physical conditions in the early universe. Detailed calculations of deuterium formation in the framework of standard big-bang (SBB) models (for reviews see Walker et al. 1991) show its primordial abundance is sensitive to the baryon density in the universe. Primordial deuterium can also play an important role in the processes of gas cooling in the universe at the postrecombination epoch (Varshalovich & Khersonsky 1976; Latter 1989)

The observational results obtained with the *Copernicus* satellite (Rogerson & York 1973), the *IUE* satellite (Landsman et al. 1984), and *GHRS* on *HST* (Linsky et al. 1993) for the local interstellar medium exibit a scatter in the D/H ratio of over an order of magnitude; the average observed value for the D/H ratio is $\sim 1.5 \times 10^{-5}$. When interpreted in the context of SBB nucleosynthesis models, these measurements have to be corrected for the effects of Galactic chemical evolution which at present are not well understood.

It would seem that the most reliable measurements of primordial deuterium might be obtained through observations of extragalactic objects like the Lyα forest clouds that are observed in QSO absorption-line spectra. Some of these clouds at moderate-to-high redshift should have a primordial origin and it is, of course, reasonable to expect that, in general, gas at higher redshift has been subjected to much less chemical processing than local gas. As a case in point, Pettini, Boksenberg & Hunstead (1990) reported observations of ZnII and CrII in a high column density ($N_{HI} = 2.5 \times 10^{21}\,\mathrm{cm}^{-2}$) damped Lyα system at redshift $z=2.3$ which indicate a low metal abundance of ~ 0.04 times solar.

Here we summarize the results of a more comprehensive analysis of the requirements to detect and measure the deuterium abundance in low to intermediate velocity dispersion ($\sigma = 3.5$ and 16 km s^{-1}) QSO absorption line systems, details of which are given by Khersonsky, Briggs & Turnshek (1994a,b). Observation of the high-order Lyman lines in these systems offers the best hope for measuring deuterium. Thus, this paper considers some cloud parameters that are most favorable for making accurate measurements of the deuterium abundance. Note that a similar analysis was conducted by Webb et al. (1991) for higher velocity dispersion absorbers with Doppler parameters, $b = \sqrt{2}\sigma = 25$ and 35 km s^{-1}.

2 Synthetic HI and DI Absorption Spectra

The synthetic spectra have been computed by including Voigt profiles for the first 98 lines of the Lyman series for HI and DI. The cloud parameters considered for this contribution are: column density $2\times10^{16} \leq N_{HI} \leq 2 \times10^{20}$cm^{-2}, velocity dispersion $\sigma=9$ km s^{-1}, and D/H ratio $N_{DI}/N_{HI} = 2\times10^{-5}$ and 10^{-4}.

2.1 Effects Limiting the Detection of DI

Here we outline four main effects limiting the detection of DI lines and point out when these effects should be unimportant.

1. *Confusion of DI lines with Lyα forest lines.* Since the density of Lyα lines is lower at low redshift, confusion with forest lines statistically becomes less important with decreasing redshift.

2. *Doppler Broadening of the HI lines.* This factor is not important if the velocity dispersion of the gas is $\sigma < 57.7 (\ln \tau_{HI,0})^{-1/2}$km s^{-1}, because then the optical depth due to the HI profile will be less than one at the center of the DI line. Here $\tau_{HI,0} = \tau_{DI,0}[N_{HI}\sigma_{DI}/N_{DI}\sigma_{HI}] = \beta\tau_{DI,0}$ is the optical depth at line center of the HI line.

3. *Opacity in the Damping Wings of the HI Lyman Lines.* To avoid this effect one should study the higher order lines which have smaller damping wings for the same value of $Nf\lambda$, where $Nf\lambda \propto \tau_0$, the optical depth at line center, N is the column density, f is the transition oscillator strength, and λ is wavelength.

4. *Confusion due to high-order HI Lyman Lines.* Doppler broadening can cause the high-order HI Lyman lines to overlap with the predicted position of the DI lines, inhibiting measurement of otherwise detectable high-order DI lines. This effect will be unimportant only if the lines are narrow enough.

2.2 Summary of Synthetic Spectra

For $N_{HI} < 10^{18}$ cm^{-2} and low σ_{HI}, HI Lyα is sufficiently transparent in the damping wings that the DI line may be observed. For slightly higher column densities in the range $10^{17} \leq N_{HI} \leq 10^{19.5}$, Figure 1 demonstrates that intermediate n transitions such as the DI Lyγ line should be useful; in this range the DI Lyγ line is unsaturated or barely saturated, allowing an unambiguous determination of column density, although the precision of the measurement will depend on the actual [D/H] ratio. At still higher column densities ($N_{HI} \sim 10^{19}$ to 10^{21}cm^{-2}), the higher order lines are good candidates for study, especially since there is an increased observational efficiency in measuring many lines at once in a single integration with a high-resolution spectrograph, when the high-order lines pile up near the continuum cutoff at 912 Å.

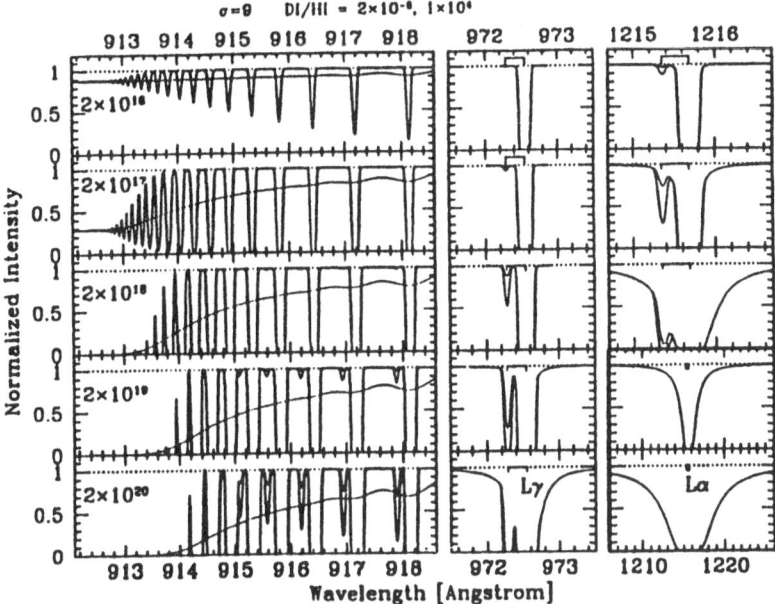

Fig. 1. Sample spectra for DI and HI Lyman absorption lines for five values of N_{HI} from 2×10^{16} to 2×10^{20} cm^{-2}. The velocity dispersion is $\sigma = 9$ km s^{-1}. Curves for both $[D/H] = 2\times10^{-5}$ and $[D/H] = 1\times10^{-4}$ are plotted. The spectral region for the Lyman continuum cut-off containing Lyman absorption transitions ending at $n = 12$ through ∞ is at the left. Panels for Lyγ and Lyα are at the right. The damping wings of Lyα are so strong for large N_{HI} that the lower two spectra for $N_{HI} \geq 2\times10^{18}$ cm^{-2} are plotted on a different scale than the spectra for $N_{HI} \leq 2\times10^{17}$ cm^{-2}. Smooth curves are drawn in the left panel to indicate the expected response of an observation with a 0.8 Å (FWHM) Gaussian resolution for the $[D/H] = 2\times10^{-5}$ case.

2.3 The Shape of the Lyman Edge and the Effect of DI

For $N_{HI} \leq 2\times10^{19}$cm^{-2} DI has no detectable effect on the shape of the Lyman edge cut-off. On the other hand, for $N_{HI} \geq 2\times10^{20}$cm^{-2} the shape of the cut-off is influenced by the presence of deuterium, but observations of the shape at moderate spectral resolution do not provide a sensitive way to measure the deuterium abundance. However, a detailed analysis of high signal-to-noise-ratio observations of the edge would need to include the DI opacity.

3 Properties of a Good Candidate QSO Absorption Line System

1. The absorption should be at relatively low redshift in order to maximize the probability of avoiding confusion due to the high density of Lyα forest absorption lines that are found at higher redshifts.
2. The best candidates will be absorption systems with a high column densities of neutral gas (damped Lyα absorption systems) in very low velocity dispersion clouds. 21cm observations and curve of growth studies aide in selecting candidates with suitably low velocity dispersion.
3. QSO absorption line systems with low metal abundance offer the possibility of observing gas that is not very chemically evolved. In good candidate systems, curve of growth studies of transitions due to, for example, CrII and ZnII in damped Lyα systems would ideally show that heavy element abundances were much less than solar, indicating that stellar nucleosynthesis minimally altered the [D/H] ratio of the gas away from the primordial abundance ratio.
4. For optimal sensitivity, it would be best if the high-order lines of the Lyman series of DI had optical depth of order unity, indicating that they would lie near the knee of the curve of growth.

References

Khersonsky, V.K., Briggs, F.H., Turnshek, D.A. (1994a): ApJ, submitted

Khersonsky, V.K., Briggs, F.H., Turnshek, D.A. (1994b): University of Pittsburgh, Department of Physics and Astronomy and Allegheny Observatory Publications

Landsman, W.B., Henry, R.C., Moos, H.W., Linsky, J.L. (1984): ApJ,285, 801

Linsky, J.L., Brown, A., Gayley, K., Diplas, A., Savage, B.D., Ayres, T.R., Landsman, W., Shore, S., Heap, S.R. (1993): ApJ, **402**, 694

Pettini, M., Boksenberg, A., Hunstead, R.W. (1990): ApJ, **348**, 48

Varshalovich, D.A., Khersonsky, V.K. (1976): Sov Astr Letters, **2**, 574

Walker, T.P., Steigman, G., Schramm, D.N., Olive, K.A., Kang, H.-S. (1991): ApJ, **376**, 51

Webb, J.K., Carswell, R.F., Irwin, M.J., Penston, M.V. (1991): MNRAS, **250**, 657

Studies of the Cloud Structure, Ionization Structure, and Elemental Abundances in High Redshift QSO BAL Region Gas

David A. Turnshek

Department of Physics and Astronomy, University of Pittsburgh

Abstract. The nature of QSO broad absorption line (BAL) gas in terms of its cloud structure, ionization structure, and elemental abundances is discussed. HST-FOS spectroscopy of the individual components of a gravitationally lensed BAL QSO is used to place constraints on BAL cloud structure, with the finding that the average line-of-sight column densities are generally constant to within ±10% across a region which is likely to be approximately the size of the continuum source. Also, HST-FOS and ground-based spectroscopy of different ions of the same element in BAL QSOs reveal information that can be used to model ionization structure and elemental abundances. Analysis indicates that the absorbing clouds have non-uniform ionization and high elemental abundances. Outflows from QSOs with BAL regions may influence the chemical evolution of the universe at high redshift.

1 Introduction

Within a few years of the discovery of moderate-to-high redshift ($z > 1.5$) QSOs observers found that some of them exhibit broad absorption lines (BALs) due to gas outflow at velocities normally up to 15,000 – 30,000 km s^{-1}. While these objects are common, representing $\sim 10\%$ of all optically selected QSOs, their role in the overall QSO scheme is still unclear. However, since the amount of time required for a BAL cloud to cross the BAL region is almost certainly shorter than a characteristic QSO lifetime, the BAL clouds are evidently continuously ejected. In addition, while most evidence suggests that BAL region covering factors are relatively small (< 0.2), indicating that most or all QSOs have BAL regions, other evidence suggests that some QSOs may have rather large BAL region covering factors. For a review and references to other work see Turnshek (1988) and Turnshek et al. (1994). Some example rest-frame spectra of BAL QSOs are shown in Fig. 1.

As discussed by Turnshek (1988), given our knowledge of the physical parameters of the BAL region (especially the distance to the central source and frequency of BALs in low-luminosity AGN) and our understanding of the initial epoch of BAL activity, QSO lifetimes and cooling times for BAL region ejecta, a significant fraction of the sky may be covered by remnant ejecta from BAL regions. The BAL outflows may, in fact, influence the chemical evolution of the universe at high redshift (Turnshek 1988), which is why I decided to make a contribution at this conference. However, the aim of my contribution is not to address these issues, but to summarize some recent advances in our understanding of the BAL region cloud structure (§2.1), ionization structure (§2.2), and elemental abundances (§2.3). These advances have resulted from HST observations, but stem from work done with large ground-based optical telescopes.

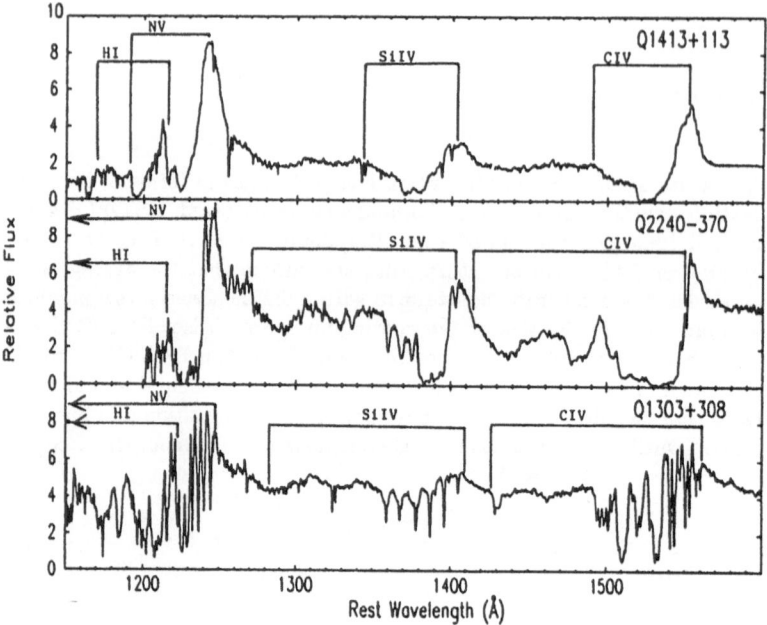

Fig. 1. Example rest-frame spectra of BAL QSOs from Turnshek (1988).

2   Results

2.1   Cloud Structure

In order to fully understand the ionization structure and elemental abundances of the BAL region gas, one must first have the ability to derive the line-of-sight column densities of ions seen in absorption as a function of outflow velocity. Procedures for doing this are discussed by Junkkarinen, Burbidge & Smith (1983), Grillmair & Turnshek (1987), and Korista et al. (1992). The procedures generally assume that the absorbing material completely covers the continuum source, that scattered emission does not fill in the bottoms of the absorption profiles, and that the absorption structure is resolved in velocity space, i.e., that absorption in the BAL profile does not get narrower and deeper at higher resolution. Kwan (1990) discusses some of these issues. A few cases have been found where there is reason for concern. These cases usually involve doublet absorption which appears saturated (i.e., nearly equal absorption equivalent widths), but for which the central intensities of the doublet are not near zero (e.g., see a SiIV doublet in the spectrum of Q0226-1024 and the discussion of it by Korista et al. 1992). However, one can always study BAL profiles in a high enough velocity interval to assure that there is little possibility of scattered emission filling in the bottoms of the absorption profiles. Moreover, as discussed below, HST observations of a gravitationally lensed BAL QSO suggest that the normal methods used to derive column densities are probably generally valid.

HST-FOS observations of the gravitationally lensed 'Cloverleaf' BAL QSO H1413+1143 (Turnshek et al. 1995) indicate that, while the BAL profiles along its four sight-lines are definitely not identical, they are very similar. Given the typical viewing angle differences of ~ 0.7 arcsec, models of the gravitational lens suggest that the sight-lines intercept the BAL region over a lateral size of $L \approx 10^{16} r_{kpc}$ cm, where r_{kpc} is the distance between the the BAL region and central source in kpc. Observational constraints reviewed by Turnshek (1988) suggest that the interval $10^{-3} < r_{kpc} < 0.5$ is likely. Photoionization equilibrium constraints and the observed column densities suggest that the thickness, x, of a BAL region cloud along a sight-line is $x < 10^{11} r_{kpc}^2$ cm. Therefore, the similar absorption indicates that *average* BAL region column densities *integrated across the continuum source* are relatively uniform (to within $\pm 10\%$) over a lateral extent which is much larger than the thickness of a cloud and approximately the size of the continuum source. This suggests that there could be some *isolated cases* (when the absorption is very weak) in BAL QSOs where the absorption doesn't completely cover the continuum source. However, in general, while the absorption appears not to be completely uniform across the continuum source along a sight-line, the average absorption profile is clearly very similar across a region which is about the size of the continuum source, which in turn suggests that the methods used to derive column densities are generally accurate enough to perform an analysis of the ionization structure and abundances.

2.2 Ionization Structure

Obtaining good data on BAL profiles of different ions of the same element had not been possible until recent HST-FOS far-UV observations of moderate-to-high redshift BAL QSOs. In the absence of far-UV data, the ionization structure can not be explored without making assumptions about elemental abundances. Single-zone photoionization models of the BAL region did result in some relevant information; however, since there was the possibility that elemental abundances were very far from solar (see below), these models were viewed as unreliable. With HST-FOS, good data on different ions of the same element in BAL region gas has been obtained for the first time. Analysis of the results indicates that a single-zone photoionization model is unable to generally explain the observed column density ratios for three ions of the same element (see below), indicating that at least a two-zone BAL cloud model is required.

For this contribution the BAL column densities derived from ground-based and HST-FOS-UV spectroscopy of Q0226-1024 (Korista et al. 1992) were considered. Column densities for different ions of the same element were available for nitrogen (NIII, NIV, and NV BAL profiles) and oxygen (OIII, OIV, and OVI

BAL profiles). Preliminary work indicates that the two-zone model gives fairly good results. Photoionization calculations using Ferland's code CLOUDY were performed with the default AGN photoionizing continuum, solar abundances, and a density of $\sim 10^8$ cm^{-3}. The inferred preliminary ionization parameters for the two zones are $\log(u_1) \approx -1.2$ and $\log(u_2) \approx -2.7$. This suggests a central density enhancement of ~ 30 (but in fact the enhancement may be larger given that there are only two zones and a two-zone cloud is unphysical). In this model the absorbing region intersected by a sight-line might consist of an inner region surrounded by a less-dense, more highly ionized outer region that has ~ 50 times the extent of the inner region. Other interpretations of the two-zone geometry are possible. Both zones contribute significantly to the observed column densities — the much larger extent of the outer region allows it to contribute to the observed column densities despite its much lower density.

2.3 Elemental Abundances

While the the two-zone model can explain the ratios of column densities of different ions of the same element for both nitrogen and oxygen with the same two ionization parameters, there are large discrepancies relative to what would be expected for solar abundances. The following preliminary results are found with CLOUDY by adopting solar abundances and the photoionizing continuum and two ionization parameters noted above. First, relative to the solar ratio, the nitrogen abundance must be ~ 3 times the oxygen abundance. In addition, relative to the *observed and computed* hydrogen column density, N(H), the *computed* metal-line column densities were much too low — N(C) was too low by a factor of ~ 20:, N(N) was too low by a factor of ~ 160, N(O) was too low by a factor of ~ 55, and N(Si) was too low by a factor of ~ 55.

3 Discussion

The results presented above can be improved upon in a number of ways. First, high-resolution studies of BAL profiles are still needed in order to understand the limitations of column density derivations. Second, the nature of the photoionization model itself, as applied to BAL column density modeling, needs to be improved. For example, the effects that far-UV BAL profiles and the outflow velocity law have on the shape of the photoionizing continuum and the effects of very enhanced abundances need to be considered carefully. Third, theoretical work is needed to explain the inferred density enhancements. Finally, observations and analysis of far-UV BAL profiles in more QSOs will allow the reliability of these results to be checked and will help pin down the systematics of abundance enhancements.

References

Grillmair, C.J, Turnshek, D.A. (1987) in *QSO Absorption Lines: Probing the Universe (Poster Papers)* (STScI publication), eds. Blades, Norman, Turnshek, 1

Junkkarinen, V.T, Burbidge, E.M., Smith, H.E. (1983), ApJ, **265**, 51

Korista, K.T., et al. (1992), ApJ, **401**, 529

Kwan, J. (1990), ApJ, **353**, 123

Turnshek, D.A. (1988) in *QSO Absorption Lines: Probing the Universe* (CUP), eds. Blades, Turnshek, Norman, 17

Turnshek, D.A., et al. (1994), ApJ, **428**, 93

Turnshek, D.A., Lupie, O.L., Espey, B.R., Sirola, C.J. (1995), ApJ, submitted

Metal Abundances in High z Absorption Lines Systems

Sandro D'Odorico

European Southern Observatory, K Schwarzschild Str.2,85748 Garching, Germany

Abstract. The abundances in high z damped $L\alpha$ systems and in metal systems with $z_{abs} \sim z_{em}$ are briefly reviewed. On the average damped systems have abundances ~ 0.1 of the solar abundances over the redshift range 1.5-3.5. The $z_{abs} \sim z_{em}$ narrow line systems on the contrary show metal abundances which are solar or higher over the same range of redshifts providing evidence for intense star formation. The status of the ESO KP "A statistical study of the IM at high z" is given. Within the program high resolution ($R \sim 30000$) spectra of QSOs at redshift $z \geq 4$ are obtained with the aim to study the statistical properties of the $L\alpha$ clouds and of the intervening metal systems.

1 . Introduction

One of the most powerful way to investigate the chemical abundances at early epochs is provided by the study of metal absorption lines detected in the absorption spectra of quasars with high emission line redshifts. These so called metallic systems are associated to galaxies or intergalactic gas in the line of sight to the QSO.

The wavelengths and the equivalent widths of the lines are used to derive the redshift and the velocity structure of the absorbing clouds, and physical parameters like ionization, temperature, chemical composition, dust content and molecular fraction.

Measurements of chemical abundances from emission lines in extragalactic ionized H II regions go back to the pre-CCD era because those lines are relatively intense. In the case of absorption systems, the most significant results have been obtained recently only. The QSOs used as background sources have typical magnitudes of 17-19 . Then telescopes of the 4 m class equipped with efficient echelle spectrograph and high performance CCD detectors are needed to carry out the observations with the required resolution (R$\succ$ 20000) and S/N ratio (≥ 5). These facilities have become available since a few years only.

In this article, & 2 summarizes the current information on abundances for high redshift damped $L\alpha$ and metal systems at redshifts close to the quasar emission line redshift and & 3 gives a brief status report of the high resolution observations of high redshift QSOs obtained within the ESO key program 149-2-01.

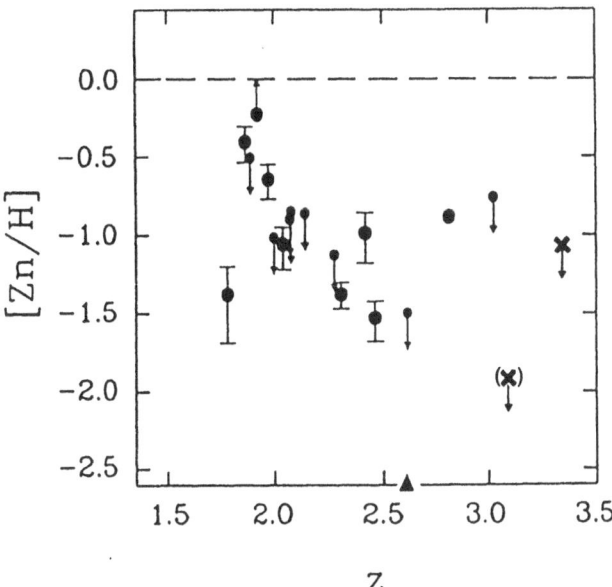

Fig. 1. A plot of the Zn/H abundances for damped systems at different redshifts. The abundances are measured on a log scale relative to solar values. The black dots are from Pettini et al. (1993), the two crosses from Savaglio et al(1994) and the triangle from X.M. Fang and Tytler (1994. All abundances by Pettini et al (1994)and the system at z=3.39 are based on Zn line measurements, the other two on different ions and ionization models.)

2 Chemical Abundances in High Z Metal Systems

Different types of metal systems are detected in the absorption spectra of QSOs. Systems at velocities well separated with from the emission redshift of the QSO are normally identified as "intervening". Among these the metal systems proper have N (HI) column densities between 10^{17} and $10^{19} cm^{-2}$and show lines of elements like C, Si, Mg, N often at different ionization stages clustered on scales of 200-300 km s^{-1}. Reliable abundance determination require the determination of the column densities of H I and of the most significant ionization stages of the different elements and must be based on lines which are neither blended or heavily saturated. A second class of intervening systems , the damped $L\alpha$ systems, have hydrogen column densities larger than 10^{20} cm^{-2} and are usually associated to dense gaseous disks of large galaxies. In these systems hydrogen is assumed to be mostly neutral and the other elements neutral or in their singly ionized state. Relatively accurate abundances can be derived from the observations of weak lines from ions like Zn II and CrII. Fig 1 gives a summary of the measurements of the metal abundances in the damped systems.

Most of the data come from the comprehensive study of Pettini et al .(1993) based on Zn abundance. The plot shows that over a wide range of redshifts, $2 \leq z \leq 4$, the abundances are one tenth of solar or lower, indicating little chemical enrichment from stellar evolution in these systems.

Absorption systems at redshifts within a few thousand km from the emission redshift of the QSO fall in two categories. Broad absorption line systems (BALs)show very high ejection velocities (up to 30000 km s^{-1}), an high degree of ionization and evidence for overabundance of metals (Hamann et al 1993). Narrow line systems with $z_{abs} \sim z_{em}$ could be associated to gas in the environment of the QSO, as it is suggested for the BAL systems, or to gas from a galaxy in a cluster to which the QSO also belongs. Table 1 gives the abundance estimates for five components of a system at $z \geq 4$ studied by Savaglio et al (1994). The metal content values, as indicated by the C/H ratio, are consistent with solar or higher than solar abundances . An equivalent result for other $z_{abs} \sim z_{em}$ systems has been obtained from high spectral resolution data by Wampler et al. (1993) on a system at z=2.32 and by Petitjean et al (1994) for six systems at $z \sim 2.1$.

Table 1. Abundances for $z_{abs} \sim z_{em}$ systems in the line of sight to 0000-26

z_{abs}	[C/H]	[Si/C]	[N/C]
4.1297	- 0.7 ÷ 0.6	-	-
4.1310	0.1 ÷ 0.8	> .2	-
4.1323	- 0.1 ÷ 0.5	-	< - 0.3
4.1334	0.1 ÷ 2	1.5	- 0.4
4.1342	- 0.4 ÷ 2.6	1.4	- 1

Line strenghts do also indicate possible departures from the solar relative abundances. Savaglio et al.(1994) observe overabundance of Silicon and an underabundance of Nitrogen with respect to Carbon. Petitjean et al. (1994) derive overabundances of Nitrogen and possibly Silicon with respect to Carbon. The presence of a significant enrichment in the metal abundances and of possible peculiar abundances in the neighbourhood of the QSO can be related to the history of star formation in the host or associated galaxy, as discussed by Matteucci and Padovani (1993). The number of studied systems is too small yet to drawn any significant conclusion from the comparison of the measured abundances and the computed models but this appear as a promising line of investigation to understand the environment and the nature of QSOs.

3 Status of the key program "Statistical Study of the Intergalactic Medium at High Redshift"

In October 1992 an ESO key program of this title was started at the ESO NTT telescope using the echelle mode of the EMMI spectrograph. With the current 2048^2 pixel,thinned CCD detector this EMMI configuration provides spectra with a resolution ~ 30000 with a one arcsec slit. The spectral range 420-900 nm is covered with high efficiency in slots of about 250 nm/exposure. The key program team includes beside the author S.Cristiani of the Astronomy Department of the Padova University, E. Giallongo and A. Fontana of the Rome Astronomical Observatory, P. Molaro of the Trieste Observatory and S. Savaglio of the Physics Department of the Calabria University.

The primary goals of the program are the study at high redshift ($z \succ 3$) of the statistical properties of the the Lα cloud population, the measure of the ionizing background and a test on the Gunn-Peterson effect.

It is planned to study 15 different line of sights through the observations of QSOs with redshift close or larger than 4. Table 1 lists the basic data on the observations obtained as of April 94, half way through our time allotment. The given range of S/N is the one to be expected over the finally reduced spectrum based on the experience of data extraction on similar data. At the time of this

Table 2. QSOs observed in the ESO Key program 149-2-013

OBJECT	m	z	Exp. No.	Total exp (sec)	R (km/sec)	coverage (Å)	Expected S/N
055-26	17.2	3.7	2	12000	12	4700-6400	5-20
BRI 0241-01	18.2	4.04	4	26100	12	4700-6400	3-10
0453-42	17.1	2.7	4	24300	10	4100-6700	10-15
BRI 952-01	18.7	4.43	2	13000	40	4700-8300	3-10
BR 1033-03	18.5	4.50	2	15200	40	4700-8300	3-10 ·
BR 1108-07	18.1	3.94	1	6000	40	5800-9500	3-15
			3	19700	40	4700-8300	5-10
			4	28800	10	4100-6700	5-10
BRI 1117-13	18	4.00	2	8900	40	5800-9500	3-10
			1	6500	40	4700-8300	
BRI 1202-07	18.7	4.7˙	2	14850	40	4700-8300	3-12
			2	15200	40	5800-9500	
1208 + 10	17.5	3.8	4	29000	10	4100-6700	5-15
BRI 2212-16	18.1	4.0	2	13600	12	4800-6400	5-15

writing (September 94) the spectra of 4 QSOs have been fully reduced and for two results have been published (Giallongo et al., 1994; Cristiani et al., 1994). While the study of the $L\alpha$ clouds has priority, we expect also to derive new valuable data on the cloud structure and on the relative ion abundances for the metallic systems at high redshifts seen in our spectra.

References

Cristiani S., D'Odorico S., Fontana A., Giallongo E. and Savaglio S. (1994) M.N.R.A.S., in press

Fan X.M.and Tytler D. (1994) Ap.J. Suppl., in press

Giallongo E., D'Odorico S.,Fontana A., McMahon R.G., Savaglio S., Cristiani S., Molaro P., Trevese D. (1994) ApJ 425,L1

Hamann F., Korista K.T., Morris S.L., 1993, Ap.J. 415,541

Matteucci F., Padovani P., 1993, ApJ 419, 485

Petitjean,P, Rauch, M., Carswell, R.F. (1994): A&A, in press

Pettini, M., Smith L.J., Hunstead R.W., King D.L. (1994) ApJ 426,76

Savaglio, S., D'Odorico S., Møller P., (1994) A&A 281,331

Wampler E.J., Bergeron J., Petitjean P., (1993) A&A 273,15

Part III

Galactic Evolution

Light Element Production in the Galactic Disc by Delayed Low Energy Cosmic Ray Flux from Intermediate Mass Stars; the ^{7}Li-Fe Dependence and Light Nuclide Ratios Explained

J E Beckman[1], E Casuso[1]

[1] Instituto de Astrofísica de Canarias, E-38200-La Laguna, Tenerife, Spain

Abstract. The mean abundance of Li in stars formed within the past 5 Gyr is $\log N(Li) = 3.2(\pm 0.2)$ while the corresponding value for the oldest stars in the galaxy is $\log N(Li) = 2.2(\pm 0.2)$. As long as the potential source of Li in the galactic disc cannot be identified there remains reasonable doubt about whether the latter represents the primordial Li abundance. Spallation of interstellar CNO nuclei by the ^4He and proton components of galactic cosmic rays (GCR) was proposed, over 20 years ago, as a source of Li but production falls short by an order of magnitude. Recently $\alpha + \alpha$ fusion was invoked to explain some Li production in the early Galaxy, but this reaction yielded, apparently, to much Li in the old disc, or too-little Li in young stars.
Failures of this sort led to the proposal of other Li-producing mechanisms: in the helium flash in AGB stars, in the nova phase, and even at supernova in specific shock mechanisms. Here we show that by assuming that the low energy GCR flux depends on the gas expulsion rate from low and intermediate mass stars rather than the supernova rate we obtain a model which explains the $\log N(Li)$ v. $\log N(Fe)$ evolution in the solar neighborhood. Incorporating interstellar depletion in the hot plasma around OB associations we can account for the observed $^7Li/^6Li$ ratios, as well as the $^{11}B/^{10}B$ ratio in the solar system, and can give a fair description of the secular decline in the D/H ratio during the disc lifetime.

1 Introduction

It is of obvious cosmological importance to be able to determine the primordial Lithium abundance, to distinguish in principle between the "population 2" value: $logN(^7Li) = 2.2(\pm 0.2)$ (Rebolo, Beckman & Molaro 1988; Spite & Spite 1991; Thorburn 1994) or the "population 1" value: $logN(^7Li) \geq 3.2$. To claim that the former value is near primordial one needs to demonstrate that population II stars have barely depleted their ^{7}Li, and that there has been a source of ^{7}Li in the galactic disc sufficient to raise the abundance by an order of magnitude. The first point has been demostrated theoretically by Pinsonneault et al. (1992) who explained quantitatively the suppression of convective transport in low metallicity stars. The second point has not lacked attention. Mechanisms adduced for disc production of ^{7}Li include cosmic ray spallation of CNO (Reeves et al. (1970), Meneguzzi et al. (1971), Walker et al. (1985)), nucleosynthesis in novae (Arnould & Norgaard (1975), Starrfield et al. (1978)) red giants (Cameron &

Fowler (1971), and in supernovae (Dearborn et al. (1989), Woosley et al. (1990)). However agreement between the observed Fe-^{7}Li dependence and model prediction based on these mechanisms is not very good (Audouze et al. (1983), Abia & Canal (1988), D'Antona & Matteucci (1991)), and the uniformity of the ^{7}Li abundance in different local sites: stars, the solar system and the interstellar medium, tells against specialized ^{7}Li sources with low population density. We show here that a GCR source can in fact explain the observations.

In observational practice it is easier to obtain reliable ^{7}Li/^{6}Li ratios than absolute ^{6}Li abundances (Lemoine et al. (1993), Meyer, Hawkins & Wright(1993)). It has been stated that the relatively high values for this ratio, greater than 10 in the solar system for example, are evidence for a stellar rather than an interstellar source for Li, but we will show here that this is not necessarily true. Finally we will use the model developed here to examine the abundances of the other light nuclides ^{9}Be, ^{10}B and ^{11}B, and very briefly that of D, where we can bring new considerations to bear.

The present study has been confined deliberately to the evolution of the Galactic disc, not because the epochs when [Fe/H] was less than -1.5 lacks interest, but because the physical assumption for a halo model are very different, and would require a major exercise rather than a mere extrapolation. A good treatment of the halo evolution of the light nuclides is found in Prantzos et al. (1993). Here we have simply taken the values of the 7Li and 9Be abundances for [Fe/H] $= -1.5$: $\log N(^7\text{Li}) = 2.2$ and $\log N(^9\dot{\text{Be}}) = 0.5$ respectively, as inputs for modelling the disc.

2 The Chemical Evolution Model

We have followed the formalism of Arimoto & Yoshii (1986) which embodies a numerical rather than an analytical approach to avoid the need for too great a reliance on simplifying assumptions such as instantaneous recycling. The model evolves the gas fraction σ_g and the abundances X_i of 6 selected nuclides: ^{4}He, ^{12}C, ^{13}C, ^{14}N, ^{16}O and ^{56}Fe. The equations used are:

$$d\sigma_g = -SFR(t) + E(t) \tag{1}$$

$$d\left(\sigma_g\right) = \int_{m_t}^{m_u} SFR\left(t - t_m\right)\phi(m) \times$$

$$\left(Q_i(m) + X_i\left(t - t_m\right)\left(R(m) - Q_i(m)\right) - R(m)X_i(t)\right) dm \tag{2}$$

with

$$E(t) = \int_{m_t}^{m_u} SFR\left(t - t_m\right)\phi(m)R(m)\,dm + P(t) \tag{3}$$

where t_m is the lifetime of a star of mass m, P(t) is the net inflow of material into the fixed volume studied, SFR(t) is the star formation rate, and $\phi(m)$ the IMF of the stars. The fractional mass expelled by a star in its lifetime is R(m) and the yield of nuclide from a star of mass m is $Q_i(m)$; E(t) is the mass added to the ISM either by stellar ejection or by inflow. The mass of a star whose lifetime is t, (the age of the zone) is m_t. We have taken m_u, the upper mass limit, as $73M_\odot$, although given the steepness of the IMF, the exact value is not critical. We have used a Schmidt (1959) law for the SFR, viz. $SFR(t) = \gamma \sigma_g^k(t)$ with k= 1 and $\gamma = 0.11\,Gyr^{-1}$, and for the IMF the Salpeter (1955) function i.e. $\phi(m) \alpha m^{-(1+x)}$ with x= 1.35, between $73M_\odot$ and $0.5M_\odot$, and flat between $0.5M_\odot$ and $0.1M_\odot$ (see eg. Scalo (1986)) and Kroupa et al. (1993)). The stellar lifetime was approximated by $t_m = 11.700/m^2$ with t_m in Myrs and m in $M_\odot$. The model takes in all relevant stellar evolutionary states, including post-main sequence phases, the planetary nebula stage, and the carbon burning phase. Gas ejection mechanisms include novae, stellar winds, and supernovae. The yields and ejected mass fractions come from Renzini & Voli (1981) for $1M_\odot \le m < 8M_\odot$ and from Arnett (1978) for more massive objects. The models were tested against the following constraints.

a) The present-day gas fraction must agree with the observational value of $\sigma_g = 0.2(\pm0.05)$ (see Rana 1991).

b) The observed trend of Fe with age in the disc must be reproduced. In Fig. 1 we compare a model prediction with observational data (Meusinger et al. (1991)).

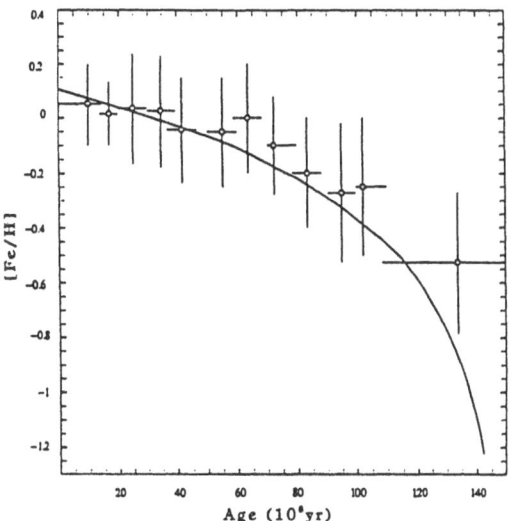

Fig. 1. Iron metallicity in the solar neighborhood as a function of age. The model prediction: the continuous curve, is compared with the observational points of Meusinger et al. (1991), showing the model used in the present paper gives a very fair account of these data.

c) The observed number of G-dwarfs as a function of Fe abundance should be predicted. In Fig. 2 we compare data from Pagel (1989b) and Norris & Ryan (1991), with model predictions. Only an infall model can yield good agreement. Infall here is given the form

$$P(t) = Ae^{-\lambda t} \tag{4}$$

following Lacey & Fall (1985) or Clayton (1985, 1986) where A takes the value $-1M_\odot\,yr^{-1}$ for the solar cylinder, of annular width 1kpc, and λ has been taken at 1.5 Gyr. The final version of the model in fact uses a positive exponent in P(t) and yields a better fit to the data.

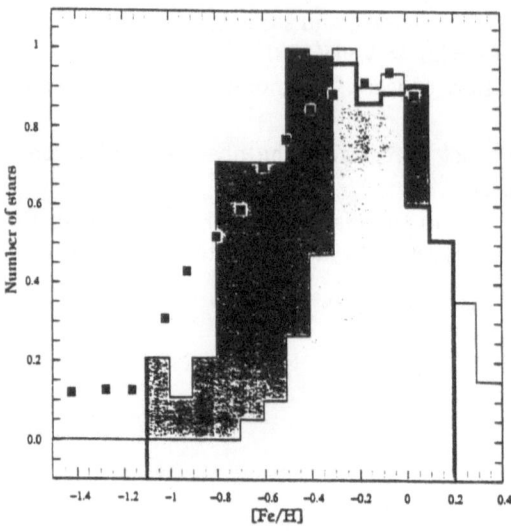

Fig. 2. The deficiency of G dwarfs at low metallicities in the solar neighborhood is predicted by this model, as is normal for models incorporating gaseous infall. The predictions here (filled squares) are not in detailed agreement with the observations, either those tabulated in Pagel (1989a) - dark shading - or in Norris and Ryan (1991) - light shading. The discrepancy could be overcome by a somewhat different treatment of infall. However the present model does not give a poorer approximation than other current literature models, and the fine tuning required to achieve a better fit would have very little effect on our present treatment of light element production.

d) The dependence of [O/Fe] on [Fe/H] should be reproducible. In Fig. 3 we combine data from Barbuy & Erdelyi-Mendez (1989), Spite & Spite (1991) and Bessell et al. (1991) to compare with the model prediction. The model also correctly predicts N/Fe (Carbon et al. (1987) and C/Fe Clegg et al. (1981) values, essentially solar in the disc metallicity range, and yields a current value of the ^{4}He mass fraction of 0.27 consistent with observations (e.g. Norris, 1971).

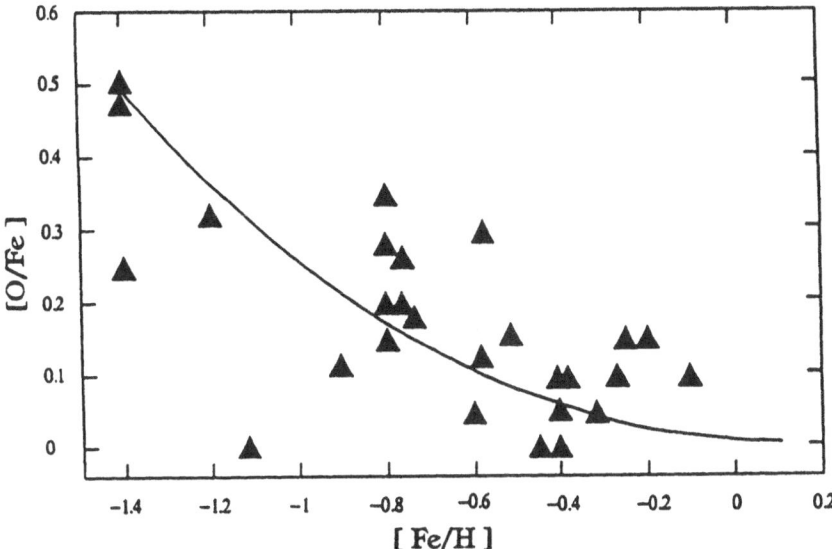

Fig. 3. The ratio oxygen: iron as a function of the iron abundance in the disk. The sources of data are described in the text. As in Fig 1, the galactic model prediction (continuous curve) gives a fair fit to the data.

3 ^{7}Li Production by $\alpha + \alpha$ Reactions of GCR

The original mechanism for producing light element nuclides in the ISM proposed by Meneguzzi et al. (1971), was spallation of CNO by GCR protons and alpha particles. We hence included this in the present model, together with a 20% contribution from spallation of GCR CNO by slow ISM protons and alphas (Meneguzzi & Reeves, 1975). However we have also included as a major source of Li, Be and B $\alpha + \alpha$ fusion between GCR and ISM alphas, whose role, though recognized (Montmerle, 1977; Steigman & Walker, 1992) has been underestimated.

The general expression used for the production process is:

$$\frac{dX_k}{dt} = \sum_j X_j^{ISM} \sum_i \int F_i^{GCR}(E,t)\sigma_{ij}^k(E)\,dE \qquad (5)$$

where $X_j(t)$ are the abundances by number of each species, j refers to ^{12}C, ^{13}C, ^{14}N, ^{16}O, or ^{4}He, while k refers to ^{6,7}Li, ^{9}Be, ^{10,11}B, and i refers to GCR protons or alphas. $F_i^{GCR}(E,t)$ is the IS GCR flux spectrum, $\sigma_{ij}^k(E)$ is the reaction cross section for i+j→k, which has an energy threshold E_T. The quantities $X_j^{ISM}(t)$ are constrained by observation, as fit by the chemical evolution model; the cross-section σ_{ij}^k are laboratory determinations (Read & Viola, 1984) with small errors. All relevant cross-sections rise from zero at E_T (near 10 MeV/nucleon) to a peak between 20-70 MeV/nucleon, and decline to a plateau value above 100 MeV/nucleon. For the $\alpha + \alpha \rightarrow ^{6,7}$Li reaction the peak is 500 times the plateau

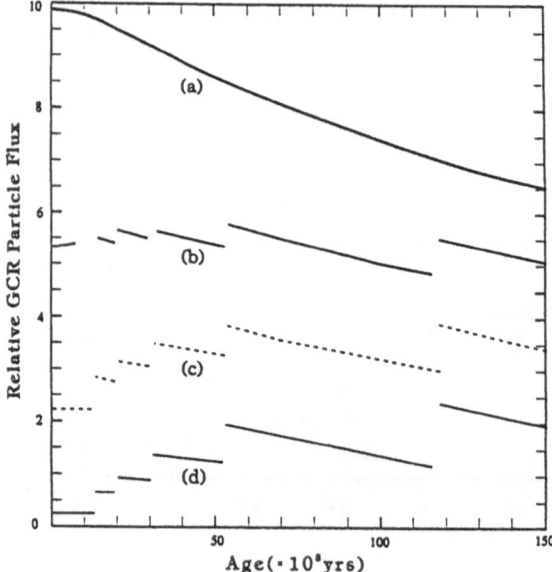

Fig. 4. Fluxes of galactic cosmic rays (GCR) in our model, when their production rate is subject to the following constraints: (a) Proportional to the star formation rate (and by implication to the type II SN rate). (b) Proportional to the total expulsion rate of gas from the stellar population. (c) Proportional to the gas expulsion rate from stars of M<8 $M_\odot$. (d) Proportional to the gas expulsion rate from stars of M<3 $M_\odot$ These plots are presented one below the other in order (a)-(d), with artificial spacings of 2 units between plots for easier visibility. They should, in practice be normalized so that their end-points coincide (at t= 1.5×10^{10} yrs). The breaks in the curves, and the cusps at the corresponding epochs and values of [Fe/H] in subsequent figures are artifacts due to finite biginning within the model, and do not significantly affect the light element predictions.

value, so that fluxes at energies above 100 MeV/nucleon can be neglected, while for all the other, spallation, reactions the peak is 5-10 times the plateau, an d the full energy spectrum must be included. This difference is important in explaining the difference between the ^{7}Li-Fe and ^{9}Be-Fe curves.

Especially for $\alpha + \alpha$, we need the low energy spectrum of the GCR α flux, which cannot be found simply from measured GR fluxes near the earth, due to solar demodulation. Here we follow Reeves & Meyer (1978) taking a value of 4 as the solar modulation factor for the whole spectrum, and another factor 3 below E= 100 MeV/nucleon, consistent with more recent Pioneer 10 measurements of low-energy α particles by McDonald et al. (1990) and McKibben (1991). Our flux spectra is proportional to $(E_T + E_o)^{-\lambda}$, where E_o is the proton dust energy and λ is 2.6 below 100 MeV/nucleon, 2.2 between 100 MeV/nucleon and 1GeV/nucleon, and 2.6 at energies >1GeV/nucleon, following Walker et al. (1985) and also Steigman & Walker (1992) We normalized to be consistent with an α/p ratio of 0.15 (Gloeckler & Jokipii, 1967) as re-examined by Webber & Lezniak (1974).

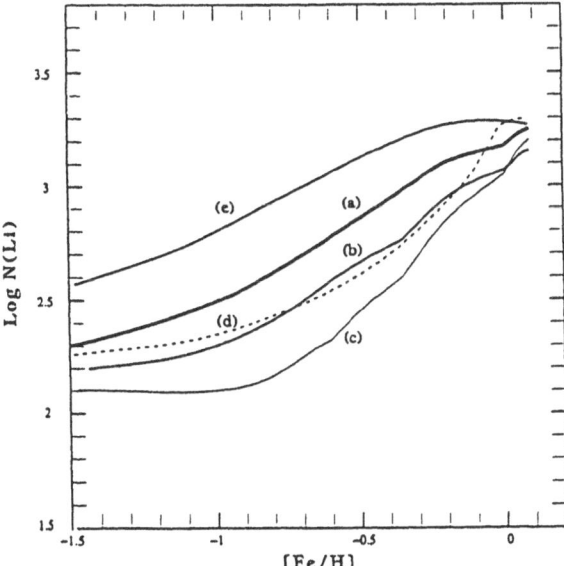

Fig. 5. Disk production curves of ^{7}Li for three models in which the GCR flux responsible for the ^{7}Li, via $\alpha + \alpha$, is proportional to the gas expulsion rate from (a) The whole stellar population. (b) The population with M<8 M$_\odot$ (c) The population with M<3 M$_\odot$. The intermediate model (b) gives a good fit to the observed envelope (curve (d)); the discrepancy here close to [Fe/H]= 0 could be improved by a more refined treatment of gaseous infall. We have also included, in curve (e), a model with GCR flux proportional to SFR, here normalized to yield the current ^{7}Li abundance, showing that in this case there would be excessive ^{7}Li production in the disk for the whole range of [Fe/H], notably at [Fe/H]= -1.5.

Since the measured GCR nuclide abundances are modulated by first ionization potentials (Cassé & Goret (1978), Meyer (1985)) their origin must be in the thermalized atmospheres of low and intermediate mass stars, whatever the subsequent acceleration mechanism. Incorporating this assumption into our chemical evolution model gives the time-dependence of the flux causing $\alpha + \alpha$ reaction shown in Fig. 4, quite different from that predicted in models where the GCR flux is proportional to the supernova rate or to the star formation rate (Steigman & Walker (1992), Steigman (1993), Prantzos et al. (1993). This difference is critical in explaining the ^{7}Li-Fe observational curve.

The final flux of GCR particles is determined not only by the injection rate, but by the acceleration mechanism, now widely accepted as collisionless plasma shocks resulting from SN explosions (Lagage & Cesarsky 1983, Blandford & Eichler 1987). Although the high energy (GeV/nucleon) GCR component is modified by the particle escape length (see e.g. Prantzos et al. (1993)) so that the high energy spectral slope has steepened with epoch, this does not materially affect the lower energy particles responsible for $\alpha + \alpha$ production of Li. Thus, as there has always been sufficient acceleration energy available, the Li produc-

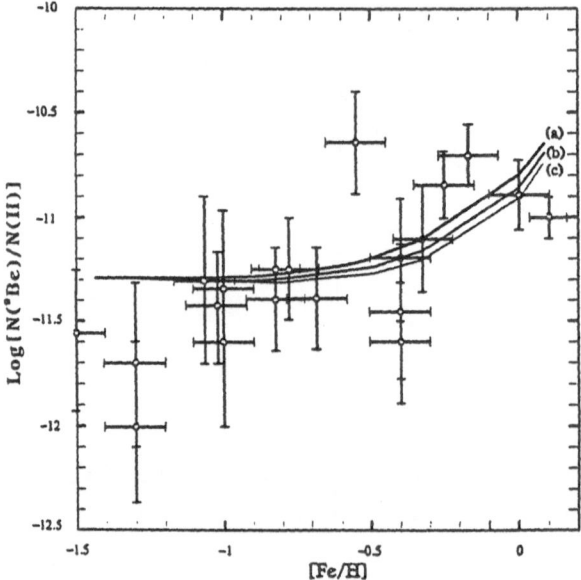

Fig. 6. ^{9}Be production for the model under three constraints: (a) GCR flux proportional to total gas expulsion from stellar population. (b) GCR flux proportional to gas expelled from stars with M<8 M$_\odot$. (c) GCR flux proportional to gas expelled from stars with M<3 M$_\odot$ Data points from Rebolo et al. (1988), Gilmore et al. (1991, 1992) plus unpublished data from our own group (Garca López et al., 1994 (in preparation)). Since the dominant process for ^{9}Be production is CNO spallation, curve (a) should give the best fit to the data, but observational scatter does not allow us to draw such a detailed conclusion here.

tion rate has evolved according to the α–particle injection rate (i.e. accumulated low-mass star density) rather than the SN rate (density of high mass stars).

4 Individual Nuclides vs Fe: Predictions vs Observations

a) **^{7}Li.** In Fig. 5 we present the results of a set of modelling exercises for the ^{7}Li-Fe curve, plus the observed curve (in fact the upper envelope of the observation of Rebolo et al. (1988) which is assumed essentially unaffected by stellar depletion). The uppermost curve shows how a model with GCR flux dependent on the SFR (and implicitly the SN rate) gives a prediction widely discrepant from the observations. The same objection would apply to any model, whether the ^{7}Li is stellar or interstellar, in which the production rate depends chiefly on the massive star population. The other production curves give better approximations to the observed envelope, and show that our hypothesis of low-energy injection for the ^{7}Li producers is consistent with observation.

b) **^{9}Be and ^{10,11}B.** In Figs. 6 & 7 we show that the model, which for Be and B gives greater weight to higher energy GCR particles, is broadly capable of reproducing the disc evolution of these elements. It is clear that for B

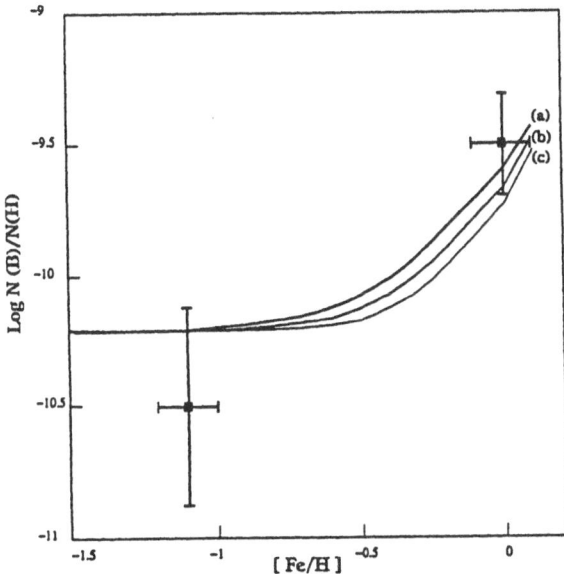

Fig. 7. $^{11}B+^{10}B$ production in the models used in the present paper: (a) GCR flux proportional to gas expelled from the whole population of stars. (b) GCR flux proportional to gas expelled from stars with M<8 $M_\odot$ (c) GCR flux proportional to gas expelled from stars with M<3 $M_\odot$ With so little data for the disk, this test of the model is of little weight; however the boron measured solar system ratio $^{11}B/^{10}B$ can be reproduced (see text), and does give support to the present model.

the data are hardly capable of discriminating this model from many others, and even for Be we would not claim more than broad consistency with the observations, so these tests cannot as yet be given high weight.

5   Isotope Ratios. The Problem of Depletion

a) $^7Li/^6Li$ After attempting to procure differential depletion of ^7Li and ^6Li via a stellar mechanism, it become clear that over practically the whole stellar mass range, gas returned to the ISM would either be undepleted or fully depleted in both isotopes. We have therefore considered depletion via destruction inside the hot gaseous envelopes of OB associations, so called "superbubbles". Tomisaka et al. (1981) showed that the temperatures of 4×10^6K should be developed in the interior of a typical superbubble. This is high enough to destroy both ^7Li and ^6Li in collisional interactions. However there is a thick boundary shell where the temperature will be in the range between 1 and 2.4×10^6K, in which ^6Li is depleted differentially. Full details of this model are described in Casuso & Beckman (1995). Application of the model yields ranges of the $^7Li/^6Li$ ratio in accord with the measured value of 12.5 for the solar system, and the range of values of 8 and below in the local ISM (Lemoine et al. (1993), Meyer, Hawkins & Wright (1993),

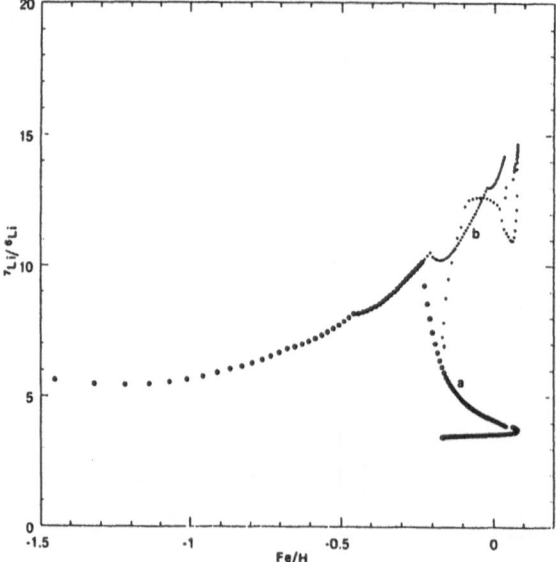

Fig. 8. Ratio of ^{7}Li to ^{6}Li v. [Fe/H] in two models where differential IS depletion in OB association envelopes is occurring. (a) Depletion acts only for the earlier 5×10^9 yrs of disc activity. (b) Depletion acts throughout the disc lifetime. The complex behaviour of the ^{7}Li/^{6}Li ratio in case (b) is caused by the non-monotonic time-dependence of the nuclides in this model with infall.

Lemoine et al. (1995, this meeting). In Fig 8 we show two possible evolutionary traces of ^{7}Li/^{6}Li, one in which IS depletion stopped after 5×10^9 yrs of disc evolution, and the other in which it has continued throughout the disc lifetime. In the second scenario we find the range of ^{7}Li/^{6}Li ratios which are observed. The non-monotonic behaviour of the ratio is consistent with the very "flat" behaviour of [Fe/H] during the past 8×10^9 years, coupled with the effect of infall. This trace of ^{7}Li/^{6}Li should clearly be taken as characteristic rather than definitive; behaviour may very spatially according to the detailed history of local OB associations. The IS mechanism for differential depletion is, however, capable of accounting for the solar system and local ISM measurements.

b) ^{11}B/^{10}B. Similar considerations apply for ^{11}B/^{10}B to those for ^{7}Li/^{6}Li. In this case the destruction temperature for ^{11}B is higher than that for ^{10}B, and to an adequate approximation the OB association envelopes leave the ^{11}B untouched, while progressively depleting ^{10}B (which is itself subject to a higher destruction temperature than ^{7}Li). In Fig. 9 we give a scenario which yields the observed solar system ^{11}B/^{10}B ratio of 4. Here we have assumed that the somewhat lower IS gas densities in the early disc gave rise to somewhat higher ISM temperatures in the superbubbles, causing relatively strong differential depletion of ^{10}B vs. ^{11}B. The formation ratio of ~ 2 was consequently enhanced to higher values, and has subsequently fallen to values close to 4. The important aspect here is not the exact scenario of

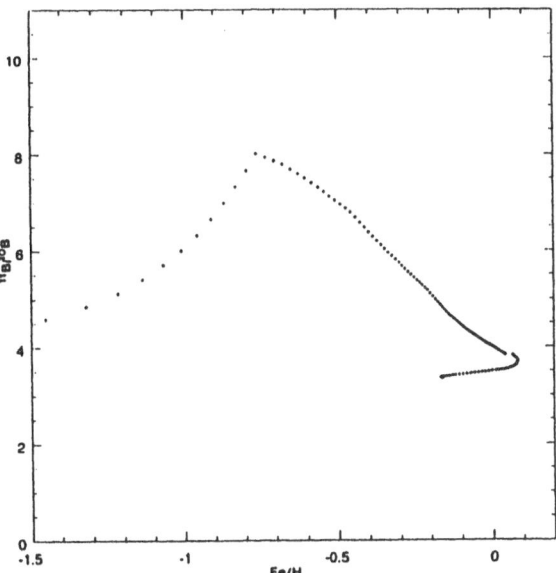

Fig. 9. One scenario for the dependence of $^{11}B/^{10}B$ on [Fe/H]. Depletion of ^{10}B in superbubbles raises the ratio (see text).

Fig. 9, but the ISM depletion of ^{10}B, which acts to enhance the $^{11}B/^{10}B$ ratio for a range of detailed evolutionary models.

6 The IS Deuterium Abundance

It is notable that in the context of the present model we can reproduce the evolution of the local D/H ratio in the ISM. If we use the primordial ^4He and ^7Li abundances to predict that primordial gas, before astration, or other depletion processes, has a D/H ratio of 7.5×10^5, and assume no subsequent sources of deuterium, the two best observational ratios: $(D/H)_\odot = 2.6 \times 10^5$ in the solar system (Geiss, 1993) and $(D/H)_{ISM} = 1.5 \times 10^5$ (Linsky et al. 1992) towards Capella (with the GHRS on HST) can be reproduced. For the detailed calculations we used an infall model with +ve exponent, and the same IS depletion mechanism required for the Lithium and Bario ratio plots. For further details of this work we refer the reader to Casuso & Beckman (1995). Clearly this exercise is one of self consistency rather than being of value in predicting the primordial D/H ratio, but it offers a consistent history of D depletion in the disc.

7 Conclusions

We are able to model the isotopic histories of the light elements: D, Li, Be and B in the Galactic disc, using a chemical evolution model developed to deal with the G-dwarf metallicity distribution and the observed evolution of CNO with Fe. The two novel elements presented are the dependence of the ^7Li and ^6Li production rates on the local spatial density of low mass stars, and the introduction of depletion in the high temperature zones of OB-association envelopes. These assumptions enable us to dispense with the requirement of stellar sources for ^7Li, although we make no comment here on whether such sources do in fact exist, or have any degree of importance.

References

Abia, C., Canal, R.: 1988, A&A **189**, 55.

Arimoto, N., Yoshii, Y.: 1986, A&A **164**, 260.

Arnett, W.D.: 1978, Ap.J. **219**, 1008.

Arnould, M., Norgaard, H. : 1975, A&A **42**, 55.

Audouze, J., Boulade, O., Martinie, G., Poilane, Y.: 1983, A&A **127**, 164.

Barbuy, B.: 1988, A&A **191**, 121.

Barbuy, B., Erdelyi-Mendez, M.: 1989, A&A **214**, 239.

Bessell, M., Sutherland, R., Ryon, K.: 1991, Ap.J. **383**, L74.

Blandford, R., Eichler, D.: 1987, Phys.Rep. **154**, 1.

Cameron, A.G.W., in Essays in Nuclear Astrophysics, C A Barnes, D D Clayton & D N Schramm eds. (CUP), p. 23.

Carbon, D.F., Barbuy, B., Kraff, R.P., Friel, E.D., Suntzeff, N.B.: 1987, P.A.S.P. **99**, 335

Cass, M, Goret, P.: 1978, Ap.J. **221**, 703.

Clegg, R., Lambert, D.L., Tomkin, J.: 1981, Ap.J. **250**, 262.

D'Antona F, Matteucci, F.: 1991, A&A **248**, 62.

Dearborn, D.S.P., Schramm, D.N., Steigman, G., Truran, J.W.: 1989, Ap.J. **347**, 455.

Geiss, J.: 1993, in Origin and Evolution of the Elements, N Prantzos, E Vangioni-Flam, M Cass eds. (CUP), p. 89.

Gloeckler, G., Jokipii, J.R.: 1967, Ap.J. **148**, 141.

Kroupa, P., Tout, Ch.A., Gilmore, G.: 1993, M.N.R.A.S **262**, 245.

Lacey, G, Fall, S.M.: 1985, Ap.J. **290**, 154.

Lagage, P.O., Cesarsky, C.: 1983, A&A, **122**, 129.

Lemoine, M., Ferlet, R., Vidal-Madjar, A., Emerich, C., Bertin, P.: 1993 A&A **269**, 469.

Linsky, J.L. et al.: 1993, Ap.J. **402**, 694.

Martn, E.L., Rebolo, R., Magazz, A., Pavlenko, Ya.V.: 1994 A&A **282**, 503.

McDonald, F.B., Lal, N., McGuire, R., von Rosenvinge, T.T.: 1990, in "Proc. XXI Int's Cosmic Ray Conference" (Adelaide) 6, 144.

McKibben, R.B.: 1991, in Physics of the Outer Atmosphere (S. Grzedzielsnij & D Page eds.) Pergamon, p. 107.

Meneguzzi, M, Reeves, H.: 1975, A&A **40**, 110.

Meneguzzi, M, Audouze, J, Reeves, H.: 1971, A&A **15**, 337.

Meusinger, H., Reimann, H., Stecklum, B.: 1991, A&A **245**, 51.

Meyer, D.M., Hawkins, I., Wright, E.: 1993, Ap.J. **411**, L61.

Meyer, J.P.: 1985, Ap.J.Suppl. **57**, 173.

Montmerle, T.: 1977, Ap.J. **217**, 878.

Norris, J.: 1971, Ap.J.Suppl. **23**, 193.

Norris, J, Ryan, S.: 1991, Ap.J. **380**, 403.

Pagel, B.E.J.: 1989a, Rev.Mex.Astr.Ap. **18**, 161.

Pinsonneault, M., Deliyannis, C, Demarque, P.: 1992, Ap.J.Suppl. **78**, 179.

Prantzos, N, Casse, M, Vangioni-Flam, E.: 1993, Ap.J. **403**, 630.

Rana, N.: 1991, Ann.Revs.Astr.Ap. **29**, 129.

Read, S., Viola, V.: 1984, Atomic Data & Nuclear Data **31**, 359.

Rebolo, R., Molaro, P., Beckman, J.E.: 1988, A&A **192**, 192.

Reeves, H., Meyer, J.P.: 1978, Ap.J. **266**, 613.

Reeves, H., Fowler, W., Hoyle, F.: 1970, Nature **226**, 727.

Renzini, A., Voli, M.: 1981, A&A **94**, 175.

Salpeter, E.E.: 1955, Ap.J. **121**, 161.

Scalo, J.M.: 1986, Fund.Cosm.Phys. **11**, 1.

Schmidt, M.: 1959, Ap.J. **129**, 243.

Spite, M., Spite, F.: 1991, A&A **252**, 689.

Starrfield, S., Truran, J.W., Sparks, W.M., Arnould, M.: 1978, Ap.J. **222**, 600.

Steigman, G.: 1993, Ap.J. **413**, L73.

Steigman, G., Walker, T.: 1992, Ap.J. **385**, L13.

Thorburn, J.A.: 1994, Ap.J. **421**, 318.

Tomisaka, K., Habe, A., Ikeuchi, S.: 1981, Astrophys.Space.Sci. **78**, 273.

Walker, T.P., Mathews, G., Viola, V.: 1985, Ap.J. **299**, 745.

Webber, W.R., Lezniak, J.A.: 1974, Ap.Sp.Sci. **30**, 361.

Woosley, S.F., Hartmann, D.H., Hoffman, D., Harton, W.C.: 1990, Ap.J. **356**, 272.

Mixing of Heavy Elements into the Interstellar Medium of Gas-rich Galaxies: Consequence for the Primordial Helium Determination

Daniel Kunth[1], Francesca Matteucci[2], Jean-Rene Roy[3]

[1] Institut d'Astrophysique de Paris, CNRS, Paris, France
[2] European Southern Observatory, Garching bei Munich,Germany
[3] Departement de Physique et Observatoire du Mont Megantic,Quebec,Canada

Abstract. Abundances fluctuations observed in the interstellar medium of gas–rich galaxies are discussed. It is argued that in all gas–rich galaxies, including large spirals, the ISM is less homogenized than expected. Processes to account for the observed amplitudes of abundance fluctuations are reviewed. O/H fluctuations measured in low mass gas–rich galaxies may arise from the loss of enriched material by galactic winds powered by the starburst events, and from the long dormant phase between successive star–forming episodes. Chemical evolutionary models including SNI and differential winds well reproduce the observed C:N:O:He abundances of the dwarf galaxy IZw18. The He/H stellar enrichment in very metal deficient galaxies must be of the order of 0.0005 by mass much lower than the observed dispersion in the Y,Z plane.

1 Introduction

Stellar and nebular abundance indicators reveal significant abundance fluctuations in the interstellar medium (ISM) of gas-rich galaxies. The ISM of the more massive galaxies appears on first examination relatively well-mixed. It is true that one never finds isolated pockets of ionized gas with O/H = 1/50 O/H$_\odot$ close to regions with O/H = 2 O/H$_\odot$; thus mixing processes appear quite effective. The chemical composition of the ISM in large spirals at a given radial distance appears not to vary much, but there are evidences that the ISM is not perfectly homogenized. O/H abundances derived from several indicators in large disk galaxies suggest that azimuthal variations could reach a factor of two (Belley & Roy 1992). The scatter in stellar metallicities itself is very likely reflecting the original inhomogeneities of the interstellar gas (François & Matteucci 1993). On the other hand low-mass galaxies have very shallow global abundance gradients, and variations of O/H from one region to another in magellanic irregulars do not exceed a factor of two (Dufour 1986), but indicate that mixing is not perfect in these systems either. The mixing in low-mass galaxies might be even less efficient. This is strongly suggested by the largest known spatial discontinuity in heavy element abundances measured recently in the dwarf galaxy IZw18 by Kunth et al.(1994a) for which they found that the oxygen abundance in the cold cloud associated with the main star–forming region is 30 times less than that of its HII region i.e. 1/1000 solar! Such a result is in agrement with Kunth and Sargent (1986) suggestion that giant HII regions are self-enriched. Not only

I Zw 18 is among the lowest abundance objects, but it also displays the largest abundance discontinuity. Local pollution effects should play a role in the $\Delta Y/\Delta Z$ derivation from HII galaxies spectroscopy as emphasized by Pagel et al. (1992) who noticed that galaxies with WR features in their spectra exhibit anomalous He overabundances at a given O/H and N/H ratio. It is precisely because we now have new informations about the metal enrichment of both the HI and HII phases in IZW18 that one can build up a coherent picture of the chemical history of this object. A chemical evolution model following the evolution of the He, C, N, O and Fe abundances of this galaxy has been derived by Kunth et al. (1994b). Their results and their implications for the primordial He are discussed in the last section.

2   The Dispersal and Mixing of the Heavy Elements

It is easy to see that if the ISM is enriched by massive stars via SN events, there have been enough of such events at any given point of the disk so as to get much smaller fluctuations than those observed (see Edmunds, 1975). The same conclusion holds even if one considers that the formation of massive stars is highly correlated in space and time. However, whenever mixing processes for the gas in galaxies are reviewed, they turn out to be very efficient at all scales: At large scale (1-10kpc) turbulent diffusion of interstellar clouds in the shear flow of galactic differential rotation wipes out azimutal O/H fluctuations in less than $\sim 5 \times 10^8$ yrs.

At intermediate scale (100 - 1000 pc) mixing occurs mainly through clouds collisions and expanding supershells driven by evolving associations of massive stars . Under continuous star-formation regime, the timescale is about 10^8 yrs.

At small scale (1 - 100 pc) turbulent diffusion is dominant in cold cloud while Rayleigh–Taylor and Kelvin–Helmhotz instabilities develop in ionised gas leading to full mixing in less than $\sim 2 \times 10^6$ yrs hence smaller than the lifetime of HII regions. This mechanism ensures the local mixing of freshly ejected nucleosynthesis products as suggested by Kunth and Sargent (1986). On the other hand turbulent diffusion is not efficient in cold phases. Roy and Kunth (1995) conclude that: (i) Because the timescales of mixing mechanisms at small scales are much shorter than those operating at larger scales, one should expect a relatively well mixed interstellar medium in disk galaxies. (ii) We reckon that the discrepancy betweeen observed and expected abundance fluctuations is significantly large, observed variations being 5 - 10 times larger than expected ones.

3 The Case for Abundance Fluctuations in Dwarf Galaxies

Roy and Kunth (1995) propose two complementary effects that may act to build up, or maintain, large abundance fluctuations in the ISM. The first effect is the of *retention of newly enriched material* in regions prone to undergo relatively quickly successive episodes of star formation. Fluctuations can be created by localized bursts of star formation. Newly produced elements are then "trapped" in SN remnants and superbubbles, which are the privileged sites for new star formation because their higher densities ensure quicker collapse under gravitational instability. The amplitude of O/H fluctuations will depend on the relative importance of *stimulated* star formation with respect to *spontaneous* star formation. In low mass galaxies, triggered star formation is less effective because *(i)* the time for gravitational collapse of an expanding spherical shell is longer than to collapse a ring (shells are likely to occur in fat dwarf galaxies), *(ii)* low metallicity results in lower compression of shells, thus again in slower collapse, and *(iii)* the loss of material by galactic wind and internal pressure through blowout or chimney quenches the piston effect in the ISM of the galaxy. The C:N:O abundance ratios in I Zw 18, as we will discuss hereafter are certainly indicative of losses through some sort of wind. The second effect, as suggested by Edvardsson et al.(1993), is *infall of relatively unprocessed gas*, where the timescale of infall events is shorter than the epicyclic mixing timescale. Some Galactic high-velocity clouds have low abundances , as demonstrated by Kunth et al.(1994a). Roy and Kunth analysis leads to obvious differences in mixing efficiency between dwarf and large galaxies. In particular, the weakness of the rotational velocity field in dwarf galaxies and selective loss through galactic winds can lead to large abundance discontinuities in the smaller galaxies. Processes associated with massive star formation are extremely efficient at mixing hot and warm ionized gas; as a corollary, mixing is much less efficient in the cold neutral and molecular gas, mixing of these latter phases being done by galactic scale stirring. The inefficiency of triggered star formation contributes to develop large abundance spatial fluctuations in dwarf galaxies undergoing long dormant phases between star forming episodes.

4 The Chemical History of the Dwarf Galaxy IZW18

The recent finding of Kunth et al. (1994a) of a very low oxygen abundance in the HI envelope of IZW 18 gives the unique opportunity to model the chemical evolution of this galaxy by reproducing the observed elements in both the HI and HII phases. A crucial question remains however on how to interpret the presence of metals in the HI zone. The issue of this question has some bearings on the very existence of young galaxies. If indeed IZW 18 is experiencing his very first episode of star-formation then one can speculate that metals in the HI were produced at an earlier epoch from a population III stars prior to the collapse of the proto galaxy. No observational evidence has been produced so far for the very existence of such an early generation of massive stars (Matteucci and Padovani,

1993). Another way to explain the presence of metals would be to assume that during the ongoing starburst, a significant fraction of metals which contribute to enrich the HII region also escape from the central region and pollute the HI gas. This can be achieved if galactic winds due to SNe transport a significant fraction of oxygen that subsequently recombines into the cold gas. Finally a third possibility is that the observed burst is not the first one and that one or few star-formation episodes have raised the HI metallicity to the present value.

A key to these scenarios is the N abundance. It turns out that in IZW18 [N/O] = -1.8 as observed from the HII. This makes it difficult to envisage a very short age for the ongoing burst if N were to be manufactured by the same star-formation episode as the O elements. Nitrogen is generally believed to be a "secondary" element - i.e. formed at the expense of carbon and oxygen already present in the star at birth - coming mainly from low and intermediate mass stars (0.8–8.0 $M_\odot$).

As noted by Matteucci (1986) and by Marconi et al. (1994), the N/O versus O/H distribution in dwarf galaxies is a puzzle in the sense that in order to get large N/O ratios even for low O metallicities requires nitrogen to be partially primary and produced on short timescales, something that stellar evolutionary model-makers had some difficulties to envisage. However, recent calculations by Weaver and Woosley (1995) seem to support the possibility that N is produced as a primary element.

Observed integrated spectra of IZW 18 can be easily reproduced by population synthesis models assuming an instant burst with an age of 3 Myrs or by a more continuous star-formation regime of 20 Myrs at most (Lequeux et al., 1981, Kunth and Mas-Hesse, 1994). This at first sight is far too short for secondary N to be released by the bulk of the intermediate mass stars whose lifetime is greater than 20 Myrs. Assuming that N originates from a previous burst, the N/O ratio should be much smaller than observed since O originates from self-pollution in the HII whereas N, produced in the previous burst/s, would have time to mix into a much larger volume. With these ideas in mind, Kunth et al. (1994b) have computed specific chemical evolution models using the code of Matteucci and Tosi (1985) which accounts for the effect of galactic winds powered by SNe and possible infalling primordial gas. This code as implemented by Marconi et al. (1994) introduces the effects of type I SNe and of a differential galactic wind in which N and He that are restored through stellar winds do not leave the star forming region whereas oxygen and the other α-elements, ejected by SN II explosions will. The details are described in Kunth et al. (1994b). The results indicate that the calculated abundance of N in single burst models is much too low and cannot reproduce the observations using standard nucleosynthesis prescriptions unless the duration of the burst is 50 Myrs. This is because N, is treated as mostly a secondary element and comes mainly from long–living stars. In another model the nucleosynthesis prescriptions treat N from massive stars as a primary element. In this case we have a good agreement with all the observed quantities, including nitrogen. The only problem with the single burst model may be the low abundance predicted for the HI gas either because nitrogen would not be ejected or because the species will not mix and recombine fast enough to be seen now.

This is the reason a two burst model has been computed in which the two episodes are separated by a temporal gap of at least 1 Gyr. The metals dispersed after the first episodes are are used as seeds for the second generation of stars in the second burst. There is no difficulty in reproducing the observed elements with either the single burst of the two burst models.

5 Implications for the Primordial Helium Determination

The chemical history of IZW18 can be modelled with very simple assumptions. All our models show that the He/H release (in the one or two bursts pictures) does not exceed 0.0005 by mass. If this is the case one needs no correction to the observed He/H ratio from any significant stellar enrichment that remains much below the observational uncertainties. This object has some objective drawbacks for the observers (see Pagel et al., 1992) making the helium value more uncertain than expected at first. However, two recent investigations have given very concordant results : Pagel et al. find 0.226 ±0.010 whereas Skillman and Kennicutt report 0.231 ±0.006. Both values are in general agrement although more accuracy would be needed for such a crucial issue. These results could indicate a provisional value of Yp=0.230. Do we need to observe IZW18 again for more than 100 hours or should one devote 10 hours to 10 more additional objects? We urge more observational concern to objects with oxygen abundances lower than 5% solar! Indeed at such a low metallicity the resulting Yp depends little on the assumption of a linear relationship between Y and Z. Our calculations suggest that the observed He/H scatter in the O/H versus He/H diagram as observed in very metal-poor galaxies is too large to be simply accounted for by chemical evolution differences(see E.Terlevich, this conference). The first task is certainly to analyse whether observational uncertainties account for most of the scatter. It remains striking that the scatter remains so large after more than 10 years of continuous improvements on both the observational and the theoretical sides (see the diagram of Kunth and Sargent(1983) for an illustration).

References

Belley, J., Roy, J.-R., 1992, ApJS, 78, 61

Dufour, R. J. 1986, PASP, 98, 1025

Edmunds, M. G., 1975, ApSS, 32, 483

Edvardsson, B., Andersen, J., Gustafsson, Lambert, D. L., Nissen, P. E., Tomkin, J., 1993, A&A, 275, 101

François, P., Matteucci, F., 1993, A&A, 280, 136

Kunth, D., Sargent, W. L. W., 1983, ApJ, 273, 81

Kunth, D., Sargent, W. L. W., 1986, ApJ, 300, 496

Kunth, D., Mas-Hesse,J.M., 1994,in preparation

Kunth, D., Lequeux, J., Sargent, W. L. W., Viallefond, F., 1994a, A&A, 282, 709

Kunth, D., Matteucci,F., Marconi, G.,1994b,A&A, submitted

Lequeux, J., Maucherat-Joubert, M., Deharveng, J. M., Kunth, D., 1981, A&A, 103, 305

Marconi, G., Matteucci, F., Tosi, M., 1994, M.N.R.A.S. in press

Matteucci, F. 1986, M.N.R.A.S., 221, 911

Matteucci, F., Padovani, P., 1993, Ap. J., 419, 485

Matteucci, F., Tosi M., 1985, M.N.R.A.S., 217, 391

Pagel, B. E. J., Simonson, E. A., Terlevich, R. J., Edmunds, M. G., 1992, MNRAS, 255, 325

Roy, J.-R., Kunth, D., 1995, A&A, in press

Skillman, E. D., Kennicutt, R. C., 1993, ApJ, 411, 655

Weaver T.A., Woosley, S.E. 1995, in preparation

Galactic Evolution of Light Elements: Theoretical Analysis

V.Chuvenkov, A.Glukhov

Department of Space Physics, Rostov University, 5 Zorge st., 344104 Rostov-on-Don, Russia

Abstract. Evolution of deuterium, helium, CNO- and heavy elements in galaxies is numerically simulated taking into account stellar activity (different in spiral and elliptical galaxies), galactic ejection and accretion on to the galaxies. Different forms of initial mass function and star formation rate are used. Observations of light and heavy elements in galaxies and intergalactic gas are used as tests for verification of the obtained theoretical results. Analysis of the model versions corresponding to these observations leads to the conclusions: i) initial mass function should be time-dependent due to enrichment of interstellar medium by heavy elements; ii) factor of deuterium astration in galaxy is $\lesssim 2$, that is mainly due to accretion on to the galaxy; iii) rates of galactic enrichment by helium and heavy elements evolve differently that leads to dependence of the value $\Delta Y/Z$ on the galaxy lifetime.

1 Introduction

Theoretical calculations of galactic chemical evolution are important both for restiction of galactic evolutionary functions and the cosmological parameters, mainly baryon density parameter Ω_b.

From this point of view the most important elements for investigation of galactic chemical evolution are the lightest isotopes produced in primordial nucleosynthesis, particularly deuterium and helium-4. It is well-known that final deuterium abundance decreases rapidly with Ω_b increasing. Since the deuterium abundance can only decrease during galactic evolution, we can restrict the upper bound of Ω_b proceeding from the deuterium evolution. Galactic evolution of helium-4 is very important for refinement of its primordial abundance and ratio $\Delta Y/Z$ in galaxies.

The aim of this work is to calculate evolution of deuterium and helium in galactic models with different forms of galactic evolutionary functions to select those ones which correspond well to recent observed data on these and other (CNO and heavy) elements. On this basis we shall restrict the above mentioned parameters.

2 Model Equations

We consider model with the following approximations: the initial chemical composition of matter is primordial and the total initial mass of the galaxy equals the initial mass of substance in intergalactic medium (IGM) normalized to one galaxy in cluster. Under such approximations, the dependences of relative contents of gas G and mass concentrations of chemical elements X_i on time t are determined by the following equations:

a) in the galaxy:

$$\frac{dG_g}{dt} = -\Psi + I_1 - E_g + A \tag{1}$$

$$\frac{d}{dt}(X_{ig}G_g) = -X_{ig}(\Psi + E_g) + I_2 + X_{ie}A, \tag{2}$$

where

$$I_1(t) = \int_{m_d}^{m_u} E_s(m)\phi(m, t - \tau_m)\Psi(t - \tau_m)dm \tag{3}$$

$$I_2(t) = \int_{m_d}^{m_u} E_{is}(m)\phi(m, t - \tau_m)\Psi(t - \tau_m)dm \tag{4}$$

b) in the IGM:

$$\frac{dG_e}{dt} = E_g - A \tag{5}$$

$$\frac{d}{dt}(X_{ie}G_e) = X_{ig}E_g - X_{ie}A. \tag{6}$$

In these equations the subscript "g" corresponds to galaxy, subscript "e" - to IGM, E_g is the rate of matter ejection from galaxy, A is the rate of accretion on to the galaxy. Integrals I_1 and I_2 describe the rate of stellar ejection into interstellar medium (ISM); in these integrals: $\phi(m, t)$ is initial mass function of stars (IMF) depending on time, Ψ is star formation rate (SFR), E_s is the share of stellar mass returned to ISM during the evolution of the star, E_{is} is the fraction of i-element in E_s, τ_m is the lifetime of the star with mass m, m_d and m_u are the down and upper stellar masses. In the present work all the parameters are normalized to the initial galactic mass and time is presented in units 10^9 years (Gyrs).

3 Model Functions and Parameters

3.1 Star Formation Rate (SFR)

Three different model dependences of SFR on time used in our calculations are shown in Fig.1. The monotoneously decreasing dependence, where $\Psi \propto G_g$ corresponds to elliptical galaxies and the others - to spiral galaxies with bursts of SFR at the periods of subsystem formation.

3.2 Initial Mass Function

Concerning the IMF, it follows to note that it should depend on galactic evolutionary time because of the decreasing of Jeans mass in the ISM enriched by heavy elements (Silk 1977). In the present work the following dependence is used:

$$\phi(m,t) = Bm^2 \exp\left[\left(-\frac{m}{m_{max}}\right)^2 \frac{Z_n - Z(t)}{Z_n} - (x+2)\frac{Z(t)}{Z_n}\ln m\right], \quad (7)$$

where $Z(t)$ is galactic metallicity at time t, and Z_n - at the present time, $x = f(m)$ is power exponent in IMF by Scalo (1986), coefficient B follows from normalization. This dependence is presented in Fig.2. It is seen, that at $t = 0$ ϕ has a maximum at $m = m_{max} = 10$ $M_\odot$. Then ϕ evolves by such a way that its maximum removes into the area of small stellar masses, and at the present time ϕ has the standard form by Miller and Scalo (1979).

3.3 Galactic Ejection and Accretion

It is assumed, that galactic ejection rate is proportional to ejection rate from II-type supernovas with $m > m_{SN2} = 8$ $M_\odot$ (Chuvenkov and Vainer 1989), and accretion rate on to galaxy is constant: $A = 0.01 \div 0.03$ Gyr^{-1}.

3.4 Stellar Ejection

Data from Renzini and Voli (1981) and Maeder (1991) are used as input data on stellar ejection.

4 Results and Conclusions

In Figs. 3, 4 evolution of deuterium abundance and value $\Delta Y/Z$ for all the considered models are presented. It is seen that in all the model versions astration of deuterium in galaxy leads to final decreasing of its pregalactic abundance by a factor of $\lesssim 2$, and the value $\Delta Y/Z$ depends on galactic age and grows with increasing of galactic lifetime.

To sum up, the main cosmological conclusion for light elements concerned with confirmation that their primordial abundances $Y_p \simeq 0.23 \div 0.24$ and $X_p(D) = (1 \div 2)\cdot 10^{-5}$ are in good argeement with galactic gaseous and chemical evolution in considered models. These results lead to the value $\Omega_b \simeq 0.1$.

In addition the following general conclusions are made: 1) The final results depend much more on time integrals of star formation, accretion and ejection rates, than on particular features of these functions. 2) Time-dependent IMF influences on the time evolution of gas and chemical elements and leads to an active enrichment of ISM by heavy elements at initial stage of evolution. At subsequent period abundance of heavy elements in galaxy evolves weakly.

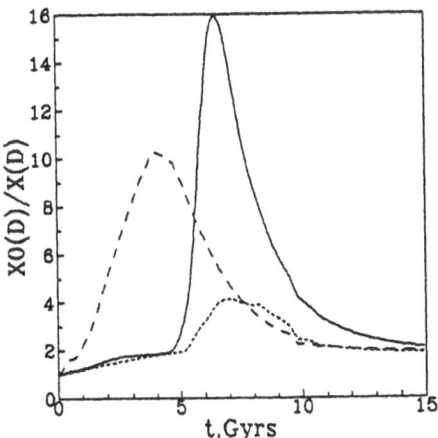

Fig. 1. SFR for models of elliptical (dashed line) and spiral (solid lines) galaxies

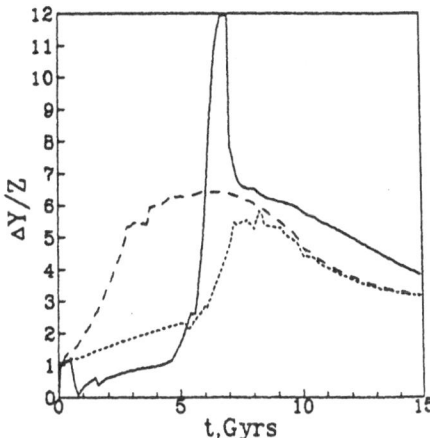

Fig. 2. IMF at $t = 0$ (solid line) and $t = 15$ Gyrs (dashed line)

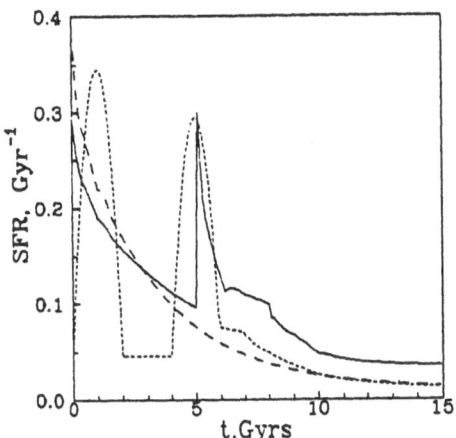

Fig. 3. Deuterium astration in galaxies with SFR presented in Fig. 1

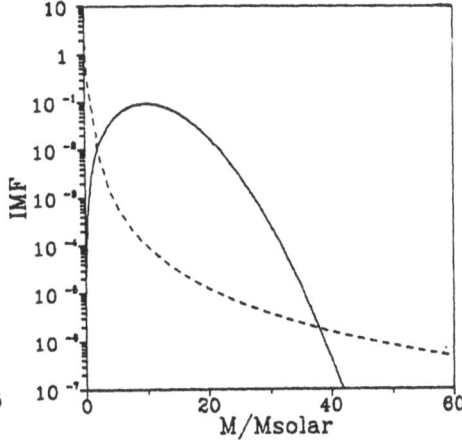

Fig. 4. Ratio $\Delta Y/Z$ in galaxies with S-FR presented in Fig. 1

References

Silk, J. (1977): Astrophys. J. **211** 638

Scalo, J.M. (1986): Fundam. Cosmic Phys. **11** 1

Miller, G.E., Scalo, J.M. (1979): Astrophys. J. Suppl. **41** 513

Chuvenkov, V.V., Vainer, B.V. (1989): Astrophys. and Space Sci. **154** 287

Renzini, A., Voli, M. (1981): Astron. Astrophys. **94** 175

Maeder, A. (1991): Publ. Observ. Geneve, **C 84**

Evolution of Light Elements in Galaxies and Intergalactic Medium

A. Glukhov, V. Chuvenkov

Department of Space Physics, Rostov University, 5 Zorge st., 344104 Rostov-on-Don, Russia

Abstract. A theoretical model of evolution of light and CNO-elements in a cluster of galaxies explaining the observations of the elements in galaxies of various types and ages and intergalactic medium is proposed. Numerical simulations take into account stellar ejection and matter exchange between galaxies and intergalactic medium with different forms of galactic ejection and accretion functions. Enrichment of intergalactic medium by heavy elements due to galactic ejection driven by supernova activity. A possibility for galactic metallicity to have inhomogeneous behaviour at redshifts $z \sim 1$ is shown, that is mainly due to accretion onto a galaxy at the latest evolutionary stage.

1   Introduction

At the present time it is well-known that galaxies interact actively with intergalactic medium (IGM) by matter ejection and accretion in varyous forms. Investigation of this process can be made by theoretical modeling of the chemical evolution on the basis of comparison the results with observed data. According to recent data, the accretion rate on to our Galaxy is $A = 1 \div 3 M_{\odot} \text{yr}^{-1}$ and the estimations of star formation rate (SFR) in it lead to practically the same value $\Psi \sim 1 \div 3 M_{\odot} \text{yr}^{-1}$. The most important elements for this investigation is helium, carbon and heavy elements. It is well-known that recent data on primordial helium yield $Y_0 \simeq 0.23 \div 0.24$ and the relative change of helium abundance in galaxies is $\Delta Y / \Delta Z \simeq 2 \div 5$. Observations of carbon and heavy elements in halo of young galaxies provide data on existence of maximum of their abundances at redshift $z \sim 1$ (Steidel 1990).

The aim of the present work is to investigate dependences of chemical evolution on parameters on galactic ejection and accretion rates. Comparing the results with set of recent observations we shall to restrict the parameters of chemical evolution of galaxy cluster.

2   Numerical Model Equations

We consider model with the following approximations: the initial chemical composition of matter is primordial and the total initial mass of the galaxy equals the initial mass of substance in intergalactic medium (IGM) normalized to one galaxy in cluster. Under such approximations, the dependences of relative contents of gas G and mass concentrations of chemical elements X_i on time t are determined by the following equations: a) in the galaxy:

$$\frac{dG_g}{dt} = -\Psi + I_1 - E_g + A \tag{1}$$

$$\frac{d}{dt}\left(X_{ig}G_g\right) = -X_{ig}\left(\Psi + E_g\right) + I_2 + X_{ie}A, \tag{2}$$

where

$$I_1(t) = \int_{m_d}^{m_u} E_s(m)\phi(m, t - \tau_m)\Psi(t - \tau_m)dm \tag{3}$$

$$I_2(t) = \int_{m_d}^{m_u} E_{is}(m)\phi(m, t - \tau_m)\Psi(t - \tau_m)dm \tag{4}$$

b) in the IGM:

$$\frac{dG_e}{dt} = E_g - A \tag{5}$$

$$\frac{d}{dt}\left(X_{ie}G_e\right) = X_{ig}E_g - X_{ie}A. \tag{6}$$

In these equations the subscript "g" corresponds to galaxy, subscript "e" - to IGM, E_g is the rate of matter ejection from galaxy, A is the rate of accretion on to the galaxy. All variables are normalized to the initial galactic mass. Integrals I_1 and I_2 describe the rate of stellar ejection into interstellar medium (ISM); in these integrals: $\phi(m,t)$ is initial mass function of stars (IMF) depending on time, Ψ is star formation rate (SFR), E_s is the share of stellar mass returned to ISM during the evolution of the star, E_{is} is the fraction of i-element in E_s, τ_m is the lifetime of the star with mass m, m_d and m_u are the down and upper stellar masses. In the present work all the parameters are normalized to the initial galactic mass and time is presented in units 10^9 years (Gyrs).

3   Model Functions and Parameters

3.1   Star Formation Rate

Two different model dependences of SFR on time used in our calculations: the monotoneously decreasing dependence, where $\Psi \propto G_g$ corresponds to elliptical galaxies and the SFR with parabolical bursts - to spiral galaxies (see Fig. 1 in our paper "Galactic Evolution of Light Elements: Theoretical Analysis" in these Proceedings).

3.2 Initial Mass Function

Concerning the IMF, it follows to note that it should depend on galactic evolutionary time because of the decreasing of Jeans mass in the ISM enriched by heavy elements. In the present work the following dependence is used:

$$\phi(m,t) = Bm^2 \exp\left[\left(-\frac{m}{m_{max}}\right)^2 \frac{Z_n - Z(t)}{Z_n} - (x+2)\frac{Z(t)}{Z_n}\ln m\right], \quad (7)$$

where $Z(t)$ is galactic metallicity at time t, and Z_n – at the present time, $x = f(m)$ is power exponent in IMF by Scalo (1986), coefficient B follows from normalization. At $t = 0$ ϕ has a maximum at $m = m_{max}$. Then ϕ evolves by such a way that its maximum removes into the area of small stellar masses, and at the present time ϕ has the present form (see Fig. 2. of our above mentioned paper in these Proceedings).

3.3 Galactic Ejection Rate

It is well-known that ejection from galaxy is initiated by different mechanisms, but depends directly on energy emission in a galaxy. The results of detailed investigations of explosion dynamics presented in the review of indicate that shock waves from II-type supernovas with masses $m \gtrsim 8\ M_\odot$ are able to provide the most intensive matter ejection from galaxy. Then, the massive supernova activity determines the following galaxy ejection rate:

$$E_g = E_0 \int_8^{m_u} E_s(m)\phi(m,t)\Psi(t)dm \qquad (8)$$

where E_0 is the model coefficient. As follows from Chuvenkov and Vainer (1989), the most optimal value is $E_0 = 3$, i.e. each supernovae should provide the ejection of galaxy mass, which is about by 3 times as much, than stellar mass ejected by itself.

3.4 Galactic Accretion Rate

At the present work we will consider the models both with constant accretion rate $A = 0.01 \div 0.03\ \text{Gyr}^{-1}$ and with exponentially decreasing rate (Clayton 1986):

$$A = A_0 exp(-\omega t) \qquad (9)$$

3.5 Stellar Ejection

Data from Renzini and Voli (1981) and Maeder (1991) are used as input data on stellar ejection.

4 Results and Conclusions

In Figs. 1-2 evolution of CNO-elements abundances, galactic metallicity and helium abundance in a spiral galaxy model and intergalactic medium are presented.

Proceeding from the results, the following conclusions are made:

1) Ratio of galactic and intergalactic metallicities $Z_g/Z_{IGM} \simeq 3 \div 3.5$.

2) Galactic metallicity Z_g and helium abundance Y_g decrease at final evolutionary stage due to accretion as both in elliptical and spiral galaxies.

3) Accretion rates are $A \sim 0.01 \div 0.03 Gyr^{-1}$ for spiral galaxies and $A \lesssim 0.01 Gyr^{-1}$ for elliptical galaxies.

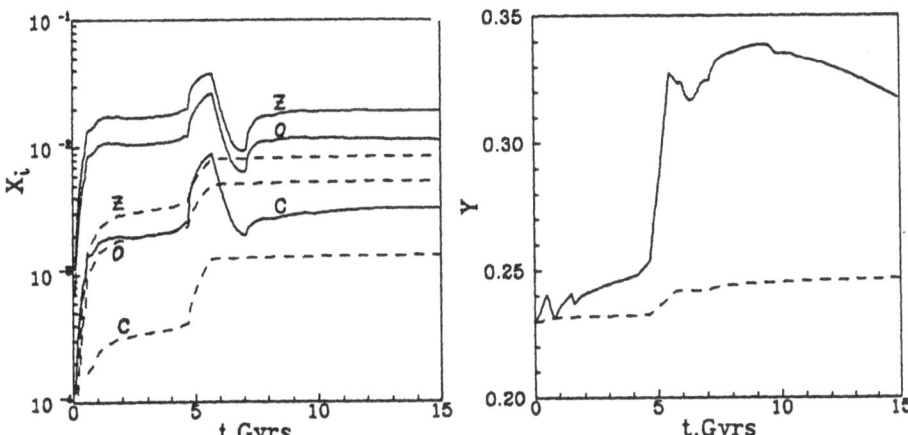

Fig. 1. Evolution of carbon, oxigen and metallicity in a spiral galaxy with bursts of SFR (solid lines) and in IGM (dushed lines)

Fig. 2. Evolution of helium in a spiral galaxy with bursts of SFR (solid line) and in IGM (dushed line)

References

Steidel, C.C. (1990): Astrophys. J. Suppl. **72** 1

Scalo, J.M. (1986): Fundam Cosmic Phys. **11** 1

Chuvenkov, V.V., Vainer, B.V. (1989): Astrophys. and Space Sci. **154** 287

Clayton, D.D. (1986): Publ. Astron. Soc. Pasif. **98** 968

Renzini, A., Voli, M. (1981): Astron. Astrophys. **94** 175

Maeder, A. (1991): Publ. Observ. Geneve, C **84**

The dY/dZ in the Chemical Evolution of Galaxies with the Multiphase Model

Mercedes Mollá[1], Angeles I. Díaz[1], and Federico Ferrini[2]

[1] Departamento de Física Teórica, Universidad Autónoma de Madrid, 28049- Cantoblanco, Madrid (Spain)
[2] Sezione di Astronomia, Università di Pisa, Piazza Torricelli 2, 56100-Pisa, Italy

Abstract. The relation between helium and oxygen abundances resulting from chemical evolution models applied to spiral galaxies is analysed. The models are of the kind referred to as "multiphase models" (Ferrini *et al.* 1992) and have been successfully applied to the Solar Neighbourhood — without and with thick disc — (Pardi & Ferrini 1994; Ferrini, Matteucci & Pardi 1994); the Galactic Disc (Ferrini *et al.* 1994); the Galactic Bulge (Mollá & Ferrini 1994) and outer spiral galaxies (Mollá, Ferrini & Díaz 1994). From the comparison between models and observations, helium production yields higher than presently calculated by factors between 5 and 10 are implied, suggesting that a revision of the nucleosynthesis prescriptions may be needed.

1 Introduction

The helium abundance at the present time, Y, has two origins: primordial and stellar (see the recent review from Wilson & Rood 1994). The primordial contribution proceeds from the Big Bang and, in the frame of the "Standard Big Bang Nucleosynthesis" (SBBN) model, its value Y_p, is found to be between 0.223 and 0.225 by mass *. In fact, model results depend on the parameter η, the ratio of baryon to photon densities. Due to this important relation between Y_p and η, which can actually govern the possible existence of dark matter, a great observational effort has been made in order to obtain a value of Y_p, and hence of η, as accurate as possible (Pagel *et al.* 1992; Skillman *et al.* 1994).

On the other hand, the stellar contribution depends mainly on stellar nucleosynthesis calculations which predict the mass of every element ejected by stars of different mass and on the Initial Mass Function (IMF), although, in the frame of given chemical evolution model, it also depends on galactic parameters such as the the star formation and the gas infall rates.

Indeed, one way of testing the consistency of a chemical evolution model is to check if results about the helium abundance are in agreement with observational data. In particular, the linear relation $Y = Y_p + \frac{dY}{d[O/H]} d[O/H]$, found for star forming regions (see *e.g* Pagel *et al.* 1992) should be reproduced.

---

* Another possibility is to use the primordial nucleosynthesis of baryonic-inhomogeneus Big Bang models, that produce significant heavy elements $[Z] \sim$ -6 to -4 (Jedamzik *et al.* 1994).

This has been done for the multiphase model of chemical evolution, widely described in Ferrini *et al.* (1992,1994), as applied to regions with different collapse times and star formation efficiencies.

2 Model Description and Results

For each galaxy we have used the model whose input parameters: collapse time scale and molecular cloud and and star formation efficiencies, best reproduce its observational constraints (Mollá, Ferrini & Díaz 1994).

Nucleosynthesis prescriptions have been taken from Renzini & Voli (1981) for low mass stars and from Woosley & Weaver (1986) for massive stars which end their lifes like type II supernovae. Also, the element ejection due to type Ia supernova events are included following Nomoto, Thielemann & Yokoi (1984) calculations. Different Y_p values between 0.23 and 0.24 have been used.

The regions we have modelled belong to the discs of five galaxies of different Hubble type (Sb to Sd): M 31, NGC 628, NGC 6946, M 33 and NGC 300. Some models of the Galactic Bulge have also been included. For each case, the abundances of O/H and Y have been computed at every time step and are plotted in Figure 1a). The regions of the spiral discs define the solid line in the figure implying a linear relation between helium and oxygen abundances. This fact means that the found relation does not depend on galactic parameters, but only on stellar nucleosynthesis (if a universal value for the IMF is assumed). The existence of this linear relation is supported by observational data, at least for low metallicities.

However, the slope, dY/d(O/H), obtained with the models is much lower than that derived from observations: dY/dZ = 0.5-0.7 against dY/dZ = 4-6 (if the conversion Z = 20 (O/H) is adopted). This result is common to other chemical evolution models: the standard stellar yields do not produce as much helium as it is needed (dY/dZ $\simeq$ 1 with Maeder (1985) massive star models). The stellar yields for helium should be multiplied by a factor from 5 to 10 in order to obtain a relation which fits the observations, as can be seen in Figure 1a). Recently, Mathews *et al.* (1994), using hierarchical clustering or accretion models for the formation of galaxies, also need ejected mass fractions of helium multiplied by a factor 3-4, to obtain an optimum fit.

The models for spirals do not reach the high values of Y observed. However, if the evolution is fast enough (high star formation rate due to high efficiencies in molecular cloud formation and collisions), a helium abundance of Y=0.28 may be obtained (see figure 1b), where the relation between O/H and Y for some galactic bulge regions is shown). But again, the shape of the increasing function is not the same as that obtained with observational data. Therefore, the fact that a chemical evolution model reproduces the solar value of Y $\simeq$ 0.28 is not sufficient to test the goodness of the model.

This high value may be reached without the observed relation being reproduced. Actually, the calculated slope has two different values: $\sim$ 0.7 for low

metallicities, a factor of 5 to 10 lower than observed, and 2.4 for high metallicities. Taking all the obervational data together, the opposite behaviour is found:
the helium abundance increases very quickly with the oxygen abundance and,
later, it reaches a saturation level or "plateau".

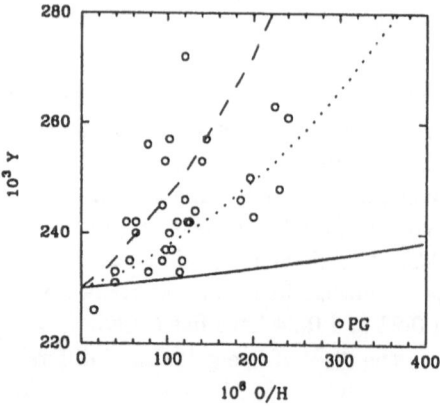

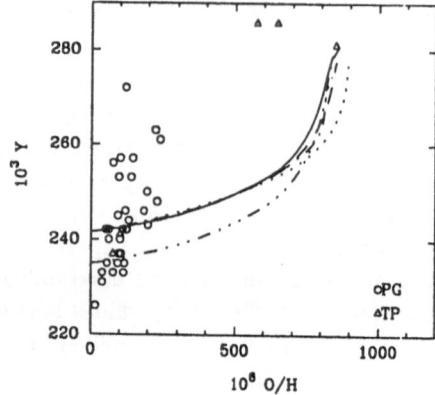

Fig. 1a. Relation between the helium
mass fraction in units of 10^3Y and the
oxygen abundance for some models of
galaxy disc regions. The solid line represents the locus of all the models. The
dotted and dashed lines show the same
models but calculated with an helium
production yield multiplied by a factor
of 5 and 10 respectively. Observational
data are from Pagel *et al.* (1994) (PG).

Fig. 1b. Same as in Figure 1a for galactic bulge regions with Y_p from 0.23 to
0.24 values. Solid line represents the
model with a collapse time of 1 Gyr,
and standard star formation efficiencies
for the best bulge model (Mollá & Ferrini, 1994). Other models have differen-
t collapse times (dot-dashed, and long
dashed lines); or differents star formation efficiencies (dotted and three dot-
dashed lines). Observational data are
from Pagel *et al.* (1994) (PG) and from
Torres-Peimbert, Peimbert & Ruiz
(1994) (TP) for low and rich metallicity
regions, respectively.

We have also run models with different collapse times, τ_{coll}, from the halo
(infall rates) or star formation efficiencies. With these models, something interesting occurs: after reaching an oxygen abundance of $\sim 500 \times 10^6$, models start
to separate (Figure 1b), and models with higher τ_{coll} or efficiency values start
to increase the helium abundance before — *i.e.* at a lower oxygen abundance—
than those with lower values. Actually, every galaxy or region may evolve in a

different way and reach a different value. Therefore, the set of these final values may show a correlation which is not related with the evolutionary track followed to reach these final values.

Finally, we can also note in the figures that the slope $dY/d(O/H)$ does not change if another value of the primordial abundance is used as input. Although the value Y_p is important to define other aspects of the chemical evolution, it does not change the final results concerning the abundances of helium and other elements.

3   Conclusions

Regions of the discs of galaxies of different Hubble types seem to follow the same increasing function of helium with oxygen abundance. This is interpreted as evidence that this relation depends mainly on stellar nucleosynthesis with galactic parameters having a secondary importance.

However, the models do not fit the data for helium abundances as derived from the observations: the slope of the $dY/d(O/H)$ linear relation obtained for low metallicity regions is not reproduced. Helium production yields higher by a factor of 5 to 10 are needed. Furthermore, if higher metallicity regions are considered, the observed functional relation is not in agreement with model resuls, either. Therefore, the nucleosynthesis prescriptions should be reanalysed.

The value of the primordial helium abundance used as input for the models, $Y_p = 0.23$ or $Y_p = 0.24$, does not change the slope calculated for low metallicity regions and a value of Y as high as 0.28 can be reached in either case. The fact that a solar abundance for helium is obtained by a model does not constitute a proof of its "goodness" since the observed relation between helium and oxygen abundances might not be reproduced.

This work has been partially supported by the Spanish DGICYT through project PB90-0182.

References

Ferrini,F., Matteucci,F., Pardi,C., Penco, U.1992, Astrophys. J. 387138

Ferrini, F, Mollá, M.,Pardi,M.C., Díaz, A.I.1994,Astrophys. J. 427,745

Ferrini, F., Matteucci, F., Pardi, C.1994,in preparation

Jedamzik,K., Fuller, G.M., Mathews,G.J., et al. 1994,Astrophys. J. 422,423

Mathews, G.J., Boyd, R.N., Fuller, G.M.1993,Astrophys. J. 403,65

Maeder, A.1985Astr. Astrophys. ,120,113

Mollá, M., Ferrini, F.,,1994, in preparation.

Mollá, M., Ferrini, F., Díaz, A.I.,1994., in preparation.

Nomoto, K., Thielemann, F.J., Yokoi, K.,1994,Astrophys. J. 286,644

Pagel, B.E.J., Simonson, E.A., Terlevich, R.J., Edmunds, M.G.,1992,Mon. Not. R. astr. Soc. 255,325.

Pardi, C. & Ferrini, F.1994,Astrophys. J. 420,87

Renzini,A., Voli, M.1981, Astr. Astrophys. 94,175.

Skillmann, E.D., Terlevich, R.J, Kennicutt, R.C., Garnett D.R., Terlevich, E., 1994 Astrophys. J. , accepted.

Torres-Peimbert,S., Peimbert,M. & Ruiz,M.T.1994, private communicatio.

Wilson, T.L., Rood, R.T1994,Ann. Rev. Astr. Astrophys. in press.

Woosley, S.E. & Weaver,P.1986I.A.U. Coll. nº 89. Radiation Hydrodynamics in Stars and Compact Objects.Eds. D.Mihalas & K.H.A. Winkler Springer-Verlag. P. 91

Chemical Evolution Models with a New Stellar Nucleosynthesis

A.Giovagnoli[1], M.Tosi[2]

[1] Dipartimento di Astronomia, Via Zamboni 33, I-40126 Bologna, Italy
[2] Osservatorio Astronomico, Via Zamboni 33, I-40126 Bologna, Italy

The chemical evolution of galaxies is mostly regulated by the production (or destruction) of the elements inside the stars. Maeder (1992, hereinafter M92) has examined this problem and presented the results of nucleosynthesis in stars with two initial metallicities, Z=0.001 and 0.02. An important conclusion of M92 is that the nucleosynthetic production strongly depends on the stellar initial metallicity. In particular, for incresing Z: ^{16}O is strongly depleted, ^{12}C is highly enhanced in stars more massive than 25 $M_\odot$ but fairly reduced in smaller stars and ^4He is highly enhanced in very massive stars and roughly constant in the others. We have computed numerical models for the chemical evolution of the Galaxy adopting M92 stellar nucleosynthesis and compared the model predictions with the corresponding observational constraints and with the predictions of one of Tosi's (1988) best models (Giovagnoli and Tosi 1995).

The table lists the models described here: model 1 represents Tosi's (1988) best model but calculated with M92 nucleosynthesis. We have adopted the low metallicity yields from the disc formation up to the time when the ISM reaches the solar metallicity (an epoch different for different galactocentric distances) and the Z=0.02 yields afterwards. In model 2 we have assumed no infall after disc formation, thus removing any dilution of the ISM. In model 3 we have investigated the impact of assuming that stars more massive than $11.6 M_\odot$ (as suggested in M92) become black holes and therefore do not explode as SNe II. Thus for these stars the ISM enrichment is due only to the stellar wind contribution. Model 4 has been calculated with a black hole mass limit $M_{BH} = 22.5 M_\odot$.

Figure 1 shows the oxygen abundance distribution in the galactic disc. The dots with average error bars represent the observations of HII regions derived from Peimbert (1979) and Shaver et al. (1983). The lines are the model predictions: the thick solid one represents the *standard* model by Tosi (1988), calculated with the parameters as in model 1 but using the nucleosynthesis computations by Renzini and Voli (1981) for low and intermediate stars and by Arnett (1978) revisited by Chiosi and Caimmi (1979), for massive stars. To take into account also the effect of massive stars winds, the latter yields are combined with Maeder's (1981, 1983) values.

Models with M92 nucleosynthesis

Model number	SFR	Infall $(M_\odot kpc^{-2} yr^{-1})$	IMF	Mass limit for black hole formation
1	$\tau = 15\ Gyr$	$B = 4 \cdot 10^{-3}$	Tinsley	-
2	$\tau = 15\ Gyr$	$B = 0$	Tinsley	-
3	$\tau = 15\ Gyr$	$B = 4 \cdot 10^{-3}$	Tinsley	$11.6 M_\odot$
4	$\tau = 15\ Gyr$	$B = 4 \cdot 10^{-3}$	Tinsley	$22.5 M_\odot$

Model 1 (thin solid line) shows a good agreement with the observational data and with the *standard* model. The separation between the solid lines at large radii is due to the differences between the Z=0.001 and Z=0.02 oxygen yields. The former are higher than the latter, the outer ring reaches the solar metallicity much later than the inner ones, therefore it is enriched for longer times by the higher O-yields and shows a slightly higher oxygen abundance. In model 2 (long-dashed dotted line) the absence of infall allows to reach early the solar metallicity at all radii. As often found for no infall models, the slope of the abundance gradient is not reproduced. The introduction of a low mass limit for black hole formation doesn't allow to reproduce the oxygen observed abundance. Model 3 (short-dashed line) shows an oxygen abundance in the ISM of the whole disc more than one order of magnitude lower than that observed in HII regions. In model 4 (long-dashed line) the higher M_{BH} mass limit allows more contribution to oxygen but the model predictions are still underabundant.

The carbon abundance distribution in the galactic disc is shown in Figure 2. The vertical line corresponds to the range of values derived by Laird (1985) from observations of 116 stars in the solar neighbourhood and its length includes both his quoted observational error ($\pm 25\ dex$) and the abundance spread due to the different ages and/or initial metallicity of the examined stars. The line symbols are as in Fig. 1. All the model predictions are in agreement with the observations except for the dotted line (model 2) where the total absence of infall causes higher carbon values. The models calculated with M92 nucleosynthesis show a rapid decrease in the outer regions of the disc. This behaviour is due to the differences in the carbon yields at different metallicity.

Figures 3a and 3b show the fraction of G-dwarf stars in the solar neighbourhood as a function of their oxygen abundance. The histogram corresponds to Pagel's (1989) observational data: we have plotted the stellar distribution as a function of the oxygen abundance rather than the overall Z metallicity, since the later is the sum of elements produced in too different sites and at too different epochs. The line symbols are as in Fig.1.

The predicted frequency distributions with [O/H] of the G-dwarfs from the *standard* model and from model 1 are consistent with the data although model 1 predicts too many G-dwarfs with low oxygen. Model 2 (closed box model)

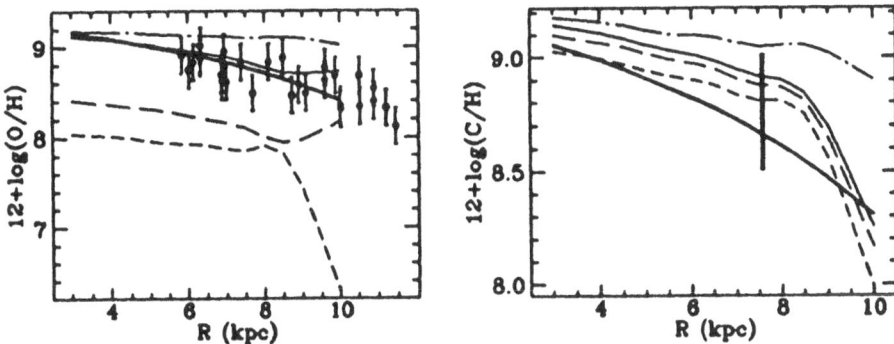

Fig. 1 and 2. Oxygen and carbon abundance distribution in the galactic disc. The thick solid line reprensents the *standard* model, the thin solid line model 1, the dotted line model 2, the short-dashed model 3 and long-dashed line model 4, in both the figures. See the text for a description of the observational data.

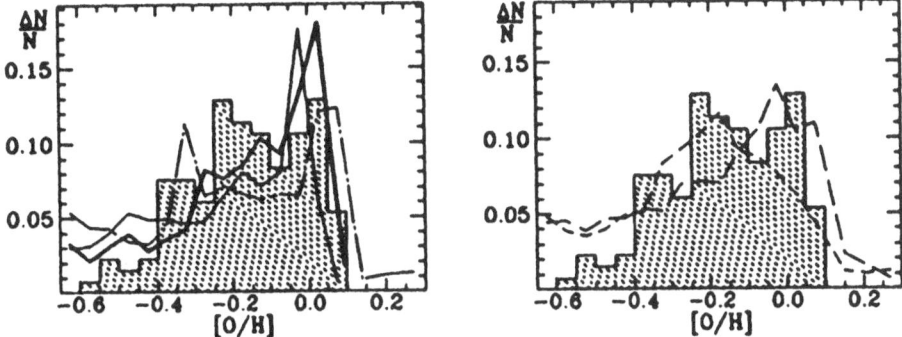

Fig. 3a and 3b. Fraction of G-dwarfs stars in the solar neighbourhood as a function of their oxygen abundance. The histogram corresponds to Pagel's (1989) observational data; the symbols are as in Fig.1.

predicts even more stars with low [O/H] which is the historical reason for introducing infall in chemical evolution models. From the predictions of models 3 and 4 it is clear that there are too many stars with [O/H]$\leq$ -0.4 in comparison with the observational data. This is due to the low production of oxygen when we introduce the mass limit for the black hole formation, in particular at Z=0.001.

We can summarize our results by saying that the oxygen predictions based on M92 yields are in agreement with the observational constraints. The introduction of low mass limits for black hole formation would however lead to a strong decrease in the ISM oxygen abundances which makes the predicted values inconsistent with the data (see also Prantzos, 1994).

All the infall models predict carbon abundances in agreement with the observations. This is due to the fact that the bulk of carbon derives from intermediate stars that are not eliminated by the introduction of the mass limit for the black hole formation. Contrary to Prantzos et al. (1994), we think that there is no need for more contribution from massive stars carbon to reproduce the observational constrains.

The helium metallicity relation $\Delta Y/\Delta Z$ has important implications since it provides the primordial value of ^{4}He by extrapolation to Z=0. Model 1 provides $\Delta Y/\Delta Z$=3.25, the *standard* model 3. Since the observational data do not provide directly $\Delta Y/\Delta Z$ but $\Delta Y/\Delta(O/H)$, we prefer however to compare this latter ratio. Pagel et al. (1992) derive from extragalactic HII regions $\Delta Y/\Delta(O/H)$=125±40; model 1 predicts $\Delta Y/\Delta(O/H) \sim$72 wich seems too small but may be consistent with the empirical one if we take into account the fact that observational data refer mostly to metal poor galaxies.

References

Arnett, W.D. 1978, ApJ , **219**, 1008

Chiosi, C., Caimmi, R. 1979, A&A , **80**, 234

Giovagnoli, A., Tosi, M., 1995, MNRAS , in press

Laird, J.B., 1985, ApJ , , **289**, 556

Maeder, A. 1981, A&A , **102**, 401

Maeder, A. 1983, A&A , **120**, 113

Maeder, A. 1992, A&A , **264**, 105

Pagel, B.E.J. 1989, in Beckman, J., Pagel, B.E.J. eds. Evolutionary Phenomena in Galaxies, p. 201

Pagel, B.E.J., Simonson, E.A., Terlevich, R.J., Edmunds, M.G., 1992, MNRAS , **255**, 325

Peimbert, M., 1979, in W.B. Burton ed. The Large Scale Characteristics of the Galaxy (Reidel, Dordrecht), p. 307

Prantzos, N., 1994, A&A , **284**, 477

Prantzos, N., Vangioni-Flam, E., Chauveau, S., 1994, A&A , **285**, 132

Renzini, A., Voli, M. 1981, A&A , **94**, 175

Shaver, P.A., McGee, R.X., Newton, L.M., Danks, A.C., Pottasch, S.R. 1983, MNRAS , **204**, 53

Tosi, M. 1988, A&A , **197**, 33

Chemical Enrichment of Mergers by Violent Star Formation

Everton Lüdke[1][2]

[1] Manchester University, Nuffield Radio Astronomy Laboratories, Jodrell Bank, Cheshire SK11 9DL, UK
[2] Universidade Federal de Santa Maria, NEPAE-Centro de Tecnologia, Santa Maria, RS, 97119, Brazil

Abstract. In this work, an analysis of the chemical abundances along the main body of the startburst galaxy NGC5253 is presented in order to study the gas enrichment due to the stellar wind. Observational evidence suggests that past interaction with the giant spiral M83 induced violent star formation towards massive stars. The observed excess of primary N and Ne in the external regions in comparison with the nucleus are consistent with simple contamination models with strong mass loss, with an upper limit of $\sim 55 M_\odot$ for the IMF of the second generation of stars formed about 500 Myrs ago.

1 Introduction

The issue of galactic interaction has received plenty of attention throughout the literature, since gravitational forces present in encounters are strong enough to change the dynamical processes in the main bodies of individual objects(Binney, and Tremaine, 1987). In particular, tidal interactions between massive and compact galaxies are important since objects do not exchange mass but the star formation history and the gas chemical enrichment can be affected if massive stars are formed. An interesting example of starburst events induced by such interactions is pair M83/NGC5253, in the Centaurus local group of galaxies(Sersic, Carranza, and Pastoriza, 1972). The main evidence for such interactions include a neutral hydrogen projection from the halo of M83 towards NGC5253(Rogstad et al.,1974), and the presence of very blue and red star clusters surrounding a giant HII nuclear complex(van den Bergh, 1982). Multiple generations of fast starburst events are a natural explanation for the starcluster colours of NGC5253, from which ages can be inferred from the stellar population synthesis technique(Bica and Alloin, 1986). Their contribution to the gas metalicity have to be investigated, to understand the chemical evolution of the gas due to multiple bursts of star formation.

In a previous work(Lüdke, Pastoriza, and Bica, 1990), measurements of the equivalent widths of the absorption lines were reproduced by fitting standard star cluster continuum spectra to the optical continuum measured at several regions along NGC5253 (nuclear, 7", 14" and 24" south). 60% of stars in the inner 7" radius are younger than ~ 100 Myr and are associated with a gas-rich substrate. The external regions (24", 38" south) show roughly equal contributions of 1 Gyr and 500 Myr stellar populations, suggesting that different bursts have taken place towards the inner regions.

3 Considerations on Contaminations by Stellar Ejecta

Electron temperatures, densities and chemical abundances of the gas were calculated at each region using the standard formulae (Peimbert, 1990, see also M. Peimbert, this proceedings. The He, N and O abundances vary with the radial distance to the nucleus, as expected from chemical enrichment although the gas is metal-poor with nuclear metallicity $logZ = -0.25 \pm 0.05$ and the helium content is almost primordial. An extrapolation to zero metallicity gives the primordial abundance Yp= $3(He/H) = 0.22 \pm 0.02$ which is consistent with the best values obtained from samples of dwarf and compact galaxies, as discussed by B.E.J. Pagel in this conference.

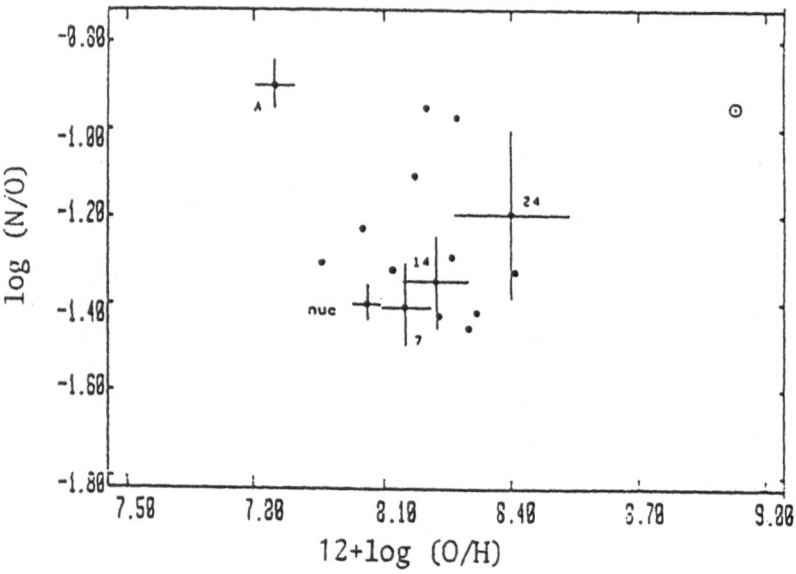

Fig. 1. Nitrogen enrichment along NGC5253

Correlations between abundance ratios show gradients of O and N along the object while both nitrogen content (fig 1) and Ne/O show a trend with the oxygen abundance (fig 2), which is higher at central regions. An overabundance of S with respect to the O content was also detected with (S/O) exceeding the solar value, but which is not observed for N/O and Ne/O. The nitrogen content appears to be of primary origin since log(N/O) correlates weakly with log(O/H) and log (N/H) with log(O/H). The apparent independence of metal abundances with the local helium abundance suggests that N and O were produced by the bi-cycle CNO mechanism, which is known to be independent on the initial helium stellar content(Maeder, 1983). In addition, log(Ne/H) is independent of log(N/H) implying that these elements were ejected from stars formed in the same epoch, consistent with the idea of primary origin.

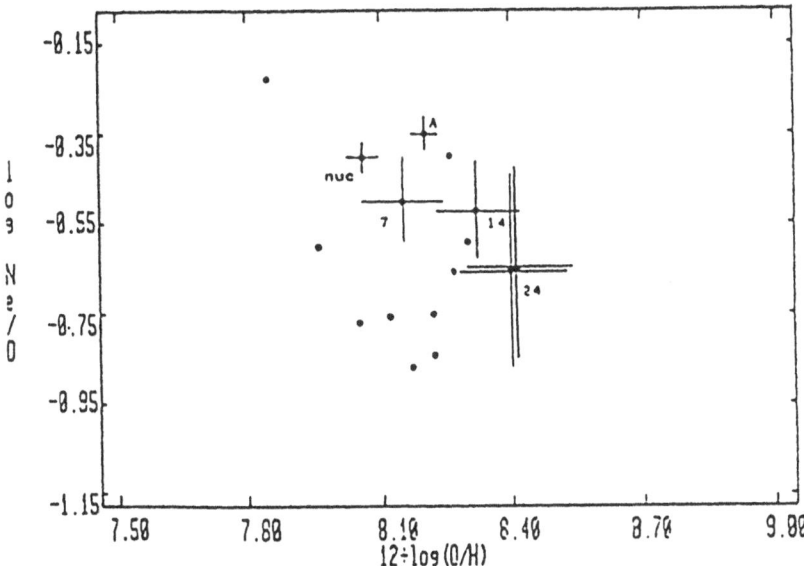

Fig. 2. Production of neon with respect to the oxigen content

The chemical evolution of the gas can be quantitatively described if a given initial chemical composition of the original gas (X_j) which fed the first generation of stars in NGC5253, will be enriched by matter ejected during the star cluster lifetime. If instantaneous and complete mixing of the ambient and ejected gas occurs, the initial abundances will then be converted to new species X_i, with individual stellar contributions weighted by the normalized Salpeter initial mass function $\Phi(m) \propto m^{-\gamma}$. Talbot and Arnett(1973) defined the yield matrix q_{ij}, which specifies the mass fraction of the initial element j which is converted to the element i, for a given burst event.

$$q_{ij} = \int_0^\infty \Phi(m) Q_{mij} dm$$

And the quantities of interest are the final mass fractions returned to the diffuse medium:

$$p_i = \frac{\sum_j q_{ij} <MX_j>}{1 - \sum_{k,j} q_{k,j} <MX_j>}$$

where $< MX_j >$ is the original mass fraction (in $M_\odot$ units) available for the element X_j. The overall mass fraction in the form of metals can then be estimated by the definition $p = \sum_i p_i$. The present calculations consider a burst towards massive stars with flat IMF between 50 and 100 $M_\odot$ and the O, N and Ne yields predicted by the stellar models given by Prantzos et al.(1986). Hence, the fraction of the required stars with masses of 50, 80 and 100 $M_\odot$ which can produce the excess of the observed oxygen mass $< M_O >$ with respect to the nucleus can be computed within the age constraints from the stellar population synthesis, giving predictions for the $< M_N >$ and $< M_{Ne} >$ mass contents. At 24"S the observational values of $< M_O > \sim 409$, $< M_N > \sim 20$ and $< M_{Ne} > \sim 44$ agree with the calculations if a cut-off in the IMF at about $55 M_\odot$ is included since higher limits on the IMF would require overabundances of nitrogen and neon which are not observed. Similar limit is also required at the regions 7" ($< M_O > \sim 81$, $< M_N > \sim 3$) and 14" south ($< M_O > \sim 270$, $< M_N > \sim 8$ and $< M_{Ne} > \sim 64$).

If the observed abundances at the present stage are due to a single event occurred at about ~ 100 and 500 Myrs ago, the total yield predicted by the model, $p \sim 3.8 \cdot 10^{-3}$, is also consistent with the empirical fit to other compact and irregular galaxies(Lequeux et al.(1979).

References

E. Bica, D. Alloin : Astr. Ap. **162** 21 (1986)

J. Binney, S. Tremaine: Galactic Dynamics, Princeton Press (1987)

J. Lequeux et al: Astr. Ap. **80** 155 (1979)

E. Lüdke, M.G. Pastoriza, E. Bica: Rev. Mex. Astr. Ap. **21** 187 (1990)

A. Maeder: Astron. Ap. **120** 113 (1983)

M. Peimbert: Rep. Prog. Phys.**53** 1559 (1990)

N. Prantzos et al: Ap. J. **304** 695 (1986)

D.N. Rogstad et al.: Ap. J. **193** 309 (1974)

J.L. Sersic, G. Carranza, M.G. Pastoriza: Ap Sp Sc, **19**, 469 (1972)

R.J. Talbot, W.D. Arnett:Ap. J. **186** 51 (1973)

S. van den Bergh : PASP **92** 122 (1982)

Ultra-Metal-Poor Stars for the 21st Century

Timothy C. Beers

Department of Physics and Astronomy, Michigan State University, E. Lansing, MI 48824, USA

Abstract. I present a summary of the status of spectroscopic and photometric followup observations of Galactic metal-deficient stars identified to date from the HK objective-prism/interference-filter survey of Beers and collaborators. I also discuss the directions we intend to go with the followup efforts over the course of the next five years, and speculate on the final total numbers of Extremely-Metal-Poor (EMP) stars ([Fe/H] $\leq$ −3.0) and Ultra-Metal-Poor (UMP) stars ([Fe/H] $\leq$ −4.0) which are likely to be identified.

1 Introduction – The Search Goes On

As is evident from other talks presented at this conference, stars of low metallicity provide crucial tests of the production and evolution of light elements from the Big Bang to the present. It is sometimes overlooked that the most metal-deficient stars in the Galaxy are apparently as primordial as the most distant quasars, and thus represent a powerful probe of the early Universe right in our own backyard. Just in the past year, observations of the lithium content in extremely-metal-poor stars has revealed the possible presence of a mild slope in the Spite plateau, a slope consistent with models of enhancement of lithium from the primordial value by cosmic-ray spallation (Thorburn 1994; Norris, Ryan, and Stringfellow 1994). The search for a beryllium plateau, or constraints on the slope of Be abundance with declining metallicity, also requires large numbers of (ideally, bright) stars of low metal abundances. As the next generation of large telescopes and blue/UV sensitive CCD chips emerges, the way will open for detailed spectroscopic inspection of stars as faint as V = 18 − 19. Stars of magnitude V = 14.5 − 15.5, which are challenging with present instrumentation, will become trivial. Thus, now is the time to marshal efforts to find the stars we will be studying well into the next century.

The search for stars with ever-lower abundances of heavy metals (parameterized here relative to the sun, using the notation [Fe/H] = $\log (Fe/H)_*$ − $\log (Fe/H)_\odot$) has engaged numerous astronomers and their telescopes over the four decades since the discovery by Chamberlain and Aller (1951) that stars with metal content less than [Fe/H] = −1.0 exist in the Galaxy. Based on his own extensive searches, Bond (1981) expressed some doubts as to the existence of significant numbers of stars with abundances below that of the lowest-abundance globular clusters ([Fe/H] < −2.5). In short order, this paper was followed up by several theoretical explanations why this must be the case. Even at the time of Bond's paper, however, there were several examples of stars that had been found to exhibit quite extreme paucity of heavy metals – the dwarf G64-12 ([Fe/H] = −3.5; Carney and Peterson 1981), and the giant CD-38° 245 ([Fe/H] = −4.5; Bessell and Norris 1984), thus begging the question: "Are these stars flukes, or the tip of a low-metallicity iceberg?"

In the search for stars of the lowest metallicity, the trend in recent years has been to cast as wide a net as possible, probing for ever-fainter stars, thus increasing dramatically the likelihood of successful fishing. This approach resulted in the publication of a list of some 134 stars by Beers, Preston, and Shectman (1985), selected on the basis of their appearance in a modified objective-prism survey described below, which appeared to have abundances [Fe/H] < −2.0. This sample included 10 stars apparently below [Fe/H] = −3.0, and a few at a level near [Fe/H] = −4.0. With the availability of an abundance calibration based on theoretical spectra provided by Norris, Beers (1987) was able to derive reasonably accurate iron abundances for these stars, and more importantly, was able to show that the distribution function of stellar metallicity for the lowest abundance stars appears "flat" (i.e. the number of stars below a given metal abundance declines in proportion to the logarithmic [Fe/H] index, thus yielding a derivative of with respect to the linear parameter $Z/Z_\odot$ consistent with zero). Such results are expected if the so-called Simple Model of chemical evolution applied throughout the Galaxy (see Pagel 1994 for a modern discussion). Regardless of the physical interpretation of the Simple Model, the flat metallicity distribution function opened the possibility that stars of arbitrarily low metallicity might be found, if one were willing to make the considerable observational effort. We have been looking for these stars ever since.

2   Where We Are Now

Owing to the shortage of IIaO/103aO emulsions, the last plates in the HK objective-prism/interference-filter survey are presently being taken. The final plate total obtained (counting both the northern and southern surveys) will be approximately 300, covering some 7,500 square degrees of sky. The techniques involved with obtaining the plates have been described in detail elsewhere (Beers, Preston, and Shectman 1985, 1992) and will not be repeated here.

The metal-poor stars discovered in the HK survey cover the range of apparent magnitudes $11 < V < 16.0$, and the color range $0.35 < B - V < 1.2$. Roughly half of the candidates are located near the main-sequence turnoff of an old Galactic field population; the other half are distributed along the giant and asymptotic giant branches. The turnoff stars provide the best probes of the light elements, while the later-type stars are most suitable for examination of intermediate (CNO) and heavy element production in the Galaxy.

All but the last few plates have been visually inspected for low-metallicity candidates, an effort which we expect to have completed by the time this article is published. Figure 1 is a cartoon of the appearance of stars on the HK prism plates. The spectrum marked "solar" shows the strength of the CaII H and K lines for the vast majority of stars on the plates which appear to have abundances similar to the sun. The spectrum marked "metal-poor" is typical of the 1% or so of the images which are classified as low-metallicity candidates. The spectrum marked "extremely-metal-poor" is typical of an even smaller fraction of the stars which are so metal-weak that their prism spectra appear continuous (corresponding to CaII H and K equivalent widths of roughly 4 Å or less). Moderate-resolution (1-2 Å) digital spectroscopy is obtained for as many of the metal-poor and extremely-metal-poor candidates as possible.

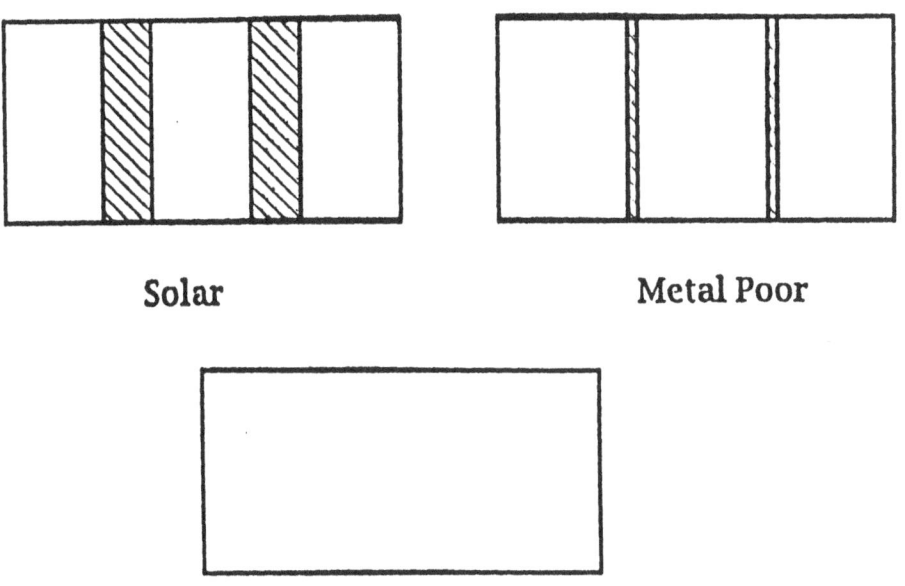

Fig. 1. Depiction of the appearance of spectra on the HK survey plates

High-resolution observations of several of the HK stars noted in the 1985 paper were obtained by Nissen (1989), Molaro and Castelli (1990), and Molaro and Bonifacio (1990). These important observations confirmed the extremely low metallicities claimed on the basis of moderate-resolution spectroscopy.

The first large sample of HK candidates with available moderate-resolution spectroscopic followup was published by Beers, Preston, and Shectman (1992). This sample (which included the 1985 data as a subset) includes some 1000 s-tars with metallicities [Fe/H] < −0.5. Included among these stars are 700 with [Fe/H] < −1.0, 450 stars with [Fe/H] < −2.0, and 70 stars with [Fe/H] < −3.0. Between 5 and 10 stars in this sample are likely to have [Fe/H] < −4.0. The uncertainty arises because of the need for testing and refinement of the calibration of our metallicity estimation technique, especially for the giants in our sample. The lack of available standards meant we first have to find suitable s-tars, then observe them at high resolution to obtain estimates of the true iron abundances prior to using them for calibrators. Fortunately, this work is well underway. Norris, Peterson, and Beers (1993) and Primas, Molaro, and Castelli (1994) present estimates of [Fe/H] based on high-resolution spectroscopy of four HK stars. Preston, McWilliam, and Sneden (1994) are preparing a comprehensive high-resolution study of some 20 HK stars. Norris, Beers, and Ryan (1995) will discuss a smaller list of stars observed at high-resolution, including the first detailed inspection of stars identified in the northern HK survey. These observations should provide more than sufficient material to verify/refine our extant calibration of metal abundance.

3   Where We Are Going

The final list of metal-poor candidates from the HK survey totals on the order of 10,000 stars. We have only obtained the moderate-resolution spectroscopy which we require in order to estimate metal abundances (and measure radial velocities) for some 2,000 of these stars. Clearly, much work remains.

There are several groups which are actively pursuing the acquisition of this data, particularly in the southern hemisphere. One group (including Beers, Cayrel, Nissen, F. and M. Spite, Andersen, and Nordstrom) is using the ESO 1.5m telescope on La Silla, Chile. To date, some 250 spectra have been obtained. We hope to observe some 750 additional stars over the next two years. Already some 10-15 stars have been identified which appear to have metal abundances [Fe/H] < −3.0. One example is shown in figure 2. Another major southern effort is underway using the 2.3m telescope at Siding Springs, Australia. Norris, Ryan, and Beers have obtained spectra (and in many cases, broadband UBV photometry) for some 450 HK candidates to date, and plan to observe some 600 more over the course of the next two years. At least 20 of the stars observed to date have been shown to have [Fe/H] < −3.0. A sample spectrum of one such star is shown in figure 2.

The followup effort of the northern hemisphere sample is just getting underway. One group (Beers, Twarog, Anthony-Twarog, Hawley, and Sarajedini) plans to observe a list of some 1,500 metal-poor candidates from the northern HK list over the next three years. As a result of discussions at the present meeting, I am hopeful that a similarly large effort can be initiated shortly with facilities on La Palma.

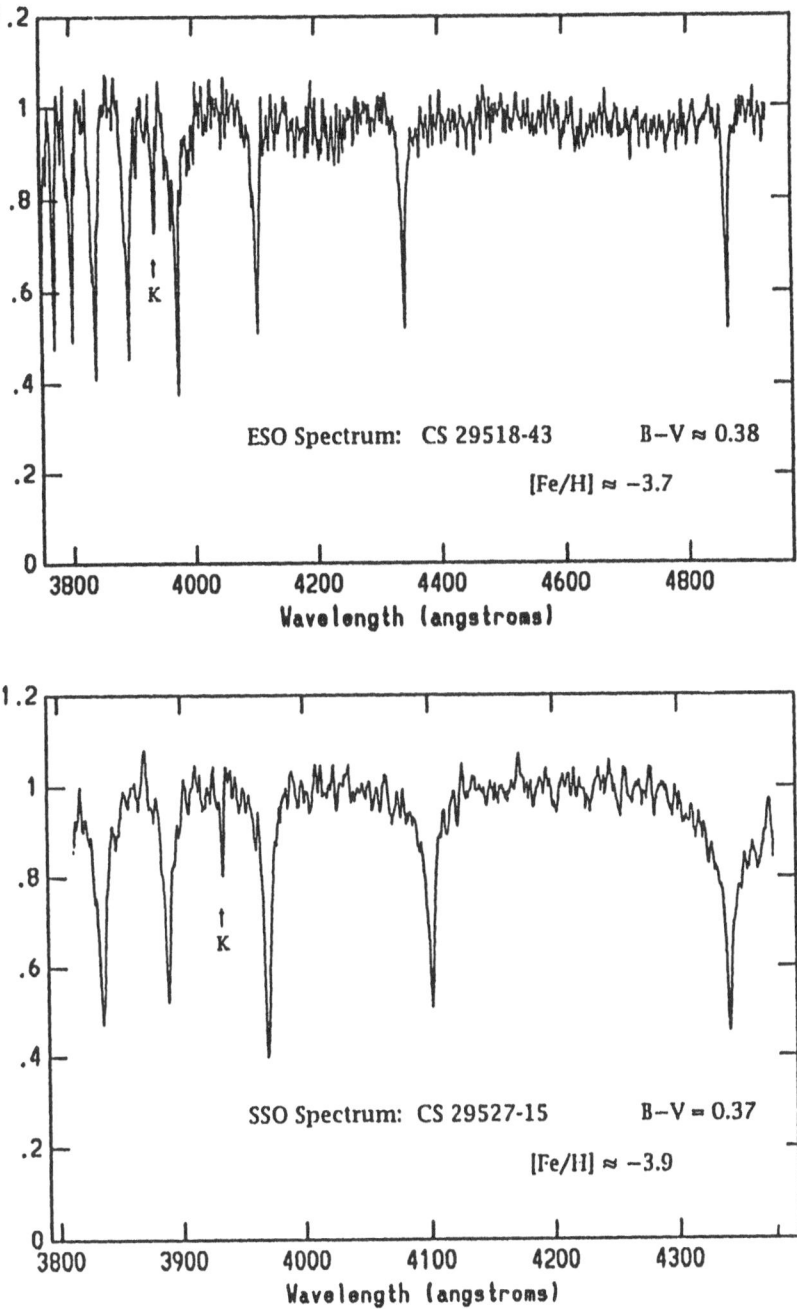

Fig. 2. Example moderate-resolution spectra of metal-poor stars from ESO and Siding Springs. Note the very weak line of CaII K at 3933 Å . All of the other strong absorption lines are due to hydrogen

4 What We Expect to Find

If all of the followup efforts now underway, or envisioned to be starting soon, are successful in reaching their goals, what might we expect to find based on our results to date ? Our presently-available sample of stars, and our intermediate- and long-range goals are summarized below.

Table 1. Numbers of HK survey stars below limiting [Fe/H]

[Fe/H]	Now (1994)	Soon (1997)	Someday (2000)
−3.0	100	250	500
−4.0	10	25	50
−5.0	0	2	5

Note that we have yet to discover our first star with metallicity [Fe/H] < −5.0. If the metallicity distribution function remains flat, as our present data indicate, then this is no surprise. We have simply not yet discovered enough stars below [Fe/H] = −3.0 that 0 stars below [Fe/H] = −5.0 (found) is statistically different than 1 star (expected). As we enlarge our sample of EMP stars over the next few years, we will no longer be able to hide behind the shield of low-number statistics. Either we will find the predicted numbers of HMP (Hyper-Metal-Poor; [Fe/H] < −5.0) stars or we will not. If they are found, then we may well have reached the limit of the lowest-abundance (surviving) stars which can be found in the Galaxy. As emphasized by Primas et al. (1994), the expected Fe output from a single 25 $M_\odot$ supernova (0.2-0.3 $M_\odot$) dispersed over 10^6 $M_\odot$ of hydrogen gas results in a metallicity of roughly [Fe/H] = −5.0. If NO stars are found below [Fe/H] = −5.0, then we will have learned something fundamental about the early chemical evolution of our Galaxy, and possibly even that of the Universe.

References

Beers, T.C. (1987): "The Metallicity Distribution Function of the Extreme Halo Population", in Nearly Normal Galaxies From the Plank Time to the Present, ed. by S.M. Faber (Springer-Verlag, New York), pp. 41–44

Beers, T.C., Preston, G.W., Shectman, S.A. (1985): "A Search for Stars of Very Low Metal Abundance. I.", Astronomical Journal, Vol. 90, pp. 2089–2102

Beers, T.C., Preston, G.W., Shectman, S.A. (1992): "A Search for Stars of Very Low Metal Abundance. II", Astronomical Journal, Vol. 103, pp. 1987–2034

Bessell, M.S., Norris, J. (1984): "The Ultra-Metal-Deficient (Population III?) Red Giant CD-38 245", Astrophysical Journal, Vol. 285, pp. 622–636

Bond, H.E. (1981): "Where is Population III ?", Astrophysical Journal, 248, pp. 606–611

Carney, B.W., Peterson, R.C. (1981): "Abundance Analyses of Subdwarfs of the Remote Halo ", Astrophysical Journal, Vol. 245, pp. 238–246

Chamberlain, J.W., Aller, L.H. (1951): "The Atmospheres of A-Type Subdwarfs and 95 Leonis", ApJ, Vol. 114, pp. 52–72

Molaro, P., Bonifacio, P. (1990): "Abundances in Two New Ultra-Metal-Poor Giants," Astronomy and Astrophysics, Vol 236, pp. L5–L8

Molaro, P., Castelli, F. (1990): " A New Ultra Metal-Deficient Star: CS 22876-32", Astronomy and Astrophysics, Vol. 228, pp. 426–442

Pagel, B.E.J. (1994): "Chemical Evidence on Galaxy Formation and Evolution", in IAC Winter School on Galaxy Formation and Evolution, preprint

Primas, F., Molaro, P., Castelli, F. (1994): "Abundances of Four Very Metal-Poor Stars of the BPS Survey", Astronomy and Astrophysics, in press

Nissen, P.E. (1989): "CASPEC Observations of the Most Metal Deficient Main-Sequence Star Currently Known", ESO Messenger, Vol. 58, pp. 40–41

Norris, J.E., Beers, T.C., Ryan, S.G. (1995): in preparation

Norris, J.E., Peterson, R.C., Beers, T.C. (1993): "Abundances of Four Ultra-Metal-Deficient Stars," Astrophysical Journal, Vol. 415, pp. 797–810

Norris, J.E., Ryan, S.G., Stringfellow, G.S. (1994), "Lithium Abundances in the Most Metal-Deficient Stars", Astrophysical Journal, Vol. 423, pp. 386–393

Preston, G.W., McWilliam, A., Sneden, C. (1994): in preparation

Thorburn, J.A. (1994): "The Primordial Lithium Abundance From Extreme Subdwarfs: New Observations", Astrophysical Journal, Vol. 421, pp. 318–343

Part IV

Helium

Helium in HII Regions and Stars

B.E.J. Pagel

NORDITA, Blegdamsvej 17, Dk-2100 Copenhagen Ø, Denmark

Abstract. Following pioneering work in the 1960's and 70's, many efforts have been made to find the cosmologically and astrophysically significant parameters Y_P and dY/dZ as precisely as possible by observing emission lines in extragalactic HII regions. With the accumulation of improved data, several controversial issues have arisen and observational limits on Y_P and dY/dZ may have widened somewhat. New, accurate parallaxes from the HIPPARCOS satellite promise to revive interest in estimating Y_P and dY/dZ from main-sequence stars with differing metallicities, but even HIPPARCOS parallaxes will be marginal for fitting extreme subdwarfs.

1 Introduction and Historical Perspective

Helium is the second most abundant element in the visible universe and accordingly there is plenty of data from optical and radio emission lines in nebulae, optical emission lines in the solar chromosphere and prominences and absorption lines of hot stars. Yet more data are derived more indirectly by applying the theory of stellar structure, evolution and pulsation. Despite all this, there are still significant uncertainties in the important parameters Y_P and dY/dZ, partly owing to the high precision that is needed in order to derive significant results and partly because their determination involves extrapolations whose validity cannot be taken for granted.

Nevertheless, we have come a long way in the 40-odd years during which I have been aware of the subject and it is of some interest to take a look back at the early history (*cf.* Danziger 1970; Pagel 1982). In the 50's, there were conflicting estimates of the helium abundance in B-stars ranging from $y \equiv$ He/H by number of atoms = 5 to $y = 15$ per cent, eventually settling down to somewhere near 10 per cent in agreement with preliminary estimates in the planetary nebula NGC 7027 by Menzel, Aller and Goldberg (*cf.* Aller 1956). In 1957, B²FH in their classic paper on stellar nucleosynthesis assumed for the most part that low-metallicity objects would also have low helium, but they wisely left open the possibility that there might be "some helium in the initial matter of the Galaxy" (Burbidge *et al.* 1957) and Burbidge (1958) noted the implications if the large amount of helium observed results from purely stellar production. Hoyle & Tayler (1964), in a remarkably prescient paper, pointed out the cosmological importance of the presence or absence of a universal "floor" to the helium abundance noting the dependence on the number of neutrino types (the ν_μ had just been discovered) and Tayler in later papers pointed out also the significance of uncertainties in the neutron free-decay lifetime, which then was thought to be 1175s compared to the modern value of 889s (*cf.* Tayler 1990). Discovery of the microwave background in 1965 led to more detailed Big-Bang nucleosynthesis calculations (Peebles 1966; Wagoner, Fowler & Hoyle 1967, etc.) which in turn stimulated active searches for evidence for or against such a floor.

Investigations of line blanketing in the UBV colours (Eggen & Sandage 1962) initially suggested a very low helium abundance for low-metallicity subdwarfs, but Faulkner (1967) using a revised distance for the Hyades and more accurate quasi-homology relations showed that their data were more consistent with a primordial helium mass fraction $Y_P \geq 0.2$ and $\Delta Y/\Delta Z \simeq 3.5$ resulting in a lower mass for globular cluster stars and an improved fit to their HR diagrams (Faulkner & Iben 1966)*. Cayrel (1968) using colour systems less affected by blanketing also deduced a high helium abundance for cool subdwarfs. In the meantime, similarly high helium abundances in low-metallicity objects had been derived from the mass of the visual binary 85 Peg (Smak 1960), from the planetary nebula K-648 in the globular cluster M 15 (O'Dell, Peimbert & Kinman 1964) and from the blue edge of the RR Lyrae instability strip (Christy 1966).

About this time, attention was drawn to the lack of helium in the atmospheres of subdwarf B stars, on the blue extension of the horizontal branch (Sargent & Searle 1966; Greenstein & Münch 1966), and this caused much confusion until Sargent & Searle (1967) noted other chemical peculiarities reminiscent of chemically peculiar stars of Population I, supporting the explanation by gravitational settling (Greenstein, Truran & Cameron 1967). Catchpole, Pagel & Powell (1967), in an abundance analysis of μ Cas (an astrometric binary with a good parallax, but with a very faint secondary), used a mass estimate of $0.75 M_\odot$ for the primary by Wehinger & Wyckoff (1966) to derive (provisionally) a near-solar helium abundance in this mild subdwarf. Iben (1968) pointed out that the relative numbers of stars on the horizontal and red-giant branches of globular clusters provide robust evidence for a reasonably high initial helium abundance in those systems.

However, the controversy was not yet over. μ Cas figured once again in a report by Hegyi & Curott (1970) who deduced a large separation and hence a large mass and a low helium content from a tentative detection of the secondary using a photoelectric scanning system, and this paper (unlike ours!) attracted considerable attention. Faulkner (1971) pointed out serious mistakes and uncertainties in their work* and there have been numerous investigations of μ Cas in the intervening years. The latest one, using speckle interferometry, gives a mass for the primary of $0.73 \pm 0.05\ M_\odot$ and a helium mass fraction $Y = 0.23 \pm 0.05$ (Haywood, Hegyi & Gudehus 1992).

---

* I thank John Faulkner for reminding me of the significance of his 1967 paper and for mentioning that his interest in the subject had been stimulated by his collaboration with Hoyle & Tayler.

* He also pointed out that Wehinger and Wyckoff, whose mass estimate we had used, gave a wrong position angle. However, I have since been assured by Peter Wehinger that their separation measurement was a real one, even if they were not sure about the orientation of their telescope!

The controversy was finally settled, to all intents and purposes, when Searle & Sargent (1972) measured a near-normal helium abundance in the two famous Zwicky blue compact galaxies IZw18 and IIZw40, which they had identified as extragalactic HII regions with a very low heavy-element content (IZw18 still holds the record for low abundances in HII regions, with about 1/50 of the solar O/H ratio). Shortly afterwards, Peimbert and Torres-Peimbert (1974, 1976) used their accurate He/H determinations in the Magellanic Clouds and Orion to study the correlation between Y and the heavy-element mass-fraction Z (essentially proportional to O/H) and initiate a research programme to determine Y_P and the astrophysically significant parameter dY/dZ by plotting regressions of helium against oxygen in HII regions.

Table 1. Primordial Helium (ESO 1983 *et al.*)

	Y_P	Method	First author	Problems
Sun	$< .28 \pm .02$	Interior	Turck-Chiéze	κ; eq of st; ν problem
	$< .28 \pm .05$	Prom. HeI	Heasley	Lev.pops.
B-stars	$< .30 \pm .04$	Abs. lines	Kilian	Precision
Subdw.	$.19 \pm .05$	Main seq.	Carney	Plx.; T_{eff}; conv.
Glob-ular clus-ters	$.23$:	RR, Δm	Caputo	Physical basis of stellar evolution
	$.23 \pm .02$	N(HB)/N(RG)	Buzzoni	
	$.20 \pm .03$:	"	Cole	
	$.21 \pm .03$:	HB in uv	Bohlin	
Gal. neb.	$.22 \pm .02$	Plan. neb.	Peimbert	enr.
	$.22$:	HII regions	Mezger	He^0; enr.
Ex-gal. HII reg.	$.233 \pm .005$	Irr. + BCG	Lequeux	He^0; data
	$< .243 \pm .010$	BCG	Kunth	IIZw40

The ESO Workshop on "Primordial Helium" (Shaver, Kunth & Kjär 1983) collected together a variety of data from different sources. Table 1 gives a selection of results reported there, with some updatings. The different objects give a quite consistent picture, but the best data come from emission lines in nebulae, especially extragalactic HII regions where low metallicities occur giving the twin advantages of hot ionizing stars (so that neutral helium is unimportant) and a short extrapolation to pregalactic abundances. The globular clusters and especially field subdwarfs can include still less processed material, but the errors due to uncertainties in the basic physics of stellar evolution are hard to quantify. Field subdwarfs have the additional problem of parallaxes, to which I return later.

Coming back to the extragalactic HII regions, there was some disagreement in 1983 between the results of Lequeux *et al.* (1979), who found a substantial slope $dY/dZ \simeq 3$ leading to an extrapolated $Y_P = 0.23$, and Kunth & Sargent (1983) who with generally better data but some bias due to the high weight given to $\lambda 5876$ in IIZw40 which is absorbed by Galactic sodium (French 1980), detected no slope and therefore presented only an upper limit to Y_P.

2   More Recent HII Region Studies

Having the good fortune to be working with Roberto and Elena Terlevich during the later 1980s, it seemed to me that their spectrophotometric survey of HII galaxies (Campbell, Terlevich & Melnick 1986; Terlevich *et al.* 1991) would be a good starting point for attempts to improve the situation on primordial and non-primordial helium. Davidson & Kinman (1985) had given a very careful survey of many of the difficulties involved, but I thought I had answers to most of them. Pagel, Terlevich & Melnick (1986) noted that there seemed to be real scatter in helium at a given oxygen abundance and we attributed this to local pollution by winds of hydrogen-burned material from embedded Wolf-Rayet stars that sometimes show up in the form of broad HeII λ 4686 and accordingly introduced the idea of plotting a regression against nitrogen as well as oxygen, in the vain hope that this effect would cancel out to first order. Later we added new data that we obtained with IPCS and CCD detectors using the INT and AAT; some (not enough) of this was of unprecedentedly good quality (especially AAT CCD data taken with a long-focus camera) and the results have been described in detail in Simonson's thesis (1990) and in Pagel *et al.* (1992) where we combined our observations with selected (and, if necessary, corrected) data from the literature. The main problems in this and other work are listed below.

(i) **Data quality and reddening corrections**

Much effort was spent in checking the detector linearity and flux calibration, reddening being deduced in the usual way from the Balmer decrement. The reddening seriously affects only λ 5876, since λ 4471 and λ 6678 are close to Balmer lines, but the dynamic range is large, e.g. Hα/λ 6678 $\simeq$ 100.

(ii) **Theoretical recombination coefficients**

In common with previous investigators we used coefficients calculated by Brocklehurst (1972), but later calculations by Smits (1991ab) show small though non-negligible differences. Smits (1994) vindicates Brocklehurst's values except in the case of λ 7065 which is sensitive to radiative transfer and especially collisional effects.

(iii) **He^0 and underlying absorption lines**

Neutral helium is not a problem if the ionizing stars are hot enough, which we were able to judge by comparing O^+/O^{++} and S^+/S^{++} ionization ratios. Weak [OII] lines implied that interference from a superposed low-ionization HII region would be insignificant. Underlying absorption lines can be a problem, especially for λ 4471, but could be safely neglected when the weaker λ 4388 line was seen.

Because of underlying or intervening absorption effects, it is dangerous to use one helium line alone.

(iv) **Radiative transfer and electron collision effects**

Because of the metastability of the 2^3S state, helium lines can be excited by fluorescence and electron collisions, as well as by pure recombination. Collisional effects were calculated from the electron density deduced from the [SII] line ratio using the formulae of Clegg (1987) and were very small, so we are not too worried by the suggestion that they are overestimated by up to a factor of 2 in planetary nebulae (Peimbert & Torres-Peimbert 1987). The agreement with λ 7065 seemed good enough to exclude significant fluorescence effects.

(v) **Cosmic scatter in the He,O or He,N relation**

There are some signs of a real dispersion in the helium abundance at a given "metallicity". The evidence is marginal compared to possible errors, but it is interesting that Roelfsema, Goss & Mallik (1992) have found local helium excesses from radio observations of the Galactic HII region W3A. The most prominent case among our objects is that of NGC 5253, a nearby HII galaxy that is well resolved spatially and shows local variations in both nitrogen and helium abundance with a very narrow helium peak in the central nucleus which also shows a broad WR feature (Campbell, Terlevich & Melnick 1986). A correlation between excess helium and visible WR features is far from being proven, but in order to find an objective criterion for excluding such effects, we excluded all spectra showing a definite WR feature from our solutions. Fig 1 shows our results from Pagel *et al.* (1992) with some later data superposed.

The result from this was $Y_P = 0.228 \pm 0.005$ (s.e.), or, assuming a systematic error of up to 0.005, $Y_P \leq 0.242$ with 95 per cent confidence, and $dY/dZ = 4 \pm 1$ for O/H $\leq 2.4 \times 10^{-4}$ or 0.3 solar oxygen abundance, not significantly different from the results of Lequeux *et al.* (1979). We derived the same Y_P from the regression against nitrogen, taking into account the contribution of "secondary" nitrogen production, which led to some discussion but no serious subsequent change (Pagel & Kazlauskas 1993; Balbes, Boyd & Mathews 1993). The large dY/dZ, already implicit in earlier work (Faulkner 1967; Perrin *et al.* 1977; Lequeux *et al.* 1979), and supported by studies of planetary nebulae (Chiappini & Maciel 1994), has led to much discussion, since it is about twice the value expected from conventional stellar nucleosynthesis, and basically three different hypotheses have been put forward:

(i) A low upper limit ($\simeq 25 M_\odot$) to the initial mass of stars undergoing supernova explosions rather than going into black holes (Maeder 1992, 1993; Brown & Bethe 1994). Apart from questions about the initial mass function and resulting absolute yields assumed by Maeder, objections have been raised by Prantzos (1994) on grounds of overproduction of carbon.

(ii) Preferential loss of oxygen and other components of Z by selective galactic winds following bursts of star formation in dwarf galaxies (Pilyugin 1993; Marconi, Matteucci & Tosi 1994). Pilyugin also includes temporary increases of both Y and Z by self-pollution of HII regions, for which Kunth *et al.* (1994) have provided independent evidence in the case of IZw18 from HST observations

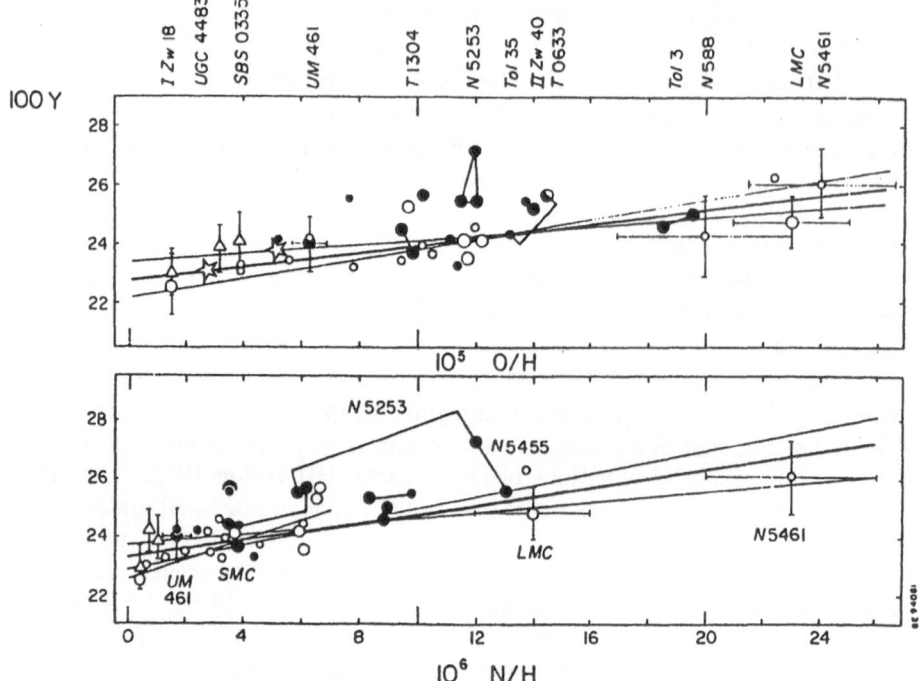

Fig. 1. Regressions of helium against oxygen and nitrogen in HII galaxies. Filled and open circles (representing objects repectively with and without definite WR features and with different sizes according to weight) are from Pagel *et al.* (1992); full and broken lines show their maximum-likelihood regressions and equivalent $\pm 1\sigma$ error limits. (The short line in the lower panel is the regression preferred for nitrogen, since there is some indication of a flattening at higher abundances.) The stars represent SBS 0335-052 recalculated from data by Melnick *et al.* (1992), while triangles show IZw18 from Skillman & Kennicutt (1993) and UGC 4483 from Skillman *et al.* (1994) and Izotov *et al.* (1994).

giving an extremely low oxygen abundance in the neutral gas. Such differences cannot, however, be a major factor in objects like the Magellanic Clouds and the Galaxy, where no HII regions are found to be overabundant relative to the stellar population or the neutral gas, and the scatter in helium abundance that they are supposed to explain only marginally exceeds experimental errors.

(iii) Our oxygen abundances are underestimated because we assumed a uniform electron temperature. This seemed to be the right thing to do in 1992, but recently evidence has been found from recombination lines and Balmer and Paschen discontinuities in planetary nebulae, Galactic HII regions and even a low-abundance extragalactic HII region that our abundances could be too low by as much as a factor of 2 (González Delgado *et al.* 1995). For Galactic HII regions, with a generous allowance for temperature fluctuations ($t^2 = 0.04$), Peimbert (1993) has derived $\Delta Y/\Delta Z = 3.2$ for Orion, 3.5 for M8 and 2.5 for M17. An upper IMF slope of 1.7 and an upper mass limit of $50M_\odot$ for SN progenitors gives $\Delta Y/\Delta Z = 3$ (Maeder 1992), so this may be the solution.

3 Still More Recent HII Region Studies

At present, efforts are being made towards still further refinements of helium abundance determinations, concentrating on improvements in the following areas: Discovery of more low-metallicity systems (i.e. more low-luminosity galaxies), better signal:noise and resolution using various telescopes and calibration routines, and improvements in atomic data.

Skillman & Kennicutt (1993) have re-observed IZw18 and their result is shown in Fig 1, together with the result for UGC 4483 by Skillman *et al.* (1994) who used spectra taken at McDonald Observatory, the MMT, and the INT and WHT on La Palma. Further results from this group (Skillman *et al.* 1994ab) have been exhaustively analyzed, together with our data (Pagel *et al.* 1992), by Olive & Steigman (1994); from various subsets of the data, including most of the definite WR objects which we excluded, they find $Y_P = 0.232 \pm 0.003$ and $Y_P^{2\sigma} + \sigma_{syst} \le 0.243$.

Izotov, Thuan & Lipovetsky (1994) have observed and analysed ten HII galaxies from the Second Byurakan Survey, using the 4m Mayall Telescope on Kitt Peak. They compare results using recombination coefficients by Brocklehurst (1972), Smits (1991ab) and Smits (1994), resulting in some confusion, since Smits (1994) confirms Brocklehurst (and so outdates Smits 1991) except in the case of λ 7065. Using Brocklehurst coefficients, they argue that the [SII] electron densities used by the rest of us are inappropriate for He^+ and derive much higher densities from λ 7065 and accordingly larger corrections for collisions; but when this exercise is repeated with Smits (1994) coefficients, they find low electron densities that are actually in fair agreement with the ones from [SII]. Using Smits (1994) recombination coefficients and omitting one deviant galaxy, they find $Y_P = 0.239 \pm 0.007*$ and $dY/dZ = 5.8^\dagger \pm 4.4$, the large error in dY/dZ coming from the rather small range in oxygen abundances among their objects. The Y_P value derived by Olive & Steigman is within 1σ of both Pagel *et al.* and Izotov *et al.*, but radiative transfer effects not previously taken into account could apparently raise the upper limit, perhaps by as much as 0.01 (Sasselov & Goldwirth 1994).

---

* With Brocklehurst coefficients the agreement with Pagel *et al.* is very much better, but, as they themselves remark, this is probably just fortuitous.

† With the corrections applied by Pagel *et al.*, this raw value is reduced to about 4.

4   Location of Stars on the Main Sequence

HII regions provide the most accurate helium abundances, but it is useful to check them independently from considerations of stellar structure. Some results from globular clusters were briefly quoted in Table 1, and here I just comment further on studies of the main sequence as a function of metallicity.

The zero-age main sequence follows approximately quasi-homology relations of the form

$$L = (X + 0.4)^a (Z + Z_0)^b f(T_{\text{eff}}) \qquad (1)$$

(*cf.* Faulkner 1967), where the first factor represents the molecular weight and the second the combined effect of opacity and nuclear energy generation, while $f(T_{\text{eff}}) \propto T_{\text{eff}}^7$, approximately. Models for low-metallicity stars aged 15 Gyr and with $\log T_{\text{eff}} = 3.72$ (e.g. VandenBerg 1983) fit this relation with $a \simeq 2.67$, $Z_0 \simeq 0.003$, $b \simeq 0.64$. As a result of this type of relation, correlated changes in helium and heavy-element abundances push the main sequence in opposite directions and tend to cancel each other out at the higher metallicities where, however, uncertainties in opacity, convection theory and corrections for evolution introduce major uncertainties. Perrin *et al.* (1977) were unable to find any metallicity-correlated dispersion among disk stars and from this with the aid of evolutionary tracks by Hejlesen they deduced $\Delta Y/\Delta Z = 5 \pm 3$ (see also Cayrel de Strobel in this volume), which looks like a fair estimation of the uncertainties, and their result gives modest support to the HII region data.

Carney (1983) used modified Yale isochrones to estimate the helium abundance in extreme subdwarfs, noting the difficulties that arise from systematic effects in the evolutionary tracks. Assuming that this problem can ultimately be solved, one can differentiate eq (1) to derive an error budget for the helium abundance:

$$\delta Y \simeq 0.44 \frac{\delta L}{L} + 3.1 \frac{\delta T_{\text{eff}}}{T_{\text{eff}}}. \qquad (2)$$

Assuming $\delta T_{\text{eff}}/T_{\text{eff}} = 1$ per cent, even a perfect parallax gives an error in Y of 0.03. Among the stars used by Carney, ground-based parallaxes are accurate to 5 per cent (contributing 0.04 to δY) in only one case, that of Groombridge 1830, the next best being HD 25329 with $\delta L/L \simeq 0.2$. There is thus considerable interest in forthcoming parallaxes from HIPPARCOS, but, disappointingly, it turns out that even with the HIPPARCOS accuracy of 2 m.arcs., $\delta L/L$ for the next best candidate with extreme metal deficiency is 10 per cent, contributing an error in Y of 0.04. HIPPARCOS will undoubtedly lead to enormous improvements in the search for fine structure in the main sequence of nearby stars belonging to the thin and thick disks, but submilliarcsec precision is needed in order to study the main sequence of extreme subdwarfs; this may be attainable with HST, if the correction to absolute parallax can be established well enough, but must otherwise await still more advanced space missions (Kovalevsky 1995).

References

Aller, L.H. 1956, *Gaseous Nebulae*, London: Chapman & Hall, p. 199.

Balbes, M.J., Boyd, R.N. & Mathews, G.J. 1993, *Ap. J.*, **418**, 229.

Bohlin, R.C., Cornett, R.H., Hill, J.K., Smith, A.M. & Stecher, T.P. 1985, *Ap. J.*, **292**, 687.

Brocklehurst, M. 1972, *M.N.R.A.S.*, **157**, 221.

Brown, G.E. & Bethe, H.A. 1994, *Ap. J.*, **423**, 659.

Burbidge, G.R. 1958, *Pub. Astr. Soc. Pacific*, **70**, 83.

Burbidge, E.M., Burbidge, G.R., Fowler, W.A. & Hoyle, F. 1957, *Rev. Mod. Phys.*, **29**, 547.

Buzzoni, A., Fusi-Pecci, F., Buonanno, R. & Corsi, C.E. 1983, *Astr. Ap.*, **128**, 94.

Campbell, A., Terlevich, R.J. & Melnick, J. 1986, *M.N.R.A.S.*, **223**, 811.

Caputo, F., Martínez Roger, C. & Páez, E. 1987, *Astr. Ap.*, **183**, 228.

Carney, B.W. 1983, in *Primordial Helium*, p. 179.

Catchpole, R.M., Pagel, B.E.J. & Powell, A.L.T. 1967, *M. N.R.A.S.*, **136**, 403.

Cayrel, R. 1968, *Ap. J.*, **151**, 997.

Chiappini, C. & Maciel, W.J. 1994, *Astr. Ap.*, submitted.

Christy, R.F. 1966, *Ann. Rev. Astr. Ap.*, **4**, 353.

Clegg, R.E.S. 1987, *M.N.R.A.S.*, **229**, 31P.

Cole, P.W., Demarque, P. & Green, E.M. 1983, in *Primordial Helium*, p. 235.

Danziger, I.J. 1970, *Ann. Rev. Astr. Ap.*, **8**, 161.

Davidson, K. & Kinman, T.D. 1985, *Ap. J. Suppl.*, **58**, 321.

Eggen, O.J. & Sandage, A.R. 1962, *Ap. J.*, **136**, 735.

Faulkner, J. 1967, *Ap. J.*, **147**, 617.

Faulkner, J. 1971, *Phys. Rev. Lett.*, **27**, 206.

Faulkner, J. & Iben, I., Jr. 1966, *Ap. J.*, **144**, 995.

French, H.B. 1980. *Ap. J.*, **240**, 41.

González Delgado, R.M. *et al.* 1995, *Ap. J.*, in press.

Greenstein, G.S., Truran, J.W. & Cameron, A.G.W. 1967, *Nature*, **213**, 871.

Greenstein, J.L. & Münch, G. 1966, *Ap. J.*, **146**, 618.

Haywood, J.W., Hegyi, D.J. & Gudehus, D.H. 1992, *Ap. J.*, **392**, 172.

Heasley, J.N. & Milkey, R.W. 1978, *Ap. J.*, **221**, 677.

Hegyi, D. & Curott, D. 1970, *Phys. Rev. Lett.*, **24**, 415.

Hoyle, F. & Tayler, R.J. 1964, *Nature*, **203**, 1108.

Iben, I., Jr. 1968, *Nature*, **220**, 143.

Izotov, Y.I., Thuan, T.X. & Lipovetsky, V.A. 1994, *Ap. J.*, in press.

Kilian, J. 1992, *Astr. Ap.*, **262**, 171.

Kovalevsky, J. (ed.) 1995, IAU Symp. 166: *Astronomical and Astrophysical Objectives of Submilliarcsecond Astrometry*, to appear.

Kunth, D., Lequeux, J., Sargent, W.L.W., & Viallefond, F. 1994, *Astr. Ap.*, **282**, 709.

Kunth, D. & Sargent, W.L.W. 1983, *Ap. J.*, **273**, 81.

Lequeux, J., Peimbert, M., Rayo, J.F., Serrano, A. & Torres-Peimbert, S. 1979, *Astr. Ap.*, **80**, 155.

Maeder, A. 1992, *Astr. Ap.*, **264**, 105.

Maeder, A. 1993, *Astr. Ap.*, **268**, 833.

Marconi, G., Matteucci, F. & Tosi, M. 1994, *M.N.R.A.S.*, in press.

Melnick, J., Haydari-Malayeri, M. & Leisy, P. 1992, *Astr. Ap.*, **253**, 16.

Mezger, P. & Wink, J.E. 1983, in *Primordial Helium*, p. 281.

O'Dell, C.R., Peimbert, M. & Kinman, T.D. 1964, *Ap. J.*, **140**, 119.

Olive, K.A. & Steigman, G. 1994, Preprint, UMN-TH-1230/94; OSU-TA-6/94.

Pagel, B.E.J. 1982, *Phil Trans. R. Soc. London A*, **307**, 19.

Pagel, B.E.J. & Kazlauskas, A. 1993, *M.N.R.A.S.*, **256**, 49P.

Pagel, B.E.J., Simonson, E.A., Terlevich, R.J. & Edmunds, M.G. 1992, *M.N.R.A.S.*, **255**, 325.

Pagel, B.E.J., Terlevich, R.J. & Melnick, J. 1986, *Pub. Astr. Soc. Pacific*, **98**, 1005.

Peebles, P.J.E. 1966, *Ap. J.*, **146**, 542.

Peimbert, M. 1983, in *Primordial Helium*, p. 267.

Peimbert, M. 1993, *Rev. Mex. Astr. Astrofis.*, **27**, 9.

Peimbert, M. & Torres-Peimbert, S. 1974, *Ap. J.*, **193**, 327.

Peimbert, M. & Torres-Peimbert, S. 1976, *Ap. J.*, **203**, 581.

Peimbert, M. & Torres-Peimbert, S. 1987, *Rev Mex. Astr. Astrofis.*, **15**, 117.

Perrin, M.-N., Hejlesen, P.M., Cayrel de Strobel, G. & Cayrel, R. 1977, *Astr. Ap.*, **54**, 779.

Pilyugin, L. 1993, *Astr. Ap.*, **277**, 42.

Prantzos, N. 1994, *Astr. Ap.*, submitted.

Roelfsema, P.R., Goss, W.M. & Mallik, D.C.V. 1992, *Ap. J.*, **394**, 188.

Sargent, W.L.W. & Searle, L. 1966, *Ap. J.*, **145**, 652.

Sargent, W.L.W. & Searle, L. 1967, *Ap. J. Lett.*, **150**, L33.

Sasselov, D. & Goldwirth, D. 1994, preprint.

Searle, L. & Sargent, W.L.W. 1972, *Ap. J.*, **173**, 25.

Shaver, P.S., Kunth, D. & Kjär, K. (eds.) 1983, *Primordial Helium*, ESO, Garching.

Simonson, E.A. 1990, PhD Thesis, Sussex University.

Skillman, E.D. & Kennicutt, R.C., Jr. 1993, *Ap. J.*, **411**, 655.

Skillman, E.D., Terlevich, R.J., Kennicutt, R.C., Jr., Garnett, D.R. & Terlevich, E. 1994a, *Ap. J.*, in press; 1994b, in prep.

Smak, J. 1960, *Acta Astr.*, **10**, 153.

Smits, D.P. 1991a, *M.N.R.A.S.*, **248**, 193.

Smits, D.P. 1991b, *M.N.R.A.S.*, **251**, 316.

Smits, D.P. 1994, *M.N.R.A.S.*, in press..

Tayler, R.J. 1990, *Q. J. R. A. S.*, **31**, 371.

Terlevich, R.J., Melnick, J., Masegosa, J. & Moles, M. 1991, *Astr. Ap. Suppl.*, **91**, 285.

Turck-Chièze, S. & Lopez, I. 1993, *Ap. J.*, **408**, 347.

VandenBerg, D.A. 1983, *Ap. J. Suppl.*, **51**, 29.

Wagoner, R.V., Fowler, W.A. & Hoyle, F. 1967, *Ap. J.*, **148**, 3.

Wehinger, P.A. & Wyckoff, S. 1966, *Astr. J.*, **71**, 185.

The Helium to Heavy Elements Enrichment Ratio

M. Peimbert

Instituto de Astronomía, UNAM, Apdo. Postal 70-264, Mexico D.F. 0450, México

Abstract. A review on $\Delta Y/\Delta Z$ and $\Delta Y/\Delta O$ determinations is presented. Results based on galactic and extragalactic gaseous nebulae are analyzed. The temperature structure of the gaseous nebulae is taken into account as well as the fraction of O atoms tied up in dust grains. The best galactic and extragalactic determinations of $\Delta Y/\Delta O$ are compared with chemical evolution models of the solar vicinity and of metal poor galaxies. The relevance of this comparison to the presence of: black holes, different types of galactic outflows and temperature inhomogeneities in H II regions is discussed.

Keywords: Planetary nebulae, H II regions, chemical evolution of galaxies

1 Introduction

Work previous to 1983 on $\Delta Y/\Delta Z$ is presented in the ESO workshop on "Primordial Helium" (Shaver, Kunth & Kjär 1983). I presented a review on $\Delta Y/\Delta Z$ in 1986 (Peimbert 1986). Excellent reviews on the pregalactic helium abundance by mass, Y_p, and the $\Delta Y/\Delta Z$ ratio derived from stars have been presented in this conference (Pagel 1995, Cayrel de Strobel 1995). I will concentrate mainly on the $\Delta Y/\Delta Z$ and $\Delta Y/\Delta O$ determinations from gaseous nebulae and their consequences.

Observers usually determine the oxygen abundance by mass, O, and $\Delta Y/\Delta O$. To compare with results derived from stellar evolution models observers usually assume that O makes a fraction of the total Z ratio (usually between 43% and 62%) and transform the observed $\Delta Y/\Delta O$ into $\Delta Y/\Delta Z$. The fraction of Z due to O varies from object to object mainly due to changes in the C/O and N/O ratios with O/H. I will base the discussion mainly on the $\Delta Y/\Delta O$ ratio because its determination is more accurate than that of $\Delta Y/\Delta Z$.

The main problem in the $\Delta Y/\Delta O$ and $\Delta Y/\Delta Z$ determinations is provided by the temperature structure of a gaseous nebula which, to a second approximation, can be characterized by an average electron temperature, T_0, and a mean square temperature variation, t^2 (Peimbert 1967).

Photoionized models of H II regions and PNe predict t^2 values around 0.01 (e.g., Harrington et al. 1982; Garnett 1992; Gruenwald & Viegas 1992), these values do not affect significantly the determination of chemical abundances and isothermal nebulae can be adopted. Most abundance determinations assume $t^2 = 0.00$, the isothermal case. There is a growing body of evidence that indicates that typical values of t^2 are around 0.04, these values do indeed produce higher

O abundances than those derived under the isothermal assumption. Moreover an explanation for these high t^2 values has to be sought.

2 Abundances

2.1 Helium Abundance

The helium abundance by mass of a gaseous nebula depends on: (a) the recombination coefficients of H^+, He^+, and He^{++} that are well known (Hummer & Storey 1987, Brocklehurst 1972, Smits 1994), the H^+ data are needed because to obtain Y it is also necessary to determine the hydrogen and heavy elements by mass, X and Z respectively, (b) the helium ionization degree, since inside a gaseous nebula $N(\text{He}) = N(\text{He}^0) + N(\text{He}^+) + N(\text{He}^{++})$, ($c$) the collisional excitation coefficients from the 2^3S level of He I that are also well known (Sawey & Berrington 1993, Kingdon & Ferland 1994), (d) the radiative transfer due to the optical depths of the 2^3S and 1^1S levels of He I (Robbins 1968; Robbins & Bernat 1973; Almog & Netzer 1989; Sasselov & Goldwirth 1994), (e) departures from case B due to absorption of hydrogen Lyman line photons by internal dust (Cota & Ferland 1988; Baldwin et al. 1991), leading to an overestimation of the helium abundance, this effect should increase with the amount of dust present and hence with metallicity leading to an overestimation of $\Delta Y/\Delta Z$, (f) the temperature structure, since the higher the t^2 value the lower the Y value for a given $T_e(4363/5007)$, this effect increases with decreasing $T_e(4363/5007)$.

Since we are interested in ΔY, the difference in Y between two H II regions to a first approximation is independent of errors in the recombination coefficients. The amount of neutral helium inside an ionized gaseous nebula can be estimated by observing objects with a varying degree of ionization at different lines of sight (*e.g.*, Peimbert, Torres-Peimbert & Ruiz 1992), or can be neglected by observing objects of high degree of ionization chosen based on the value of the radiation softness parameter (Vílchez & Pagel 1988). The He I radiative transfer effects for the triplet series can be estimated by comparing the intensity of a line strongly affected, like $\lambda 7065$, with another weakly affected, like $\lambda 4472$; while for the singlet series the $\lambda\lambda 5016$ and 6678 lines can be used; these effects are usually negligible. The collisional effects from the 2^3S level can be easily computed if T_e and N_e are known. For O-poor H II regions N_e is considerably smaller than the critical density for depopulating the 2^3S level by collisions and the collisional effects are often negligible. For solar vicinity H II regions T_e is relatively small and the collisional effects become negligible. Alternatively for Type I PN the collisional effects are considerable. Departures from case B for O-poor H II regions are expected to be negligible due to the very small amount of dust present inside the H II regions.

2.2 Oxygen Abundance

The gaseous component of the O abundance in PN and H II regions is given by

$$\frac{N(O)}{N(H)} = \frac{N(O^+ + O^{++})}{N(H^+)} \left[\frac{N(He^+) + N(He^{++})}{N(He^+)}\right]^a , \tag{1}$$

where most people have used $a = 1$, but recently Kingsburgh & Barlow (1994) based on ionization structure models have recommended a value for $a = 2/3$. For most H II regions the He^{++}/He^+ ratio is negligible. High degree of ionization H II regions have more than 75% of their O in the O^{++} stage.

The O abundance derived from equation (1) does not include the fraction of O embedded in dust grains, which for the solar vicinity amounts to about 0.08 dex (Meyer 1985); while for O-poor extragalactic H II regions, based on the Si depletion (Dufour et al. 1994) amounts to about 0.04 dex.

The main problem with the O determination is the value of t^2. The effect on the O abundance determination for a constant t^2 value decreases with decreasing O/H. For an increase in t^2 from 0.00 to 0.04 the O/H increase amounts to about 0.3 dex for an H II region with solar abundances and to about 0.18 dex for an H II region with one fourth of the solar abundances.

3 Planetary Nebulae

From different methods to determine the electron temperature it has been found that t^2 varies from 0.01 to 0.08 with a typical value around 0.035 (e.g., Peimbert 1971; Dinerstein, Lester & Werner 1985; Liu & Danziger 1993; Peimbert, Storey & Torres-Peimbert 1993; Liu et al. 1994, 1995).

Possible explanations for the large t^2 values observed are: (a) the deposition of kinetic energy by shocks or by subsonic turbulence produced by mass loss from the central star (e.g., Peimbert, Sarmiento & Fierro 1991 and references therein; Bohigas 1994), (b) the presence of chemical abundance inhomogeneities (e.g., Jacoby & Ford 1983; Torres-Peimbert, Peimbert & Peña 1992), (c) in some cases erroneous T_e (4363/5007) and t^2 determinations due to the presence of high density clumps (Viegas & Clegg 1994).

Table 1 includes three $\Delta Y/\Delta O$ and $\Delta Y/\Delta Z$ determinations for different values of t^2 based on PN of the solar vicinity (Peimbert & Serrano 1980). Table 1 also includes the $\Delta Y/\Delta O$ and $\Delta Y/\Delta Z$ values derived by Chiappini & Maciel (1994) by combining solar vicinity PN with O-poor H II regions under the assumption that $t^2 = 0.00$. In the PN determinations the amount of O in dust grains has not been considered.

4 H II Regions

4.1 Solar Vicinity

The most studied H II region of the solar vicinity is the Orion nebula. From optical and radio observations it has been found that the $N(\text{He}^+)/N(\text{H}^+)$ ratio decreases with distance from the Trapezium (*e.g.*, Peimbert & Torres-Peimbert 1977; Peimbert et al. 1988; Pogge, Owen & Atwood 1992), this change has been interpreted as due to the presence of neutral helium inside the Orion nebula. From ionization structure models of the Orion nebula Rubin et al. (1991) find a substantial amount of neutral helium present, while Baldwin et al. (1991) find a negligible amount; this discrepancy needs further investigation. The correction due to the presence of neutral helium introduces a relatively large uncertainty in the $\Delta Y/\Delta Z$ determination for the Orion nebula (Peimbert 1993).

M17 has the smallest fraction of neutral helium among the brightest H II regions of the solar vicinity, therefore it is the best H II region to determine the $\Delta Y/\Delta Z$ ratio. Peimbert et al. (1992) determined $\Delta Y/\Delta Z$ for $t^2 = 0.04$ and by assuming that 0.08 dex of O was trapped in dust grains (see Table 1); the t^2 value was derived from a comparison of the Balmer continuum temperature with T_e (4363/5007). The high t^2 value was confirmed by comparing the O^{++} abundances derived from recombination lines with those derived from forbidden lines (Peimbert et al. 1993).

4.2 O-poor H II Regions

From a selected group of well observed H II regions, where those with W-R features were omitted, Pagel et al. (1992) found $\Delta Y/\Delta Z = 6.1 \pm 2.1$ (see Table 1) by assuming that: (*a*) $t^2 = 0.00$, (*b*) no O trapped in dust grains and (*c*) no deviations from case B of the hydrogen lines. Pagel et al. recommended a $\Delta Y/\Delta Z \sim 4$ value by considering: (*a*) an increase of 0.04 dex due to $t^2 \neq 0.00$, (*b*) 0.08 dex of O trapped in dust grains and (*c*) possible deviations of the H lines from case B for the metal richer H II regions.

Olive & Steigman (1994) obtain $\Delta Y/\Delta Z = 5.7 \pm 0.5$ (see Table 1) from a sample of 49 extragalactic objects, including some with W-R features, and without any of the three corrections mentioned above.

Carigi et al. (1994) from a sample of 10 well observed extragalactic objects derived $\Delta Y/\Delta Z = 3.8 \pm 0.9$ for $t^2 = 0.00$ without considering the amount of O in dust grains nor deviations from case B. The main difference in the $\Delta Y/\Delta Z$ values between Carigi et al. and Pagel et al. (1992) is due to the use by Carigi et al. of recent results for the two metal poorest objects in their sample: IZw18 and UGC 4483 (Skillman & Kennicutt 1993; Skillman et al. 1994). Carigi et al. derived $\Delta Y/\Delta Z = 2.4 \pm 0.6$ for $t^2 = 0.035$ (t^2 based on the studies by Campbell 1988, and McGaugh 1991), and adopting an increase of 0.04 dex in O due to the fraction trapped in dust grains (increase based on the work by Dufour et al. 1994); these results are presented in Table 1.

Additional arguments in favor of large t^2 values in O-poor H II regions are those by González-Delgado et al. (1995) who find a considerably smaller T_e from the Paschen continuum than that derived from the [O III] lines for NGC 2363. Moreover, Esteban & Peimbert (1994) find that the scattering in the Y versus N diagram of the H II regions with W-R features can be explained if they have larger values of t^2 than objects without W-R features; they also find that self-enrichment is not responsible for the scattering in the Y versus N diagram of H II regions with W-R features because self-enrichment moves the points in a direction parallel to that defined by the objects without W-R features.

Table 1. Helium to heavy elements ratio from observations

Object	$\Delta Y/\Delta Z$	$\Delta Y/\Delta O$	O/Z	t^2	Reference
PN (SV)	3.6 ± 0.8	8.0 ± 1.8	0.45	0.00	1
PN (SV)	3.0 ± 0.7	6.7 ± 1.6	0.45	0.02	1
PN (SV)	2.2 ± 0.5	4.9 ± 1.1	0.45	0.035	1
PN + H II	5.2 ± 1.1	12.1 ± 2.6	0.43	0.00	2
H II (M17) + Y_P	2.5 ± 0.5	5.7 ± 1.1	0.44	0.04	3
H II (O-poor)	6.1 ± 2.1	10.2 ± 3.5	0.60	0.00	4
H II (O-poor)	5.7 ± 0.5	9.5 ± 0.8	0.60	0.00	5
H II (O-poor)	3.8 ± 0.9	7.1 ± 1.6	0.54	0.00	6
H II (O-poor)	2.4 ± 0.6	4.5 ± 1.0	0.54	0.035	6

1 Peimbert & Serrano 1980; 2 Chiappini & Maciel 1994; 3 Peimbert et al. 1992; 4 Pagel et al. 1992; 5 Olive & Steigman 1994; 6 Carigi et al. 1994.

5 Models of Chemical Evolution of Galaxies

5.1 Solar Vicinity

The large $\Delta Y/\Delta Z$ values determined for $t^2 = 0.00$ and the low O/H values in irregular and blue compact galaxies led Maeder (1992) to suggest that massive stars above a given threshold, $m(BH)$, produce black holes that prevent them from enriching the interstellar medium with heavy elements at the time of the SN explosion.

Based on the stellar evolution models by Maeder (1992) at least four groups have computed chemical evolution models for the solar vicinity (Carigi 1994; Peimbert, Sarmiento & Colín 1994b; Traat 1994, Giovagnoli & Tosi 1994). They find $\Delta Y/\Delta Z$ ratios in the $2 - 2.5$ range without the use of black holes, in agreement with the observed values of M17 (see Table 1). The agreement is obtained with the Scalo (1986) or with the Kroupa, Tout & Gilmore (1993) IMF's (hereinafter KTG), alternatively the Salpeter (1955) IMF predicts $\Delta Y/\Delta Z$ values smaller than observed. Notice that the suggestion by Maeder is not supported not because the stellar evolution models are at fault but because a different value of the observational restriction is used.

The $(Z\text{-O})/\text{O}$ observational constraint can be fitted by chemical evolution models of the solar vicinity without the production of black holes by massive stars (Peimbert et al. 1994b), which supports the result derived from $\Delta Y/\Delta Z$. Moreover $(Z\text{-O})/\text{O}$ is almost independent of t^2, unlike $\Delta Y/\Delta Z$.

5.2   Irregular Galaxies

5.2.1 Closed Models Peimbert, Colín & Sarmiento (1994a) based on the stellar evolution models by Maeder (1992) compute chemical evolution models for O-poor irregular galaxies and find that with black holes they can adjust $\Delta Y/\Delta Z$ but not $(Z\text{-O})/\text{O}$; alternatively without black holes they can adjust $(Z\text{-O})/\text{O}$ but not $\Delta Y/\Delta Z$. Marconi, Matteucci & Tosi (1994) from their chemical evolution models for irregular galaxies also find that they can not explain $\Delta Y/\Delta Z$ without assuming black holes.

Carigi et al. (1994), based on the abundances of 10 well observed irregular galaxies, define a "typical" irregular galaxy and derive the M_{gas}/M_{total}, O/H, C/O, $(Z\text{-C-O})/\text{O}$ and $\Delta Y/\Delta O$ values associated with it. Carigi et al. based on the stellar evolution models by Maeder (1992) and Renzini and Voli (1981) compute chemical evolution models for their typical irregular galaxy and find that by introducing a $m(\text{BH})$ in the 40 to 50 $M_\odot$ range they can fit $\Delta Y/\Delta O$ but not C/O and $(Z\text{-C-O})/\text{O}$. Alternatively without black holes they are able to fit C/O and $(Z\text{-C-O})/\text{O}$ but not $\Delta Y/\Delta O$. The C/O and $(Z\text{-C-O})/\text{O}$ values are almost independent of t^2, particularly C/O because the lines used to derive the C and O abundances have similar excitation energies and to a very good approximation the temperature structure cancels out (Garnett et al. 1994). Carigi et al. conclude that a different solution to that of the production of black holes by massive stars is needed to explain the large $\Delta Y/\Delta Z$ ratios.

5.2.2 Outflow of Well-mixed Material Matteucci & Chiosi (1983) and Chiosi & Matteucci (1984) suggested the presence of galactic outflows of well mixed material to explain the low O/H values observed in irregular galaxies. Matteucci & Tosi (1985) have computed chemical evolution models that explain the observed low O/H values. Marconi et al. (1994) with well mixed outflow models find that it is not possible to explain the observed $\Delta Y/\Delta Z$ values. Similarly Carigi et al. (1994) find that the $\Delta Y/\Delta Z$ values for well mixed outflow models are slightly higher than those for closed models but not large enough to explain the observed values.

5.2.3 Outflow of O-rich Material It has been proposed that an outflow of O-rich material is present, at least in some galaxies, to explain: (a) the small yield in heavy elements seen in irregular galaxies (Lequeux 1989; Tosi 1994; Peimbert et al. 1994a), (b) the large helium abundances derived in some irregular and blue compact galaxies (Aparicio, García-Pelayo & Moles 1988), (c) the large $\Delta Y/\Delta O$ ratios derived from samples of galaxies (Aparicio et al. 1988; Lequeux 1989; Pilyugin 1993; Tosi 1994; Peimbert et al. 1994b), (d) the high Fe/O ratio in the

Magellanic Clouds (Russell, Bessell & Dopita 1988a, b; Lequeux 1989) and (e) the Z-Mass relation present in irregular galaxies (De Young & Gallagher 1990).

The O-rich outflow can be characterized by the fraction of the mass of SNe of Type II, γ, which is ejected to the intergalactic medium without mixing with the interstellar gas (*e.g.*, Pilyugin 1993; Peimbert et al. 1994a; Carigi et al. 1994). Since practically all the O is produced by Type II SNe and only a fraction of the He, it is possible to obtain increasing values of $\Delta Y/\Delta O$ just by increasing γ.

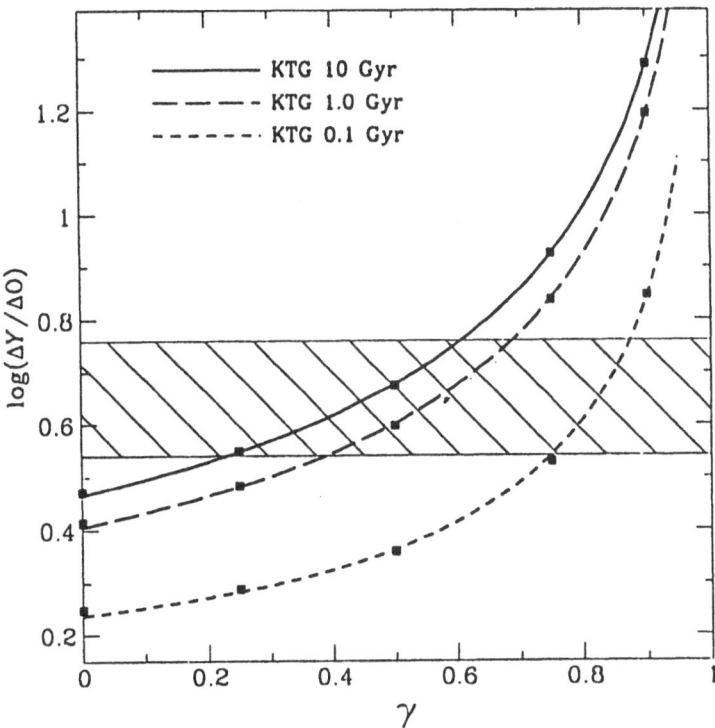

Fig. 1. $\Delta Y/\Delta Z$ versus γ for three different galactic ages (Carigi et al. 1994). The shaded band indicates the region allowed by observations. The best fit with the $\Delta Y/\Delta O$, C/O and (Z-C-O)/O constraints is obtained for the 10 Gyr models.

To explain the scatter in the Y versus O and Y versus N diagrams of irregular galaxies Pilyugin (1993) and Marconi et al. (1994) have constructed models with bursts of star formation. They are able to fit the $\Delta Y/\Delta Z$ values of all the observed irregular galaxies by a different bursting model and a different γ value for each galaxy. One of the problems with these models is that they predict higher N/O values than observed. It is not clear that the scatter in the Y versus N and Y versus O diagrams is mainly due to a different bursting activity and a different γ value for each galaxy. It is possible that most of the scatter is due to errors in the abundance determinations (*e.g.*, Olive & Steigman 1994; Esteban & Peimbert 1994); there are at least three sources of error in the abundance determinations: (*a*) errors in the observed line intensities, (*b*) different t^2 values for each object and (*c*) problems with the N ionization correction factor.

Carigi et al. (1994) find that to fit the $\Delta Y/\Delta O$ value for $t^2 = 0.035$ present in Table 1 they need $0.22 < \gamma < 0.87$ (see Figure 1) in their models they assume that the star formation rate is proportional to the gas mass. On the other hand to fit the C/O and (Z-C-O)/O constraints they require $0 \leq \gamma \leq 0.24$. They conclude that there is a marginal fit for an age of 10 Gyr at $\gamma = 0.23$. It is also possible to have a γ value closer to cero and adjust all the observational constraints if t^2 is higher than 0.035. Carigi et al. are able to fit M_{gas}/M_{total} and O/H by adopting an IMF similar to the KTG IMF for masses larger than $1 M_\odot$, but with a variable fraction of stars less massive than $1 M_\odot$ for different values of γ; the fraction of stars less massive than $1 M_\odot$ for $\gamma < 0.7$ is higher than the amount given by the KTG IMF.

6 Conclusions

The value of t^2 plays a paramount role in the determination of $\Delta Y/\Delta O$ as can be seen from Table 1. The errors in Table 1 do not include errors associated with the determination of t^2. There is increasing evidence that $t^2 \sim 0.04$ for PN, solar vicinity H II regions and O-poor H II regions. This value of t^2 decreases significatively the $\Delta Y/\Delta O$ and the $\Delta Y/\Delta Z$ ratios.

From chemical evolution models of the solar vicinity and of irregular galaxies it has been found that there is no evidence for a mass cut-off above which massive stars do not enrich the interstellar medium with heavy elements during the SN event.

From the chemical evolution of irregular galaxies it is found that $\gamma = 0.23$ marginally fits the $\Delta Y/\Delta O$, C/O and (Z-C-O)/O observational restrictions. Larger values of γ fit $\Delta Y/\Delta O$ but fail to fit C/O and (Z-C-O)/O; alternatively lower values of γ fit C/O and (Z-C-O)/O but fail to fit $\Delta Y/\Delta O$. A larger value of t^2 would produce a smaller value of $\Delta Y/\Delta O$ and a better agreement among the three observational restrictions.

A $t^2 = 0.04$ for O-poor H II regions starts affecting the third decimal place in a Y_p determination in the sense of reducing Y_p by about 0.003 relative to the Y_p value for $t^2 = 0.00$.

It is a pleasure to acknowledge several fruitful discussions on this subject with: L. Carigi, P. Colín, D. Garnett, B.E.J. Pagel, A. Sarmiento and S. Torres-Peimbert.

References

Almog, Y. & Netzer, H. 1989, *M.N.R.A.S.*, **238**, 57.

Aparicio, A., García-Pelayo, J.M. & Moles, M. 1988, *A.& A.S.S.*, **74**, 375.

Baldwin, J.A., Ferland, G.J., Martin, P.G., Corbin, M.R., Cota, S.A., Peterson, B.M. & Sletteback, A. 1991, *Ap.J.*, **374**, 580.

Bohigas, J. 1994, *A. & A.*, **288**, 617.

Brocklehurst, M. 1972, *M.N.R.A.S.*, **157**, 211.

Campbell, A. 1988, *Ap.J.*, **335**, 644.

Carigi, L., 1994, *Ap.J.*, **424**, 181.

Carigi, L., Colin, P., Peimbert, M. & Sarmiento, A. 1994, *Ap.J.*, submitted.

Cayrel de Strobel, G. 1995, these proceedings.

Chiappini, C. & Maciel, W.J. 1994, *A. & A.*, **288**, 921.

Chiosi, C. & Matteucci, F. 1984, in *Stellar Nucleosynthesis*, eds. C. Chiosi & A. Renzini, (Dordrecht: Reidel), p. 359.

Cota, S.A. & Ferland, G.J. 1988, *Ap.J.*, **326**, 884.

De Young, D.S. & Gallagher, J.S. 1990, *Ap.J.*, **356**, L15.

Dinerstein, H.L., Lester, D.F. & Werner, M.W. 1985, *Ap.J.*, **291**, 561.

Dufour, R.J., Walter, D.K., Garnett, D.R., Skillman, E.D., Peimbert, M., Torres-Peimbert, S., Terlevich, R.J., Terlevich, E., Shields, G.A. & Mathis, J.S., 1994, in preparation.

Esteban, C. & Peimbert, M. 1994, *A.&A.*, submitted.

Garnett, D.R. 1992, *A.J.*, **103**, 1330.

Garnett, D.R., Skillman, E.D., Dufour, R.J., Peimbert, M., Torres-Peimbert, S., Terlevich, R.J., Terlevich, E. & Shields, G.A. 1994, *Ap.J.*, submitted.

Giovagnoli, A. & Tosi, M. 1994, this conference.

González-Delgado, R.M., Pérez, E., Tenorio-Tagle, G., Vílchez, J.M., Terlevich, E., Terlevich, R.J., Telles, E., Rodríguez-Espinosa, J.M., Mas-Hesse, M., García-Vargas, M.L., Díaz, A.I., Cepa, J. & Castañeda, H.O. 1995, *Ap.J.*, in press.

Gruenwald, R.B. & Viegas, S.M. 1992, *Ap.J.S.*, **78**, 153.

Harrington, J.P., Seaton, M.J., Adams, S. & Lutz, J.H. 1982, *M.N.R.A.S.*, **199**, 517.

Hummer, D.G. & Storey, P.J. 1987, *M.N.R.A.S.*, **224**, 801.

Jacoby, G.H. & Ford, H.C. 1983, *Ap.J.*, **266**, 298.

Kingdon, J. & Ferland, G.J. 1994, *Ap.J.*, submitted.

Kingsburgh. R.L., & Barlow, M.J. 1994, *M.N.R.A.S.*, in press.

Kroupa, P., Tout, C.A. & Gilmore, G. 1993, *M.N.R.A.S.*, **262**, 545.

Lequeux, J. 1989, in *Evolution of Galaxies-Astronomical Observations*, eds. I. Appenzeller, H.J. Habing & P. Lena, (Springer), p. 147.

Liu, X. & Danziger, J. 1994, *M.N.R.A.S.*, **263**, 256.

Liu, X.W., Storey, P.J., Barlow, M.J. & Clegg, R.E.S. 1994 *M.N.R.A.S.*, in press.

Liu, X.W., Storey, P.J., Barlow, M.J. & Clegg, R.E.S. 1995, in Physics of Emission Lines, ed. R. Williams (Cambridge University Press), in press.

Maeder, A. 1992, *A.&A.*, **264**, 105.

Marconi, G., Matteucci, F. & Tosi, M. 1994, *M.N.R.A.S.*, **270**, 35.

Matteucci, F. & Chiosi, C. 1983, *A.&A.*, **123**, 121.

Matteucci, F. & Tosi, M. 1985, *M.N.R.A.S.*, **217**, 391.

McGaugh, S.S. 1991, *Ap.J.*, **380**, 140.

Meyer, J.P. 1985, *Ap.J.S.*, **57**, 151.

Moles, M., Aparicio, A. & Masegosa, J. 1990, *A.&A.*, **228**, 310.

Olive, K.A. & Steigman, G. 1994, preprint.

Pagel, B.E.J. 1995, these proceedings.

Pagel, B.E.J., Simonson, E.A., Terlevich, R.J. & Edmunds, M.G. 1992, *M.N.R.A.S.*, **255**, 325.

Peimbert, M., 1967, *Ap.J.*, **150**, 825.

Peimbert, M., 1971, *Bol. Obs. Tonantzintla y Tacubaya*, **6**, 29.

Peimbert, M., 1986, *P.A.S.P.*, **98**, 1057.

Peimbert, M., 1993, *Rev. Mexicana Astron. Astrof.*, **27**, 9.

Peimbert, M., Colín, P. & Sarmiento, A. 1994a, in *Violent Star Formation, from 30 Doradus to QSO's*, ed. G. Tenorio-Tagle (Cambridge University Press), in press.

Peimbert, M., Sarmiento A. & Colín P. 1994b, *Rev. Mexicana Astron. Astrof.*, in press.

Peimbert, M., Sarmiento, A. & Fierro, J. 1991, PASP, **103**, 815.

Peimbert, M. & Serrano, A. 1980, *Rev. Mexicana Astron. Astrof.*, **5**, 9.

Peimbert, M., Storey, P.J. & Torres-Peimbert, S. 1993, *Ap.J.*, **414**, 626.

Peimbert, M. & Torres-Peimbert, S. 1977, *M.N.R.A.S.*, **179**, 217.

Peimbert, M., Torres-Peimbert, S. & Ruiz, M.T. 1992, *Rev. Mexicana Astron. Astrof.*, **24**, 155.

Peimbert, M. & Ukita, M., Hasegawa, T. & Jugaku, J. 1988, *P.A.S.J.*, **40**, 581.

Pilyugin, L.S. 1993, *A.&A.*, **277**, 42.

Pogge, R.W., Owen, J.M. & Atwood, B. 1992, *Ap.J.*, **399**, 147.

Renzini, A. & Voli, M. 1981, *A.&A.*, **94**, 175.

Robbins, R.R. 1968, *Ap.J.*, **151**, 511.

Robbins, R.R. & Bernat, A.P. 1973, *Mémoires Societé Royale des Sciences de Liége, 6ᵉ série, tome V*, 263.

Rubin, R.H., Simpson, J.P., Haas, M.R. & Erickson, E.F. 1991, *Ap.J.*, **374**, 564.

Russell, S.C., Bessell, M.S. & Dopita, M.A. 1988a, in *Galactic and Extragalactic Star Formation*, eds. R.E. Pudritz & M. Fich, (Dordrecht: Kluwer), p. 106.

Russell, S.C., Bessell, M.S. & Dopita, M.A. 1988b, in *The Impact of Very High S/N Spectroscopy*, eds. G. Cayrel de Strobel & M. Spite, (Dordrecht: Kluwer), p. 545.

Salpeter, E.E. 1955, *Ap.J.*, **121**, 161.

Sasselov, D.D. & Goldwirth, D. 1994, *Ap.J.*, submitted.

Sawey, P.M.J. & Berrington, K.A. 1993, *Atomic Data and Nuclear Data Tables*, **55**, 81.

Scalo, J.M. 1986, *Fund. Cosmic Phys.*, **11**, 1.

Shaver, P.S., Kunth, D. & Kjär, K. (eds.) 1983, *Primordial Helium*, (ESO, Garching).

Skillman, E.D. & Kennicutt, R.C. 1993, *Ap.J.*, **411**, 655.

Skillman, E.D., Terlevich, R.J., Kennicutt, R.C., Garnett, D.R. & Terlevich, E. 1994, *Ap.J.*, **431**, 172.

Smits, D.P. 1994, *M.N.R.A.S.*, in press.

Torres-Peimbert, S., Peimbert, M. & Peña, M. 1990, *A.&A.*, **233**, 540.

Tosi, M. 1994, preprint.

Traat, P. 1994, this conference.

Viegas, S.M. & Clegg, R.E.S. 1994, *M.N.R.A.S.*, submitted.

Vílchez, J.M. & Pagel, B.E.J. 1988, *M.N.R.A.S.*, **231**, 257.

Primordial Helium from Extremely Metal-Poor Galaxies

Elena Terlevich[1], Evan D. Skillman[2], Roberto Terlevich[3]

[1] Institute of Astronomy, Madingley Rd., Cambridge CB3 0HA, UK
[2] Department of Astronomy, 116 Church Street S.E., University of Minnesota, Minnesota 55455, USA
[3] Royal Greenwich Observatory, Madingley Rd., Cambridge CB3 0EZ, UK

Abstract. Primordial helium determination relies heavily in the value obtained from extremely metal defficient galaxies ($Z \leq 1/20 \ Z_\odot$) but they have been almost impossible to find. In the past few years, we have been successful in discovering them. We are now embarked in a programme for obtaining very high S/N spectra of these objects with linear detectors, in order to derive He abundances to better than the 5% per object needed to constrain Big Bang models of the origin of the universe. We will discuss some results, problems encountered in this quest, and future plans.

1 Introduction

Three observational findings sustain the Big Bang model of the origin of the Universe: 1) the relic 3 K microwave background radiation; 2) the expansion of the Universe; and 3) the relative abundances of the light elements (H, D, ^{3}He, ^{4}He, Li); (see the reviews by Pagel and by Steigman, presented at this conference). Accurate measurements of these observables provide better constraints on the details of the model.

Even though an accurate measurement of the primordial light element abundance and in particular of He, is critical to our understanding of the origin of the universe it has not been the target of an observational effort comparable to that for the other cosmological observables: the microwave background and the expansion of the universe. Experimental progress has been made on the number of neutrino species and the half-life of the neutron, which together with the nucleon mass density, relate directly to the relative abundances of the light elements just after the Big Bang. Therefore, a determination of the primordial helium abundance, under the assumption of the standard hot Big Bang model of nuclesynthesis, will eventually provide a fundamental parameter in cosmology: the nucleon density in the early universe.

We will discuss in what follows the process for measuring helium abundance and hence inferring the primordial value; we will also discuss uncertainties in the method.

2 Measuring the Primordial Helium Abundance

2.1 The Method

Empirical evidence of the existence of primordial He, that is that He was present before the heavy metals were synthesized in the interior of stars, comes from the early discovery of a plateaux in the relation of the abundance by mass of He versus metals (Y vs. Z) in extragalactic HII regions and HII galaxies (Figure 1). Y tends to a finite value as Z tends to zero. Therefore one can gauge the relative contributions of primordial and post-primordial helium by measuring the abundance of a heavier element (e.g. oxygen) that was not produced in the Big Bang.

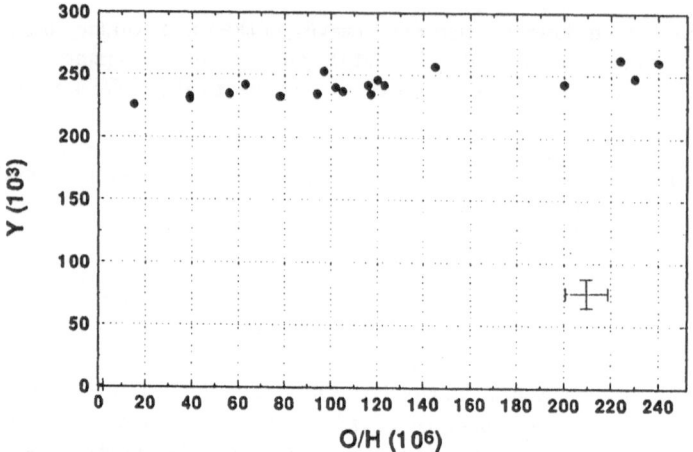

Fig. 1. Abundance by mass of He (Y) vs O/H for a sample of low metallicity giant HII regions and HII galaxies taken from the literature. Typical error bars are indicated. Notice the plateaux in the relationship.

The most accurate method of measuring helium abundance consists of observing diffuse gas which has been ionized by the ultraviolet photons of an inbedded cluster of hot stars in the so called giant HII regions. The goal is to measure the fraction of helium to hydrogen present in ionized gas of low heavy elements abundance (in practice, oxygen).

Early studies (Peimbert and Torres-Peimbert 1974, 1976; Lequeux et al. 1979) pioneered the method of extrapolating the He/H vs. O/H relation to a value of the helium abundance where the oxygen abundance is zero. The smaller the oxygen abundance, the smaller the extrapolation, the more accurate the value obtained for the primordial helium abundance.

This method has been refined by Pagel and his collaborators in a series of papers culminating with Pagel et al. (1992; P92) for which they only used best quality data of Y vs. O/H and also Y vs. N/H for high excitation, high surface brightness HII regions not including those contaminated by Wolf-Rayet stars. Figure 2 serves as an illustration.

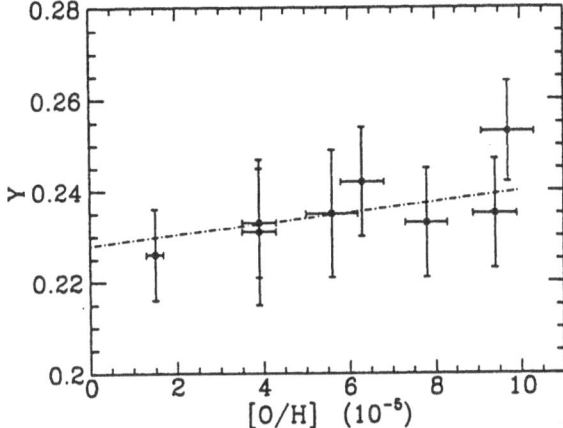

Fig. 2. Abundance by mass of He (Y) *vs* O/H for a sample of low metallicity giant HII regions and HII galaxies from P92, shown in an enlarged vertical scale.

Clearly, objects of very low oxygen abundance are crucial.

2.2 The Search for Very Low Abundance Objects

Unfortunately, several attempts in the past to find such regions have failed. Candidates have been obtained amongst the highest excitation objects from surveys capable of detecting intense bursts of star formation. As of 1985, only I Zw 18 (with an oxygen abundance of 2 % of that of the sun) was known, and no one seemed to know how to find more (Shields 1986).

It has been known for more than a decade that low mass irregular galaxies are much less evolved chemically than our own galaxy (Lequeux et al. 1979, Talent 1980); in the last few years, searches for very low metal-abundance regions in extremely low mass galaxies have been very successful (Skillman et al. 1988,1989; Terlevich et al. 1992; Izotov et al. 1992). Searching for these metal poor galaxies Stepanian & Terlevich (private communication) have compiled a list of low luminosity candidates from the Second Byurakan Survey (SBS) of ultraviolet excess galaxies, which have been studied spectroscopically by Lipovetskii and Stepanian. We obtained blue spectra of these galaxies with the 2.5m INT in order to select a subsample of the objects that are the brightest, lowest oxygen abundance and of highest excitation.

This was followed-up by high precision, high spatial and spectral resolution, high signal-to-noise spectroscopy with linear detectors with the 4.2m WHT in La Palma (Skillman et al. 1994; Skillman, Terlevich & Terlevich, in preparation). ISIS, the double spectrograph at the WHT, allows simultaneous observations in the blue (3600–5100 Å) and red (6300–6800 Å) regions of the spectrum. Figure 3 shows a typical spectrum. While we have discovered about ten objects with oxygen abundances in the range 2% to 5% of the solar value, none has an oxygen abundance lower than that of I Zw 18.

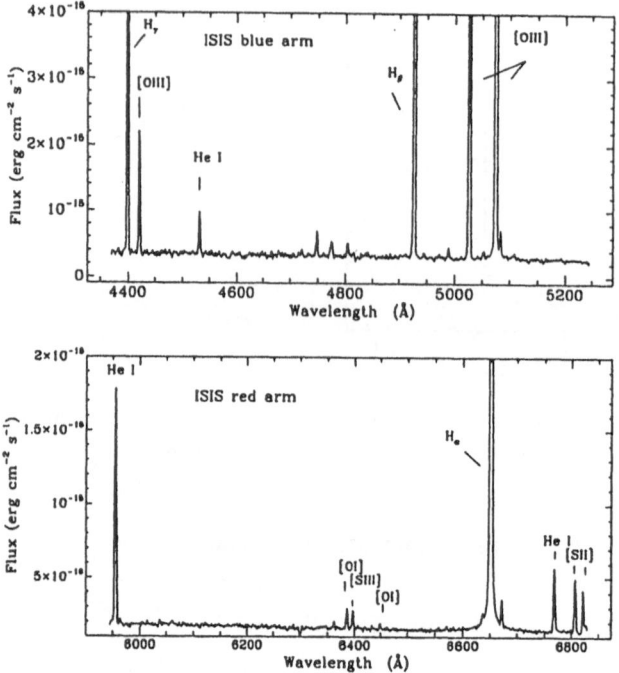

Fig. 3. ISIS blue and red spectra obtained with the WHT, of one of the newly discovered very low abundance (O/H< 5% solar) HII galaxies. Notice the temperature sensitive [OIII] λ 4363 Å line and the density sensitive [SII] λλ 6717,31 Å doublet, as well as the HeI lines λλ 4471 and 6678 Å.

3 Abundances Derivation

Ionized gaseous nebulae are fairly well understood, and their physical parameters (temperature, density, chemical abundance) can be deduced by analysing relative intensities of strong emission lines (Aller 1984; Osterbrock 1989).

Electron densities (n_e) for our sample were obtained from the ratio of the [SII] λλ6717,6731 Å or [OII] λλ 3726,3729 Å emission lines, corresponding in all cases to the low density limit ($n_e \sim 100 \, \mathrm{cm^{-3}}$).

Electron temperatures for the O^{++} zone ($T_e(O^{++})$) have been estimated following standard procedures, from all three [OIII] λλ5007, 4959, 4363Å lines observed simultaneously. We followed P92 method for estimating the low ionization zone temperatures, based on Stasińska's (1990) photoionization models.

Total oxygen abundances were then derived from the ionic abundances as discussed in detail in Skillman et al. (1994).

Assuming that hydrogen and helium lines are produced from recombination, the He^+ abundance can be determined using the strength of the HeI $\lambda\lambda$ 4471,5876,6678 Å lines relative to the reference Balmer lines. Careful attention has been given to the corrections for neutral helium and for collisional excitation (Skillman et al. 1994).

3.1 Uncertainties in the He Abundance Determination

Constraints to cosmological models need for the abundance of He to be determined to better than 5%. Uncertainties in its derivation have therefore to be well understood and corrected for or, better still, avoided whenever possible.

One of the problems that trouble the determination of the He abundance is neutral helium that, if present, remains undetected leading to an underestimation of He; the so called ionization correction factor (ICF) is very difficult to estimate. That is why we select high excitation giant HII regions for which ICF is expected to be negligible.

Stellar population from the galaxy or from the HII region itself may produce underlying absorptions in the H and He nebular lines. This problem can be particularly severe in the HeI λ4471 Å line and can lead to an underestimation of the value of He.

The strength of helium lines might be affected by fluorescent and collisional excitation effects which need to be estimated and corrected for (Clegg 1987, Peimbert & Torres-Peimbert 1987). Stratification of the nebula constitutes a potential source of error because it could lead to wrong electron density determination and hence to miscalculation of the collisional excitation effects. Disregarding them leads to overestimation of the helium abundance. Data obtained with high spatial resolution is hence of paramount importance.

The HeI λ 5876 Å line intensity might be reduced by galactic NaI absorption in λ 5890 Å in objects with redshifts between 0.002 and 0.004. This was for many years the case for II Zw 40, a typical object used for these studies.

Due to the large intensity difference between He and the H reference lines ($4471/H\gamma$ is typically about 0.1 while $6678/H\alpha$ is about 0.01) spectra to this high precision requirements, have to be obtained with linear detectors. Also, due to the weakness of the lines involved, and to the need to disentangle emission from underlying absorption, the observations need high spectral resolution.

The question of T_e fluctuations that shocks, for instance, can produce in the nebulae (e.g. Peimbert & Costero 1969), needs to be addressed. If present and ignored, these fluctuations cause O abundances to be systematically underestimated (González-Delgado et al. 1994).

Objects that show clear evidence for shocks – e.g. W-R and supernova signatures – are better not included in the quest for primordial helium (P92). Incidentally, both phenomena can also produce local pollution effects.

The uncertainty in recombination coefficients for He has been discussed recently (Smits 1991; P92); and its effect on He abundance determinations (through discrepancies for the λ 5876 and λ 6678 Å lines of up to 5% for $T_e \sim 20\,000$ K) was analysed in Skillman & Kennicutt (1993) where low O/H abundance data from P92 and our own preliminary data, are plotted using old Brocklehurst (1972) and new Smits (1991) coefficients.

Since then Smits (private communication) has discovered an error in his 1991 paper and his new results agree with Brocklehurst's (Smits 1994, in preparation). The whole subject of the recombination coefficients for helium needs revising.

Problems inherent to the observation and reduction processes, are to be considered as well, when seeking line ratios to better than 2%. At this level all aspects are suspect; from the standard stars calibration to the reduction procedures – flat fields, response curve fitting, wavelength calibration, etc – they have been found by us potentially able to introduce errors at levels much higher than 2% $r.m.s.$ (Skillman et al. 1994 presents a detailed discussion).

Is in this context that we have developed a method that involves ideally the combination of repeated observations with different systems (telescope, spectrograph, standard stars, reduction procedures, etc.) and the final result is achieved only when the different methods agree to better than 2% in the final line ratios.

4 Discussion and Conclusions

We have obtained new He abundance for very low metallicity giant HII regions (O abundances in the range 2% to 5% of the solar value). These are all high excitation regions with negligible ICF. Underlying absorption and collisional excitation effects have been carefully corrected for. He abundance has been calculated from the HeI $\lambda6678$ Å line, measured from high S/N spectra obtained with linear detectors. Figure 4 represent our addition to P92 Y $vs.$ O/H. A maximum likelihood linear fit to these new data (not showed in the figure) combined with that of P92 yields a slightly higher value of the intercept (0.238±0.003), bridging the small gap between the theoretically predicted one (Walker et al. 1991) and the best observationally derived one (P92). Both regressions – with O/H and N/H – give identical values for Y_p.

Note that the dispersion in He at a given metal abundance is larger than the uncertainty in the intercept. The dispersion in He/H may be due to systematic effects and therefore it is probably unwise to take the uncertainty in the maximum likelihood fit as equal to the uncertainty in the primordial helium abundance.

We are now in the position of determining observationally He abundances to better than 5%, by obtaining high quality, high dispersion, high spatial resolution repeated spectrophotometric observations of low metal abundance, high excitation compact HII galaxies, and avoiding the cases showing clear signs of local contamination by Wolf-Rayet stars and supernovae.

Linear detectors are a must, and the best line to use is HeI λ 6678Å, because it has a very small dependence on electron temperature and almost none on reddening correction (thanks to its proximity to Hα) and is only slightly affected by underlying absorption and collisional excitation effects. Crucial to deriving

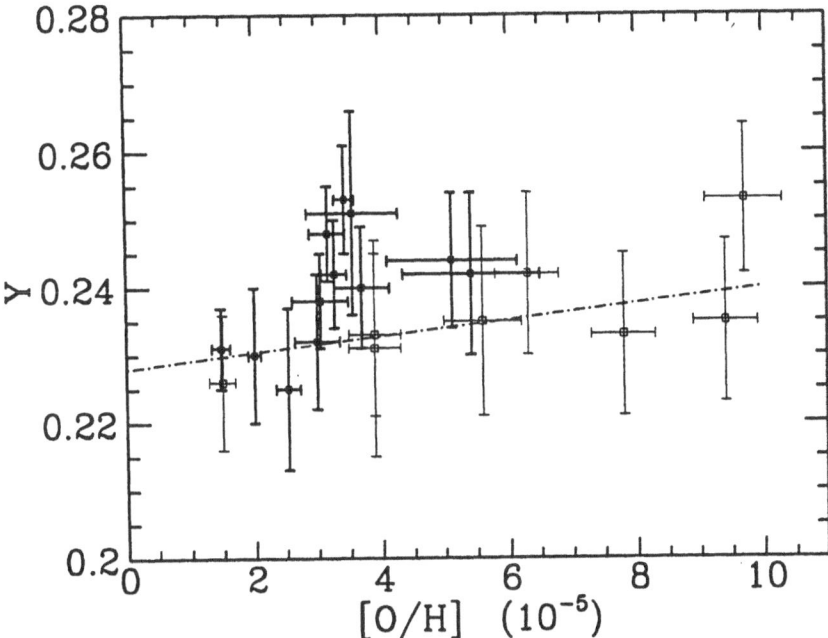

Fig. 4. Abundance by mass of Y *vs.* O/H of low metallicity giant HII regions and HII galaxies; fit and open squares from P92, filled squares and darker trace, our new sample.

primordial helium abundance now, is the adoption of the best He emissivity coefficients (as discrepancies in their theoretical derivations are larger than observational errors) and the improvement of the models to include the effects of dust, winds and shock heating. The challenge continues to be to increase the number of objects at the low metallicity end of the sample, and, most important for chemical evolution models, the search for these elusive objects less metallic than I Zw 18.

We wish to thank Lipovetskii and Stepanian for allowing us to select the objects from their unpublished survey. We acknowledge productive discussions with Bernard Pagel, Don Garnett, Max Pettini, David Hummer and Derck Smits, and financial support from a NATO grant for collaborative research.

References

Aller, L. H. (1984): *Physics of Thermal Gaseous Nebulae*, Reidel, Dordrecht, Holland.

Brocklehurst, M. (1972) *MNRAS* **157**, 211.

Clegg, R. E. S. (1987) *MNRAS* **229**, 31p.

González-Delgado, R.M., Pérez, E., Tenorio-Tagle, G., Vilchez, J.M., Terlevich, E., Terlevich, R., Telles, E., Rodríguez Espinosa, J.M., Mas-Hesse, M., García Vargas, M.L., Díaz, A.I., Cepa, J., Castañeda, H. (1994) *Astrophys. J.* in press.

Izotov, Yu. I., Lipovetsky, V. A., Guseva, N. G. & Kniazev, A. Yu. (1992) in *The Feedback of Chemical Evolution on the Stellar Content of Galaxies*, eds. D. Alloin & G. Stasińska, Publication de l'Observatoire de Paris 134–137.

Lequeux, J., Peimbert, M., Rayo, J. F., Serrano, A. & Torres-Peimbert, S. 1979 *Astr. and Astrophys.* **80**, 155.

Osterbrock, D. E. (1989) *Astrophysics of Gaseous Nebulae and Active Galactic Nuclei*, University Science Books, Mill Valley, California.

Pagel, B. E. J., Simonson, E. A., Terlevich, R. J. & Edmunds, M. G. (1992) *MNRAS* **255**, 325.

Peimbert, M. & Costero, R. (1969) *Bol. Obs. Ton. y Tac.* **5**, 3.

Peimbert, M. & Torres-Peimbert, S. (1974) *Astrophys. J.* **193**, 327.

Peimbert, M. & Torres-Peimbert, S. (1976) *Astrophys. J.* **203**, 581.

Peimbert, M. & Torres-Peimbert, S. (1987) *Rev. Mex. Astron. Astrof.* **15**, 117.

Shields, G. A. (1986) *Publ. astr. Soc. Pacif.* **98**, 1072.

Skillman, E. D. & Kennicutt, R. C. (1993) *Astrophys. J.* **411**, 655.

Skillman, E. D., Kennicutt, R. C. & Hodge, P. W. (1989) *Astrophys. J.* **347**, 875.

Skillman, E. D., Melnick, J., Terlevich, R. & Moles, M. (1988) *Astr. and Astrophys.* **196**, 31.

Skillman, E. D., Terlevich, R. J., Kennicutt, R. C., Garnett, D. R. & Terlevich, E. (1994) *Astrophys. J.*, 431, 172.

Skillman, E. D., Terlevich, R. & Melnick, J. (1989) *MNRAS* **240**, 563.

Smits, D. P. (1991) *MNRAS* **251**, 316.

Stasinska, G. (1990) *Astr. Astrophys. Suppl.* **83**, 501.

Talent, D. L. (1980) Ph. D. Thesis, Rice University.

Terlevich, E., Terlevich, R., Skillman, E., Stepanian, J. & Lipovetskii, V. (1992) *Elements and the Cosmos*, eds. M.G. Edmunds & Terlevich, R.J., C.U.P. 21–27.

Walker, T. P., Steigman, G., Schramm, D. N., Olive, K. A. & Kang, H.-S. (1991) *Astrophys. J.* **376**, 51.

Δ Y/ Δ Z from Fundamental Stellar Parameters

G. Cayrel de Strobel, F. Crifo

Observatoire de Paris, section de Meudon, 92195 Meudon Cedex, France

Abstract. The position of unevolved low mass stars in the (log T_{eff}, M_{bol}) diagram, having relatively well determined bolometric magnitudes, effective temperatures and metal abundances is discussed in terms of helium and metal content of the objects. It is shown that the narrow width of the observational main sequence implies the relation: Δ Y/Δ Z = 5 $\pm$ 2.

1 Introduction

The preceding papers by Terlevich, Pagel, and Peimbert have all stressed the paramount importance to astrophysics of a precise determination of the two helium abundance parameters: Yp, the primordial helium mass fraction, and Δ Y/Δ Z, the amount of additional helium ejected by dying stars relative to the ejected amount Z of heavy elements in the Galaxy or in other galaxies. Cosmology will certainly take advantage of a new determination of Y_p having its third decimal digit known with great precision, (Pagel & Kazlauskas 1994), as well as our knowledge of the structure of the Galaxy will improve by a more firm value of Δ Y/Δ Z, no more fluctuating between 2 and 5 among different authors or from different methods, as is currently the case. The paper written by Bernard Pagel (1994) for the IAU Symposium 166 on "Fine structure in the Main Sequence: primordial helium and Δ Y/Δ Z" is a good introduction to this paper. We advise the reader to consult that paper before reading the present one.

In what follows we shall shortly describe the method used by Marie Noel Perrin in determining Δ Y/Δ Z by means of a sample of unevolved F, G and K disk stars (Perrin 1975; Perrin et al. 1977), and then discuss our new results, obtained with the help of a very similar sample of stars to that of M. N. Perrin, but based on observational material derived not from photographic, but from very precise solid state spectroscopic high resolution observations.

2 The Intrinsic Width of the Zero Age Main Sequence

When it is possible to directly determine by spectroscopy the abundance ratio He/H in early type stars, in low mass solar type stars, and cooler, the spectra of which do not contain He-lines, this ratio can be determined only indirectly. Such an indirect method was employed by M.N. Perrin. In a star model if one varies simultaneously and in the same way both Y and Z, one gets effects of opposite sign and therefore it is possible to get the same ZAMS for a He-rich and a metal-rich model and for a He-poor and metal-poor model, yet having different masses at the same location in the HR diagram. This dependence of the ZAMS with respect to Y and to Z enables to estimate the He abundance of solar and later type stars from their position in the observational (log T_{eff}, M_{bol}) diagram, provided that they are not yet evolved from their original main sequence, and that their metal abundances are well known. This helium estimation in low mass dwarfs requests that their luminosities and effective temperatures are very accurately determined. In the mid- seventies, M.N. Perrin (1975) has interpreted with the help of theoretical evolutionary tracks, computed by Hejlesen, and afterwards published in Hejlesen (1980), an observational (log T_{eff}, M_{bol}) diagram composed of unevolved low-mass disk stars of different metal contents. The chosen stars had to have reliable absolute bolometric magnitudes, precise bolometric corrections, effective temperatures and metal abundances from high resolution detailed spectroscopic analyses. This sample of stars has been carefully chosen in one of the first editions of the [Fe/H] - Catalogue (Morel et al. 1975). The author has shown that there was some evidence that the narrow width (small dispersion) of the observational main sequence, limited to disk stars having well known metal abundances between : $-0.60 < $ [Fe/H] $ < +0.50$, and effective temperatures smaller than 5500 K, is not well explained by a variation of the metal to hydrogen abundance alone. For that it seemed necessary to assume in these stars the existence of a simultaneous variation of the He-content which acts in the opposite way than the metal content, thus narrowing the observational ZAMS. The narrow width of the observational main sequence determined by M.N. Perrin implied the relation : $\Delta Y / \Delta Z \approx 5 \pm 3$.

Eighteen years later, we are interpreting again the observational (log T_{eff}, M_{bol}) diagram, constituted by a sample of nearby unevolved, low mass, disk stars. Some of the stars have been chosen in the "1991 edition of the [Fe/H] Catalogue" (Cayrel de Strobel et al, 1992), and some have been observed and analysed by Cayrel de Strobel (unpublished). Note that a few of these stars have known masses, being members of multiple visual systems. The inclusion of binaries in the sample is interesting because they represent small fragments of ZAMS's or main sequences and can be used for calibrating theoretical ZAMS's. The new observational material is essentially based on high resolution, high S/N Reticon and CCD spectra taken at ESO, CFHT and OHP. The results are more accurate by almost an order of magnitude than those based on the 1975 photographic spectra. The reduction of data has been refined, and the spectra of the stars have been interpreted with theoretical line computations using adequate grids of model atmospheres. Three grids of up-to-date internal structure models have been used for interpreting the observations: Noels 1993 (private communication), Chieffi and Straniero 1993 (private communication), Morel (1994). We show in fig.1 the effect of varying Y using the code of Morel (1994), computed using the OPAL opacities (Iglesias et al. 1992). In a forthcoming paper, to appear in A&A, the complete grid will be given.

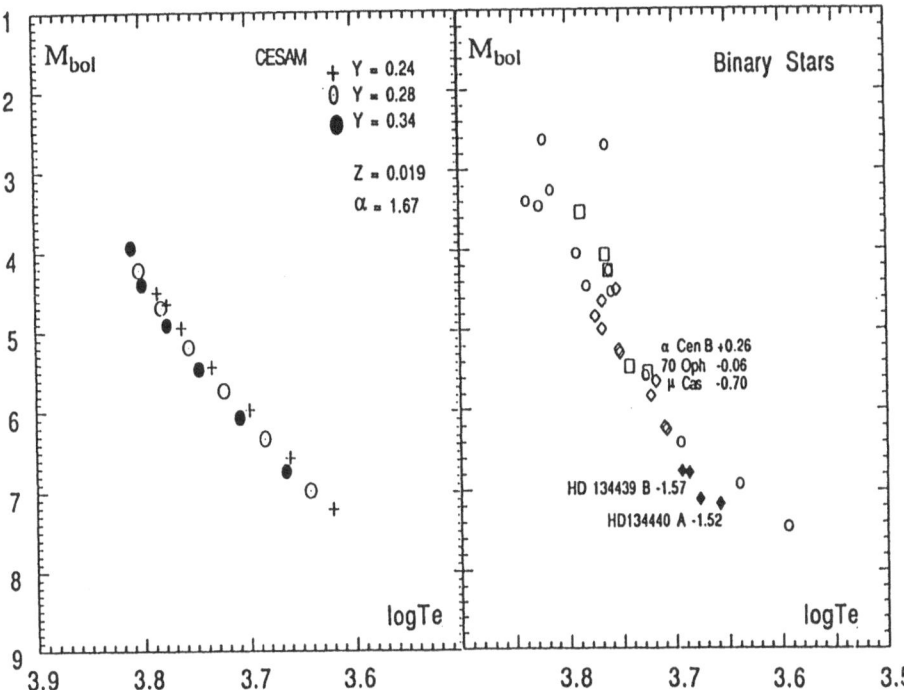

Fig. 1. (left): Example of three ZAMS computed with the CESAM (Code Evolution Stellaire Adaptative et Modulaire) code (P. Morel 1994) keeping constant Z and varying Y and taking: $\alpha = 1.67$

Fig. 2. (right): Observational HR (logTeff, M_{bol}) diagram for the visual binary stars of our sample. The symbols according to their [Fe/H] ratio are the following: rectangles = metal enriched binaries $+0.45 > [Fe/H] > +0.12$, circles = metal normal binaries $+0.12 > [Fe/H] > -0.12$, diamond shaped = metal deficient binaries $-0.12 > [Fe/H] > -0.80$. The filled diamond shaped symbols represent Halo binaries: these very metal-deficient stars do not follow the linear relationship between Y and Z (section III).(Some of the more massive stars are already evolved, and their position has not been taken into account in determining the relationship.)

Helium is correlated with the heavy element abundance Z by a cosmological concomitance relationship, which in its simplest form can be written:

$Y = (\Delta Y/\Delta Z) Z + Y_p,$

where $\Delta Y/\Delta Z$ is the relative helium to heavier element enrichment and Y_p, the primordial helium abundance. If the observational ZAMSs, constructed with unevolved low mass stars of very different metal content, **do not segregate** (according to their metal content), it seems necessary to assume the **existence** in these stars of a simultaneous variation of the helium content which acts in the opposite way than the metal content on the position of the observational ZAMS. We have studied **eight** observational ZAMSs composed of nearby F, G, and K stars. **Two** of them are metal-rich, **one** metal-normal, and **five** metal-poor. Some of these stars belong to binary or multiple systems. According to J. Fernandes et al. (1994):

$\Delta M_{bol} = 2.5 \, \Delta Y - 14.2 \, \Delta Z$ or, $\Delta M_{bol} = 2.5 \, (\Delta Y - 5.7 \, \Delta Z)$.

- If, $\Delta Y = 5.7 \, \Delta Z$, there is a single ZAMS.
- If, $\Delta Y = 2.5 \, \Delta Z$, as suggested by the global enrichment from: $(Y = Y_p = 0.23, Z \approx 0)$ to $(Y = Y_o = 0.28, Z \approx 0.02)$, the ZAMSs should be different for metal-rich and metal-poor stars. Following our observations this does not seem to be the case (see fig. 2).

The above linear relationship between Y and Z is valid only for : $0.01 < Z < 0.04$. It is not applicable between ΔM_{bol} and Z, when Z varies by several orders of magnitude. For example, $\Delta M_{bol} / \Delta Z$ becomes very large for small values of Z ($\approx 10^{-3}$). It is for this reason that subdwarfs ($Z \approx 0.0006$) segregate in the ($\log T_{eff}$, M_{bol}) plane. For them Y varies insignificantly around $Y=Y_p$, but their M_{bol} varies by several tenths of magnitude between : $Z=10^{-3}$ and 10^{-4}.

3   Conclusions

From figure 2, which displays the ($\log T_{eff}$, M_{bol}) diagram for all the binary stars we have observed and analyzed ourselves, we can conclude that:

$Y = 0.32$ for α Cen ($Z = 0.033$), $Y = 0.28$ for 70 Oph ($Z = 0.018$), $Y = 0.25$ for μ Cas ($Z = 0.004$), in spite of their proximity in the observational ($\log T_{eff}$, M_{bol}) diagram. These conclusions will become stronger, or weaker, when the Hipparcos results become available.

References

Cayrel de Strobel, G. Hauck, B., François, P., Thevenin, F., Friel, E., Mermilliod, M., Borde, S.: 1992, A&A Suppl. **95**, 273

Fernandes, J., Lebreton, Y., Baglin,A.: 1994 A&A in press

Hejlesen, P.M.: 1980, A&A **84**, 135

Iglesias, C.A., Rogers, F.J., Wilson, B.J.: 1992, Ap.J. **397**, 717

Morel, P.: 1994, A&A in press

Morel, M., Bentolila, C., Cayrel de Strobel, G., Hauck, B.: 1975 in IAU Symp. 72, ed. B. Hauck and P.C. Keenan, Reidel

Pagel, B.E.J.: 1994, in IAU Symp 166, ed. J. Kovalevsky, Kluwer

Pagel, B.E.J., Kazlauskas, A.: 1994, M.N.R.A.S., in press

Perrin, M. N.: 1975,Thesis, Obs.Paris

Perrin, M. N., Hejlesen, P.M., Cayrel de Strobel, G., Cayrel, R. : 1977, A&A **54**, 779

On the $\Delta Y/\Delta Z$ Determination

Leonid S. Pilyugin

Main Astronomical Observatory, Ukrainian Academy of Sciences, Goloseevo, 252022 Kiev, Ukraine

Abstract. An origin of the scatter of the points in the Y versus Z_O diagram is briefly discussed. It has been shown that this scatter can be caused by the fact that the ratio of relative helium to oxygen enrichments in any galaxy is defined not only by the stellar yields of helium and oxygen but also by the efficiencies of enriched galactic winds taking place in that galaxy. In addition, the giant HII regions can have a temporary and local reduction of the helium to oxygen abundances ratio due to the self-enrichment of star formation region.

1 Introduction

For a few HII regions, Peimbert and Torres-Peimbert (1974,1976) found that the helium mass fraction, Y, has a tendency to increase together with the heavy-element mass fraction, Z. From this, they proposed to derive the pregalactic helium abundance, Y_P, from an empirical correlation between the abundances of helium and heavy elements observed in HII regions with different metallicities. It has often been assumed that Y and Z are linearly related

$$Y = Y_P + Z\,(\Delta Y/\Delta Z). \tag{1}$$

Almost always the Z values have been based on the O abundances alone, Z being taken as proportional to the oxygen abundance, $Z \approx 2\,Z_O \approx 23\,(O/H)$, where Z_O is the oxygen mass fraction. Eq.(1) can be replaced by

$$Y = Y_P + Z_O\,(\Delta Y/\Delta Z_O) = Y_P + O/H\,(\Delta Y/\Delta(O/H)). \tag{2}$$

Within the framework of the simple closed-box model of chemical evolution the ratio $\Delta Y/\Delta Z_O$ of relative helium to oxygen enrichments is defined by the stellar yields of helium, P_Y, and oxygen, P_O,

$$\Delta Y/\Delta Z_O = P_Y/P_O. \tag{3}$$

Both the pregalactic helium abundance, Y_P, and the ratio $\Delta Y/\Delta Z_O$ of relative helium to oxygen enrichments can be derived from the empirical relationship between Y and Z_O observed in HII regions with different metallicities. The scatter of the points in the Y versus Z_O diagram is strong, Fig.1. To clarify the origin of this scatter is very important. If the scatter of the points is due only to uncertainties in the element abundance determinations then all the galaxies evolve along the single track in the Y versus Z_O diagram. If the part of this scatter is real cosmic scatter then the evolutionary tracks in the Y versus Z_O plot can differ from galaxy to galaxy.

The second case seems to be more realistic.

2 The Possible Reasons of Scatter

The selective element loss via enriched galactic winds is one of the possible reasons the scatter in the Y versus Z_O diagram (Pilyugin 1992a, 1993, 1994). The bulk of oxygen is produced by massive stars. The fraction of newly-produced oxygen produced by massive stars of star generation, f_O, is close to 1, $f_O \approx 1$. The contribution of massive stars to the amount of helium produced by all the stars of star generation is much less, $f_Y \approx 0.55$ (Pilyugin 1994). Due to the type II supernova driven galactic wind both some mass of ambient interstellar matter (ordinary galactic wind) and the fraction λ_E of supernova ejecta (enriched galactic wind) leave the galaxy. Since a fraction of newly produced element leaves the galaxy instead of being mixed to the interstellar medium we must use the effective yield of ith element, P_i^*,

$$P_i^* = P_i \left(1 - \lambda_E f_i\right), \tag{4}$$

in model of chemical evolution of galaxies. In this case the ratio $\Delta Y/\Delta Z_O$ of relative helium to oxygen enrichments can be expressed (Pilyugin, 1994)

$$\frac{\Delta Y}{\Delta Z_O} = \frac{P_Y}{P_O} \frac{1 - \lambda_E f_Y}{1 - \lambda_E f_O}. \tag{5}$$

A similar relationship has been also derived by Peimbert et al. (1994).

Thus, the $\Delta Y/\Delta Z_O$ observed in any galaxy is defined not only by the stellar yields of helium and oxygen but also by the efficiencies of enriched galactic winds taking place in that galaxy. Since the efficiency of enriched galactic wind, λ_E, can vary from galaxy to galaxy the various galaxies can have different values of the $\Delta Y/\Delta Z_O$. Campbell (1992) and Marconi et al. (1994) have also concluded that the observed dispersion in the Y versus Z_O diagram can be explained by chemical evolution models in which a fraction of the newly-synthesized elements is lost through type II supernova driven wind.

The evolutionary tracks of galaxies with three different values of efficiencies of enriched galactic winds are shown in Fig.1 by solid lines. The effect of increasing of efficiencies of enriched galactic winds is to make evolutionary track of galaxy steeper. The high values of λ_E which is necessary to explain the entire scatter do not seem to be very credible. However, it must be taken into account that a part of scatter is certainly due to the uncertainties in the element abundances.

The occurence of the selective element loss via type II supernova driven galactic wind in irregular galaxies has received further support through the recent study of the oxygen and iron abundances in the Large Magellanic Cloud. Luck and Lambert (1992) have shown that oxygen is less abundant in the Large Magellanic Cloud than in Galactic stars of the same iron abundance. The conclusion that the relative contribution of type Ia supernova to the enrichment of the Large Magellanic Cloud is larger than for the Galactic disk has been made by de Freitas Pacheco et al. (1993).

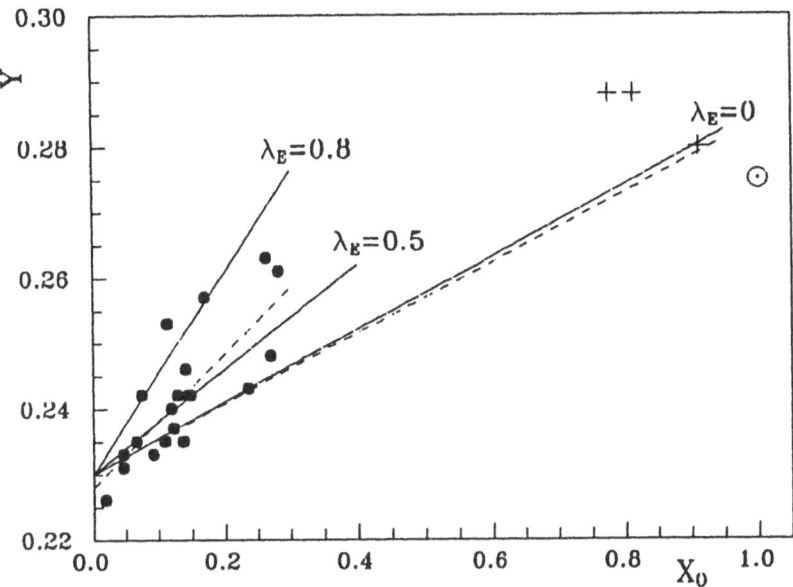

Fig. 1. Y as a function of $X_O = (O/H)/(O/H)_\odot$ for the data from Pagel et al. (1992) (fulled circles) and from Peimbert (1993) (crosses). Dashed lines correspond to $\Delta Y/\Delta(O/H) = 122$ (Pagel et al. 1992) and to $\Delta Y/\Delta(O/H) = 64$ (Peimbert 1993). Solid lines are models with $\lambda_E = 0$, 0.5, 0.8.

The self-enrichment of star forming regions is another possible reason for the observed scatter in the Y versus Z_O plot (Pilyugin 1992a,b, 1993). The star forming regions can have a temporary and local excess of oxygen and helium due to pollution by supernova ejecta. On a short time scale the oxygen abundance in star formation region (and associated H II region) is enhanced more than the helium abundance due to the fact that the part of helium is produced by stars living longer than those producing the bulk of oxygen. It leads to a decrease of $\Delta Y/\Delta Z_O$ derived from this H II region.

In the last few years the observational evidences of the self-enrichment of star formation regions have been obtained. Cunha and Lambert (1992) have found that the stars in the youngest subgroup in the Orion star formation region are enriched in oxygen relative to the abundance of two oldest subgroups. Kunth et al. (1994) have obtained that most of the heavy elements in the H II region in I Zw 18 have been produced in the present burst of star formation.

3 Conclusions

The $\Delta Y/\Delta Z_O$ observed in any galaxy is defined not only by the stellar yields of helium and oxygen but also by the efficiencies of enriched galactic winds taking place in that galaxy. Since the efficiency of enriched galactic wind, λ_E, can vary from galaxy to galaxy the various galaxies can have different values of the $\Delta Y/\Delta Z_O$. The temporary deviation of H II region from real evolutionary track in the Y versus Z_O diagram can be caused by the self-enrichment of H II region. The underlying reason both of change of the evolutionary track of galaxy in the Y versus Z_O diagram under the action of the enriched galactic wind and of temporary deviation of H II region from real evolutionary tack under the action of the self-enrichment of H II region is that part of helium but not oxygen is produced by long-lived stars.

Acknowledgements. I wish to thank the Organizing Committee of the ESO/EIPC workshop and the ESO C&EE Programme for the invitation to participate in this conference and for the financial support (Grant D-04-004).

References

Campbell, A. (1992): ApJ **401** 157

Cunha, K., Lambert, D.L. (1992): ApJ **399** 586

de Freitas Pacheco, J.A., Costa, R.D.D., Maciel, W.J. (1993): A&A **279** 567

Kunth, D., Lequeux, J., Sargent, W.L.W., Viallefond, F. (1994): A&A **282** 709

Luck, R.E., Lambert, D.L. (1992): ApJS **79** 303

Marconi, G., Matteucci, F., Tosi, M. (1994): MNRAS *in press*

Pagel, B.E.J., Simonson, E.A., Terlevich, R.J., Edmunds, M.G. (1992): MNRAS **255** 325

Peimbert, M. (1993): Rev. Mexicana Astron. **27** 9

Peimbert, M., Colin, P., Sarmiento, A. (1994): in *Violent star formation from 30 Doradus to QSO's* (ed. G. Tenorio-Tagle) Cambridge University Press, *in press*

Peimbert, M., Torres-Peimbert, S. (1974): ApJ **193** 327

Peimbert, M., Torres-Peimbert, S. (1976): ApJ **203** 581

Pilyugin, L.S. (1992a): in *The feedback of chemical evolution on the stellar content of galaxies* (ed. D. Alloin, G. Stasinska) Meudon, p.153

Pilyugin, L.S. (1992b): A&A **260** 58

Pilyugin, L.S. (1993): A&A **277** 42

Pilyugin, L.S. (1994): A&A *in press*

$\Delta Y/\Delta Z$ - No Real Controversy Between Theory and Observations

Peeter Traat

NORDITA, Blegdamsvej 17, DK-2100 Copenhagen Ø, Denmark
Institute of Astrophys. and Atm. Phys., EE2444 Tõravere, Estonia

For several years a controversy has been claimed to exist between observational determinations of $\Delta Y/\Delta Z$ and theoretically computed values from nucleosynthesis in stellar models. More specifically, since neither observational nor theoretical estimates are very numerous, the essentially controversial results are the value $\Delta Y/\Delta Z \sim 6$ obtained by Pagel et al. (1992) from observations of emission lines in HII galaxies and the theoretical results following from the extensive stellar model grids by Maeder et al.: $\Delta Y/\Delta Z = 1.1 \div 1.8$ for metal-deficient systems $Z = 0.001$ or $\Delta Y/\Delta Z = 1.5 \div 2.2$ when $Z = 0.02$ (Maeder 1992). Still another series of observational $\Delta Y/\Delta Z$ determinations, published by Peimbert et al. over the years, converges to a value $\Delta Y/\Delta Z = 2.4 \div 2.5$ (e.g. the latest publications: Peimbert (1993) – $\Delta Y/\Delta Z =2.5$, Torres-Peimbert et al. (1994) – $\Delta Y/\Delta Z =2.41$). This value is based on selected uncontaminated HII regions in the Galaxy and nearby dwarfs with the correction of abundances from collisionally excited lines for temperature variations inside nebulae. It also agrees well with the Sun's value of $\Delta Y/\Delta Z \sim 2.3$.

The discrepancy between all existing evidence on $\Delta Y/\Delta Z$ somewhat weakens with the correction by Pagel et al. for the possible lock-up of oxygen on dust grains and for the influence of possible dust absorption on the derived He abundances - the $\Delta Y/\Delta Z$ range estimated by the authors for this case extending from 3 to 6, with the preferred value $\Delta Y/\Delta Z = 4 \div 5$. But even with these corrections it still exceeds Maeder's values by a factor $2 \div 4$. This circumstance led Maeder to suggest that there may be an upper cut-off in stellar masses contributing to heavy-element synthesis with the more massive stars swallowing all their products in the supernova event as black holes.

However, yet another fundamental problem is buried behind the diverging $\Delta Y/\Delta Z$ evidence. Namely, if taken at face value, both the theoretical and observational data mentioned indicate different $\Delta Y/\Delta Z$ slopes for object samples with different compositions, so questioning the existence of a universal linear regression between Y and Z, first claimed by Peimbert & Torres-Peimbert (1974).

To investigate the apparent controversy between theoretical nucleosynthetic predictions and observations more closely, I have used Maeder's (1992) stellar nucleosynthesis data to recompute wide tables of $\Delta Y/\Delta Z$, $\Delta Y/\Delta O$ and $\Delta C/\Delta O$ as functions of the IMF slope x and the upper cut-off mass of black holes, M_{cut}, since Maeder's original computations included only two values of the initial mass function slope, $x = 1.35$ and $x = 1.7$. These tables contain also all of Maeder's separate cases A,B,C,D,F, with no black-hole cut-off in case A and $M_{cut} = 27.5$,

22.5, 17.5 and 11.6 $M_\odot$ for cases B÷F, respectively. For stars with masses above M_{cut}, elements lost by stellar wind were taken into account, but not the products from the final SN explosion. As to the nucleosynthetic products themselves, there is, of course, some additional uncertainty in Maeder's data from the use of presupernova models by Woosley and Weaver (1986) for the estimates of element contributions from the final explosion, because these models had been evolved without mass loss. As testified by some new models evolved with mass loss (Woosley et al. 1993), this really has a great influence on the final configuration of stars with masses above $\sim 30 \div 40$ $M_\odot$. Originally I hoped to use in computations also these new SN models to get alternative/improved estimates of the final nucleosynthesis, but the published data set is not sufficient for that purpose (yet).

The results for my computed $\Delta Y/\Delta Z$ are displayed in table 1. The general behaviour is as expected, with $\Delta Y/\Delta Z$ values increasing with increasing x and decreasing M_{cut}, since in both cases the relative content in the stellar population of massive stars, the main Z-contributors, becomes smaller. It is evident from the table, that the canonical power-law IMF-s with values $x = 1.35 \div 1.7$ really do fall somewhat short of giving the observed $\Delta Y/\Delta Z$ values without some reduction of Z production through locking into black holes, although the discrepancy is not so great with Peimbert's observational value, being of the order of 2 and, in fact, nearly reaching the observed value in the case of standard mass loss and $Z = .02$, if $x = 1.7$. The difference with the results of Pagel et al. in these cases remains large.

The last column of the table, labeled by Φ_1, however, is the most intriguing one. It is NOT computed with a power-law IMF, but with the lognormal IMF derived by Miller and Scalo (1979) for the solar cylinder. The values obtained with that IMF are practically coincident for both compositions $Z = .001$ and $Z = .02$ ($\Delta Y/\Delta Z$ =2.49 and 2.56, respectively) and also most perfectly fitting the observational values $\Delta Y/\Delta Z = 2.4 \div 2.5$ by Peimbert et al. (1993,1994) WITHOUT ANY FUTURE NEED FOR ARTIFICIAL BLACK HOLES. The case of reduced stellar winds with $Z = .02$, although giving the somewhat smaller $\Delta Y/\Delta Z$ value 2.16, also fits the observational results of Peimbert within error limits (or, in fact, also coincides with the value derived from SMC HII regions and M17 by Torres-Peimbert et al. (1994), $\Delta Y/\Delta Z = 2.23 \pm 0.5$, for $t^2 = 0.00$). As to the overproduction of carbon in Maeder models with low metallicity, claimed by Prantzos (1994) from comparison with carbon–to–oxygen abundance ratios measured in field halo dwarf stars by Tomkin et al. (1992), our computations with the Miller-Scalo IMF and the IMF upper mass value 120 $M_\odot$ (i.e. without black holes) gave $\Delta C/\Delta O = 0.19$ ([C/O]= —0.7), in good accordance with the observational constraint $[C/O]_{HALO} = -0.6 \pm 0.2$. This aspect together with other nucleosynthetic results will be considered in more detail in a forthcoming paper (Traat 1995).

The main conclusions, obtained from my computations and concerning $\Delta Y/\Delta Z$, can be summarized as follows:

Table 1. $\Delta Y/\Delta Z$ ratio and its dependence on the assumed IMF high-mass cut-off M_{cut} of black holes, for coeval stellar populations with different IMF slopes and metallicities. Helium production is accounted for in the entire mass interval $1 \div 120$ $M_\odot$.

$M_{cut}\backslash x$	0.6	1.	1.2	1.35	1.5	1.7	2.	2.4	3.	Φ_1
Z=.001, nucleosynthesis and mass loss from Maeder (1992)										
120.00	0.56	0.77	0.95	1.13	1.36	1.80	2.88	5.79	18.44	2.49
80.00	0.76	0.98	1.17	1.36	1.61	2.07	3.18	6.15	18.91	2.63
50.00	1.11	1.35	1.56	1.77	2.05	2.56	3.78	6.95	20.22	2.96
35.00	1.54	1.80	2.04	2.30	2.62	3.22	4.61	8.14	22.45	3.47
27.50	2.01	2.28	2.56	2.85	3.24	3.93	5.52	9.49	25.08	4.06
22.50	2.60	2.88	3.22	3.56	4.02	4.83	6.69	11.21	28.49	4.82
17.50	4.03	4.28	4.71	5.17	5.78	6.86	9.29	15.08	36.35	6.55
15.00	5.71	5.89	6.42	7.01	7.78	9.16	12.25	19.47	45.25	8.54
11.60	12.32	12.05	12.93	13.99	15.41	17.96	23.60	36.47	80.40	16.14
Z=.02, nucleosynthesis and mass loss from Maeder (1992)										
120.00	1.15	1.24	1.36	1.51	1.71	2.11	3.15	6.11	19.56	2.56
80.00	1.18	1.26	1.38	1.52	1.72	2.12	3.17	6.13	19.59	2.57
50.00	1.23	1.32	1.44	1.58	1.79	2.20	3.26	6.26	19.88	2.63
35.00	1.29	1.39	1.51	1.66	1.88	2.30	3.40	6.50	20.52	2.75
27.50	1.39	1.51	1.66	1.83	2.07	2.55	3.77	7.17	22.28	3.05
22.50	1.51	1.68	1.87	2.08	2.38	2.96	4.43	8.46	25.86	3.63
17.50	1.60	1.83	2.07	2.34	2.72	3.47	5.37	10.59	32.65	4.52
15.00	1.63	1.89	2.17	2.48	2.93	3.80	6.09	12.50	39.73	5.25
11.60	1.64	1.93	2.25	2.62	3.16	4.27	7.39	17.03	62.55	6.71
Z=.02, reduced stellar wind models from Maeder (1992)										
120.00	0.47	0.63	0.78	0.93	1.14	1.53	2.53	5.33	18.24	2.16
80.00	0.51	0.67	0.82	0.97	1.18	1.58	2.58	5.40	18.34	2.18
50.00	0.59	0.76	0.91	1.07	1.29	1.70	2.74	5.65	18.86	2.30
35.00	0.66	0.86	1.03	1.21	1.44	1.89	3.01	6.11	20.01	2.51
27.50	0.71	0.93	1.13	1.33	1.60	2.11	3.38	6.83	21.98	2.83
22.50	0.74	1.00	1.23	1.47	1.79	2.40	3.91	8.00	25.45	3.33
17.50	0.75	1.05	1.32	1.60	2.00	2.76	4.69	9.96	32.20	4.11
15.00	0.75	1.06	1.34	1.66	2.10	2.98	5.25	11.69	39.18	4.75
11.60	0.73	1.03	1.33	1.68	2.19	3.24	6.21	15.64	61.33	5.94

1. The true value of $\Delta Y/\Delta Z$ is probably around 2.5, with some allowance for observational/theoretical errors.

2. Theoretical nucleosynthetic results for two Maeder stellar model sets with normal and metal-deficient compositions studied here converge to the practically unique $\Delta Y/\Delta Z$ value of ~ 2.5 and perfectly fit the observations of Peimbert, if the correct IMF for the solar neighbourhood is used.

3. The basic contradiction is NOT the contradiction between the theoretical and observational values of $\Delta Y / \Delta Z$, but the discrepancy between the observationally derived values by Peimbert and by Pagel et al.

4. For a combination of reasons (contamination by WR stars, dust absorption, uniform electron temperature approximation in deriving abundances; cf. also Pagel 1994) the $\Delta Y / \Delta Z$ value published by Pagel et al. (1992) may be an overestimate. I would like to stress here first the possible WR contamination, since although the formal value of $\Delta Y / \Delta Z$ found for the whole object sample is 6.1, the regression for 9 objects definitely free from WR influence derived in the paper has much shallower slope ($\Delta Y / \Delta Z = 4.5$ assuming $Z = 20(O/H)$, or 3.9 if $Z = 23(O/H)$ as in the Sun). These values can be rather easily further reduced to 3 and even lower by corrections due to temperature variations.

5. Should the high values of $\Delta Y / \Delta Z$ for metal-poor galaxies be later confirmed, there might exist an alternative explanation of these results by having a steeper IMF with effective slope at the high-mass end $x \sim 2 \div 2.4$ and a shortage of massive stars relative to galaxies having normal composition now. Since there is no obvious reason why the IMF has to be unique and the same in the galaxies with low and high volume densities of matter, this would also help to explain the very slow chemical evolution of dwarf galaxies.

6. The power-law simplification of the IMF, although very simple in use, should be avoided in practical applications, since its deficiencies (infinite mass in low-mass stars when $m \to 0$, the deviation from the observed mass spectrum in high-mass stars) are well known. In this case it seems to have been the main source of previous erronous conclusions on the incapability of theory to match observations.

The author wishes to express his sincere thanks to Prof. B.Pagel for multiple stays and discussions at Nordita, Denmark, where this work was carried out.

References

Maeder, A.: Astron. Astrophys. **264** 105 (1992)

Miller, G.E., Scalo, J.: ApJ Suppl. **41** 513 (1979)

Pagel, B.E.J., Simonson, E.A., Terlevich, R.J., Edmunds, M.G.: M.N.R.A.S. **255** 325 (1992)

Pagel, B.E.J.: Lect. Notes Phys., *in press* (*Nordita Prep.* No. 94/33A) (1994)

Peimbert, M., Torres-Peimbert, S.: ApJ **193** 327 (1974)

Peimbert, M.: Rev. Mex. Astr. Astrofis. **27** 9 (1993)

Prantzos, N.: Astron. Astrophys. **284** 477 (1994)

Tomkin, J., Lemke, M., Lambert, D., Sneden, C.: AJ **104** 1568 (1992)

Torres-Peimbert, S., Peimbert, M., Ruiz, M.T.: Lect. Notes Phys., *in press* (1995)

Traat, P., in preparation (1995)

Woosley, S.E., Weaver, T.A.: Lect. Notes Phys. **255** 91 (1986)

Woosley, S.E., Langer, N., Weaver, T.A.: ApJ **411** 823 (1993)

The Helium Abundance of the Galactic Bulge

Dante Minniti

ESO Garching

Abstract. We present new measurements of the He abundance of the Galactic bulge based on IR photometry of several bulge fields. The fields cover a wide range of distances from the Galactic center, allowing to study possible effects of disk contamination and the He gradient in the bulge. The findings are: 1) the mean He abundance of the Galactic bulge is $Y = 0.28 \pm 0.02$, and 2) there is no significant He gradient in the bulge.

1 Introduction

The absolute He abundance of the inner regions of the Galaxy is a key ingredient for models of Galactic chemical evolution. However, previous He abundance determinations in the Galactic bulge give contradictory results: $Y = 0.30$ (Terndrup 1988), $Y \leq 0.22$ (Davidge 1991), and $Y = 0.30$–0.35 (Renzini 1994). In this work we present new measurements of the He abundance of the Galactic bulge based on IR photometry of several bulge fields.

2 The Data

We have selected several "windows" towards the Galactic bulge, covering a wide range of Galactic latitudes and longitudes, with Galactocentric distances ranging from 0.3 to 1.7 kpc. The reasons to study several windows are to have better statistics, to search for a He gradient in the bulge, and to account for the possible effects of disk contamination on the computed He abundances. The positions of the fields are plotted in Figure 1.

The observations were taken with the 256×256 Nicmos 3 array at the Steward Observatory 90" telescope in several observing runs in 1990, 1992 and 1993. The scale was 0.6 arcsec pixel^{-1}, giving a field of view of 2.5 arcmin square. Typically, several hundred frames were taken per field. The total size of the final mosaics were typically 25 square arcmin.

The photometry was obtained with DAOPHOT. Completeness corrections were estimated for the fainter magnitudes by adding artificial stars. The photometry reaches 2–3 magnitudes below the HB (horizontal branch). Typical errors are $\sigma_K = 0.03$, 0.07, and 0.22 for $K = 10$, 13, and 15, respectively. The luminosity functions are constructed down to magnitudes where the completeness corrections are less than about 50%. The fainter, more incomplete data are not considered. The luminosity functions for all the fields observed are shown in Figure 2.

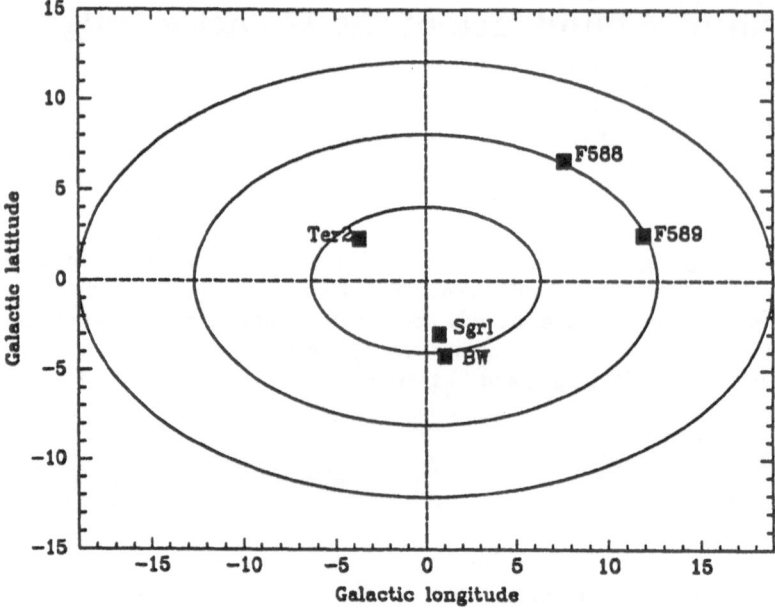

Fig. 1. Location of the fields projected on the Galactic plane.

3 Measuring Helium Abundances

The relative numbers of HB and GB stars can be used to measure the He content of a stellar population (Iben 1968, Caputo et al 1983). Here we will compute He abundances using the R' method, where the ratio R' is defined as $N_{HB}/N_{RGB+AGB}$. The number of HB stars is obtained by subtracting a power-law fit to the luminosity function outside the magnitude range of the HB. This fit is performed after accounting for 25% AGB stars for magnitudes brighter than the HB.

The He abundance determinations using the R' method are insensitive to age for stellar populations with ages in excess of ~1 Gyr (Renzini 1994). The value of R' itself is independent of reddening and distance, which can be very uncertain. Also, the number statistics for R' are very good for the bulge fields, which is not the case for globular clusters, where small number statistics is the major source of error. The correction required for extremely blue HB stars (which would be too faint in the IR) is negligible in bulge fields (e.g. Terndrup & Walker 1994).

Other factors determining the errors in the bulge He abundance are the calibration of R', and disk contamination. We adopt the calibration of R' given Caputo et al. (1983). We also correct for the presence of the RGB "bump", as advised by Ferraro et al. (1993). This correction is estimated to be 10% of the total number of HB stars, and it is in the sense of lowering the measured value of R'.

Disk contamination is a problem because of the different range of ages compared with the bulge, and the different distribution of stars along the line of

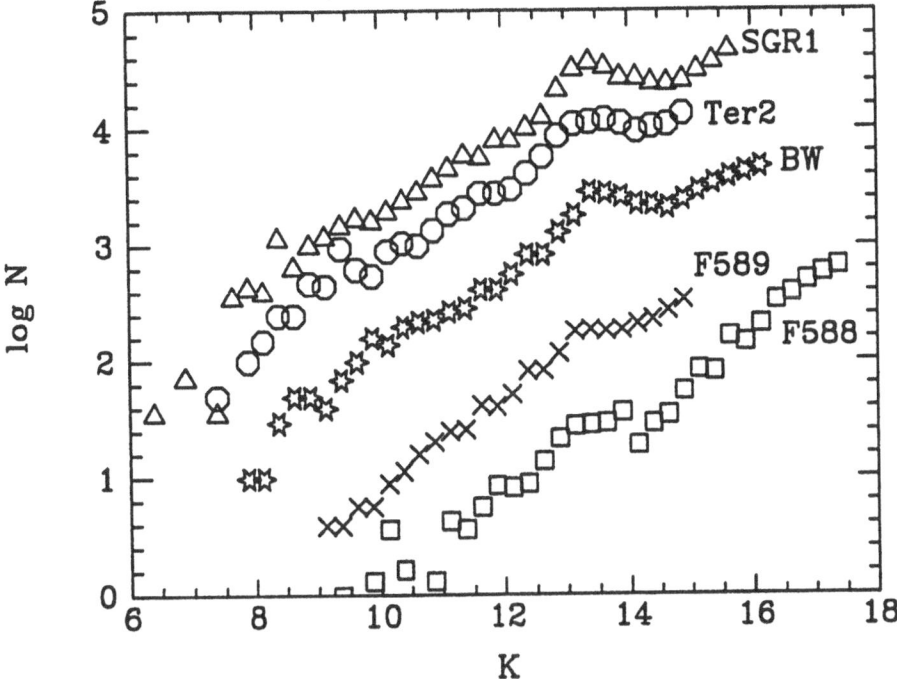

Fig. 2. Luminosity functions on a logarithmic scale for all the fields observed. These are corrected for completeness at the faint end. The scale corresponds to the counts in the Sgr1 field. The other fields have been displaced in subsequent steps of $log\ N = 0.5$ to make the plot clear.

sight. To check the effect of disk contamination on R', and thus He abundance, we observed the low–latitude field F589. We estimate the disk to bulge ratio to be D/B $\sim$ 40% in this field, based on a simple Galactic model. The R' value of field F589 is smaller than that of the field F588, located at similar Galactocentric distance. Thus, the effect of disk contamination is to lower the measured He abundance.

Buzzoni et al. (1983) used the R (and R') method to study the Helium abundances for several globular clusters as function of metallicity. In figure 3 we present their results compared with our Y values for two bulge fields. The bulge fields plotted are Baade's window and F588, which are the only fields in our sample with accurate metallicity determinations: [Fe/H] = -0.25 (from McWilliam & Rich 1994) and -0.50 (from Minniti 1994), respectively. The bulge population follows the general trend given by the most metal rich globulars (e.g. Buzzoni et al. (1983) derive $Y = 0.28 \pm 0.03$ for 47 Tuc which has $[Fe/H] = -0.7$).

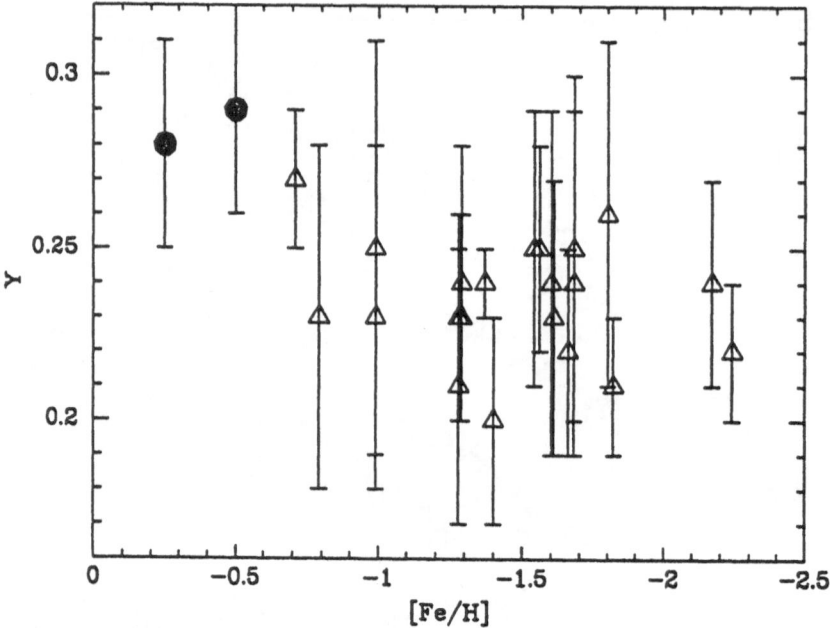

Fig. 3. Helium abundances *vs* metallicity for globular clusters from Buzzoni et al. (1983) –triangles– compared with bulge fields –circles–.

4 Conclusions

• The mean He abundance of the Galactic bulge is $Y = 0.28 \pm 0.02$, slightly lower than the value derived by Renzini (1994). Helium abundances as low as found by Davidge (1991) are inconsistent with the present observations.

• The bulge He abundance is comparable to those for metal rich globular clusters, and higher than those for metal poor globulars. The high He abundance of the Galactic bulge compared with the metal poor globular clusters suggests that the Galactic bulge is younger and more chemically evolved than the Galactic halo, in agreement with the formation scenario described by Minniti (1994).

• We do not find a significant He gradient in the Galactic bulge, between 0.3 and 1.6 kpc.

References

Buzzoni, A., Fussi Pecci, F., Buonanno, R. & Corsi, C. 1983, A&A, 128, 94

Caputo, F., Martinez Roger, C., & Paez, E. 1987, A&A, 183, 228

Davidge, T. J. 1991, ApJ, 380, 116

Ferraro, F. R., Fussi Pecci, F., Guarnieri, M. D., Moneti, A., Origlia, L., & Testa, V. 1994, MNRAS, 266, 829

Iben, I. 1968, Nature, 220, 143

Minniti, D. 1994, ApJ submitted

Renzini, A. 1994, A&A, 285, L5

Terndrup, D. M., 1988, AJ, 96, 884

Terndrup, D. M., & Walker, A. R. 1994, AJ, 107, 1787

Part V

Deuterium and ^3He

The Quest for the Cosmic Abundance of ^{3}He

R. T. Rood[1], T. M. Bania[2], T. L. Wilson[3], D. S. Balser[2]

[1] Department of Astronomy, University of Virginia, Charlottesville, VA 22903
[2] Department of Astronomy, Boston University, Boston, MA 02215
[3] Max Planck Institut für Radioastronomie, D-53121 Bonn, Germany

Abstract. We give a progress report on our observations of the $\lambda = 3.46$ cm hyperfine line of ^{3}He$^+$ and our first efforts at realistic models for the H II regions in which the line is observed. Both of these are required to derive an abundance. The abundances for the H II regions range from ^{3}He/H $= 1$–5×10^{-5}, and we argue that the abundance differences are real, i.e., the consequence neither of observational error nor of the source modeling. In three planetary nebula we find high abundances ^{3}He/H $\sim 10^{-3}$ consistent with the predictions of standard stellar models. The implications both for standard Big Bang nucleosynthesis and for chemical evolution are discussed.

1 Introduction

Determining the primordial abundance of any isotope involves three steps:

1) *Making a measurement.* In many cases the measurement itself is so difficult, the following steps are often overlooked.

2) *Converting the measured quantities to abundances.* The necessary quantity is x/H. As emphasized by Linsky at this meeting and by us below, making the transition from x to x/H is nontrivial.

3) *Inferring a primordial value from the observed abundance ratios.* There are sources and/or sinks for all of the isotopes of interest. One must account for these in arriving at the primordial abundance. There are two major components involved: first one must have a theoretical understanding of the production/destruction and evolution of each isotope; second one must have enough observations to both calibrate the free parameters of the model and to provide a consistency check.

In the case of ^{3}He each of these three steps presents a formidable obstacle making the analysis of ^{3}He potentially the most difficult of the isotopes of cosmological interest.

In the Solar System ^{3}He can be measured in the solar wind and in meteorites. The protosolar value ^{3}He/H $= (1.5 \pm 0.3) \times 10^{-5}$ is inferred by observing different components of helium in carbonaceous chondrites (e.g., Geiss 1993). Outside the solar system the only way the ^{3}He abundance can be determined is via the 3.46 cm hyperfine line of ^{3}He$^+$ (Rood, Bania, & Wilson 1984a [RBW]; Bania, Rood, & Wilson 1987 [BRW]). Our latest results in this series have recently been published (Balser, Bania, Brockway, Rood, & Wilson 1994 [BBBRW94]) and are summarized below.

In converting the observed hyperfine line parameters to an abundance one is faced with the problem that the strength of the hyperfine line from an H II region is proportional to the source density and the hydrogen abundance depends on the square of the density. Thus the H II region must be modeled. In our earlier work H II regions were assumed to be homogeneous spheres. Crude efforts at more detailed modeling suggested that in a few cases ^{3}He/H could be a factor of two higher but never lower than the homogeneous sphere case. However various lines of reasoning led us to be highly suspicious of these early models. Only recently have we had the resources to attempt more accurate models and to acquire the high resolution continuum data they require. These results will be presented in Balser, Bania, Rood & Wilson (1995) (BBRW95) and we give preliminary results below.

The chemical evolution of ^{3}He at first seemed simple. Rood, Steigman and Tinsley (1976) suggested that ^{3}He was produced in significant quantities by stars of 1–2 $M_\odot$. The ^{3}He is produced in the outer half of the star during main sequence hydrogen burning. The amount produced basically depends on the main sequence lifetime. It is mixed to the surface in the first "dredge-up" on the lower red giant branch (RGB). Once in the convective envelope it is difficult to destroy because the base of the convective zone retreats when the temperature rises. Rood et $al.$ (1976) hypothesized that the ^{3}He survives the thermal pulsing phase on the asymptotic giant branch (AGB). Recently models of Vassiliadis & Wood (1993) have confirmed this for stars with $M < 5\,M_\odot$. For stars of mass 1–2 $M_\odot$ they find the surface ^{3}He/H is $\sim 3 \times 10^{-4}$. Thus the RGB and AGB winds, and planetary nebulae (PNe) of stars $M < 2\,M_\odot$ should be substantially enriched in ^{3}He compared to the protosolar value. Because the PNe and their preceding winds are a major source of mass input to the interstellar medium one might expect ^{3}He/H to grow with time and to be higher in those parts of the galaxy where there has been substantial stellar processing.

^{3}He is destroyed in the interiors of more massive stars. The same stars convert primordial D into ^{3}He throughout the star. Dearborn, Schramm & Steigman (1986) showed that when weighted over reasonable mass functions, stars were net producers of ^{3}He even if the Rood et $al.$ (1976) production scheme is ignored. Hence it has become common to take the lowest observed abundance of ^{3}He as an upper limit to the primordial value, in which case it places a lower limit on the baryon to photon ratio η. The pre-solar D + ^{3}He abundance can be used to place a lower limit to η (Yang et $al.$ 1984)

Despite the simplicity of the synthesis of ^{3}He in low mass stars chemical evolution models (e.g., Steigman & Tosi 1992; Vangioni-Flam, Olive, & Prantzos 1994) have until recently included only processed primordial D as a source of ^{3}He. These did confirm the validity of taking the current ^{3}He as an upper limit to the primordial value. Models including stellar production of ^{3}He are now appearing (Tosi & Steigman 1995; Vangioni-Flam & Cassé 1994; Olive et $al.$ 1995). These give results consistent with the expectations of Rood et $al.$ (1976) and, as we shall see below, in disagreement with the abundances we derive.

2 Current Status of the 3-Helium Experiment

2.1 Observations of H II Regions

Our observations of H II regions from 1984–1991 have been described in detail in BBBRW94. The results can be summarized:

- *Much longer integration times:* During this period we have concentrated on obtaining extremely accurate line parameters for a few H II regions. In several cases we have now acquired more than a 100 hours of integration.

- *Multi-epoch observations:* We have found that the phase of the instrumental spectral baseline structure varies with sky frequency and thus with the time of the year. By obtaining smaller amounts of integration time scattered around the year the magnitude of the baseline structure in the averaged data can be reduced.

- *Accuracy of baseline removal:* The spectral baseline structure is real, arising primarily from standing waves in the telescope. It *must* be removed to get accurate line parameters. To monitor the accuracy of our baseline removal we introduce the concept of "baseline noise." Plots of the parameter $Q = \sqrt{t_{\text{intg}}} RMS/T_{\text{sys}}$ against $\log t_{\text{intg}}$ demonstrate the necessity of removing a very high order baseline. Criteria have been developed which indicate when baseline removal might introduce a substantial error.

- *Accounting for weak recombination lines:* At the levels we are now reaching the spectrum is beginning to fill up with weak recombination lines which were not considered in our earlier work. In particular the H213ξ ($\Delta N = 14$!) line lies close to the ^{3}He$^+$ line. Not accounting for it can distort the baseline shape in the region of the ^{3}He$^+$ line. Unfortunately these lines reduce the amount of "blank" spectrum that can be used for baseline modeling.

For cosmological purposes our most interesting sources are those with the lowest abundances, i.e., the most difficult to observe. In Fig. 1 we show our current spectrum for W49. Note that the observed features for the C171η and He171η lines are consistent with what we expect from scaling the observed C and He91α lines with the observed H91α and H171η intensities. (The lines cannot be "fit" uniquely because of their weakness and large combined line width.) Since these features are weaker than the ^{3}He$^+$ it gives us some confidence in our result.

We would like to stress that the difference in abundance between our sources *is not due to observational error*. To demonstrate this we plot as a dotted line the ^{3}He$^+$ feature as it would appear if W49 had the same abundance as in W3. We would detect such a line in a single 12 minute ON/OFF pair and have a precision determination of line parameters after a few hours integration scattered among 3 observing epochs. If the ^{3}He abundance found in W3 were commonplace, this project would have been completed more than a decade ago.

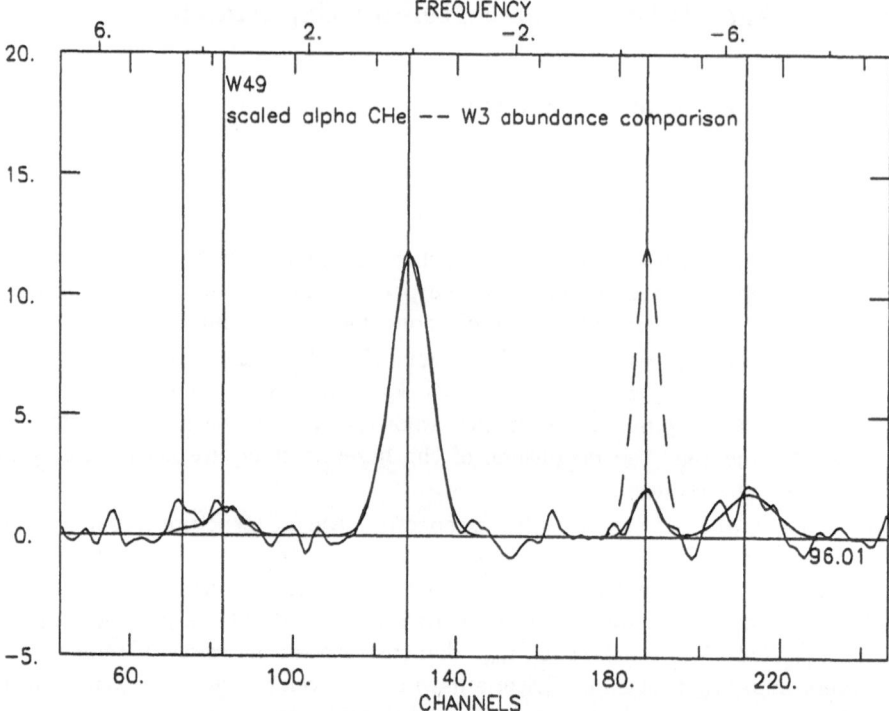

Fig. 1. The spectrum of W49. The vertical scale is antenna temperature in mK. The vertical marks show respectively the expected positions of the C171η, He171η, H171η, ^{3}He$^+$, and H213ξ lines. The best gaussian fits to the H171η, ^{3}He$^+$, and H213ξ lines are shown. The "fit" to the C171η and He171η lines is scaled from the 91α lines. The dashed line shows what the ^{3}He$^+$ line would look like if the ^{3}He abundance were the same in W49 as in W3. The total integration time is 106 hr.

2.2 Abundances

In BBBRW94 we give abundances assuming that the H ɪɪ regions are homogeneous spheres in local thermodynamic equilibrium, and that the H$^+$ and He$^+$ Strömgren spheres are identical. Slightly updated values for these abundances are given in Table 1. We have made considerable progress toward more accurate source models which will be described in detail in BBRW95. These include the following improvements:

- *High resolution continuum modeling:* This is required because the strength of the hyperfine line from an H ɪɪ region is proportional to $\int n_e \, dl$ through the source (where n_e is the electron density and l is the path length through the source) whereas the hydrogen abundance goes like the emission measure,

Table 1. ^{3}He in H II Regions

Source	$\left(\frac{^3\text{He}^+}{\text{H}^+}\right)_{hs}$ 10^{-5}	f_{dens}	f_{noPB}	f_{ion}	$\left(\frac{^3\text{He}}{\text{H}}\right)$ 10^{-5}	$M(\text{H II})$ $M_\odot$	R_{gal} kpc
W3	4.22	1.18	1.39	1.00	$5.42^{+0.75}_{-0.75}$	25	10.1
G133.8	2.73	1.01	1.02	1.54	$2.77^{+1.50}_{-0.42}$	59	10.1
S206	2.93	1.44	1.44	1.22	$5.16^{+0.51}_{-0.62}$	76	11.5
S209	1.03	0.87	0.87	1.00	$0.90^{+0.13}_{-0.13}$	1373	16.2
S311	4.28	0.51	0.51	1.19	$2.61^{+0.29}_{-0.29}$	295	11.0
SGRB2	2.49	1.09	1.28	1.00	$2.95^{+2.30}_{-0.34}$	2507	0.1
M17S	1.70	1.05	1.12	1.00	$1.84^{+0.17}_{-0.17}$	320	6.4
W43	1.13	1.25	1.50	1.00	$1.55^{+0.19}_{-0.19}$	3089	4.6
W49	0.68	1.37	1.97	1.00	$1.14^{+0.23}_{-0.23}$	7776	8.1
W51	2.29	0.93	1.31	1.00	$2.56^{+0.45}_{-0.45}$	1718	6.5
NGC7538	1.87	0.95	1.00	1.00	$1.82^{+0.21}_{-0.21}$	98	10.3

$\int n_e^2\, dl$. We have obtained higher resolution 8.7 GHz continuum maps with the 100 m radio telescope of the Max Planck Institut für Radioastronomie and with the VLA. These have been used along with our Green Bank continuum observations to produce nested multiple core/ halo models for each source. This correction alone will always increase the value of ^{3}He/H as compared to the homogeneous sphere value.

- *Recombination lines:* During the course of our observations we observe many recombination lines of H and He. Earlier we used these as a monitor of system performance. We now use them to determine 1) the extent of non-LTE effects; 2) the level of micro-clumping not resolved by the VLA which we incorporate via a volume filling factor in the core components; and 3) the expected width of the ^{3}He$^+$ line. Adopting the latter can lead to a reduced abundance ratio compared to the homogeneous sphere value where we always use the best fit to the observed ^{3}He$^+$ line width.

- *Pressure broadening:* In the densest environments pressure broadening can lead to very wide wings for recombination lines originating from levels with large principle quantum number. In particular, the wings of the H213ξ line can extend into the region of the ^{3}He$^+$ line. This can lead to ^{3}He/H abundances higher than the true value. The combined effects of the density structure, calculated line width, and pressure broadening are reported as in Table 1 as f_{dens}. At least some of the H213ξ line "wing" is probably removed during baseline fitting, so we also give a correction factor f_{noPB} which includes no pressure broadening. The true structure correction is bracketed by f_{dens} and f_{noPB}.

- *Ionization affects:* We have used Ferland's program CLOUDY (Ferland 1993) to estimate how much the underionization of He as compared to H reduces the observed abundance. These correction factors are reported as f_{ion} in Table 1. We are still assessing some apparent inconsistencies between the observed $^4He^+/H^+$ and f_{ion}.

For the abundances given in Table 1 we adopt the average of the f_{dens} and f_{noPB} corrections – the error bars have been accordingly adjusted. We take $f_{ion} = 1$ for G133.8 because of the high observed $^4He^+/H^+$ and increase the upper error bar to account for the calculated f_{ion}. For S206 we adopt the calculated f_{ion} which seems consistent with the observed $^4He^+/H^+$ but increase the lower error bar. For Sgr B2 we increase the upper error bar to account for a low observed $^4He^+/H^+$.

2.3 Planetary Nebulae

Rood *et al.* (1976) suggested that planetary nebulae might be rich in ^{3}He. Rood, Wilson, & Steigman (1979) reported on early efforts to measure ^{3}He in PNe, which were not successful because the sensitivity and stability of the receivers were not adequate. Finally, Rood, Bania, & Wilson (1992) reported the first detection of ^{3}He in the PN NGC 3242. They cautioned that while the detection seemed solid, line parameters from a single observing epoch could be inaccurate. We report here on our continuing efforts.

We have now observed NGC 3242 during four observing epochs. The integration time is almost doubled. The measured line intensity is less than half our first reported value. The spectrum is shown in Fig. 2. From the point of signal/noise and baseline uncertainties the result is very solid. The presence of an H171η much stronger than anticipated is the only worrisome matter. We also have another detection, a tentative detection, and upper limits for other PNe. The detection in NGC 6720 is shown in Fig. 3. While the signal to noise is quite high, the line does not appear at quite the velocity expected. The velocity flags are set on the basis of optical data which might not be appropriate for the $^3He^+$ line. Further, the complex baselines at the 100 m telescope often lead to small velocity offsets in data comparable to that for NGC 6720. "Homogeneous sphere" abundances and limits are given in Table 2. The upper limits have been obtained by adjusting the baseline fit so that the maximum possible feature is created near the expected $^3He^+$ line. The derived abundances depend on the adopted distances to the PNe which are typically uncertain at the factor of 2 level. Basically at this point one should not pay too much attention to the specific PNe abundance values but simply conclude that in some, perhaps most, PNe the ^{3}He/H abundance is enhanced by at least a factor of 10 and perhaps by a factor up to 100 over the protosolar value.

Table 2. ^{3}He$^+$ in Planetary Nebulae

Source	T_L (mK)	Δv (km s^{-1})	t_{intg} (hr)	^{3}He$^+$/H$^+$ (10^{-3})
IC 289	2.75	37.73	38.5	1.16
NGC 3242	4.16	47.24	36.8	1.42
NGC 6543	<4.82	56.50	16.8	<3.46
NGC 6720	2.85	36.55	39.1	0.27
NGC 7009	<3.64	43.82	11.5	<0.89
NGC 7662	<6.98	40.73	10.9	<3.32

3 Conclusions

The current state of the ^{3}He experiment can be summarized:

• *The source to source abundance variations are not due to observational error.* Further, it is unlikely that some factor such as density structure corrections vary systematically in the low abundance sources such that they uniformly have high abundances. It makes no sense at all to average our results to obtain a cosmological value, a value for the present local ISM, etc.

• *The observed abundance pattern does not make sense.* Chemical evolution models including ^{3}He (Vangioni Flam & Cassé 1994; Tosi & Steigman 1995; Olive *et al.* 1995) lead to 3 "expectations:" 1) the present ^{3}He/H abundance should be substantially greater than the primordial value; 2) the current value of ^{3}He/H in the local ISM should be greater than the primordial value; and 3) the gradient of ^{3}He/H with galactocentric distance (R_{gal}) should be negative. Applying rather generous error limits to the primordial D, ^{4}He, & ^{7}Li one finds that $\eta \lesssim 4 \times 10^{-10}$ which requires (^{3}He/H)$_{primordial} \gtrsim 1.3 \times 10^{-5}$. This is marginally consistent with currently observed values. There is certainly no room for substantial galactic production of ^{3}He. As far as the expected gradient, the most obvious "feature" in the data is a collection of high abundance sources in the Perseus Arm, and it is not clear that there is any smooth variation with R_{gal}. In a plot of ^{3}He/H against R_{gal} as in Fig. 4 ^{3}He/H increases with R_{gal} more or less orthogonal to expectations. Given this and the fact that none really sample the local ISM we are not in a position to assess whether ^{3}He/H has increased since the formation of the Solar System.

• *Is ^{3}He made in low mass stars?* Given the confusing abundance pattern noted above, one might wonder whether low mass stars are indeed a source of ^{3}He. Indeed, Galli *et al.* (1994 & at this meeting) show that a large low energy

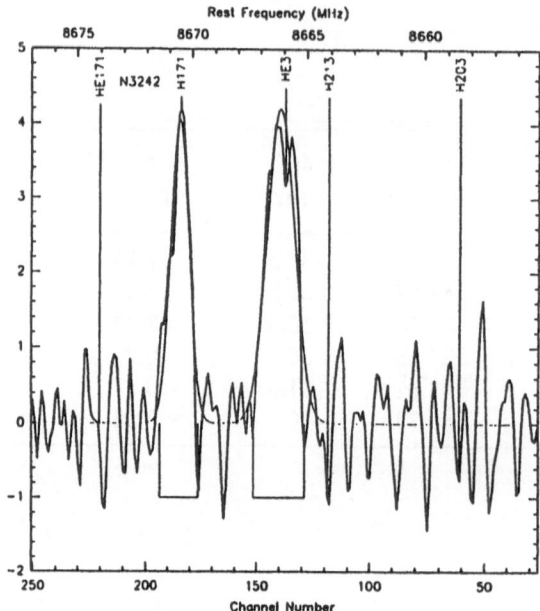

Fig. 2. The spectrum of the PN NGC 3242. The vertical scale is brightness temperature in mK. The vertical marks show respectively the expected positions of the He171η, H171η, ^{3}He$^+$, H213ξ, and H203μ lines. The best gaussian fits to the H171η and ^{3}He$^+$, lines are shown.

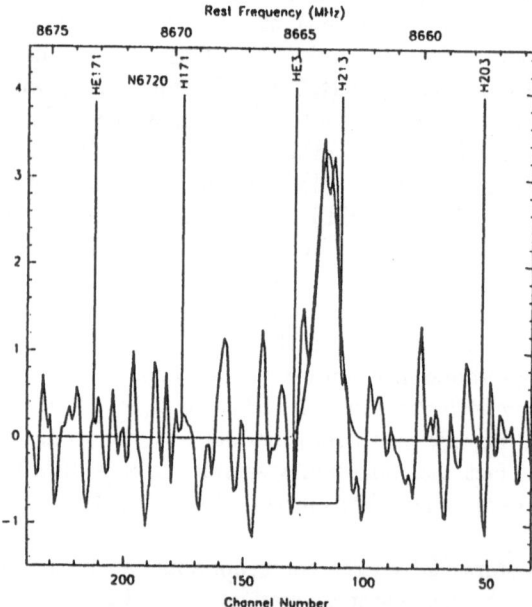

Fig. 3. The spectrum of the PN NGC 6720.

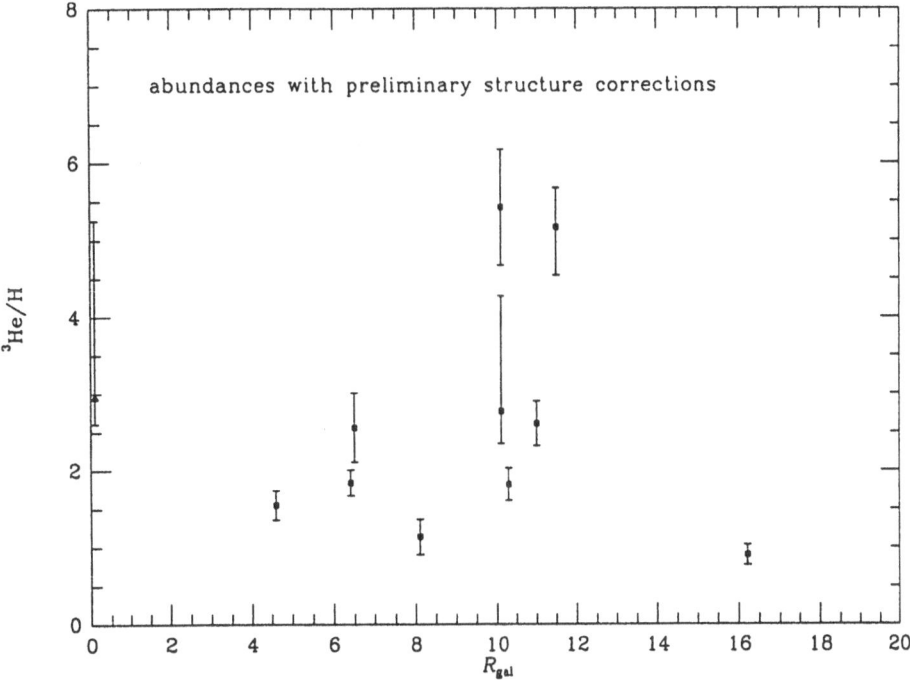

Fig. 4. ^{3}He/H abundance in units of 10^{-5} as a function of galactocentric radius in kpc.

resonance in the ^{3}He (^{3}He, $2p$)^{4}He reaction can suppress the production of ^{3}He. Another possibility is that some non-convective mixing process either prevents the buildup of ^{3}He on the main sequence (see §V(c) of RBW) or destroys ^{3}He along the upper red giant branch (Hogan 1995; Deliyannis at this meeting). Both of these possibilities are at odds with our observations of ^{3}He in PNe: the Galli *et al.* hypothesis simply requires that we are wrong; the mixing requires that our PNe sample is atypical.

• *Are the abundances of our sample not a simple measure of the cosmic abundance?* Perhaps the most compelling evidence in this direction is shown in Fig. 5 where the ^{3}He/H abundance is plotted as a function of the mass of the H II region observed. There is a strong inverse correlation. This is suggestive of local pollution. The trouble with this scenario at first glance is that the stars which might plausibly pollute H II regions are massive, i.e., ^{3}He sinks. But a moderate ^{3}He enhancement could be achieved as follows: It is generally agreed that H II regions are ionized by massive stars and that massive stars have very substantial winds which can carry away most of the stellar mass within their lifetimes. As far as we are aware no calculations have been published which give the ^{3}He abundance in massive star winds. However, it is plausible that the very earliest winds are slightly enriched in ^{3}He from the initial (D + ^{3}He). From Maeder (1990) it seems possible that the first few $M_\odot$ of wind is ^{3}He rich. The later winds would be depleted in ^{3}He, becoming first enhanced in N, then ^{4}He, and finally C & O. Thus, in a young H II region whose ionized gas was composed primarily by the young winds of massive stars ^{3}He could be enhanced. Allowing even a small dispersion in formation times, since the ^{3}He rich winds are a small fraction of the integrated wind mass loss, the combined winds of many stars would be low in ^{3}He. Only those regions containing a very few (perhaps 1 or 2) stars would have high ^{3}He. W3, the H II region with the highest observed ^{3}He could fit this model. W3A is a bubble like structure with two embedded IR sources shaping the region (Harris & Wynn Williams 1976). The region observed by BBBRW94 (W3A plus some surrounding gas) is estimated to contain about 15–25 $M_\odot$ of ionized gas. So a significant fraction of the observed gas could be composed of slowed winds. W3 shows one other sign of local pollution. Roelfsema, Goss, & Mallik (1992) have observed substantial variations in the ^{4}He abundance in W3. Yet averaged over the entire source ^{4}He/H in W3 is "normal" (BBBRW94) suggesting that winds in the W3 stars have just reached the ^{4}He rich layers and that the ^{4}He rich blobs are slowed winds not yet mixed into the nebula as a whole. This scenario is further developed in Olive et al. (1995) who also suggest that the massive, low ^{3}He abundance H II regions have been significantly polluted by the late winds of many massive stars and thus have an atypically low ^{3}He/H ratio. The lowest abundance H II regions can have masses up to $\sim 1000\, M_\odot$, and self pollution at this level might violate other observational constraints.

4 Possible Relation to ^{13}C

Interpreting ^{3}He abundances is more difficult than the other light isotopes. For example, ^{7}Li reaches a constant abundance as the metallicity of Pop II stars decreases, suggesting that this plateau value is the primordial value. Likewise, since O and ^{4}He have similar stellar sources one can use the O abundance in low metallicity H II regions in dwarf galaxies to extrapolate the ^{4}He to its primordial value. Unfortunately ^{3}He cannot be observed in low metallicity objects nor is there any other easily observable species thought to be produced predominantly in the same stars that make ^{3}He.

Observations of the isotope ^{13}C can however offer some clues to ^{3}He production and evolution. From the stellar evolution point of view the possible connection between ^{3}He and ^{13}C is clear cut. Basically, as one goes radially inward in a solar analog star (these include globular cluster stars) one progressively finds that: (1) D is converted to ^{3}He all the way to surface; (2) Li is depleted; (3) Be is depleted; (4) B is depleted; (5) there is a ^{3}He peak; (6) there is a ^{13}C peak; (7) ^{14}N is enhanced; and (8) ^{16}O is depleted. If ^{13}C is mixed to the surface ^{3}He must also be mixed. ^{13}C (and other CNO) "anomalies" are common in solar analog stars. If these anomalies arise from some nonconvective mixing processes affecting ^{13}C, the surface ^{3}He must also be altered from its "standard" value. Indeed when we first mulled over the idea that ^{3}He might be destroyed in stars (RBW), we suggested that it might be related to ^{13}C anomalies.

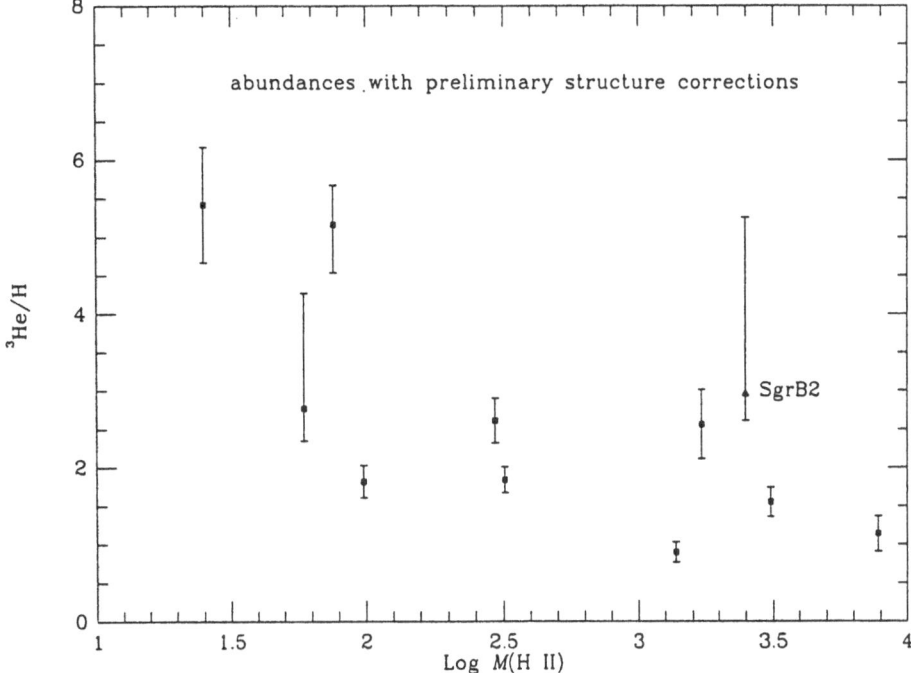

Fig. 5. ^{3}He/H abundance in units of 10^{-5} as a function of the mass of the H II region observed. SGR B2 is marked since the galactic center region is generally thought to have an evolutionary history different from the rest of the Galaxy.

It is nontrivial to concoct a mixing scheme which leads to very low surface ^{12}C/^{13}C. Inside the ^{13}C peak, where ^{12}C/^{13}C ~ 4, both isotopes have been converted almost completely to ^{14}N. Thus when the tooth fairy mixes, if he/she is not careful a little low ^{12}C/^{13}C but also low C stuff is diluted with lots of high ^{12}C/^{13}C high C stuff resulting in very little change. To get lots of ^{13}C one has to process just a little bit and then put the ^{13}C in a cool, safe environment. This happens during He shell flashing on the AGB where one gets incomplete CN processing because the burning is taking place in a C rich but p poor environment. There is a nice discussion of the necessity of partial CN processing in Lambert & Ries (1981) where they show that in ^{13}C rich stars the processing is less than one proton per initial ^{12}C.

We suggested a mechanism (RBW; Rood, Bania & Wilson 1984b) which would suppress the ^{3}He buildup on the main sequence and possibly increase ^{13}C at the surface of red giants. At this time the ^{13}C observations seemed most consistent with a main sequence origin as well (Lambert & Ries 1981). The idea was that the solar global overstability suggested by Dilke & Gough (1972) and confirmed by Christensen-Dalsgaard, Dilke, & Gough (1974) would lead to some sort of slowish mixing which would maintain a "critical" gradient in the ^{3}He. As far as we aware, this overstability "still exists" but is ignored in stellar model calculations.

Recently the observational evidence has suggested that the stars with the lowest $^{12}C/^{13}C$ are produced on the upper RGB (Brown 1987; Gilroy 1989; Gilroy and Brown 1991). Charbonnel (1994) argues that the stars observed on the lower RGB have $^{12}C/^{13}C$ consistent with "standard models" with ^{13}C reaching the surface at the first dredge-up. She points out that the lowest $^{12}C/^{13}C$ stars are all observed above the so called RGB "luminosity function bump" (e.g., Fusi Pecci *et al.* 1990). At luminosities less than that of the bump there is a molecular weight gradient between the base of the convective envelope and H burning shell which can suppress slow mixing processes. Above the bump processes akin to meridional circulation could operate more efficiently.

It is not clear how much ^3He can be destroyed in such a scheme. The problem is that there is a significant reservoir of ^{12}C and ^3He in the convective envelope. The bottom of this envelope is set by the temperature where the opacity rapidly drops; this region is too cool to burn either species. Nonconvective mixing below the base of the convective envelope must be present to get any burning. To change the surface $^{12}C/^{13}C$ abundance substantially one must imagine a situation analogous to taking a glass of water from a barrel, processing the water somehow, returning it to the barrel, mixing it well, and repeating. As pointed out by Hogan (1994), any such process which changes ^{13}C will also deplete ^3He. We suspect that ^3He will not go to zero abundance and guess that in luminous RGB stars $(^3He)_{final}/(^3He)_{prebump} \sim (^{12}C/^{13}C)_{final}/(^{12}C/^{13}C)_{prebump}$. Even factors of a few could not make the observed ^3He consistent with chemical evolution models.

Observations of Kraft *et al.* (1993 and references therein) show that the level of CNO processing in different RGB stars varies quite a bit from star to star. The same would probably be true for ^3He. This could allow for some ^3He rich PNe.

From the point of view of chemical evolution the one possible corollary isotope is again ^{13}C. In the early days of CNO chemical evolution ^3He and ^{13}C were thought to share a common source (e.g., Fig. 3 of Audouze & Tinsley 1976). The work of Renzini & Voli (1981) suggests a larger ^{13}C contribution from intermediate mass stars. There is the further complication that ^{13}C in low mass stars would be a "secondary" nucleosynthesis product while ^3He is primary. For whatever it is worth, observations of ^{13}C in the ISM show that $^{12}C/^{13}C$ increases with R_{gal} more or less as expected from chemical evolution models (Wilson & Rood 1994).

5 What's in the Works

We note with some amusement the "confusion" created by the high D observations in large redshift galaxies (Songaila *et al.* 1994; Carswell *et al.* 1994). Things were so simple when there were only the observations in the local ISM (e.g., Linsky *et al.* 1993). If we had observations of only one H II region ^{3}He would look simple too. Unfortunately we have been able to make detections in several places and confusion reigns. There is some promise, however, that some pattern will emerge allowing us to both determine the primordial ^{3}He and to understand its evolution. We hope our situation is a bit like that of ^{7}Li when it had only been observed in the Sun and a few nearby stars. Before the Spite's found ^{7}Li in extreme Pop II stars while hunting for aluminum, no one would have guessed that it would be possible to determine the primordial ^{7}Li.

In the next few years we can look forward to:

- *Outer Galaxy sources:* In the hope of finding more nearly primordial abundances we are observing more outer Galaxy sources. We have a tentative detection in S252.
- *W3/S206-like sources:* We have started trying to identify and to observe "small," bubblelike sources to test the self-pollution hypothesis.
- *Planetaries:* Our observations of planetary nebulae using the MPIfR 100 m telescope continue. We plan to increase the sample size and to get more observing epochs for currently observed sources. We will try to confirm our observations with the VLA.
- *The Green Bank Telescope:* The new off-axis 100 m telescope under construction in Green Bank almost appears to have been designed specifically for this project. We look forward to its completion.

We wish to thank Dave Schramm, Keith Olive, Gary Steigman, Monica Tosi, Craig Hogan, and Elisabeth Vangioni-Flam for helpful conversations and providing information in advance of publication. This research was supported in part by NSF grant AST–9121169 and NATO Travel Grant CRG 900482. TLW's research is partially supported by the Max Planck Forschungpreis of the A. v. Humboldt-Stiftung.

References

Audouze, J. and Tinsley, B.M. 1976, ARAA, 14, 43

Balser, D. S., Bania, T. M., Brockway, C. J., Rood, R. T., & Wilson, T. L. 1994, ApJ, 430, 667 [BBBRW94]

Balser, D. S., Bania, T. M., Rood, R. T., & Wilson, T. L. 1995, in preparation [BBRW95]

Bania, T.M., Rood, R.T., & Wilson, T.L. 1987, ApJ, 323, 30 [BRW]

Brown, J. A. 1987, ApJ, 317, 701

Carswell, R. F., Rauch, M., Weymann, R. J., Cooke, A. J., and Webb, J. K., 1994, MNRAS, 268, L1

Charbonnel, C. 1994, A&A, 282, 811

Christenson-Dalsgaard, J., Dilke, F. W. W., & Gough, D. O. 1974, MNRAS, 170, 589

Dilke, F. W. W., & Gough, D. O. 1972, Nature, 240, 262

Fusi Pecci, F., Ferraro, F., Crocker, D. A., Rood, R. T., & Buonanno, R. 1990, A&A, 238, 95

Lambert, D. L. & Ries, L. M. 1981, ApJ, 248, 228

Dearborn, D. S. P., Schramm, D. N., & Steigman, G. 1986, ApJ, 302, 35

Ferland, G. J. 1993, U. Kentucky, Dept. Phys. Internal Report

Galli, D., Palla, F., Straniero, O., & Ferrini, F. 1994, ApJ, 432, L101

Geiss, J. 1993, in *Origin and Evolution of the Elements* eds. N. Prantzos, E. Vangioni-Flam, and M. Cassé (Cambridge:Cambridge University Press), p. 89

Iben, I. & Truran, J. W. 1978, ApJ, 220,980

Kraft, R. P., Sneden, C., Langer, G. E., & Shetrone, M. D. 1993, AJ, 106, 1490

Maeder, A. 1990, A & A Supp, 84, 139

Olive, K. A., Rood, R. T., Schramm, D. N., Truran, J., & Vangioni-Flam, E. 1995, submitted to ApJ

Renzini, A. & Voli, M. 1981, A&A, 94, 175

Roelfsema, P. R., Goss, W. M., & Mallik, D. C. V. 1992, ApJ, 394, 188

Rood, R. T., Bania, T. M., & Wilson, T. L. 1984a, ApJ, 280, 629 [RBW]

Rood, R. T., Bania, T. M., & Wilson, T. L. 1984b, in Observational Tests of Stellar Evolution Theory, ed. A. Maeder, A. Renzini (Dordrect: Reidel), 567

Rood, R. T., Bania, T. M ., & Wilson, T. L . 1992, Nature, 355, 618

Rood, R. T., Steigman, G. & Tinsley, B. M. 1976, ApJ, 207, L57

Songaila, A., Cowie, L. L., Hogan, C. & Rugers, M. 1994 Nature, 368, 599

Steigman, G. & Tosi, M. 1992, ApJ, 401, 150

Vangioni-Flam, E. & Cassé, M. 1994, ApJ (submitted)

Vangioni-Flam, E., Olive, K. A., & Prantzos, N. 1994, ApJ, 427, 618

Vassiliadis, E. & Wood, P. R. 1993, ApJ, 413, 641

Wilson, T. L., & Rood, R. T. 1994, ARAA, 32, 191

Yang, J., Turner, M. S., Steigman, G., Schramm, D. N., & Olive, K. A. 1984, ApJ, 281, 493

Hubble Observations of D/H in the Local ISM and Consequences for Cosmology

J.L. Linsky[1][2], A. Diplas[3], T.R. Ayres[4], B. Wood[1],
A. Brown[1]

[1] Joint Institute for Laboratory Astrophysics, University of Colorado and N.I.S.T., Boulder CO 80309-0440
[2] Staff Member, Quantum Physics Div., National Institute of Standards and Technology
[3] CASS, University of California at San Diego, La Jolla, CA 92093-0111
[4] CASA, University of Colorado, Boulder CO 80309-0389

Abstract. An accurate measurement of the primordial value of D/H would provide one of the best tests of nucleosynthesis models for the early universe and the baryon density. We first summarize previous observations of atomic D and deuterated molecules, and then report on our ongoing program of HST/GHRS observations of the interstellar neutral H and D Lyman-α lines along the lines of sight toward nearby stars. The first stars studied in this program were Capella (12.5 pc) and Procyon (3.5 pc). We compare the derived properties of the local interstellar medium (D/H ratio, temperature, turbulent velocity) for Procyon and Capella at two orbital phases. Our data are consistent with a single value for D/H (about 1.6×10^{-5} by number) in the local ISM. We estimate the primordial value of D/H using Galactic chemical evolution models and compare this value with recent standard cosmological models. We find that $\Omega_B h_{50}^2$ lies in the narrow range 0.06-0.08. Finally we comment on recent reports of extragalactic D/H measurements.

1 The Importance of an Accurate Measurement of the Primordial D/H Ratio

The Hubble expansion, microwave background, and light element abundances are the main observational pillars upon which the standard Big Bang cosmology now rests. Of these three tests, the light element abundances provide the main constraint on the total baryon density (luminous and dark matter), and the D/H ratio provides the tightest constraint. This follows from (1) the absence of any known significant sources of deuterium after about 10^3 seconds in the early universe, (2) the subsequent destruction by nuclear reactions in the cores of stars where D is the most fragile species, and (3) the steep monotonic slope between the primordial D/H ratio and the baryonic density in contemporary Big Bang nucleosynthesis models (e.g. Walker et al. 1991). Since none of the other light elements (^{3}He, ^{4}He, ^{6}Li, ^{7}Li, Be, or B) share these properties, their abundances provide more uncertain estimates of the baryon density of the universe. We have therefore established an HST observing program to obtain both accurate values of the D/H ratio in local interstellar gas and estimates of the destruction of D over the lifetime of the Galaxy to infer the primordial D/H ratio.

The importance of measuring D/H has led to searches in a wide variety of environments from terrestrial seawater (using HDO/H_2O), atmospheres of the giant planets (using HD/H_2 and NH_2D/NH_3), cold interstellar clouds (using deuterated molecules and the as yet unsuccessful search for the 92-cm hyperfine transition of atomic D), and warm interstellar gas (using the Lyman series lines of D and H). This work has been reviewed by Boesgaard & Steigman (1985) and most recently by Wilson & Rood (1994). Deuterated molecules have been observed in many sources but their abundances relative to the corresponding undeuterated molecules are generally larger (often orders of magnitude larger) than expected for sensible values of D/H. This overabundance is usually explained by the slightly larger binding energy of the deuterated molecules and the cold environments in which they formed (e.g. Geiss & Reeves 1981).

The ratio of D to H column densities in warm interstellar gas ($T \approx 7000$ K) as inferred from absorption in the Lyman series lines for the lines of sight toward stars is now thought to be the most accurate method for inferring the D/H ratio in the Galaxy. Although this gas has been chemically processed over the lifetime of the Galaxy and the D/H ratio must be lower than primordial, this method should not suffer from other systematic errors because the relative ionization fraction, molecular association fraction, and degree of condensation onto dust grains should be the same for D and H in this environment. For Galactic lines of sight only the Lyman-α line can be studied by IUE and HST, but the higher Lyman lines were observed toward a few hot stars by Copernicus. Unfortunately, H absorption overlaps the D line (-0.33Å from the H line). This limits the use of the Lyman-α line to relatively nearby stars where log $N_{HI} <$ 18.7. A reanalysis of the best available Copernicus and IUE data led McCullough (1992) to estimate that the mean value of D/H by number in the local ISM is $(D/H)_{LISM} = 1.5 \pm 0.2 \times 10^{-5}$. Since these data have rather low S/N, the H line is very saturated, and the intrinsic shapes of the D and H lines are unresolved at the spectral resolutions of IUE and Copernicus, we initiated an observing program with HST to obtain more accurate values of $(D/H)_{LISM}$.

2 Hubble Space Telescope Observations of D/H in the Local ISM

Since 1990 the Goddard High Resolution Spectrograph (GHRS) on the Hubble Space Telescope has been acquiring ultraviolet spectra with unprecedented spectral resolution and signal-to-noise. (See Brandt et al. 1994 and Duncan 1992 for descriptions of the GHRS and its capabilities.) The high quality of these data permits us to measure very accurate column densities of interstellar D, H, and the heavier elements, together with the thermal and dynamic properties of interstellar gas, in the region of space near the Sun that includes the local warm cloud and surrounding hot substrate plasma and additional warm clouds. (See

Linsky et al. 1993, hereafter Paper I, and below for a discussion of the local interstellar medium properties.)

2.1 The Line of Sight Toward Capella

Our program to measure the D/H ratio and the gas properties in the local interstellar medium (LISM) began on 1991 April 15 with our echelle observations of the resonance lines of H and D (Lyman-α at 1216Å), the FeII multiplet UV1 (at 2599Å), and the MgII h and k lines (at 2796Å and 2803Å) toward the Capella binary system. The first observations were made at orbital phase 0.26 when the G8 III star in the Capella system had a radial velocity of +55.6 km s^{-1} and the G1 III star had a radial velocity of +2.0 km s^{-1}, very close to their maximum radial velocity separation. The broad interstellar H absorption and narrower D absorption centered at –0.33Å (–81.6 km s^{-1}) from the H line center are superimposed upon Capella's chromospheric H Lyman-α emission line (see Fig. 1). Detailed analysis of this spectrum and the FeII and MgII spectra in Paper I provided very accurate measures of the temperature ($T = 7000 \pm 200$ K) and nonthermal broadening ($\xi = 1.66 \pm 0.03$ km s^{-1}), which characterize all of the gas in this line of sight and which may be representative of the gas properties in the local cloud. A careful analysis of these Capella spectra, including systematic errors associated with the uncertain intrinsic Lyman-α emission line of the stars, shows that the neutral H column density is $N_{HI} = (1.7–2.1) \times 10^{18}$ cm^{-2} and $(D/H)_{LISM} = 1.65$ (+0.07, -0.18) $\times 10^{-5}$ for this line of sight.

A major systematic error in our analysis of the Capella phase 0.26 observations was the uncertain intrinsic Lyman-α emission line profiles of the two stars in the Capella system, especially those portions of the emission lines within 0.5 Å of line center but outside of the dark core of the interstellar H Lyman-α line. These portions of the intrinsic line profiles are critically important because they form the "continuum" against which the observed profile is compared to determine the interstellar column densities and broadening parameters for H and D. It was difficult to solve for both the shape of the intrinsic stellar emission lines and the interstellar gas properties with spectra obtained at only one orbital phase, although IUE spectra at other phases were helpful. We therefore reobserved Capella on 1993 September 19 at orbital phase 0.80 close to the opposite orbital quadrature to verify our results in Paper I by analyzing the (assumed constant) interstellar absorption against the background of a somewhat different intrinsic emission line from the Capella system. At phase 0.80 the radial velocities of the G8 III and G1 III stars were +4.7 and +55.5 km s^{-1}, respectively, nearly the reverse of their velocities at phase 0.26. We used the G160M grating to obtain a moderate resolution spectrum of the Lyman-α region, because the echelle-A grating of the GHRS was not available for use at that time. Analysis of both Capella data sets together with a more accurate representation of the instrumental point spread function by Linsky et al. (1994) led to a small increase in temperature and decrease in turbulent velocity of the interstellar gas but essentially the same D/H ratio, $(D/H)_{LISM} = 1.60 \pm 0.08 \times 10^{-5}$.

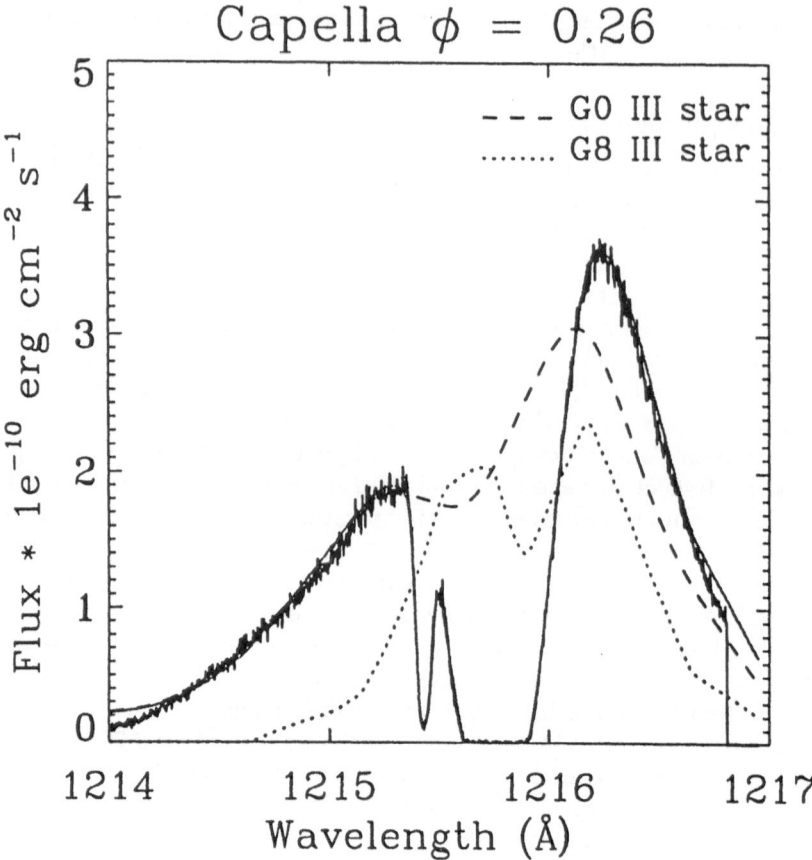

Fig. 1. The observed Lyman-α spectrum of Capella at phase 0.26, intrinsic profiles of the two stars, and a best fit model which is nearly indistinguishable from the data. From Linsky et al. 1993.

2.2 The Line of Sight Toward Procyon

Our second target was Procyon, an F5 IV-V star located 3.5 pc along a line of sight about 54 degrees from Capella. We observed this star on 1992 December 21 in the same way as we observed Capella at phase 0.26, except that the Lyman-α line (see Fig. 2) was observed with the G160M grating through the small science aperture (SSA). The spectral resolution at Lyman-α was only 20,000 (15 km s^{-1}) instead of 84,000 (3.57 km s^{-1}) when we used the echelle-A grating. The MgII and FeII lines were observed through the SSA with echelle-B, providing the same high spectral resolution as for Capella. These observations and their analysis will be described in more detail by Linsky et al. (1994).

We first assumed that there is only one velocity component for the Procyon line of sight, but were unable to match accurately the Mg II and Fe II line profiles. We then tried a variety of two-component models. We obtained excellent fits to the spectra with a two-component model in which the velocities differ by 2.7 km s^{-1}, slightly less than the instrument resolution of about 3.4 km s^{-1}. The value of the reduced χ^2 for the Mg II lines decreased from 3.23 to 1.32 when we changed from the best one-component model to the best two-component model.

Since the interstellar H absorption is saturated and broad, we had to infer a model for the intrinsic stellar line profile against which to measure the interstellar H and D absorption. Fortunately Procyon (unlike Capella) is similar to the Sun in spectral type, low rotational velocity, and weak chromospheric emission. We therefore first assumed that the intrinsic chromospheric emission line is similar in shape to that of the Sun. Since the Lyman-α lines of solar-type stars broaden with increasing luminosity (Landsman & Simon 1993), we broadened the solar line profile (about line center) by a range of multiplicative factors to best match the observed line profile outside of the interstellar absorption. We found that a broadening factor of 1.4 leads to a very good fit to the observed profile, except in the region of the D line.

With the assumed intrinsic Lyman-α line profile, our fit to the shapes of the H and D line profiles leads to a H broadening parameter $b_{HI} = 10.83 \pm 0.15$ km s^{-1}, which is slightly smaller than the value $b_{HI} = 11.22$ km s^{-1} for Capella. The thermal and turbulent contributions to this broadening differ somewhat from those for the Capella line of sight with a slightly higher temperature ($T \approx 7,200$ K) and significantly lower turbulence ($\xi \approx 1.2$ km s^{-1}). These quantities are assumed to be the same in both components as we cannot derive values for b_{HI} separately for the two components. The smaller value of ξ is a result of the identification of the two closely spaced velocity components. In our best one-component fit the value of ξ increases to 2.15 km s^{-1}.

We have explored a range of parameters N_{HI} and D/H to minimize χ^2 characterizing the difference between the observed and computed line profiles. Since no set of parameters provided an excellent match to the shape of the observed profile in the region of the D line, we concluded that the broadened solar line profile must be modified by a small amount. We assumed that the interstellar D/H ratio is the same as for the Capella line of sight [(D/H)$_{LISM} = 1.60 \times 10^{-5}$] and modified the broadened solar line profile in order to minimize χ^2 between the observed line profile and the computed line profile folded through the interstellar medium and the instrument. Figure 2 shows the modified intrinsic Lyman-α line profile and the observed and computed line profiles. Since the changes made to the intrinsic line profile do not result in a profile that is very different from the solar line, we believe that this intrinsic profile is plausible and the Procyon data are thus consistent with but do not prove that (D/H)$_{LISM} = 1.60 \times 10^{-5}$. Since the derived interstellar parameters are not very different from the values obtained using the scaled solar Lyman-α profile, the interstellar parameters are nearly independent of our assumed stellar Lyman-α line profile.

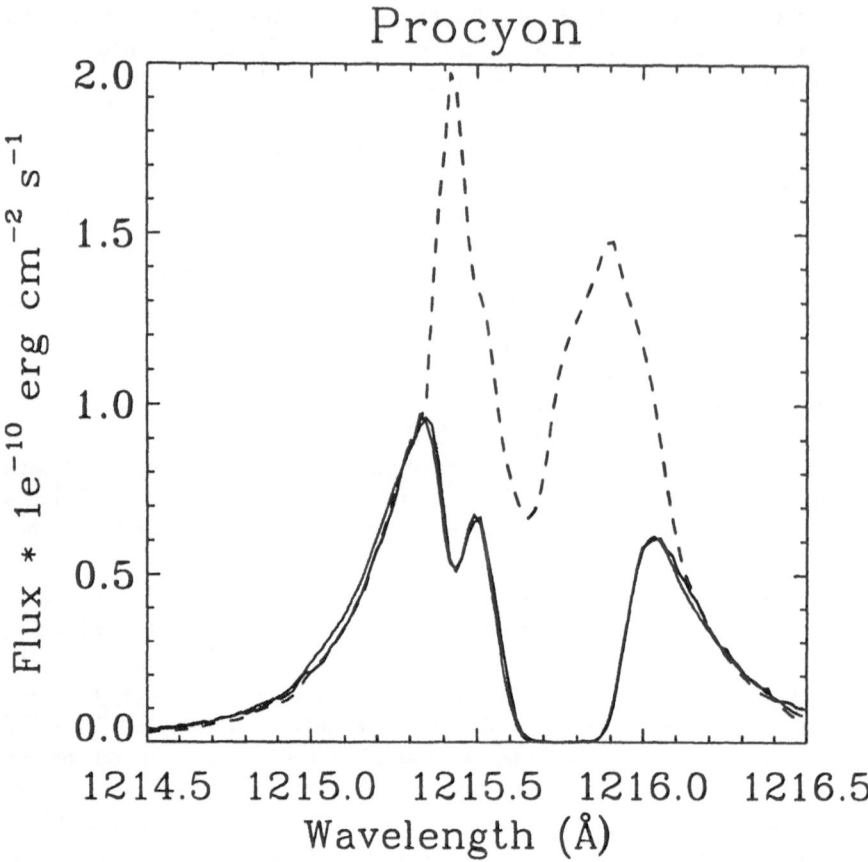

Fig. 2. Observed Lyman-α spectrum of Procyon compared with the instrinsic stellar emission line (assumed to be a slightly modified broadened solar profile) and the computed profile through the local ISM from Linsky et al. 1994. The computed and observed profiles are nearly indistinguishable.

2.3 Other Lines of Sight

The GHRS is being used to measure the D/H ratio toward other nearby stars. In addition to testing the accuracy of the D/H ratio determined for the Capella and Procyon lines of sight, observations of stars in other directions and at other distances are needed to determine whether the D/H ratio varies in our region of the Galaxy. In a poster at this meeting Lemoine describes the analysis of his GHRS spectra of the hot white dwarf G191-B2B and the inferred value of D/H for this 50 pc line of sight. We have requested GHRS spectra to observe the closest (1.3 pc) stars, α Cen A and B, and the binary system HR 1099 (33 pc).

Observations at both quadratures in the orbit of HR 1099 should help remove the uncertainty of the intrinsic stellar Lyman-α emission line. Other lines of sight should be explored using the GHRS echelle-A grating. HST programs to obtain D/H ratios for extragalactic lines of sight have been approved, but as yet there have been no reports of results from these difficult observations.

3 The Range of Ω_B Implied by D/H Measurements

An accurate determination of the primordial number ratio, $(D/H)_p$, should tell us the number density of baryons during the period 100–1000 seconds after the Big Bang when the temperature became low enough for the light nuclei to form. This follows from the density sensitivity of nuclear reaction rates that yield a higher abundance of ^{4}He and lower abundance of D for larger densities at that time. Since the Hubble expansion relates the baryon densities then and now, $(D/H)_p$ also determines the mean baryon density in the universe today and the ratio Ω_B of the local baryon density to the critical density needed to eventually halt the expansion. Thus $(D/H)_p$ is a critical parameter for experimental cosmology.

Although our data do not allow us to measure $(D/H)_p$ directly, we can infer its value from our measurement of $(D/H)_{LISM}$ and chemical evolution calculations for the Galaxy. Steigman & Tosi (1992) have calculated the survival fraction of D as the primordial D is converted to heavier elements in the cores of stars and this deuterium-depleted gas is dispersed into the interstellar medium from which later generations of stars are formed. Their calculations indicate that $(D/H)_p$ = $(1.5$–$3.0) \times (D/H)_{LISM}$, so that $(D/H)_p = (2.2$–$5.2) \times 10^{-5}$. Comparison of the Capella value for $(D/H)_p$ with recent Big Bang nucleosynthesis calculations (Walker et al. 1991) indicates that $\eta_{10} = 3.8$–6.0, where η_{10} is 10^{10} times the ratio of nucleons to photons by number. This range in η_{10} leads to the very important result that $0.06 \leq \Omega_B h_{50}^2 \leq 0.08$, where h_{50}^2 is the Hubble constant in units of 50 km s^{-1} Mpc^{-1}. Thus, no matter what value one assumes for the Hubble constant, $\Omega_B \ll 1$ and the universe must be open if the cosmological constant is zero and if only baryons are present. Tremaine (1992) and others, however, argue that $\approx 90\%$ of the universe consists of dark nonbaryonic matter. Thus whether $\Omega = 1.0$ or not remains an open question.

4 Comments on Recent Reports of·Extragalactic D/H

One way to avoid the uncertainties in Galactic chemical evolution models is to measure D/H in warm gas which has very low metallicity. In warm gas H and D are mainly neutral and not tied up in molecules. The measured D/H ratio should therefore be close to and can be extrapolated to zero metal abundance, the primordial value. Songaila et al. (1994, and independently, Carswell et al. 1994) has reported on observations of the line of sight toward the $z_{em} = 3.42$ quasar Q0014+813, which has a metallicity $< 10^{-3.5}$ solar (Donahue & Shull 1991). Precise measurements of D/H in such environments would provide an unambiguous test of Big Bang nucleosynthesis models, and as a byproduct would also test the accuracy of Galactic chemical evolution models.

The major uncertainty in the estimates of D/H in absorbing clouds toward Q0014+813, and by implication other distant lines of sight, is the possibility that a low column density H-absorbing cloud at the predicted D velocity is masquerading as D. This possibility was recognized by Songaila et al. (1994), Carswell et al. (1994), and by Bahcall (1994) in his comments on the initial discovery paper. Let us consider this point in more detail as the search for primordial D will surely lead to new observations and new "results" in the literature.

For each line of the Lyman series, D absorbs at -82 km s^{-1} relative to H but with a column density $10^{-4} - 10^{-5}$ that of H. It is critical therefore to determine velocity components (hereafter called "clouds") with neutral H columns $10^{-4} - 10^{-5}$ that of the most opaque H clouds. By analyzing the unsaturated higher Lyman line profiles, Songaila et al. (1994) identified five clouds in the Q0014+813 line of sight ranging from $N_{HI} = 5.5 \times 10^{16}$ cm^{-2} down to $N_{HI} = 6.5 \times 10^{14}$ cm^{-2}, nearly a factor of 100, with relative velocities of -17.5 to +155.0 km s^{-1}. The proposed D features lies at -99.5 km s^{-1} (-82 km s^{-1} from the most opaque H cloud). The column density of this feature is estimated to be 2×10^{12} cm^{-2}, a factor of 2.5×10^{-4} that of the most opaque H cloud. Is this feature D or H? If it is D, then $(D/H)_p \approx 2.5 \times 10^{-4}$, a factor of 5-10 times larger than $(D/H)_{LISM}$, whereas Galactic chemical evolution calculations (e.g. Steigman & Tosi 1992) indicate that this factor should lie in the range 1.5 – 3.

To answer this critical question one must guess what clouds may be present with H column densities a factor of 300 below the least opaque cloud measured. We have no real guidance for this other than that the local ISM is very complex with individual clouds on a scale of a few parsecs. On the other hand, warm gas on the line of sight to a quasar may have a different structure as a consequence of smaller cooling rates (due to lower metal abundance) and the larger ionizing radiation field from the quasar. Consider therefore a crude model of the warm gas region with two simple assumptions and the testable predictions of this model. We assume that (1) an increasing number of clouds are present in the line of sight with decreasing values of N_{HI}, and (2) the velocity centroid of the cloud distribution is centered on the mean velocity of the observed clouds ($< v >= +70$ km s^{-1} for Q0014+813).

These very simple assumptions lead to the prediction that for some quasars low N_{HI} but detectable clouds may appear just outside the saturated core of the Lyman-α line. The probability that this occurs depends on how rapidly the number of clouds increases with decreasing N_{HI}. Inspection of the Lyman-α line of Q0014+813 (Fig. 3 in Songaila et al.) shows that just outside the Lyman-α saturated core there is absorption centered at -99.5 km s^{-1}, which they interpreted as D Lyman-α absorption, and a weak depression near +230 km s^{-1}, which they did not discuss. According to our model both features could be weak H clouds. One way to tell whether one is observing D or H masquerading as D is to observe a large number of quasars and look for the absence of absorption at velocities more negative than the blue edge of the saturated core. For such systems one could estimate an upper limit to $(D/H)_p$. Another possible way is to search for an excess of absorption components blueward compared to redward of the saturated core in a large sample of quasars.

This work is supported by Interagency Transfer S-56460-D from NASA to the National Institute of Standards and Technology.

References

Bahcall, J.N. (1994): Nature **368** 584

Boesgaard, A.M. & Steigman, G. (1985): ARAA **23** 319

Brandt, J.C. et al. (1994): PASP **106** 890

Carswell, R.F., Rauch, M., Weymann, R.J., Cooke, A.J., & Webb, J.K. (1994): MNRAS **268** L1

Donahue, M. & Shull, J.M. (1991): ApJ **383** 511

Duncan, D.K. (1992): *Goddard High Resolution Spectrograph Instrument Handbook*, Version 3.0, Space Telescope Science Institute

Geiss, J. & Reeves, H. (1981): A&A **93** 189

Landsman, W. & Simon, T. (1993) ApJ **408** 302

Linsky, J.L., Brown, A., Gayley, K., Diplas, A., Savage, B.D., Ayres, T.R., Landsman, W., Shore, S.N., & Heap, S.R. (1993): ApJ **402** 694 [Paper I]

Linsky, J.L., Diplas, A., Wood, B., Brown, A., & Savage, B.D. (1994), in preparation

McCullough, P.R. (1992): ApJ **390** 213

Songaila, A., Cowie, L.L., Hogan, C.J., & Rugers, M. (1994): Nature **368** 599

Steigman, G. & Tosi, M. (1992): ApJ **401** 150

Tremaine, S. (1992): *Physics Today* **45** 28

Walker, T.P., Steigman, G., Schramm, D.N., Olive, K.A., & Kang, H.-S. (1991): ApJ **376** 51

Wilson, T.L. & Rood, R.T. (1994): ARAA **32** 191

Galactic Evolution of D and ^{3}He

D. Galli[1], F. Palla[1], F. Ferrini[2], and O. Straniero[3]

[1] Osservatorio Astrofisico di Arcetri, I-5015 Firenze, Italy
[2] Dipartimento di Fisica, Università di Pisa, I-56100, Italy
[3] Osservatorio Astronomico di Collurania I-64100 Teramo, Italy

Abstract. The evolution of D and ^{3}He is considered in the framework of a numerical model for the chemical evolution of the Galaxy. The destruction of D in the course of galactic evolution is found to be modest, of the order of a factor ~ 2. Conversely, the evolution of ^{3}He remains one of the major problems in this field: results based on the predictions of updated stellar models are shown to lead to an overproduction of ^{3}He by a factor $\sim$5–7 at the time of formation of the Sun, and by a factor $\sim$5–20 with respect to measured abundances in galactic HII regions. Some possibilities to reduce this discrepancy are presented and discussed quantitatively.

1 A Model for the Evolution of the Light Elements

According to the standard hot Big Bang model of nucleosynthesis, the abundances of D, ^{3,4}He, ^{6,7}Li, ^{9}Be, and ^{10,11}B in the primordial gas depend on one cosmological parameter only, the baryon-to-photon ratio $\eta \equiv n_b/n_\gamma$. The window of consistency for η_{10} ($= \eta \times 10^{10}$) is bounded from below by the upper limit on the primordial (D+^{3}He/H) and from above by the upper limit to the primordial ^{4}He (see e.g. Smith et al. 1993).

Since the abundance of D and ^{3}He is known only at the time of formation of the Sun ($t_\odot = 8.5$ Gyr) and at the present epoch ($t_0 = 13$ Gyr), the determination of the primordial abundance of D and ^{3}He requires some degree of theoretical modeling to evaluate the amount of D destruction during galactic evolution and the amount of ^{3}He destruction/production by stellar processes.

The equations that describe the evolution of the abundance of the chemical species i are

$$d(X_{i,h}g_h)/dt = -\psi_h X_{i,h} - fg_h X_{i,h} + W_{i,h},$$

and

$$d[X_{i,d}(g_d + c_d)]/dt = -\psi_d X_{i,d} + fg_h X_{i,h} + W_{i,d},$$

(see Ferrini et al. 1992), where the subscript h and d refer to the halo and disk regions, respectively. Here, ψ is the star formation rate, f the infall parameter, g and c are the mass fractions in the gas and cloud phases, respectively, and $W_i(t)$ is the restitution rate. These quantities are computed in the framework of the matrix formalism introduced by Talbot & Arnett (1976). Each star is subdivided into one or more concentric mass shells, where different nucleosynthesis processes have occurred. The mass of the element i ejected by a star of mass m after a time $\tau(m)$ from birth is assumed to be a linear combination of the masses of elements

j initially present in the star, $m_i^{\text{ej}} = \sum_j Q_{ij} m_j^{\text{in}}$, where the matrix elements Q_{ij} depend only on the mass of the star. The restitution rate then results

$$W_i(t) = \int_{m(t)}^{m_{\text{upp}}} \sum_j Q_{ij}(m') X_j[t - \tau(m')] \psi[t - \tau(m')] \phi(m') dm'.$$

The matrix elements $Q_{ij}(m)$ are computed from updated stellar evolutionary models (see Ferrini et al. 1992 and Galli et al. 1994a for details).

1.1 Stellar Nucleosynthesis of D and ^{3}He

Deuterium can only be destroyed in stars. Recently, Palla & Stahler (1991, 1993) have shown that D is completely destroyed during the protostellar and Pre-Main Sequence (PMS) phases in stars of all masses. Therefore, a mass equal to 3/2 of the initial mass of D is added to the initial mass of ^{3}He. In terms of the matrix elements Q_{ij}, this means $Q_{22} = 0$ and $Q_{32} = 3(1 - q_3)/2$, where q_3 is the stellar mass fraction within which ^{3}He has been burnt into ^{4}He and ^{7}Be.

^{3}He is produced in a star by the burning of the initial D during the PMS phase and via the pp chain in the MS phase, during which new ^{3}He is created in any sufficiently cool H burning zone independently of the initial D and ^{3}He abundances. When the star rises the Red Giant Branch for the first time, this freshly produced ^{3}He is uniformly distributed over the region covered by the convective envelope at its maximum inward extent. Thus, ^{3}He enrichment of the ISM from stars of low and intermediate mass is an unavoidable consequence of stellar evolution (in this context, a particular relevance assumes the first detection of ^{3}He in the ejecta of the planetary nebula NGC3242, recently reported by Rood et al. 1992, with $X_3^{\text{surf}} \simeq 3 \times 10^{-3}$).

The net production of ^{3}He in our model is expressed by the function $w_3(m)$. If $X_3^{\text{surf}}(m)$ is the abundance of ^{3}He brought to surface, then $w_3(m) = [X_3^{\text{surf}}(m) - X_3^{\text{in}}(m) - 3X_2^{\text{in}}(m)/2] \times [1 - d(m)]/X^{\text{in}}$, where $d(m)$ defines the mass fraction which remains as a stellar remnant upon the death of the star. We have adopted $X_3^{\text{surf}}(m) = 1.063 \times 10^{-3}(m/M_\odot)^{-2}$ (Schatzman 1987) which represents a good fit to Vassiliadis & Wood (1993) results for intermediate masses. This case will be referred to as the "Standard Model".

Massive stars ($m \geq 8M_\odot$) destroy rather than produce ^{3}He (Dearborn et al. 1986). Dearborn et al. (1986) compute the survival fraction $g_3(m)$ of ^{3}He at selected stellar masses (8, 15, 25, 50 and 100$M_\odot$); their quantity $g_3(m)$ is thus equivalent to our quantity $1 - q_3(m)$, and $Q_{33}(m) = g_3(m)$.

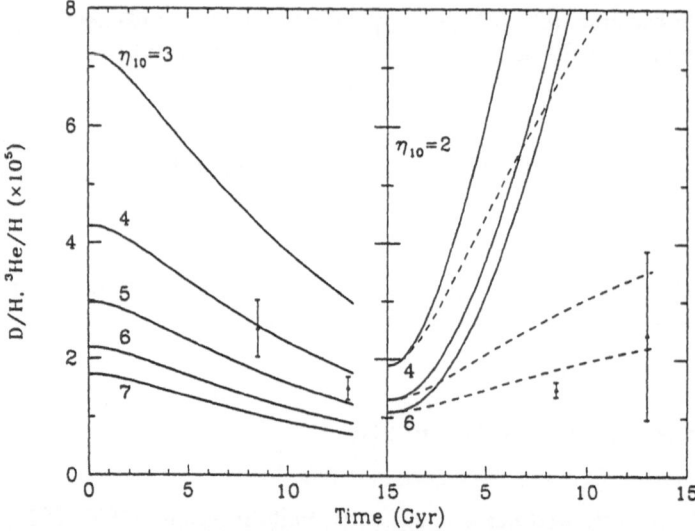

Fig. 1. (*a*) Evolution of (D/H) vs. time, and (*b*) evolution of (^{3}He/H) vs. time, for selected values of η_{10} (*solid* curves: Standard Model; *dashed* curves: $w_3 = 0$). Filled triangles with error bars represent the observed pre-solar and ISM abundances.

2   Results

The evolution of X_2 and X_3 for various values of η_{10} is shown in Fig. 1a,b. The astration of D is in all cases relatively modest: the surviving D is less than the primordial value by a factor ~ 1.5 at $t = t_\odot$ and ~ 2.5 at $t = t_0$. The abundance of D in the pre-solar material and in the ISM set stringent bounds on the value of η_{10}. If the values adopted in this paper are taken at face value, then the bounds on η_{10} from $(D/H)_\odot$ and from $(D/H)_0$ are $3.8 \leq \eta_{10} \leq 4.8$ and $4.2 \leq \eta_{10} \leq 4.9$, respectively.

The run of the ^{3}He abundance vs. time of Fig. 1b clearly shows a serious conflict with observations, unless stellar production of this isotope is zero, i.e. $w_3 = 0$. Even in this favourable circumstance, however, it is not possible to reproduce the solar value unless $\eta_{10} \simeq 7$, too high a value to provide acceptable results for D (and ^{4}He). We are thus left with the fundamental problem of finding an effective mechanism capable of eliminating, or at least reducing, the inconsistency.

In our opinion, there are only two possible solutions to this dilemma: (*i*) stellar models give correct values for the ^{3}He yield, *but* its abundance in the protosolar nebula and in HII regions has been reduced by about one order of magnitude by some unknown process; or, conversely, (*ii*) observed abundances of ^{3}He are representative of the (uniform) composition of the ISM at the corresponding epochs, *but* the predicted stellar yields of ^{3}He need to be drastically reduced.

Even though the first possibility cannot be completely dismissed, possibility (*ii*) is more appealing, although it appears to be in conflict with the ^{3}He abundance measured in the ejecta of the planetary nebula NGC3242. Should this value be substantially reduced by future observations, the possible existence of a mechanism able to destroy ^{3}He in stellar envelopes during the latest stages of stellar evolution would gain credit.

If this is the case, the solution to the problem of the galactic evolution of ^{3}He might come from nuclear physics. For instance, Galli et al. (1994b) have considered the effects that a resonance at low energies ($E_r \simeq 100$ keV) in the ^{3}He(^{3}He,2p)^{4}He reaction might have on the stellar nucleosynthesis of ^{3}He (Fowler 1972, Fetysov & Kopysov 1972, Castellani et al. 1993). To evaluate quantitatively the effect of a low energy resonance on the ^{3}He yield, we used the results from a set of stellar models computed with the FRANEC code, in which the rate for the ^{3}He(^{3}He,2p)^{4}He reaction was modified to include the effect of a resonance $E_r = 10$ keV or $E_r = 3.0$ keV. The time evolution of the ^{3}He abundance obtained with the stellar yields evaluated in the resonant case gives a starting result: the ^{3}He abundance remains almost constant during galactic evolution and the pre-solar value may be easily accounted for by either values of the resonance parameters (see Galli et al. 1994b for more details). Experiments currently in progress in underground laboratories should be able to extend cross-section measurements for the ^{3}He(^{3}He,2p)^{4}He reaction down to $E_r \simeq 10$ keV with the required sensitivity (Arpesella et al. 1991).

References

Arpesella, C, et al. 1991, Nuclear Astrophysics at Gran Sasso Laboratory, INFN-Laboratori Nazionali del Gran Sasso, int. rep.

Castellani, V., Degl'Innocenti, S., & Fiorentini, G. 1993, A&A, 271, 601

Dearborn, D. S. P., Schramm, D. N., & Steigman, G. 1986, ApJ, 302, 35

Ferrini, F., Matteucci, F., Pardi, C., & Penco U. 1992, ApJ, 387, 138

Fetysov, V. N., & Kopysov, Y. S. 1972, Phys. Lett., B40, 602

Fowler, W. A. 1972, Nature, 238, 24

Galli, D., Palla, F., Ferrini, F., & Penco, U. 1994a, ApJ, in press

Galli, D., Palla, F., Straniero, O., & Ferrini, F. 1994b, ApJL, in press

Palla, F., & Stahler, S. W. 1991, ApJ, 375, 288

Palla, F., & Stahler, S. W. 1993, ApJ, 418, 414

Rood, R. T., Bania, T. M., & Wilson, T. L. 1992, Nature, 355, 618

Schatzman, E. 1987, A&A, 172, 1

Smith, M. S., Kawano, L. H., & Malaney, R. A. 1993, ApJS, 85, 219

Talbot, R. J. & Arnett W.D. 1973, ApJ, 186, 51

Vassiliadis, E., & Wood, P. R. 1993, ApJ, 413, 641

Chemical Evolution Models of D and ^{3}He: Problems ?

M.Tosi[1], G.Steigman[2], D.S.P.Dearborn[3]

[1] Osservatorio Astronomico, Via Zamboni 33, I-40126 Bologna, Italy,
[2] Institute of Astronomy, University of Cambridge, Madingley Road, Cambridge CB3 0HA, U.K.
[3] Lawrence Livermore National Laboratory, Livermore, CA 94550, USA

1 Introduction

The importance of studying the chemical evolution of the Galaxy to interpret the observed light element abundances and their cosmological implications was already recognized in the early seventies (Reeves *et al.*,1973) and only recently reconsidered by several different groups. Our aim is to check if *standard* (i.e. non *ad hoc*) chemical evolution models can reproduce the observed amounts of D and ^{3}He and, if so, for what range of primordial abundances.

In this framework, a few years ago we applied (Steigman and Tosi 1992, hereinafter ST92) two numerical models to D and ^{3}He which had proven successful (Tosi,1988) in reproducing all the major observational constraints of our galactic disk (see e.g. Giovagnoli and Tosi in this volume). In ST92 we did not assume the instantaneous recycling approximation thus taking into account properly the stellar lifetimes. We took the primordial abundances of D and ^{3}He as free parameters and for their stellar nucleosynthesis we assumed that all deuterium which enters a star is burnt to ^{3}He in the pre-main sequence phase, whereas a significant fraction of the ^{3}He present on the main sequence (which thus corresponds to the combination of the initial ^{3}He with the initial D) survives stellar processing. For the survival fractions g_3 in stars of different initial mass, we followed Dearborn et al. (1986); this corresponds to $g_3=1$ for M/M$_\odot$ <3, $g_3=0.7$ for $3 \leq$ M/M$_\odot$ ≤ 8, and $0.2 < g_3 < 0.5$ for M/M$_\odot$ >8. By comparing the model predictions with the D and D+^{3}He abundances observed in the solar system and in the interstellar medium (ISM), we were able to put the following limits to the primordial abundances by mass of D and ^{3}He: $3.9 \leq 10^5 X_{2p} \leq 9.0$ and $1.9 \leq 10^5 X_{3p} \leq 4.0$. For standard BBN (Walker et al. 1991), they correspond to $3.7 \leq \eta_{10} \leq 4.2$.

In the meantime, the early suggestion by stellar evolution theorists of a net production of ^{3}He in low mass stars (e.g. Iben 1967, Rood, Steigman and Tinsley 1976) has become popular again. We have now recomputed our chemical evolution models adopting the new stellar evolution computations by Dearborn (1994), which do take ^{3}He production into account. These stellar models cover the entire mass range from 0.65 to 100 M$_\odot$ and are computed for several initial metallicities Z. We adopt the yields for Z=0.002 from the formation of the galactic disk up to the time (different at different galactocentric distances) when the

ISM reaches the solar abundance and then adopt the Z=0.02 yields. According to Dearborn's new results, the smaller the stellar mass the larger the ^{3}He net production. For stars heavier than 2.5 M$_\odot$ the temperature is too high and there is, instead, a net destruction, which is more significant than earlier estimated by Dearborn et al.(1986).

2 Models and Results

We have compared the observational data with the predictions of our chemical evolution models based on the above nucleosynthesis prescriptions for ^{3}He and assuming, as in ST92, that all stellar D is immediately converted into ^{3}He. Figure 1a shows the D abundances by mass X_2 in the solar ring as a function of time. The vertical bars represent the values (with a 2 σ error) adopted by ST92 and derived from observations in the solar system (taken as representative of the local ISM 4.5 Gyr ago) and in the present ISM.

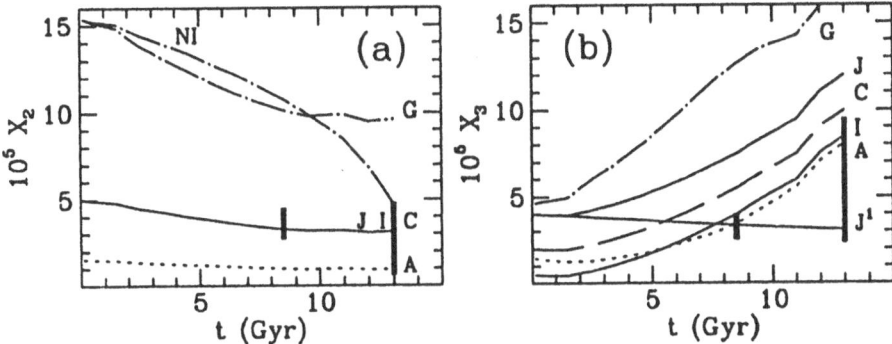

Fig. 1a. Time distribution of deuterium in the solar ring.

Fig. 1b. Same for ^{3}He. See text for symbol description.

All curves refer to model 1 of ST92 (with almost constant SFR and infall of primordial gas) except for the top dot-dashed one in Fig.1a, labeled NI, which shows the predictions of a model assuming no galactic infall after the disk formation. This model does not reproduce the major observational constraints but is useful to show what could be the maximum D destruction in these *standard* chemical evolution models. The different curves are labeled according to the various combinations of primordial abundances of D and ^{3}He examined by ST92 (see their Table 1) with the addition of cases I and J, assuming $X_{2p}=5\ 10^{-5}$ and $X_{3p}=5\ 10^{-6}$ or $X_{3p}=4\ 10^{-5}$, respectively. Since the overall depletion of D turns out to be only a factor of 1.6 (3 in the NI case), it is apparent that only with low primordial abundances of D ($X_{2p}\simeq5\ 10^{-5}$) can our models reproduce both sets of data: if the primordial value is as large as the upper limit suggested by Songaila et al. (1994) our models have no chance of predicting consistent D abundances. Qualitatively the same results are obtained also with chemical evolution models (see model 25 in ST92) with more rapidly decreasing SF and infall rates.

Figure 1b shows the analogous plot for the ^{3}He abundance by mass X_3. The vertical bars correspond respectively to the observational range derived by Steigman (1991) for the solar system and to the whole range of abundances in the ISM measured by Bania et al. (1987) in HII regions. All curves are labeled as in Fig.1a and refer again to model 1 of ST92 but with the new ^{3}He nucleosynthesis described above. The effect of the net production is apparent in the rise of all curves (except case J^1, see below) from the epoch when stars smaller than 2.5 $M_\odot$ start to die and contribute to the ISM enrichment. The observed abundances can be reproduced only if the primordial X_{3p} is lower than $1.5 \ 10^{-5}$.

It is somewhat reassuring that at least one of the cases (labeled I) which fits the observed X_3 abundances also reproduces very well the observed X_2 values. However, one should never restrict such analyses only to the solar neighbourhood, which is not representative of the whole Galaxy. Models of the chemical evolution of the Galaxy should always be compared with all available constraints.

Figure 2 shows the present distribution with galactocentric distance of the abundances by mass of the four lightest elements as predicted by model 1. Hydrogen and deuterium are the only elements always depleted by any stellar process and they show a positive gradient with galactic radius. This is because in the inner galactic regions there is (and has been) more star formation activity, therefore more stellar processing, and a smaller fraction of the primordial amount of these elements is still present in the ISM. From the same argument, elements produced by stellar processes show instead a negative galactocentric gradient, as indicated in the figure by the ^{3}He and ^{4}He distributions and confirmed by observations of heavier elements.

In the case of ^{3}He, we can compare its predicted distribution with galactocentric distance with that derived by Bania et al. (1993) from HII regions observations. In Figure 3 the dots and relative error bars correspond to the observed data and the solid curve refers to case I of model 1, which predicts a time behaviour of D and ^{3}He in agreement with solar neighbourhood data, as shown in Figs.1. The space behaviour predicted by this model is however totally inconsistent with the observed distribution. Whatever the uncertainty attributable to the empirical abundances, they would never show the steep negative gradient predicted by the theoretical models. The slope of this gradient can depend on the chemical evolution model but not its sign, which is the natural consequence of assuming net stellar production of ^{3}He. In fact, if we arbitrarily remove any such production from the stellar models, by simply assuming that all D+^{3}He which enters in stars smaller than 2.5 $M_\odot$ is ejected unaltered (case J^1), we obtain the dashed line, which is in amazingly good agreement with the observed data. It is worth noticing that case J^1 also fits very well the data observed in the solar neighbourhood (Fig. 1).

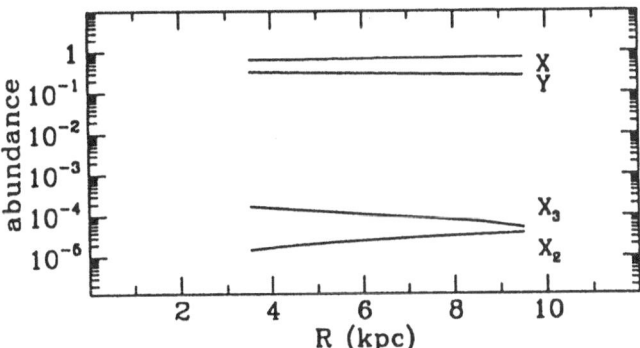

Fig. 2. Predicted present distribution of X, X_2, X_3 and Y as a function of galactocentric distance.

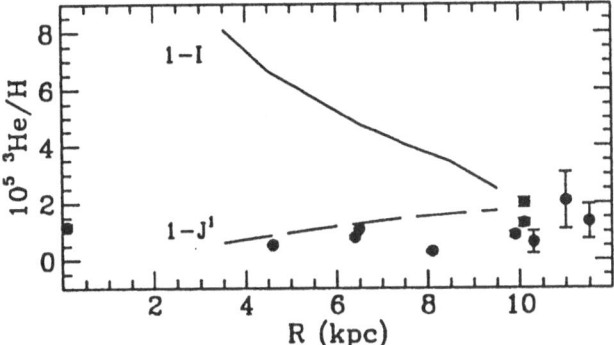

Fig. 3. Present radial distribution of ^{3}He from models and HII region data. Curves and labels are as in Fig.1.

3 Summary

We can summarize the current status of our understanding of the evolution of D and ^{3}He by saying that if stellar nucleosynthesis indeed follows the suggestions of Dearborn's(1994) models (as well as others, see Galli et al. in this volume), *standard* chemical evolution models cannot reproduce *all* the observed data on D and ^{3}He. However, the available stellar models have not followed in full detail the ^{3}He nucleosynthesis, specially in the late evolutionary phases. We thus hope that some process to burn ^{3}He into heavier elements or some mechanism to inhibit its production will be found by future stellar evolution studies. In that case, our chemical evolution models may reproduce both the time and the space observed behaviours of D and ^{3}He.

References

Bania, T.M., Rood, R.T., Wilson, T.L. 1987, Ap.J. 323, 30

Bania, T.M., Rood, R.T., Wilson, T.L. 1993, in *Origin and Evolution of the Elements*, N.Prantzos, E.Vangioni-Flam, M.Casse eds (Cambridge Univ. Press), p.107

Dearborn, D.S.P., 1994 in preparation

Dearborn, D.S.P., Schramm, D.N., Steigman, G. 1986, Ap.J. 203, 35

Galli, D., Palla, F., Ferrini, F. 1994 this volume

Giovagnoli, A., Tosi, M. 1994, this volume

Iben, I.Jr. 1967, Ap.J. 147, 624

Reeves, H., Audouze, J., Fowler, W.A., Schramm, D.N. 1973, Ap.J. 179, 909

Rood, R.T., Steigman, G., Tinsley, B.M. 1976, Ap.J. 207, L57

Songaila, A., Cowie, L.L., Hogan, C.J., Rugers, M. 1994, Nature 368, 599

Steigman 1991, in *Prim. Nucl. and Evol. of Early Universe*, K.Sato and J.Audouze eds (Dordrecht, Kluwer) p.3

Steigman, G., Tosi, M., 1992, Ap.J. 401, 150

Tosi, M., 1988, A.A. 197, 33

Walker, T.P., Steigman G., Schramm, D.N., Olive, K.A., Kang, H.S., 1991, Ap.J. 376, 51

The Interstellar D/H Ratio Toward G191–B2B

Martin Lemoine[1], Alfred Vidal–Madjar[1], Roger Ferlet[1], Philippe Bertin[1], Cécile Gry[2], Rosine Lallement[3]

[1] Institut d'Astrophysique de Paris, CNRS, 98 bis boulevard Arago, 75014 Paris, France
[2] Laboratoire d'Astronomie Spatiale, CNRS, Marseille, France
[3] Service d'Aéronomie du CNRS, Verrieres-Le-Buisson, France

Abstract. We report the results of our HST–GHRS observations of the DA white dwarf G191–B2B aiming at deriving the interstellar D/H ratio. Our data suggest possible variations of the local interstellar D/H ratio and that a refined estimate of the D/H ratio toward α Aur (Linsky et al., 1993) might lie closer to 1.5×10^{-5} than to 1.7×10^{-5}.

1. Introduction

The primordial (D/H) ratio is well-known for providing the most sensitive direct observable to the single parameter of standard Big–Bang nucleosynthesis, viz. the baryonic density Ω_B. The interstellar (D/H) ratio is representative of the present epoch, and should allow to put constraints on the primordial abundance *via* use of models of chemical evolution. This is however not the case, since simple models predict a factor of depletion around 2–3 for D during the galactic evolution (there is no known astrophysical site where D may be produced in significant amounts), and the interstellar (D/H) ratio is $(D/H)=1.65^{+0.07}_{-0.18} \times 10^{-5}$ (Linsky et al., 1993) whereas the predictions of BBN give $(D/H)_p \sim 10^{-4}$ and recent estimates of the primordial ratio in a QSO line of sight yield $(D/H)_p \lesssim 2.5 \times 10^{-4}$ (Songaila et al., 1994; Carswell et al., 1994). In order to account for a depletion factor of $\simeq 7$, some specific models of chemical evolution have to be put at work (see for instance Olive, 1994; Tosi, 1994). If a value as high as $(D/H)_p \simeq 2.5 \times 10^{-4}$ was confirmed, then a very specific model based on galactic winds issued by SNII and a bimodal star formation rate might have to be considered (Cassé, 1994). The interstellar (D/H) ratio is thus also a sensitive probe of galactic chemical evolution.

The history of interstellar (D/H) ratios measurements is long, and we refer the reader to reviews by Vidal–Madjar (1990), Ferlet (1992). These have been performed following mostly two observational methods: one is to observe the absorption of interstellar DI against the chromospheric emission line of cool stars, and the other is to observe the DI absorption against the continuum of hot stars. Both methods suffer from high systematics. In the first case, the (D/H) ratio will somehow depend on the modelisation of the stellar Lyman α emission line, while in the second, hot stars are usually located more than 50 pc away, and as a result the line of sight is often complex and the column density of HI is too high to allow to detect DI at Lyman α.

Here we present the results of our HST–GHRS observations using a new observational strategy that avoids the problems mentioned above. We observed the nearby DA white dwarf G191B2B, whose stellar continuum in the Lyman α region is smooth (Lorentzian profile of stellar absorption), so that the line of sight was not too complex (see below), and the stellar continuum could be normalized to unity without difficulty.

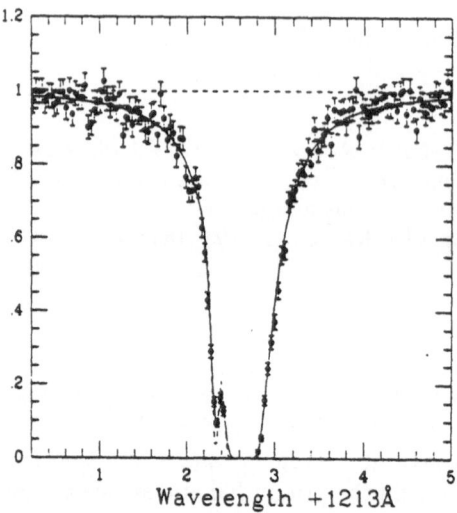

Fig. 1. Spectrum of G191B2B in the Lyman α region. The stellar absorption profile has been removed. The interstellar DI absorption is seen in the blue wing of the interstellar HI profile. The solid line shows the fit to the HI and DI profiles. The missing data points at the bottom of the HI line correspond to the geocoronal HI emission line which has been removed.

We have observed and analyzed using a profile fitting procedure the following interstellar absorption lines: FeII (2344Å), MgII (2800Å, doublet) at high resolution ($\simeq$ 5 km/s); SiII (1190Å doublet, 1260Å, 1304Å), SiIII (1206Å), CII (1334Å), NI (1200Å triplet), OI (1302 Å), and HI Lyman α (1216Å) at medium resolution ($\simeq$ 18 km/s). The main idea is to reach an agreement on the structure on the line of sight among the different lines analyzed by comparing the fitting of a line with that of the others and iterating, so that at the end, the remaining unknown parameters involved in the profile fitting of the DI and HI Lyman α lines are precisely the DI and HI column densities. We detect 3 absorbing components with the following characteristics:

N(HI) (cm^{-2})	$N_A = 0.40 \times 10^{18}$	$N_B = 1.85 \times 10^{18}$	$N_C = 0.01 \times 10^{18}$
Heliocentric Velocity (km/s)	$V_A = 9.9$	$V_B = 20.6$	$V_C = -1.7$
Temperature (K)	$T_A \simeq 9000$	$T_B \simeq 7000$	$T_C \sim 2500$
Turbulence (km/s)	$\sigma_A \simeq 1.5$	$\sigma_B \simeq 1.6$	$\sigma_C \simeq 1.5$

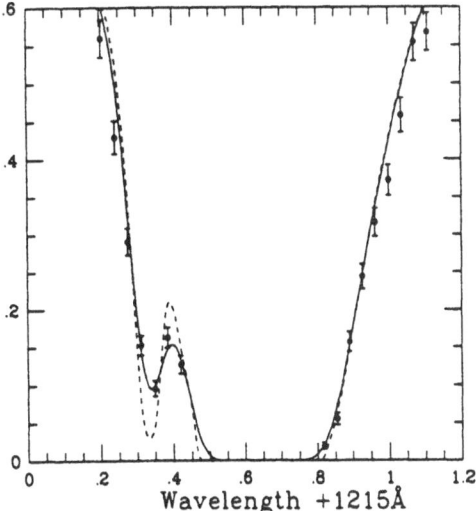

Fig. 2. Enlarged portion of the spectrum shown in Fig.1. As before, the solid line shows the fit to the profile and the dashed line the theoretical profile before convolution with the instrumental profile.

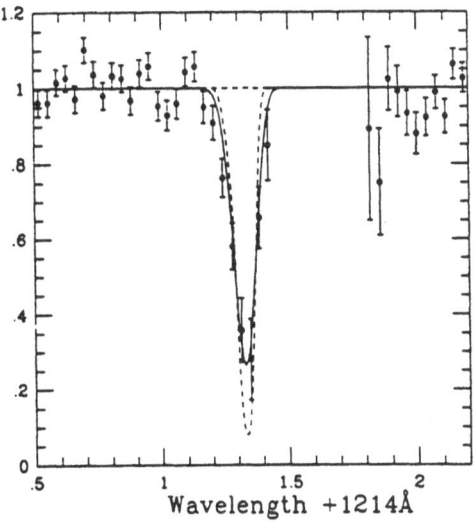

Fig. 3. The interstellar DI line is shown here as residuals of the previous fit by subtracting to the observed profile the calculated HI absorption of the three interstellar clouds. The fit to the DI profile calculated in Fig.1 and Fig.2 is shown as the solid line.

We derive an average D/H ratio:

$$N(DI)_{Total}/N(HI)_{Total} = 1.3 \times 10^{-5}$$

and the individual D/H ratios:

$$(D/H)_A = 1.0^{+0.4}_{-0.1} \times 10^{-5}$$
$$(D/H)_B = 1.4^{+0.1}_{-0.3} \times 10^{-5}$$
$$(D/H)_C = 1.5^{+1.0}_{-1.0} \times 10^{-5}$$

Since the DI line is clearly detected in the wing of the HI Lyman α line, the average D/H ratio is strongly constrained to its value of 1.3×10^{-5}. The unfortunately large error bars associated with individual (D/H) ratios are due to the lack of spectral resolution, the high resolution grating Echelle–A being out of order by the time of our observations. It appears however that our cloud B may be exactly identified with the cloud detected by Linsky et al. (1993) in their analysis of the D/H ratio toward α Aur: same turbulence velocity and temperature, same radial velocity, same column densities of FeII and HI, and a separation of only $7°$ between both targets for a distance less than 50pc to the further G191-B2B. This cloud corresponds in fact to the local cloud in which the solar system is embedded, and this similarity was to be expected.

This means that, if we enforce our $(D/H)_B$ ratio to the 1σ lower limit of the D/H ratio found by Linsky et al. (1993), $(D/H)_{\alpha Aur} = 1.65^{+0.07}_{-0.18}$, then our $(D/H)_A$ ratio in cloud A is constrained to $(D/H)_A \leq 1.1 \times 10^{-5}$.

This data set thus adds its name to the list of interstellar D/H ratios measurements that suggest the possible existence of variations of this ratio on short distances in the ISM, a longstanding and yet unresolved problem....

References

Carswell, R. F., Rauch, M., Weymann, R. J., Cooke, A. J., Webb, J. K. W.: 1994, preprint

Cassé, M., Vangioni–Flam, E.: 1994, these proceedings

Ferlet, R.: 1992, in IAU Symposium 150, p.85

Linsky, J. L., Brown, A., Gayley, K., Diplas, A., Savage, B.D., Ayres, T.R., Landsman, W., Shore, S. W., Heap, S. R.: 1993 ApJ **402**, 694

Olive, K. A.: 1994, these proceedings

Songaila, A., Cowie, L. L., Hogan, C., Rugers, M.: 1994, Nature **368**, 599

Tosi, M.: 1994, these proceedings

Vidal-Madjar, A.: 1991, Adv. Sp. Res. vol **11**, n° 11, p. 1197

Part VI

Lithium

Observational Status of Lithium in Stars

F. Spite

Observatoire de Paris, section de Meudon, 92195 Meudon Cedex, France

Abstract. Some of the new data about lithium abundance in stars are reviewed here. The data, that seem more directly linked with the problem of the primordial nucleosynthesis, are briefly discussed. In order to reach a clear conclusion, progresses in the stellar atmosphere theory are necessary.

1 Introduction

I will try here to update previous reviews on the subject (e. g. Boesgaard & Steigman, 1985, Rebolo 1989, D'Antona 1990, Deliyannis 1990, Michaud & Charbonneau 1991, Spite's 1993, 1994), but I am unable in a limited space to quote all the numerous works made recently in this field; I beg the pardon of the authors I will not quote here. I will concentrate on the works more closely related to the two questions which appear to have far reaching consequences :

1) - is the lithium abundance in the old Pop II dwarfs identical to the primordial lithium abundance ? or is the lithium abundance (observed in the superficial layers of these stars) severely altered ? by depletion ? by enrichment of the material forming the star (cosmic rays) ?

2) - is lithium efficiently produced in the Galaxy ? how ?

The answer to these questions has to be sought in the careful comparison between the observations and the predictions of the physical processes of depletion and production of lithium in stars of different masses, temperatures, structures, ages, metallicities, rotations etc.

Let us recall that lithium may be depleted at the surface of stars by stellar wind (Faulkner, Reeves), by microscopic diffusion (Michaud, Vauclair), by turbulent diffusion induced by rotation (Charbonnel, Vauclair), by meridional circulation (Charbonneau), by simultaneous transport of (Li-poor) matter and angular momentum (Zahn, Demarque, Deliyannis, Pinsonneault), by mixing due to internal waves (Schatzman, Montalbán), by convection in the Pre-Main Sequence phase (Faulkner, D'Antona & Mazzitelli), by convection in the Main Sequence phase for low mass, low temperature stars, etc.

Even if we are essentially interested in the behavior of lithium in Pop II dwarfs, the discussion of other types of stars is important for testing the theories. Let us begin by a brief look at the behavior of the Li abundance in the Pop I stars.

2 Lithium in Pop I Stars

2.1 T Tauri Stars and Very Young Stars

The abundance of lithium in the T Tauri stars is a difficult problem : the absorption lines in the spectra of T Tauri stars are often affected by some veiling, which has to be taken into account, and the veiling also affects the colors. In some samples, the Li and the K resonance lines are enhanced simultaneously (Padgett 1990) suggesting an atmospheric effect rather than a real abundance enhancement. In fact, Li has been found more abundant in the T Tauri phase by some authors, but not by other ones, using more sophisticated methods (Duncan using a subordinate line, Martín E. L. et al. 1992, Martín E. L. et al. 1994, using NLTE analysis). Some very young stars (King 1993 for PMS stars in Orion, Favata 1993 for X sources) have been found Li-rich (log NLi = 3.6 dex or more).

Is this enhancement real ? Is there a difference of Li abundance between different star forming regions ? And even (King 1993) within the Orion region ? The theory of Li depletion in PMS stars do not seem to agree with the observations. Let us note that several recent works (e. g. Edwards 1994) suggest a non negligible accretion of material from the disk to the surface of the T Tauri stars. Is the stellar surface fed by Li-rich interstellar matter? Is it clearly established that the WTTS are Li depleted relatively to the Classical TTS ? Detailed observations should provide the answers to these questions : it is essential to understand wether there is (or not) some Li depletion between the T Tau phase and the Main Sequence phase in the Population I stars or wether a real depletion is masked by an accretion. On this point (and this remark is also valuable for the following points, I will not repeat it) see the contributions in this Workshop.

2.2 Open clusters

A considerable amount of work has been done in this field, especially by Soderblom and collaborators. Let us note for the Hyades, the experimental work of Thorburn et al. (1993), one of their results (the scatter of the Li abundance) and the excellent fit with the PMS depletion computed by Swenson et al. (1994), using a major advance accomplished on the theoretical side : the computation of better opacities.

2.3 The Boesgaard-Tripicco's Dip

One important feature of the lithium abundance in field and cluster stars is the lithium dip : the work of Balachandran in the field stars(1990a) and in the old cluster M 67 (1990b) shows that the dip should be due to lithium destruction in the main sequence phase. However, a dip is marginally found for Be (Michaud and Charbonneau 1991) and O (García López 1993) in the Hyades, favoring, if confirmed, the interpretation by diffusion. The depletion of Li, Be, B in Procyon is possibly not linked with the process of the Dip, but with another process (Lambert et al. 1991).

2.4 Brown Dwarfs

Magazzú et al. (1993) emphasize that Li is not destroyed in brown dwarfs, so that Li in Pop II brown dwarfs could provide directly an upper limit of the primordial abundance. However, the limits for the brown dwarfs have to be better understood : for example, at the Keck telescope, Marcy et al. (1994) do not detect Li in the candidates of the Pleiades, and the problem is not yet understood.

2.5 Active Stars

A considerable work has been done by Pallavicini and collaborators on the RS CVn stars : Li is statistically (but only statistically) more abundant in RS CVn, and in active stars in general. Is there some inhibition of Li depletion ? The RS CVn stars are tidally locked binaries, we will see hereafter that the theory of Zahn (1994) predicts the inhibition of the Li destruction in some cases of binarity.

Is the higher Li rather linked with the fact that the progenitors are stars with a higher mass and therefore a shallower convective zone ?

2.6 Lithium in Giants

The observed lithium abundance in the giants is generally lower that the abundance predicted by the theory, but Charbonnel & Vauclair (1992), and Charbonnel (1994) explain this lower abundance by a lithium depletion on the main sequence (in the deep layers, not on the surface) followed by an extra-mixing on the giant branch.

A few giants have a high Li, in spite of a strong mixing attested by the isotope ^{13}C. Fekel & Balachandran (1993) suggested some Li production : in the spin-down of the envelope of the massive progenitor the rotating core produces a quick dredge-up of ^7Be (and therefore a corresponding ^7Li production). Wallerstein and Morell (1994) propose a mini-mixing, deep enough for bringing ^7Li to the surface, shallow enough for keeping unaltered the $^{13}C/^{12}C$ ratio.

3 Population II

3.1 General Trends

Thorburn (1994) analyzes a sample large enough for providing accurate trends : she finds a scatter, a slope in the Li versus Teff plane, a slope in the Li versus Fe/H plane. She estimates that the scatter is real since it is larger than the observational errors, the scatter is explained by the progessive formation of ^7Li by cosmic rays. The slope and shape found by Thorburn for the plateau is not compatible with the microscopic diffusion model, and is strongly constraining the rotational model.

The work of Norris et al. (1994) and the work of Thorburn suggest that the extremely metal-poor stars have, in the average, a lower Li abundance, a fact difficult to understand. This is not confirmed by the analysis (Spite et al. 1993) of "primitive stars" such as the Sr-poor stars noted by Ryan et al. (1991). Could this strange feature be an artifact caused by a model atmosphere problem ?

3.2 A Digression About Temperature Determination

The scatter of the lithium abundance around the mean value of the plateau seems to be an important problem, since the tenants of the rotational model note that this scatter is the best argument in favor of this model. E. Terlevich (this workshop) recalled that "abundances are not observed, but derived from observations through long and often tortuous paths....", and it is clear that the determination of the lithium abundance depends of the choice of the effective temperature. Thorburn (1994) finds the temperature from the observed colors and assumes that there is a one to one correspondance between color and temperature, that the correspondance itself, translated in a mathematical formula, may introduce systematic errors but no random errors and therefore no scatter. This assumption is indeed suggested by the available model atmospheres, which do not predict large variations of the colors when the parameters (gravity, metallicity, microturbulence) of the model have variations. However, the model atmospheres are simplified models, far from the complexity of real stars. These models were quite satisfactory when used for the determination of abundances with an accuracy of 0.2dex. If, pushed by the curiosity of cosmologists, now abundances with an accuracy of 0.02 dex or better are needed, the details of the stellar structure can no more be ignored since we are dealing with stars of different age, degree of evolution, metallicity and structure. The same relation color-temperature does not necessarily holds for all the stars, and a part of the scatter has to be ascribed to the (up to now neglected) diversity of the color-temperature relations for the different stars. The problem is that the temperature of the model is found by fitting the predicted to the observed color, both quantities being related to deep layers of the atmospheres, whereas the absorption lines of metals depend on the temperature of the external layers. A structure difference will make that similar deep layers (similar colors) do not necessarily imply similar external layers, at the accuracy aimed for. The work of Fuhrman et al. (1993) shows how sensitive is the color of metal-poor dwarfs to the details of the convective zone in Pop II dwarfs. They suggest that there is no one to one correspondance between color and the depth of the wings of the H_α line. This important point has to be checked, but it happens that it is sometimes possible to determine the stellar temperature from the excitation equilibrium in the weak lined Pop II stars : the lithium abundances determined using this temperature show a scatter largely reduced (Spite et al. in preparation), and even so small that it could be entirely explained by measurement errors.

The slope of the plateau is probably a less important problem, but the color temperature relations used for finding the lithium abundances on the plateau are affected by systematic errors, which are not necessarily identical for cool and warm stars, so that a slope in the relation Li versus Teff could be due, at least partly, to this systematic error. Moreover, a systematic error in the zero point of the equivalent width measurement also produces a spurious slope. Finally, the NLTE effects (Carlsson et al. 1994) may reduce by half the slope found by J. Thorburn.

The numerical values of the scatter and slope derived in the work of Thorburn (1994) have to be considered with caution : both could be vanishingly small.

3.3 The ^{6}Li Isotope

The new determination of the ^{6}Li abundance in a few Pop II stars (Nissen, this Workshop) provides an important argument in favor of the non depletion of ^{7}Li in the Pop II stars : however, the interpretation of the profile of the Li line by the additional absorption due to the ^{6}Li isotope is not completely unique, and a (narrow) escape is still possible : a wider line with a higher differential redshift (Maurice et al. 1983). Moreover, even when the presence of ^{6}Li is admitted, the tenants of the rotational models emphasize that an escape is still possible by a fine tuning of these rotational models.

3.4 Subgiants

The very important work of Pilachowski et al. (1993) shows for Pop II subgiants a behavior which may agree with diffusion theory (Proffitt & Michaud 1991), however the mixing on the giant branch is stronger than predicted (similar to Pop I problem ?).

4 Pop I and II Binaries

Rotational models (Pinsonneault et al. 1992) and the theory of angular momentum loss of Zahn (1994) predict an inhibition of Li depletion in tidally locked binaries arriving on the main sequence. For Pop I stars, binaries in the Hyades (Thorburn et al. 1993) and one binary in M 67 (Deliyannis et al. 1994 and Ryan & Deliyannis 1995, preprint) have more Li than the single (or long period binary) stars. A similar case is found in the Praesepe cluster, but another binary is on the contrary Li-poor. In the Hyades, the difference is larger for the cool stars, suffering a large depletion. The sample of short period binaries is however small, and it is difficult to derive a very firm rule.

A lithium preservation in a binary may be masked by an extra-depletion by another process. In the presence of antagonistic processes, it is not possible to disprove a theory, but it is also impossible to confirm it. The case of the Pop II and old Pop I tidally locked binaries is dealt with by Pasquini (this Workshop) : partial inhibition is found only for cool stars (and not for all of them), suggesting possibly that either no depletion occurs in warm stars (and therefore an inhibition has no effect) or the inhibition does not work on warm stars, or is compensated for (by a strong depleting wind maintained for a longer time ?).

5 Li-Poor Stars in Pop I and II

Spite and Spite (1982) found a very stange feature in the Li abundance of the nearby G-type stars : the distribution of the ages is the same for the Li- rich and the Li-poor stars. Conversely, the Li abundance distributions are rather similar for the younger and the older stars. The accuracy of the ages, based on trigonometric parallaxes, was poor, and it was thought necessary to get the Hipparcos measurements for pushing the interpretation any further. Pasquini (1995) recently measured accurate Li abundances in a larger sample of nearby G-type stars, and simultaneously several (age-linked) parameters : v sini, chromospheric activity, absolute magnitude Mv : he also found a distribution of ages similar for the Li-rich and the Li-poor stars, whatever the parameter chosen for estimating the age. The proportion of stars with undetected Li is about 30%. This strange behavior of lithium in G type stars recalls a more or less similar behavior in F-type stars : Lambert et al. (1991) find that a non negligible proportion (15%) of field F-stars have a low Li abundance, owing to a fast unidentified depletion process. And in Pop II dwarfs (F and G type), Hobbs and Thorburn (1994) find 5% of stars with undetected lithium. The only link at the moment between these strange data is that the proportion of stars with undetected Li is larger where the convective zone is deeper, smaller where the convective zone is shallower. This does not help much to identify precisely the cause of the strange effect. HIPPARCOS measurements will provide more accurate ages. The data have possibly to be interpreted

-by inhomogeneities in the material which formed the stars,

-by a large proportion of unknown binaries with mass exchange

-by a large loss of angular momentum, which could be linked with the importance of the convective zone.

Moreover, let us remind that the Sun is Li-poor, and has transferred a large part of its angular momentum to the planetary system : in the theory of Zahn (1994), the presence of a planetary system could be an explanation for the low lithium abundance of the Sun. And it could be recommended to the programs aiming at the discovery of planets to concentrate on Li-poor G dwarfs. However, from what we now at present, it is not likely that 30% of the G stars have a planetary system.

6 Discussion

It appears that the lithium behavior is really complex, and not well understood, in spite of a large number of recent works devoted to the subject. The analysis of the processes existing in the Pop I are still a important clue for trying to understand the lithium behavior in the Pop II stars.

In the Pop II, progresses have been made on the observational side, the sample of observed stars is now large enough for a good determination of the main trends, the necessity of discriminating between small differences of lithium abundances is however pushing the theory of model atmospheres to its present limits, and progresses in the field of stellar atmospheres are needed.

Some results provide hints towards the non-depletion of Li in Pop II dwarfs, but there are also some hints in the other way. It seems to me that the hints towards essentially no depletion in the Pop II dwarfs are more numerous than the other ones.

7 Lithium Production

If we would accept the idea of a low primordial lithium abundance, we have to find processes to increase the lithium abundance in the Galaxy by a factor of 10. Novae are probably not significant producers of Li. Flare stars, cataclysmic variables, binaries with a compact companions (Rebolo and Martín 1993 and this workshop) coud produce some lithium. Giants are perhaps producing some Li for example by small quick mini-mixings (Wallerstein and Morell 1994) or by dredge-up of ^{7}Be by the rotating core of a massive star evolving towards the giant branch (Fekel and Balachandran 1993). Supernovae could produce a limited amount of Li. The cosmic rays, through alfa + alfa collisions, are an important source of ^{7}Li and especially of ^{6}Li. The AGB are obviously an essential source, as shown observationally (Plez et al. 1993 and this workshop) and theoretically (Boothroyd et al. 1993, 1994). The problem has obviously to be analyzed more thoroughly.

8 Conclusion

From the present lithium stellar abundances, we cannot reach a firm conclusion about the primordial Li abundance : the abundance determinations in Pop II stars have to be made more reliable and more accurate. Accurate observations of lithium in stars of various masses in old metal-poor clusters will provide some important information about the controversial depletion of Li in metal-poor dwarfs.

It has not been clearly proven that the known processes of Li production are able to produce the large amount of ^{7}Li needed for explaining the Li abundance found in the Pop I objects. Help is needed from other fields, e. g. from the interstellar abundance of ^{6}Li and ^{7}Li.

Two trends appeared recently in the research about lithium. In spite of remarkable sucesses, the first attempts made to interpret the behavior of lithium, using a single physical process, did not succeed in the interpretation of all the observed features. We are now observing attempts to use simultaneously two (or more) physical processes, and the predictions of such elaborate theories are in excellent agreement with the observations : which is not entirely unexpected, since more free parameters are available for the adjustment of the theory to the observations. On the other hand, such complex models are probably more realistic than the "single process" models, since there is absolutely no reason that only one process is operating in the stars.

The second trend appearing is related to the quality of the predictions of the theories : the quantitative results predicted are now so accurate, the tiny differences between the predictions of competing theories are so small, that the observers are urged to provide very accurate Li abundances. Unfortunately, the abundances are provided through the adjustment of stellar atmosphere models which are only approximations to the real stars, and are based on computations of the structure of the convective zone, and this structure is critical for the determination of the effective temperature (which in turn is critical for the determination of the Li abundance). The accurate parallaxes from Hipparcos, the stellar angular diameters from interferometry, a better convection theory in the stellar model atmospheres should soon provide reliable models, accurate temperature calibrations, accurate lithium abundances and therefore an accurate basis for comparison with the elaborate theories now available. Conclusions for cosmology will then be possible.

References

Balachandran S. (1990a), ApJ **354**, 310

Balachandran S. (1990b) ASP Conf. **9**, G. Wallerstein, ed., San Francisco, p. 357

Boesgaard A. M., Steigman G. (1985) Ann. Rev. Astr. Astrop. **23**, 319.

Boothroyd A. I., Sackmann I.-J., Ahern S. C. (1993), ApJ **416**, 762

Boothroyd A. I., Sackmann I. J., Wasserburg G. J. (1994) ApJ **430**, L77

Carlsson M. et al. (1994) A&A **288**, 860

Charbonnel C., Vauclair S. (1992) A&A **265**, 55

Charbonnel C. (1994) A&A **282**, 811

D'Antona F. (1990) editor of "The problem of Lithium", Mem. Soc. Ital. Astr. **62**.

Deliyannis C. P. (1990) PhD, Yale Univ.

Deliyannis C. P., King J. R., Boesgaard A. M., Ryan S. G. (1994) ApJ **434**, (in press).

Favata F., Barbera M., Micela M., Sciortino S. (1993) A&A **277**, 428

Fekel F. C., Balachandran S. (1993) Astrophys. J. **403**, 708

Fuhrman K. et al. (1993) A&A **271**, 451

Garciía López R., Rebolo R., Herrero A., Beckman J. (1993) ApJ **412**, 173

Hobbs L. M., Thorburn J. A. (1994) ApJ **428**, L25

King J. R. (1993) AJ**105**, 1087

Lambert D. L., Heath J. E., Edvardsson B. (1991) MNRAS **253**, 610

Magazzú A., Martín E. L., Rebolo R. (1993) ApJ **404**, L17

Marcy G. W., Basri G., Graham J. R. (1994) ApJ **428**, L57

Martín E. L., Rebolo R., Magazzú A., Pavlenko Y V. (1994) A&A **282**, 503

Martín E. L., Magazzú A., Rebolo R. (1992) A&A **257**, 186

Maurice E., Spite F., Spite M. (1983) in : ESO Workshop "Primordial Helium", P. A. Shaver et al., eds., Garching

Michaud G., Charbonneau P. (1991) Space Sci. Rev. **57**, 1.

Norris J. E., Ryan S. G., Stringfellow G. S. (1994) ApJ **423**, 386

Padgett D. L. (1990) ASP Conf. **9**, Wallerstein G., ed., San Francisco p. 354

Pasquini L. (1995) ESO preprint 988

Pilachowski C. A., Sneden C., Booth J. (1993) ApJ **407**, 699

Pinsonneault M. H., Deliyannis C.P., Demarque P. (1992) ApJSup **78**, 179

Plez B., Smith V. V., Lambert D. L. (1993) ApJ **418**, 812

Proffitt C. R. & Michaud G.(1991) ApJ **371**, 584

Rebolo R. (1989) Astrophys. Space Sci. **157**, 47

Rebolo R., Martín E. L. (1993) in:" Origin and evolution of the elements", N. Prantzos et al., eds., Cambridge Univ. Press, p. 149

Richer J., Michaud G. (1993) ApJ **416**, 312

Spite F.,Spite M. (1993) in :" Origin and evolution of the elements", N. Prantzos et al., eds., Cambridge Univ. Press, p. 201

Spite F., Spite M. (1994) in : "Evolution of the Universe and its observational quest", K. Sato, ed., Universal Acad. Press, Tokyo, p.81

Spite F., Spite M. (1993) A&A **279**, L9

Swenson F. J., Faulkner J., Rogers F. J., Iglesias C. A. (1994) ApJ **425**, 286

Thorburn J. A. (1994) ApJ **421**, 318

Thorburn J. A., Hobbs L. M., Deliyannis C. P., Pinsonneault M. H. (1993) ApJ **415**, 150

Wallerstein G., Morell O.(1994) A&A **281**, L37

Zahn J. P. (1994) A&A **288**, 829

The Abundances of Li in δ Scuti Stars: Can They Explain the Li Dip?

Stephen C. Russell

Dublin Institute for Advanced Studies, School of Cosmic Physics, 5 Merrion Square, Dublin 2, Ireland

Abstract. Several years ago it was suggested that mass loss from δ Scuti stars may transform them into the 'lithium dip' stars seen in older galactic clusters. Although several strands of evidence indicate that this is unlikely, there has never been a study made of the lithium abundances in delta Scuti stars themselves. This paper reports the preliminary results from a survey of lithium in these stars.

Lithium was detected in about half of the stars studied, with only upper limits being determined for the others. For those stars with measurable lithium there is evidence for a moderate depletion at the cooler temperatures, but not enough to explain the lithium dip. For those stars with no measurable lithium, the upper limits are consistent with only moderate depletion in all except one or two cases.

Introduction

Boesgaard and Tripico (1986) discovered there was a drastic depletion in lithium abundances in Population I stars in a narrow range of temperatures around 6600 K. This 'lithium dip' occurs in galactic cluster stars as old as the Hyades ($t \sim 7x10^8$ yrs) (Boesgaard and Budge 1988), but is not seen in stars as young as the Pleiades ($t \sim 8x10^7$ yrs) (Boesgaard, Budge, and Ramsay 1988). This has posed a significant problem in terms of stellar evolution theory.

Since the discovery of the lithium dip there have been many suggestions put forward to explain the observations. Michaud (1986) discussed the formation of the lithium dip in terms of diffusion processes, whereby lithium would fall below the convection zone due to gravitational settling, around temperatures of 6600 K. In hotter stars, radiative acceleration would push lithium to the surface, and there would be severe lithium overabundances on the hot side of the dip, unless a modest degree of mass took away precisely the right amount of surface material.

Most of the other explanations rely on some form of mixing of the surface layers down to depths hot enough to destroy lithium. Charbonneau and Michaud (1988), for instance, suggested that meridional circulation was the mechanism responsible for the mixing. This was supposed to be in balance with the upwards radiative acceleration on the hot side of the dip, but only at some critical rotation velocity. If the rotational velocity was too low, then overabundances would result. While if the rotational velocity was too high, underabundances would result. This was an easily verifiable prediction, and was refuted by the work of Balachandran (1990), among others.

Of particular interest to this present work, was the suggestion by Schramm, Steigman, and Dearborn (1990) (hereafter SSD) that main-sequence mass loss from stars at the red edge of the instability strip may deplete their masses and lithium abundances sufficiently to move them over into the region of the lithium dip. The feasibility of this mechanism will be addressed in this paper.

The lithium dip occurs in the temperature range 6450-6850 K (Boesgaard 1987; Boesgaard and Budge 1988). This corresponds, in the Hyades, to main-sequence masses of $1.2 \leq M \leq 1.4 \ M_{\odot}$. SSD found it suggestive that the red end of the instability strip, with a temperature of 6950 K (Abbot, Telesco, and Wolff 1984), and a mass of $1.3 \pm 0.1 \ M_{\odot}$ (Wolff 1983), intersects the main sequence just above the lithium dip. They invoked the hypothesis of Willson, Bowen and Struck-Marcell (1987, hereafter WBS) to explain the mechanism for main-sequence mass loss. WBS argue that those main-sequence stars in the instability strip that are rapid rotators, which describes most δ Scuti stars, will lose mass at a rate of $\dot{M} \geq 10^{-9} M_{\odot} yr^{-1}$.

SSD show that mass loss corresponding to 0.05 $M_{\odot}$ for a 1.3 $M_{\odot}$ star will remove all the surface lithium, while leaving the beryllium abundance largely in tact. It is necessary to remove 0.07 $M_{\odot}$ before the beryllium will be lost to the star. From observations of a lithium dip in the Hyades, but no signs of a similar beryllium dip in that cluster, they conclude that the mass loss rate should naively be expected to be around $10^{-10} M_{\odot} yr^{-1}$. This is an order of magnitude smaller than the rate preferred by WBS.

Delta Scuti stars are A or early F stars, that are lightly variable, and lie within about 2 magnitudes above the zero age main-sequence (ZAMS). They have short periods, not exceeding 0.3 days, and amplitudes ranging from 0.01 (the detection limit), up to 0.8 magnitudes. Typical amplitudes, however, are around 0.02 magnitudes. There seems to be nothing special about these stars, though for some reason, only about one third of stars in this region of the HR-diagram experience variability above the detection limit.

King (1991) discussed the work of SSD at length. He argued that the location of the blue edge of the lithium dip and the red edge of the instability strip are far from coincident, and therefore there is no connection between the two types of stars. King showed that the mean mass of δ Scuti stars is too high for mass loss to explain the lithium dip, without a consequent beryllium dip in the Hyades. Also, if mass loss is important for δ Scuti stars, there should be an enhanced IR-excess. King showed that about half as many δ Scuti stars have an IR-excess as compared to non-variable B, A, and F stars on the ZAMS.

Perhaps even more significant is the report by Deliyannis at this conference of a star with depleted beryllium, but still detectable lithium. This implies that lithium depletion can not be due to mass loss alone.

Although these arguments may seem conclusive, there are possible ways around each of them. For instance, the mass loss may increase towards the red edge of the instability strip, or only be effective at the edge itself. Clearly there is a need to study the problem directly by measuring the lithium abundances in the stars.

2 The Survey

A survey was carried out (Russell 1994) to determine the lithium abundances in a wide range of δ Scuti stars in the field, without regard for their spectral type, luminosity or rotation. In this way it was hoped that any trends in lithium abundances would become apparent.

High signal to noise, high dispersion spectra were taken of 23 of the brightest δ Scuti stars from the catalogue of de Coca $et\ al.$ (1990). Detailed abundance analyses of the spectra were made, where possible, using published Strömgren and H_β photometry, and solar gf-values for all elements except lithium. For the lithium doublet itself, the gf-value adopted by Balachandran (1990) was used.

3 Results

The main results are shown in Fig.1, where 12 plus the logarithm of the lithium abundance compared to hydrogen (Log N(Li)) is plotted against the effective temperatures of the stars. In this figure, the closed circles represent stars with a measurable lithium abundance, the open triangles are stars with only upper limits on the lithium abundances, and the open circles are data from the disk F-stars studied by Balachandran (1990) for comparison. The horizontal line is a guide to illustrate the upper envelope of lithium abundances for disk F-stars. The cross near the bottom left hand corner illustrates the expected error bars for the measurements. The well observed star HR5017 is marked in the figure at three possible positions connected by lines, according to the range of physical parameters published in the literature (see Hauck, Foy, and Proust 1985).

Most stars in the sample are already evolving off the ZAMS, so main-sequence mass loss has terminated already. If mass loss occurred at a rate of $10^{-10} M_\odot yr^{-1}$, all the lithium would have been lost within ~ 0.5 Gyrs. Since most stars in the sample are already older than this, such a high mass loss rate can not have taken place. In addition, I find no correlation between the projected rotation velocity, $v\,Sin\,i$, and the amount of lithium depletion, in contradiction to the predictions of WBS.

Our results show there to be some evidence for a modest lithium depletion at the lower temperatures. This is especially interesting for the two coolest stars, and for the well observed star HR5017 (see Fig.1). However, due to their apparent rarity, it is hard to see how the lithium dip can be populated by such stars. It can only be concluded that if mass loss through the WBS mechanism, or any other mechanism, is responsible for lithium depletion, it must have only a relatively minor rôle to play, and probably has little to do with the lithium dip.

4 Conclusions

It is concluded that a mass loss of $10^{-10} M_\odot yr^{-1}$, or greater, is not observed for the majority of δ Scuti stars in the sample. There is, however, some evidence for a moderate depletion in lithium for some stars, especially near the red edge of the instability strip, but not enough to suggest these could be the source of lithium dip stars. Finally, there is no evidence for a correlation of lithium depletion with $v\,Sin\,i$, and therefore, no support for the WBS mechanism for mass loss.

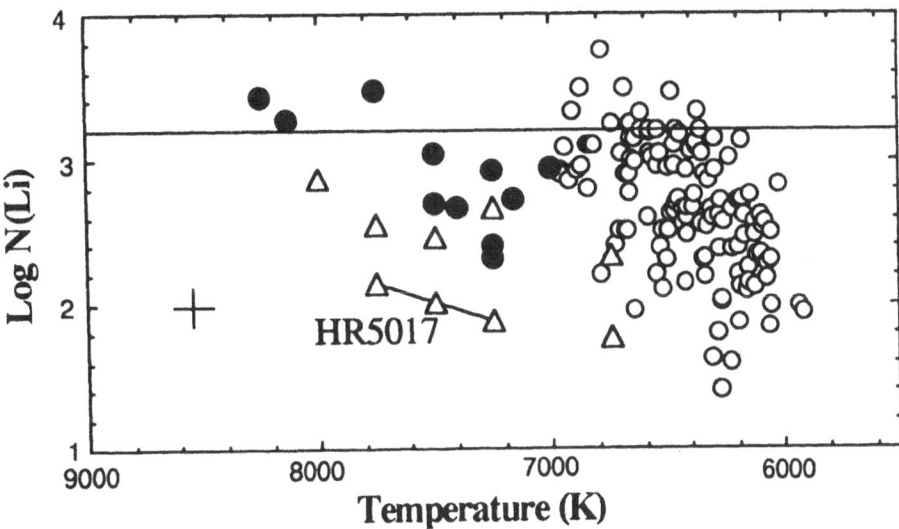

Fig. 1. Lithium abundances of δ Scuti stars and F-dwarfs (see text for details)

References

Abbot, D., Telesco, C., & Wolff, S. (1984) Ap.J. **279**, 225.

Balachandran, S. (1990) Ap.J. **354**, 310.

Boesgaard, A.M. (1987) Publ. Astr. Soc. Pacif. **99**, 1067.

Boesgaard, A.M., & Budge, K.G. (1988) Ap.J. **338**, 875.

Boesgaard, A.M., Budge, K.G., & Ramsay, M.E. (1988) Ap.J. **327**, 389.

Boesgaard, A.M., & Tripico, M.J. (1986) Ap.J. **302**, L49.

Charbonneau, P. & Michaud, G. (1988) Ap.J. **334**, 746.

de Coca, P.L., Rolland, A., Rodriguez, E., & Garrido, R. (1990) A.Ap.Sup. **83**, 51.

Hauck, B., Foy, R., & Proust, D. (1985) A.Ap. **149**, 167.

King, J.R. (1991) MNRAS **249**, 658.

Michaud, G. (1986) Ap.J. **302**, 650.

Russell, S.C. (1994), (in preparation).

Schramm, D.N., Steigman, G., & Dearborn, D.S.P. (1990) Ap.J. **359**, L55 (SSD).

Willson, L.A., Bowen, G.H., & Struck-Marcell, C. (1987), Comm. Astrophys., **12**, 17 (WBS)

Wolff, S.C. (1983) in *The A-Stars: Problems and Perspectives*, NASA SP-463.

Exploring the Lithium Dip:
A Comparison of Clusters

Suchitra Balachandran

The Ohio State University, Department of Astronomy, 174 West 18th Avenue, Columbus OH 43210, USA

Abstract. The Li dip is compared in 4 clusters ranging in age from 750 Myr to 4 Gyr and in metallicity from [Fe/H]=+0.12 to -0.15 and inferences are drawn by comparison to model predictions.

1 Introduction

The Li dip refers to the sharp drop in Li abundances of between a factor of 30 to 100 in stars in the mid-F spectral type, first seen in the Hyades (Boesgaard and Tripicco 1986). On either side of this roughly 300 K region of effective temperature, Hyades stars have their "normal" Li abundance of log ϵ(Li) = 3.0 to 3.3. The dip is seen in clusters 200 Myr and older. In order to shed light on the source of this anomalous depletion, Li abundances in clusters from the age of the Hyades (750 Myr), to M67 (4-5 Gyr) were examined. Several theoretical models have attempted to explain the formation of the Li dip by using a variety of physical mechanisms. While they all able reproduce the surface Li abundance, usually in the Hyades, they make different predictions for the interior composition. Hence determination of subgiant and giant abundances of dip stars, e.g. in M67, are particularly very useful to distinguish between the models.

2 Analysis

The Li abundances in subgiants and giants in M67 were derived from spectra we acquired at Kitt Peak, Cerro Tololo and McDonald Observatories. The remaining cluster data were taken from the literature: Hyades from Boesgaard and Tripicco (1986), Boesgaard and Budge (1988) and Thorburn et al. (1993); Praesepe from Soderblom et al. (1993); and NGC 752 from Hobbs and Pilachowski (1986) and Pilachowski and Hobbs (1988). In order to eliminate systematic differences between the various analyses, all of the stellar temperatures were re-derived from B-V colors using the Saxner and Hammarbäck (1985) calibration and all of the Li abundances recalculated using a uniform technique. Detailed reasons for the temperature calibration chosen, comparison to other temperature calibrations, and the effect of a uniform abundance analysis are provided in Balachandran (1994). Masses and ZAMS temperatures of the individual stars were calculated using isochrones of the appropriate age and metallicity interpolated from the Revised Yale Isochrone grid (Green, Demarque and King 1987).

3 Results and Discussion

With the exception of one star, no Li was detected in any of the subgiants and giants in M67 indicating that Li has effectively been destroyed in the dip stars. This provides strong evidence that microscopic diffusion of Li to a region just below the surface convective zone, as suggested by Michaud (1986) and Richer and Michaud (1993), cannot be the sole process responsible for the formation of the dip. Li abundances as a function of mass and ZAMS temperature in the Hyades, NGC 752 and M67 clusters are shown in Figure 1. Two principal facts emerge from our cluster comparisons:

a. The mass at the center of the Li dip increases with increasing metallicity.
b. The ZAMS temperature at the center of the Li dip is independent of metallicity.

Furthermore, the morphology of the Li dip is characterized by a sharp drop at the blue edge and a more gradual rise at the red edge. There is no change in the ZAMS temperature or shape of the blue edge with age. The red edge becomes less steep with age; this may either be caused by an evolution in the Li dip, or by a decrease in Li in stars cooler than the red edge and thus unrelated to the Li dip phenomenon.

Several of the models which explain the Li dip do so by recognizing that the dip occurs in the temperature domain which roughly defines the boundary between stars with and without surface convective envelopes, and utilizing the unique properties of stars which span this region. Most important among these is that all stars cooler than the dip undergo rapid rotational spin-down, while those hotter than the dip retain a large fraction of their initial angular momenta through their main sequence lives.

Charbonneau and Michaud (1988) suggested a combination of meridional circulation and microscopic diffusion to account for the Li dip. However, meridional circulation is not seen to cause Li depletion in stars hotter than the dip, in which this process should be more effective (Balachandran 1991). Vauclair (1988) and Charbonnel, Vauclair and Zahn (1992) suggested that the surface layers of stars hotter than the dip were not subject to depletion because they were shielded from mixing by a "quiet zone" which resulted from their more detailed examination of the meridional circulation mechanism. Our principal objection to their mechanism is their invocation of a constant initial rotational velocity of 100 km s^{-1} for all Hyades stars in order to fit the observed dip. Observations show that such large velocities are seen in only a small fraction of young stars. The rotational braking models of Pinsonneault et al. (1990) have accounted for the Li dip by incorporating a sharp change in the initial angular momenta of stars at the cool edge of the dip. While this does reproduce the shape of the dip, their assumption remains to be documented observationally.

Finally, the mass loss models of Schramm, Steigman and Dearborn (1990), based on the proximity of the Li dip to the instability strip, have suggested that pulsation-driven mass loss may be responsible for the dip. The principal

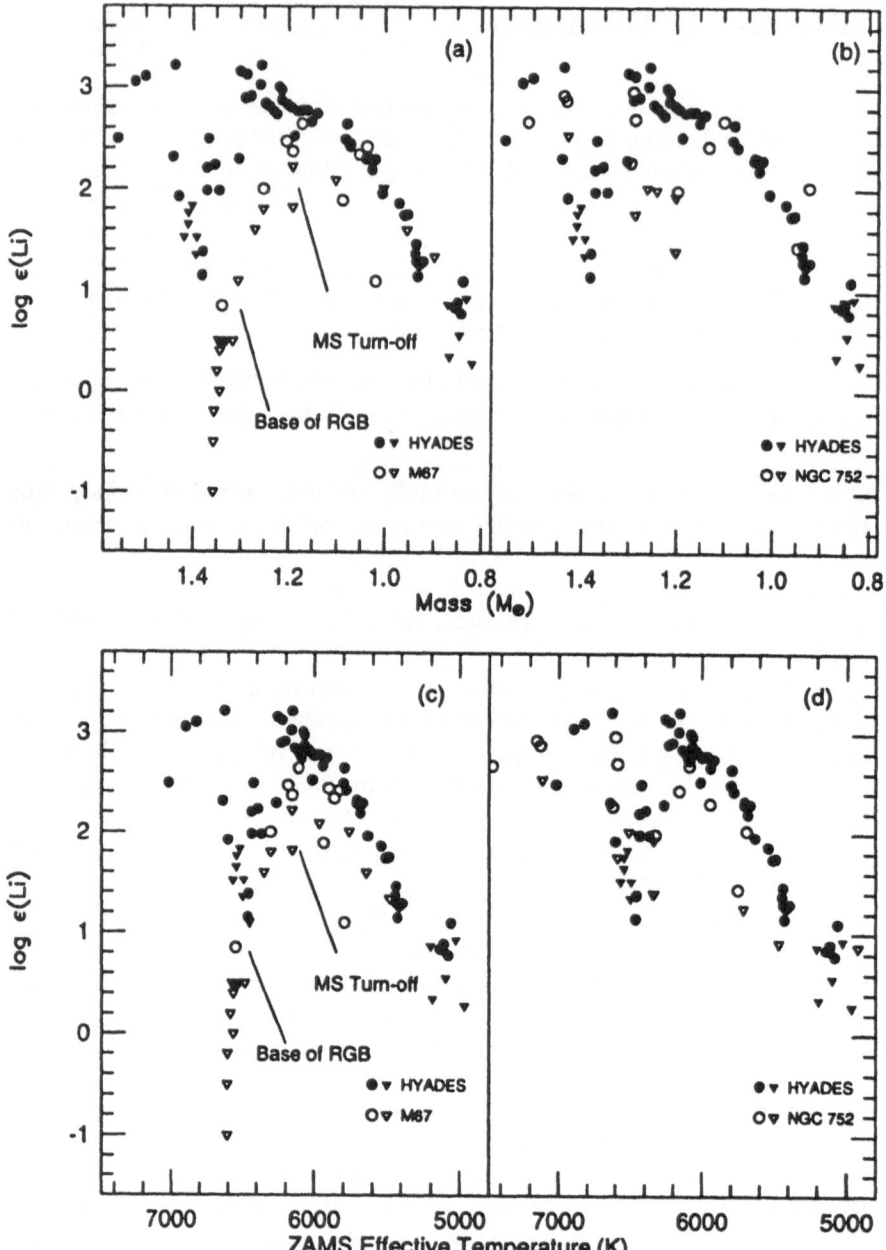

Fig. 1. Lithium abundances as a function of mass and ZAMS temperatures in the Hyades, NGC 752 and M67. In each case circles are Li measurements and triangles Li upper limits.

conflict with their suggestion is Russell's (these proceedings) findings that no Li depletion is seen in the δ Scuti stars themselves.

It appears, then, that none of the existing models adequately explains the observations. We point to one interesting observational fact which may provide a clue to this depletion phenomenon. In the Hyades, the dip stars are rotating with velocities of upto 60 km s^{-1}, while by the age of the field stars, they have spun down to much smaller velocities. Unlike G dwarfs whose surfaces appear to spin down rapidly once they arrive on the main sequence, the mid-F dip stars do so on a much longer timescale. Could such a longer spin-down timescale result in greater mixing?

References

Boesgaard, A. M., Tripicco, M. J.: ApJL, **302**, 49, (1986)

Boesgaard, A. M., Budge, K. G.: ApJ, **332**, 410, (1988)

Thorburn, J. A., Hobbs, L. M., Deliyannis, C. P., Pinsonneault, M. H.: ApJ, **415**, 150, (1993)

Soderblom, D. R., Fedele, S. B., Jones, B. F., Stauffer, J. R.,Prosser, C. F.: AJ, **106**, 1080, (1993)

Hobbs, L. M., Pilachowski, C.: ApJ, **309**, L17, 1986)

Pilachowski, C. A., Hobbs, L. M.: PASP, **100**, 336, (1988)

Saxner, M., Hammarbäck, G.: A&A, **151**, 372, 1985)

Balachandran, S.: ApJ, submitted (1994)

Green, E. M., Demarque, P., King, C. R.: The Revised Yale Isochrones and Luminosity Functions, (New Haven: Yale Univ. Observatory)

Michaud, G.: ApJ, **302**, 650, (1986)

Richer, J., Michaud, G.: ApJ, **416**, 312, (1993)

Charbonneau, P., Michaud, G.: ApJ, **334**, 746, (1988)

Balachandran, S.: in Inside the Stars, IAU Coll. 137, ASP Conf. Ser., **40**, 333, (1991)

Vauclair, S: ApJ, **335**, 971, (1988)

Charbonnel, C., Vauclair, S., Zahn, J.-P.: A&A, **255**, 191, (1992)

Pinsonneault, M. H., Kawaler, S. D., Demarque, P. ApJS, **74**, 501 (1990)

Schramm, D. N., Steigman, G., Dearborn, D. S. P.: ApJ, bf359, L55, (1990)

Convection, Opacities and Pre–Main Sequence Lithium Depletion

Francesca D'Antona

Osservatorio Astronomico di Roma, I-00040 Monte Porzio, Italy

Abstract. Different convection models applied to stellar structure computation provide different HR diagram locations and structural properties of stars. Theoreticians often avoid to remember that *calibrating the Mixing Length Theory on the Solar Model does not allow to make predictions on convection either for other masses, or for different evolutionary phases, or for a different chemistry*. Lithium depletion in pre–Main Sequence depends on the convection model and on the surface opacities in an interconnected way: it is highly dangerous to try to infer global properties on this issue until we have both very good low temperature opacities and a reliable treatment of convection.

1 Premise: CM Convection Model

The following discussion explains in some detail an aspect of the results of extensive computations of pre–Main Sequence (pre–MS) evolution published in a paper by D'Antona and Mazzitelli 1994 (hereinafter referred to as DM). These results include Lithium burning and HR diagram location of tracks having $M \leq 2.5 M_\odot$, computed by adopting two different sets of low temperature ($T \leq 6000$K) opacities (by Alexander et al. 1989 and by Kurucz 1991) and two different turbulent convection models (the Mixing Length Theory –MLT– model and Canuto and Mazzitelli (CM) 1990, 1992 model. The latter is the first model which attempts to surpass the MLT limitations, and that can be extensively applied to all phases of stellar evolution. Further, in the cases already studied, CM model has always given very good performances in stellar evolution. In particular:

- the solar T_{eff} at the solar age is reproduced within 0.5%, with the available last generation opacities (CM 1990, 1992);

- it performs better than the MLT in describing the solar spectrum of oscillations (Paternò et al. 1992);

- semiempirical solar models derived from the inversion of the oscillation spectrum agree much better with CM model than with the MLT (Basu and Antia 1994);

- when applied both to Main Sequence (MS) and giants, CM model reproduces reasonably well the HR diagram of an old open cluster (M67), and in particular it reproduces better the location of Horizontal Branch stars (D'Antona et al. 1992);

- the model has been applied to study the onset of convection in helium envelope white dwarfs (Mazzitelli and D'Antona 1991): the blue edge of the DB variables instability strip is compatible with these results;

- the location in T_{eff} of Red Supergiants has been shown to agree naturally with the results of CM model, while it requires variable l/H_p to be reproduced with MLT (Stothers and Chin 1994);

- Asymptotic Giant Branch models of mass $M \geq 5M_\odot$ naturally show the occurrence of hot bottom burning, which is necessary to understand the creation of Li–rich supergiants in the Magellanic Clouds, while the MLT required calibration (D'Antona and Mazzitelli 1995);

- CM model plus the new Rogers and Iglesias (1992) and Kurucz (1991) opacities, applied to the study of Globular Cluster stars, provide reasonable and interesting results in the whole range of GC metallicites (Mazzitelli et al. 1994).

Finally, CM model has been applied to pre–MS evolution in the computations we are going to discuss. The reason which leads us to explore the properties of these models is the following: although the MLT models, once calibrated on the solar model, provide very similar results for the Lithium depletion in pre–MS (see Tab. 1), the same is not true for CM models. Notice, however, that the final MS location is very similar for all sets. In this paper we show the physical reasons leading to this result and warn against a much too faithful use of MLT based computations concerning Li–depletion.

2 Pre–MS Lithium Depletion and HR Diagram Location

In DM we compare first of all the HR diagram location of the pre–MS of the solar model. The value of $\alpha = l/H_p$ si calibrated, in the MLT models, so that the solar radius is obtained at the solar age, as it is usually done in the stellar structure computation. The tuning provides the values listed in the first row of Tab.1.

Table 1. l/H_p values appropriate to fit the present Sun (first row) and the CM tracks location in pre–MS (second row)

	Alexander	Kurucz
MS	1.2	1.5
PRE–MS	1.6	2.0

The choice of either the "Alexander" or the "Kurucz" set of low temperature opacities ends in a different value of α for the solar tuning: this shows that we reset into α all our ignorance connected not only with convection, but also with the opacities. In the CM model, the solar T_{eff} is reproduced within 0.5% without any effort, but in DM we tuned a "second order parameter" (a) to give a more precise tuning of the solar T_{eff}, like it can be obtained by tuning α in the MLT, although our attitude is to avoid "perfect" tunings as meaningless. Therefore, the four solar tracks all coincide at the MS. This allows us to be immediately aware that the morphology of the tracks in the giant pre-MS phase is different between the CM and MLT models: in fact, *the α which in the MLT provides the solar fit does not reproduce the T_{eff} location of CM pre–MS evolution*, for which a *larger* value of α should be adopted (second row in Tab. 1). By this statement we are not implying that CM giants are located "correctly" and MLT giants are not, as we do not know whether CM model is completely adequate to describe giants envelopes. We only have shown that extension of the MLT tuned α to a different structure gives results which differ from those obtained by a different convection model, and, as such, the MLT tuning requires an act of faith to be taken seriously for structures which are not the present–day Sun itself.

Table 2. Lithium depletion from an initial $\log X_{Li} = -8$ in models of $1 M_\odot$

	Alexander	Kurucz
CM	-8.985	-8.578
MLT	-8.445	-8.439

Lithium depletion in low mass pre–MS begins towards the bottom of the Hayashi track while the star is fully convective and goes on while a radiative core develops. Therefore, the Li–depletion will be dependent on the convective model. This, in fact, is shown in Tab. 2 and in DM. As we remarked, the difference in Lithium depletion between the two MLT models is not large. On the contrary, depletion in CM model results very much dependent on the set of low–T opacities adopted. It is possible that the α MLT value which reproduces the location of CM giants would provide a depletion similar to that of the CM model: this, of course, is of no use, as we do not know a priori the correct MLT calibration! So the questions we pose are:

1. why the pre–MS giant location is different, although the present Sun is reproduced in all four sets?
2. which are the structural differences at the stage of Li–depletion?

Analysis of the models has shown that two are the inputs which play a role: 1) the convection model: CM model is less efficient than MLT *very close to the surface*, but it is much more efficient inside the envelope; 2) the opacities: the radiative atmospheric opacities provide the starting point of overadiabatic convection, and thus may change also its role on the structure.

We have examined in detail model structures along the evolutionary tracks (during the Hayashi contraction, at the phase of Lithium burning, and in the MS) to understand the differences. The comparison between the CM and MLT (Alexander) models shows that:

1) in the pre–MS (models along the Hayashi track), although the peak of overadiabaticity in much larger in CM structures, MLT convection is less efficient (the overadiabatic gradient is larger) for a larger fraction of the envelope, and therefore the MLT T_{eff} is smaller;

2) in MS, the models give similar T_{eff} values. In fact, the optical atmosphere is denser and less massive, the effect of overadiabaticity is thus less important than in pre–MS models (at least for the T_{eff} determination);

3) at the stage of Li–burning, CM models are slightly hotter and denser at the bottom of the convective envelope. The effect on Li–depletion is large, as the dominant timescale (contraction) is quite long here.

The comparison between Kurucz and Alexander CM models shows that:

- the "shape" of the pre–MS tracks is different, reflecting where Kurucz opacities are larger (smaller) than Alexander's; the difference at the stage of Li–burning is quite relevant;

- in the MS the models are very similar, although there remain differences in the behaviour of gradients in the most external envelope.

In conclusion we remark that, although pre–MS Li–burning is a very important issue, not only in itself, and for population I stars, but also to assess the importance of other possible mechanisms for Li–depletion, *it is dangerous, by now, to overemphasize the results.* Improvements are needed and will soon be included: Rogers and Iglesias are making available their new Equation of State from OPAL; Alexander is releasing new updated opacities including molecules, in any desired mixture of chemistry; Mazzitelli is now working on the convection models by understanding the effect of a small (physical) overshooting in atmosphere for the CM model description. We will have new results soon.

References

Alexander, D. R., Augason, G. C., Johnson, H. R. 1989, Astrophys. J., 345, 1014.

Basu, S. Antia, H.M., (1994): Astrophys. J., in press

Canuto, V.M., Mazzitelli, I. (1990): Astrophys. J., 370, 295 (CM)

Canuto, V.M., Mazzitelli, I. (1992): Astrophys. J., 389, 724

D'Antona, F., Mazzitelli, I. (1994): Astrophys. J. Suppl. Series, 90, 467

D'Antona, F., Mazzitelli, I. (1995): in Proceedings of 9th European Workshop on White Dwarfs, eds. D. Koester and K. Werner, in press

D'Antona, F., Mazzitelli, I., Gratton, R. (1992): Astron. Astrophys., 257, 539

Kurucz, R. L. (1991): in "Stellar Atmospheres: Beyond the Classical Models", L. Crivellari, I. Hubeny, D. G. Hummer eds., NATO ASI Series, Kluwer, Dordrecht, p. 441.

Mazzitelli, I. Caloi, V., D'Antona, F. (1994): in "Astrophysical Applications of Powerful New Atomic Databases" in press

Mazzitelli, I., D'Antona, F. (1991): in "White Dwarfs", ed. G. Vauclair and E. Sion (Kluwer, Dordrecht), p. 305

Paternò, L., Canuto, V. M., Ventura, R., Mazzitelli, I. (1993): Astrophys. J., 402, 733

Rogers, F. J., Iglesias, C. A. (1992): Astrophys. J. Suppl. Series, 79, 507 (RI).

Stothers, R.B., Chin, C. (1994): Astrophys. J., in press

The Behaviour of the Lithium Abundance Along the Pre-Main Sequence Phase

A. Magazzù[1], E.L. Martín[2], R. Rebolo[2], R.J. García López[2], Ya.V. Pavlenko[3]

[1] Osservatorio Astrofisico di Catania, Città Universitaria, I-95125 Catania, Italy
[2] Instituto de Astrofísica de Canarias, E-38200 La Laguna, Tenerife, Spain
[3] The Principal Observatory of Ukraine, Kiev, Ukraine

Abstract. Understanding of lithium burning in the pre-main sequence phase provides information on the present cosmic Li abundance, as well as on the evolution of stellar structure before the settling on the main sequence. In 1987, triggered by the scarcity of observational data, we started a long term observational programme of Li abundance in pre-main sequence stars. Our main conclusions at this stage are: the initial Li abundance of T Tauri stars, $\log N(\mathrm{Li}) \sim 3.1$, coincides with the "cosmic" Li abundance; Li depletion increases with decreasing luminosity, starting between 0.9 and $0.4\,L_\odot$, depending on the mass. A clear dependence on luminosity is seen, expecially for masses $\leq 0.4\,M_\odot$. A number of issues which deserve further analysis are outlined.

1 Introduction

Many are the reasons to study Li in pre-main sequence (PMS) stars. The comparison with model predictions leads to a better understanding of PMS evolution and stellar structure, provides constraints on parameters (mass loss, angular momentum loss), and constitutes a test for opacities and convection models. Besides the knowledge on the constitution of stars, this study can give hints on galactic evolution of Li and the initial Li abundance, with consequences on Big Bang nucleosynthesis models.

When we started our study only few determinations of Li abundance in PMS were available, due to the pioneering work by Zappala (1972). On the other hand, several important theoretical papers had already been printed (Bodenheimer 1965, D'Antona & Mazzitelli 1984), giving the guidelines to understand Li evolution during the PMS phase. In this contribution we describe the progress of our study, compared with observational and theoretical work produced in the meantime by other groups.

2 Measure of Li Abundance in PMS Stars

The first attempts to measure Li abundance in a wide sample of PMS stars are those of Magazzù & Rebolo (1989) and Strom et al. (1989). Soon it was realized that extreme care has to be taken in measuring the Li abundance in these stars. There are several aspects that make the analysis different from that of normal main sequence objects. Namely,

1. In T Tauri stars (TTS) the resonance doublet at $\lambda 6708$, the strongest feature of lithium, is often saturated and is formed in relatively high layers of the atmosphere. Departures from LTE, as well as chromospheric effects, can have non-trivial consequences in the correct determination of the abundance.
2. From their position in the HR diagram, it is clear that PMS stars cannot be considered as belonging to luminosity class V. Therefore, when deriving effective temperatures from spectral types or colours, adequate calibrations have to be used.
3. Veiling due to a continuum produced in an external envelope may reduce the equivalent width of the lines under study.

The first analysis in which all these points were taken into account was presented in Magazzù et al. (1992), who studied the Li abundance in a sample of 36 (classical and weak) TTS. They took into account veiling effects and studied non-LTE effects in the range $4000 \leq T_{\text{eff}} \leq 5000\,\text{K}$. Moreover, their effective temperatures were obtained from spectral types using a calibration for luminosity class IV (de Jager & Nieuwenhuijzen 1987). The abundances were derived using model atmospheres extrapolated to upper atmospheric levels. The results of Magazzù et al. show a rather narrow distribution of Li abundance around $\log N(\text{Li}) = 3.2$, with some indication of depletion in the most evolved TTS with $M \approx 0.8\,M_{\odot}$. These findings do not confirm the much higher abundances found by other authors (e.g. Strom et al. 1989). A refinement of this analysis has been recently presented in Martín et al. (1994), who used new 64 layers Kurucz (1992) models, extrapolated when necessary to the uppermost atmospheric levels. The effective temperatures were derived without making any a priori assumption on the luminosity class of the objects, simply relying on their position in the HR diagram. The NLTE calculations were also improved by a 20-level Li atom model. Their work is focussed on weak TTS and its results on a sample of 53 objects can be summarized as follows:

1. Clear evidence for PMS Li burning, depending on luminosity and mass;
2. initial abundance $\log N(\text{Li}) = 3.11 \pm 0.06$;
3. 0.2–$0.4\,M_{\odot}$ stars deplete earlier than predicted;
4. no high $v \sin i$ stars show low Li abundance.

The last point deserves some comment. Figure 9 in Martín et al. (1994) indicates no depletion for $v \sin i > 40\,\text{km s}^{-1}$. However, for five out of the six fast rotators the luminosity is still so high that no depletion is expected, regardless the $v \sin i$. Only one object keeps its initial Li content despite its luminosity, sufficiently low to allow a severe depletion. Nevertheless, this object

(NTTS 040012+2454N,S) is a close visual binary, whose luminosity has been divided equally between both components of the pair. We have no sufficient data to draw any definitive conclusion on the connection between rotation and Li depletion in pre-main sequence; other data are welcome. It is interesting to note that at the very end of the pre-main sequence phase (i.e. for Pleiades), in the mass range 0.7–0.9 $M_\odot$ rapid rotators show higher Li abundances than slowly rotating stars (García López et al. 1994).

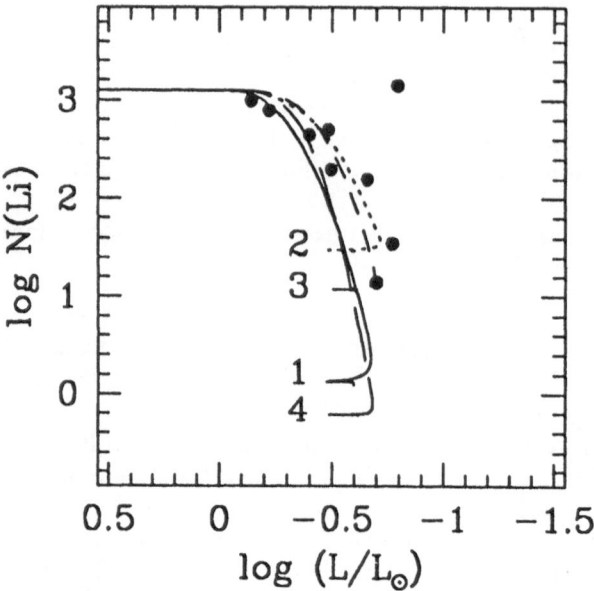

Fig. 1. Lithium abundance versus luminosity for weak TTS in the sample of Martín et al. (1994), mass range 0.7–0.8 $M_\odot$. Also shown are depletion tracks by D'Antona & Mazzitelli (1994)

In Figure 1 we plot Li abundance versus luminosity for stars in the mass range 0.7–0.9 $M_\odot$ in the sample of Martín et al. (1994). For comparison the 0.8 $M_\odot$ tracks from D'Antona & Mazzitelli (1994) are shown. The four tracks were produced with different convection models and opacities. The figure shows a general agreement between models and observations. The tracks that better fit our data seem to be 2 and 3, i.e. those obtained with mixing length theory plus Alexander opacities and CM convection treatment plus Kurucz opacities, respectively. However, the differences among the various tracks are relatively small, so these data cannot set stringent constraints on convection and opacities.

3 Open Questions

The knowledge of Li abundance evolution in pre-main sequence stars has advanced greatly in the latest years. However, we have still to clarify several important points, like the relationship between Li abundance and rotation. Other questions for which we still have to find an answer are:

1. Does Li evolution differ in classical and weak TTS? No appreciable Li depletion has been observed in classical TTS. This is mainly due to the fact that no significant sample of relatively low luminosity classical TTS has been studied to date. Observations of low luminosity classical TTS are required in order to clarify this point.

2. Does the initial Li content of stars vary depending on the forming region? The study of Magazzù et al. (1992) and Martín et al. (1994) give a quite uniform initial abundance for samples built up with stars of different associations. However, other authors have reported much higher abundancies in selected regions. A reason for these discrepancies could be found in the different analyses, but a difference in the initial composition cannot be excluded yet.

3. Do we get the correct abundance for the λ6708 doublet? The resonance doublet is the most intense Li line and usually is the only measurable line in the optical. Since in TTS this feature is often saturated and can be unpredictably affected by boundary conditions, the determination of the abundance from the subordinate line at λ6104, visible for the typical abundances of TTS, would be safer (Duncan 1991). A check with the other line at λ8126 would be also desired.

4. Is the apparent Li early depletion in 0.2–0.4 PMS stars real? If so, why do these stars deplete so early? The knowledge of the physics of these low mass stars needs still much improvement.

References

Bodenheimer, P. (1965): ApJ 142, 451
D'Antona, F., Mazzitelli, I. (1984): A&A 138, 431
D'Antona, F., Mazzitelli, I. (1994): ApJS 90, 46
de Jager, C., Nieuwenhuijzen, H. (1987): A&A 177, 217
Duncan, D.K. (1991): ApJ, 373, 250
García López, R.J., Rebolo, R., Martín, E.L. (1994): A&A 282, 518
Kurucz, R.L. (1992) private communication
Magazzù, A., Rebolo, R. (1989): Mem. S.A.It. 60, 105
Magazzù, A., Rebolo, R., Pavlenko, Ya.V. (1992): ApJ 392, 159
Martín, E.L., Rebolo, R., Magazzù, A., Pavlenko, Ya.V. (1994): A&A 282, 503
Strom, K.M., Wilkin, F.P., Strom, S.E., Seaman, R.L. (1989): AJ 98, 1444
Zappala, R.R (1972) ApJ 172, 57

Lithium Abundances Below the Substellar Limit

A. Magazzù[1], E.L. Martín[2], R. Rebolo[2]

[1] Osservatorio Astrofisico di Catania, Città Universitaria, I-95125 Catania, Italy
[2] Instituto de Astrofísica de Canarias, E-38200 La Laguna, Tenerife, Spain

Abstract. Recently we proposed a "lithium test" able to acquire information on the nature of brown dwarf candidates. Here we summarise the results of our search for the Li $\lambda670.8$ nm resonance doublet in a sample of brown dwarfs candidates. We find evidence for strong Li depletion in all the observed objects and hence we infer lower limits to their masses, and constraints on their age. The mass limits are in the range 0.08–$0.06\,M_\odot$, depending on the assumed age.

1 Introduction

We have recently proposed a spectroscopic test (the "lithium test") capable of providing direct confirmation of the substellar nature of brown dwarf candidates (Rebolo et al. 1992; Magazzù et al. 1993). Nuclear reactions destroy Li at temperatures higher than $\sim 2.0 \times 10^6$ K, which can be easily attained in the interior of very low mass stars. As a consequence, we expect no Li to be detected in these objects; conversely, below the substellar limit Li can be preserved. More precisely, preservation is expected below $0.065\,M_\odot$. According to our synthesis of the Li I doublet using Allard (1990) model atmospheres we find that this feature could be detected through spectroscopy of moderate resolution (1–2 Å). Therefore, the detection of Li in the spectrum of a brown dwarf candidate would be a confirmation of its substellar mass.

2 Search for Lithium

We have undertaken a programme aimed to apply the Li test to several of the best brown dwarf candidates known in young open clusters (α Per, Pleiades and Hyades) and in the field. We have obtained intermediate resolution spectra (FWHM= 1.5–4 Å) of several objects with spectral types M6–M9 and R magnitudes in the range 17–21. These data were obtained with the 4.2 m William Herschel telescope at La Palma, and the 3.6 m ESO telescope at La Silla. The field objects were selected from the available tables of the coolest dwarfs (Kirkpatrick et al. 1993). The open cluster objects come from very low mass proper motion studies (Bryja et al. 1994, Hambly et al. 1993). The objects are listed in Table 1.

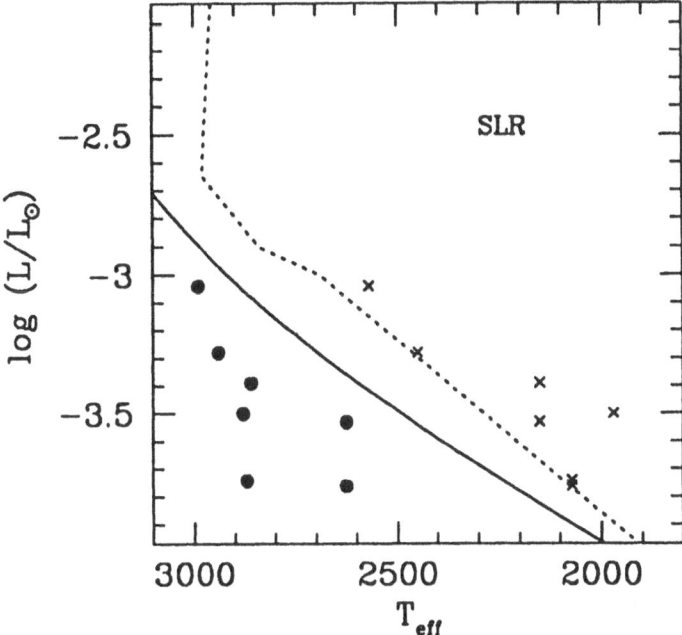

Fig. 1. H-R diagram for field BD candidates in common with Tinney et al. (1993), plotted using the luminosities and temperatures reported by them. Dots are the same objects with temperatures obtained from the calibration of Kirkpatrick et al. (1993). The continuous line is the main sequence (10^{10} yr; Burrows et al. 1993). The SLR is delimited by the dashed line

3 Results

We have not detected the LiI doublet in any of the objects of our sample. The upper limits to the equivalent width of the LiI line lead to estimate a minimum Li depletion ranging from a factor of 10 (GL 623 B) to a factor of 10^4 (LHS 2065, LHS 2924). According to calculations of Li depletion in very low mass stars and substellar objects (Pozio 1991, Magazzù et al. 1993, D'Antona & Mazzitelli 1994), the mass of the brown dwarf candidates is constrained to be larger than about $0.065\,M_\odot$.

Table 1. Effective temperatures and Li abundances

Object	T_{eff}	log N(Li)
GL 623 B	3200	<2.0
LHS 1047 B	3200	<1.8
HHJ 10	1200	<1.0
BHJ 358	3030	< -0.2
CTI 1156+28	2940	≤ 0.6
GL 644 C	2940	<-0.3
LHS 248	2940	<-0.4
TVLM 868-110639	2900	< 0.5
TVLM 513-46546	2880	≤ -0.1
ESO 207-61	2880	< -0.1
LHS 2397a	2880	≤ 1.0
LHS 2243	2880	< -0.4
LHS 2065	2630	< -1.0
LHS 2924	2630	< -0.8

Our data support membership of HHJ 10 and BHJ 358 to the Pleiades and Hyades clusters, respectively, because of their late spectral type, Hα emission and consistent radial velocities (see Table 2). The Li non detections imply masses > $0.08 \, M_\odot$ for the Pleiades member and > $0.065 \, M_\odot$ for the older Hyades member.

In order to illustrate the implications of the Li test in the H-R diagram we have defined the Substellar Lithium Region (SLR). The BD candidates are located inside or outside the SLR depending on the temperatures adopted for them (see Fig. 1). Since the observed Li depletions prevent these objects from being inside the SLR we conclude that some temperature calibrations give excessively low values of T_{eff}. Only a narrow region in the H-R diagram, between the SLR and the main sequence, is allowed for our sample of very low mass dwarfs, and the temperature calibrations should satisfy these constraints.

Table 2. Hα and heliocentric radial velocities for Pleiades and Hyades members

Object	Sp.T.	Hα (Å)	Vrad (km s^{-1})
HHJ 10	M5.5	5.7	−3
BHJ 358	M6	6.8	+42

At ages younger than about 10^8 yr the SLR comprises all the substellar domain. Thus, young objects – like the Pleiades members – that have burnt their Li are not brown dwarfs. Adopting the frequently used (I−K) vs. T_{eff} calibration for the faintest proper motion Pleiades members currently known, leads to the inconsistent result that they lie well inside the SLR in the H-R diagram. We argue that this inconsistency could be solved if these stars were about 200 K hotter than inferred from their (I−K) colours. The (R−I) colour and the spectral type of HHJ 10 suggest that such a hotter temperature may be correct. Further considerations on the validity of theoretical evolutionary tracks and isochrones have to be postponed until a reliable temperature scale is established for the very low mass Pleiades objects.

A detailed discussion will be find in Martín et al. (1994).

References

Allard, F. (1990) Ph. D. Thesis, Univ. Heidelberg.

Burrows, A., Hubbard, W.B., Saumon, D. & Lunine, J.I. (1993): ApJ, 406, 158

Bryja, C., Humphreys, R.M., Jones, T. (1994): AJ 107, 246

D'Antona, F., Mazzitelli, I. (1994): ApJS, 90, 467

Hambly, N.C., Hawkins, M.R.S., Jameson, R.F. (1993): A&AS 100, 607 .

Kirkpatrick, J.D., Henry T.J., Liebert, J. (1993): Ap.J. 406, 701.

Kirkpatrick, J.D, McGraw, J.T., Hess,.T.R., Liebert, J., & McCarthy, Jr, D.W. (1994): ApJS, in press

Magazzù, A., Martín, E.,L., Rebolo, R. (1993): ApJ 404, L17.

Martín, E.L., Rebolo, R., Magazzú, A.: (1994) ApJ, Nov 20

Pozio, F. (1991): Mem. Soc. Astron. It. 62, 171

Rebolo, R., Martín, E.L., Magazzù, A. (1992): ApJ 389, L83.

Tinney, C.G., Mould, J.R. & Reid, I.N. (1993): AJ, 105, 1045

Lithium in Nearby Main-Sequence Solar-Type Stars

F. Favata[1], G. Micela[2], S. Sciortino[2]

[1] Astrophysics Division, European Space Agency, P.O. Box 299, 2200 AG Noordwijk, The Netherlands

[2] Istituto ed Osservatorio Astronomico di Palermo

Abstract. We report on the status of an observational program aimed at determining the lithium abundance in a large volume-limited sample of nearby solar-type (with spectral-types ranging from late-F to late-K) stars. Although the results presented here are preliminary, in that the data reduction is not completed yet, the sample is already large enough to be reasonably representative of the "typical" behavior of normal stars in this mass range.

1 Introduction

The study of lithium in the last few years has been a field full of new results, with almost every observational program yielding new and interesting results. However, most if not all of the programs have concentrated on somewhat "special" stellar population, such as the active binaries (Randich et al. 1993), X-ray selected samples (Favata et al. 1994), or age-homogeneous samples from open clusters (see for example Soderblom et al. 1993). Very little attention has been given in recent times to the study, with modern detectors, of normal solar-type stars in the solar neighborhood, with the exception of the recent study of Pasquini et al. (1994), who accurately surveyed a limited range of spectral types (G0V-G5V).

On the other hand it becomes increasing difficult to assess how peculiar really are the several samples studied so far in terms, for example, of the number of objects whose other age indicators are not in agreement with their lithium-inferred age, or how common high-lithium "stragglers", such as HD 17925 (a field K0 main sequence star with an inferred lithium abundance typical of PMS stars) or GJ 182 (a field M1 main sequence star also with an inferred lithium abundance typical of PMS stars). We have therefore started a survey of lithium in nearby, apparently normal stars, covering a range of spectral types from approximately F8 down to mid-K main sequence objects.

2 The Observed Sample

The sample is composed of two sub-samples, both of which are selected from the Gliese catalog of nearby stars. Both subsamples contain about 200 stars, distributed across the whole sky. The first sub-sample has been randomly selected from the Gliese catalog within the color range of interest, extracting 200 stars. The second sub-sample is selected, with the same random procedure, from the sample of all the Gliese stars in the color range of interest which have been observed with the Einstein satellite either in pointed mode or serendipitously. While the first sub-sample is a pure random sample, the second one is likely to show some selection effects due to the criteria used to select Einstein targets. The rationale of the sub-sample definition has been to investigate, with the first sub-sample, the behavior of lithium abundance in a representative sample of normal stars, while the second sub-sample will be used to study, using both lithium and kinematics, the biases, specially in terms of age, present in the Einstein samples, on which still most of the knowledge of X-ray luminosity functions for coronal sources is based. Note that while each subsample contains 200 stars selected at random, to avoid "inverse" selection effects we have not checked for duplications, i.e. stars can appear in both samples, so that the effective observational sample size is less than 400 (actually 332).

3 Observations and Analysis

The southern part of the sample (which is discussed here) has been observed, in the course of three different observing run, using the CES spectrograph at the ESO La Silla 1.4 m CAT telescope. All of the observations in the first two runs were centered on the Li I 6707.8 Å line, using the short camera on the CES, yielding a resolution of approximately 50,000. The observations have been reduced in the standard way, except for flat fielding which has been performed using a fast rotating B star as flat field source, given that the lamps available at the CAT for external flat fields show a faint lithium emission line. The equivalent width of the Li I line has been measured using the "splot" tool in the IRAF package. When no lithium line was visible, an upper limit to the equivalent width has been computed on the basis of the measured signal-to-noise ratio of the spectrum in a nearby line-free region. The equivalent widths have been converted to lithium abundance using the curves of growth of Soderblom et al. (1993) and using effective temperatures computed from available photometric colors. In the course of the analysis we noted that some of the later-type systems (i.e. later than approximately K5V) show a quite crowded spectrum, with basically no free continuum visible. This makes it difficult to separate the (eventual) lithium line from the nearby lines, causing many of the later type system to appear to have larger upper limits than their signal-to-noise ratio would imply. Also, some of these objects show a line very close in position to the expected position of the Li I line, but again, at resolution 50,000 it is not possible to properly separate this from nearby spectral feature, and it is therefore not possible to understand if it is actually lithium or if it some yet-to-be identified line.

To better study the problem of measuring lithium in late type main sequence stars we have recently (Aug. 1994) measured some of the systems showing extreme spectral crowding and the presence of weak features at the expected position of the Li I line with the Long Camera at the CAT/CES, yielding a resolution of about 100,000. At this resolution the lines are not blended by the instrumental profile, and as soon as the data are reduced we will study in more detail these systems through spectral synthesis techniques to verify whether indeed these systems have a weak lithium line. The analysis of the R=100,000 observations will be the subject of a forthcoming paper (Favata et al. 1994b). In the meanwhile they are shown as upper limits with a relatively high value in Figure 1.

4 Results

The measured equivalent width of the Li I 6707.8 Å line for the fraction of the sample analyzed so far is shown in Figure 1. The stars belonging to the subsample of stars observed by the Einstein observatory (59 objects) are represented with crosses, while the stars coming from the purely random sample (79 objects) are represented with filled squares. All the points below the horizontal line traced at $W(Li) = 12$ mÅ are upper limits to the actual Li I equivalent width. The main points which can be seen from the plot are: a) a sudden, very sharp drop in "allowed" lithium abundances at $B - V$ approximately equal to 0.7, b) the almost complete lack of systems redder than $B - V \approx 0.7$ showing measurable amounts of lithium and c) the presence of six high-lithium K-type stars. In general, it appears that, while for systems bluer than $B - V \approx 0.7$ a range of lithium abundances is "allowed", for systems redder than this limit lithium suddenly disappears, becoming undetectable except in some rare cases.

As for eventual biases present in the sample extracted from the Einstein observed stars, among the six high-lithium K-type stars all but one come from the Einstein-selected sample. Although the statistics are obviously too small to allow any sound conclusion, it is suggestive that only one of the high lithium later type systems comes from the randomly selected sample. In particular, the two higher lithium systems are HD 17925, an already known "high-lithium straggler" in the solar neighborhood (Cayrel de Strobel and Cayrel 1989) and GJ 182, also known as a "high-lithium straggler" (Bopp 1974).

Of the four intermediate-lithium K type systems, three (GJ 879, GJ 150 and GJ 517) are variously classified as BY Dra systems in the literature, and are known as very active stars. Given that BY Dra systems are found to be kinematically quite young (Soderblom 1990), it is not surprising to find that they tend to have a higher lithium level than the general population. This result is also indicative of a bias in favor of younger and therefore more active systems present among the late-type stars observed with Einstein. The detailed analysis of these biases will be the subject of a forthcoming paper (Favata et al. 1994b).

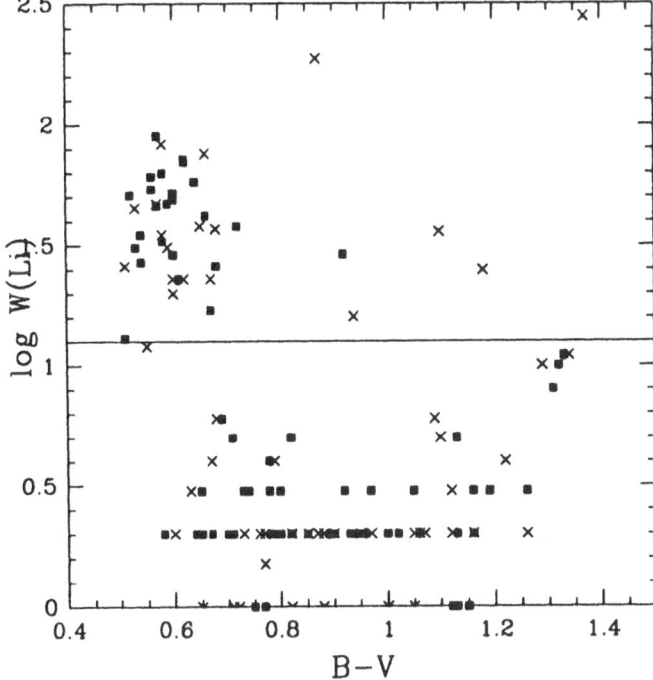

Fig. 1. The measured equivalent width of the Li I 6707.8 Å line for the sample of late-type nearby stars analyzed so far is plotted against the color. The filled squares are stars from the randomly selected sample, while the crosses are from the sample selected from the stars observed by the Einstein observatory.

Acknowledgments

We acknowledge M. Barbera's help in the sample selection procedure. G.M. and S.S. acknowledge financial support from ASI (Italian Space Agency), and MURST (Ministero della Università e della Ricerca Scientifica e Tecnologica).

References

Bopp, B.W. 1974, PASP 86, 281
Cayrel de Strobel, G., Cayrel, R. 1989, A&A 218, L9
Favata, F., Micela, G., Sciortino, S. 1994b, in preparation.
Favata, F., Barbera, M., Micela, G., Sciortino, S. 1994, A&A in press
Pasquini, L., Liu, Q, Pallavicini, R. 1994, A&A 287, 191
Randich, S., Gratton, R., Pallavicini, R. 1993, A&A 273, 194
Soderblom, D.R. 1990, AJ 100, 204
Soderblom, D.R., Jones, B.F., Balachandran, S. et al. 1993, AJ, 106, 1059

Lithium in Pleiades K Dwarfs

David R. Soderblom,[1] Burton F. Jones,[2]
Matthew Shetrone[2]

[1] Space Telescope Science Institute, 3700 San Martin Drive,
Baltimore MD 21218 USA

[2] Lick Observatory/UCO, Board of Studies in Astronomy and
Astrophysics, University of California Santa Cruz,
Santa Cruz CA 95064 USA

Abstract. Observations of Li in Pleiades K dwarfs with the Hires spectrograph of the Keck telescope show that the correlation between excess Li and rapid rotation breaks down in the coolest stars.

1 Introduction

This paper will present only a summary of observations of lithium in Pleiades K dwarfs. For a fuller discussion of Li in open cluster stars, see Soderblom (1994).

Duncan & Jones (1983) showed that solar-type stars in the Pleiades exhibit a significant spread in their Li abundances. This is very different from the tight relationship between Li and T_{eff} seen in the Hyades (Thorburn et al. 1993), for example. In a recent paper (Soderblom et al. 1993), we reported on more and better observations of Li in the Pleiades. The essence of that paper is shown in Figure 1. The scatter in N_{Li} ($= \log N(Li)$ on a scale where hydrogen is 12) for $T_{eff} > 5500\,K$ may be mostly due to observational uncertainty, but the large spread for $T_{eff} \lesssim 5300\,K$ is very real. Soderblom et al. (1993) discuss this spread and show that it probably represents real differences in Li abundances, not just apparent differences that might arise from spots or activity on these young stars.

Figures 2 and 3 show that the spread in Li is correlated with both rotation and chromospheric activity (as measured with the 8542 Å line of the Ca II infrared triplet). These illustrations are in observational coordinates to show that the abundance differences are well beyond any observational uncertainty. The fact that rotation and activity correlate with Li in the same way is not surprising since rotation and activity are so closely correlated with one another in solar-type stars. The close relation between excess Li and rapid rotation in particular has led to the speculation that rapid rotation may inhibit the internal circulation patterns that lead to Li depletion in the slower rotators.

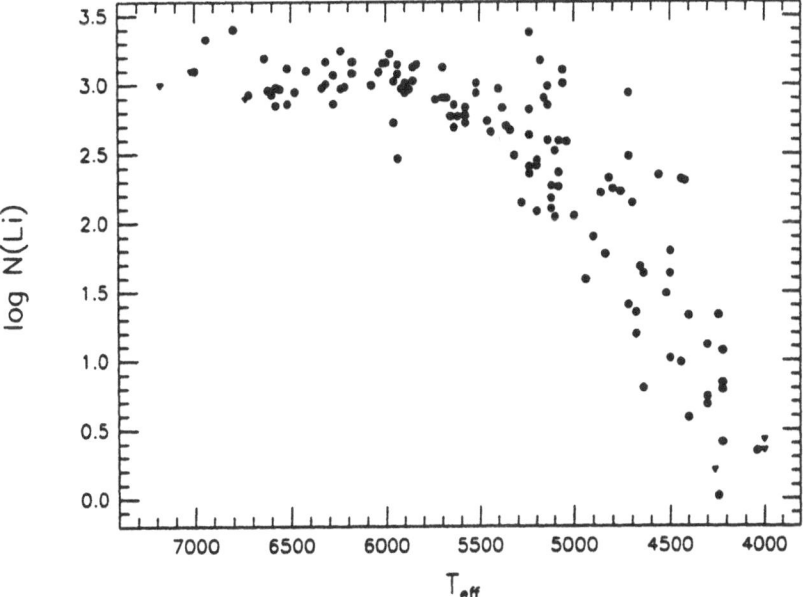

Fig. 1. Lithium abundance (logarithmic, with H = 12) versus effective temperature for stars of the Pleiades, from Soderblom et al. (1993). The open circles are data from Boesgaard et al. (1988).

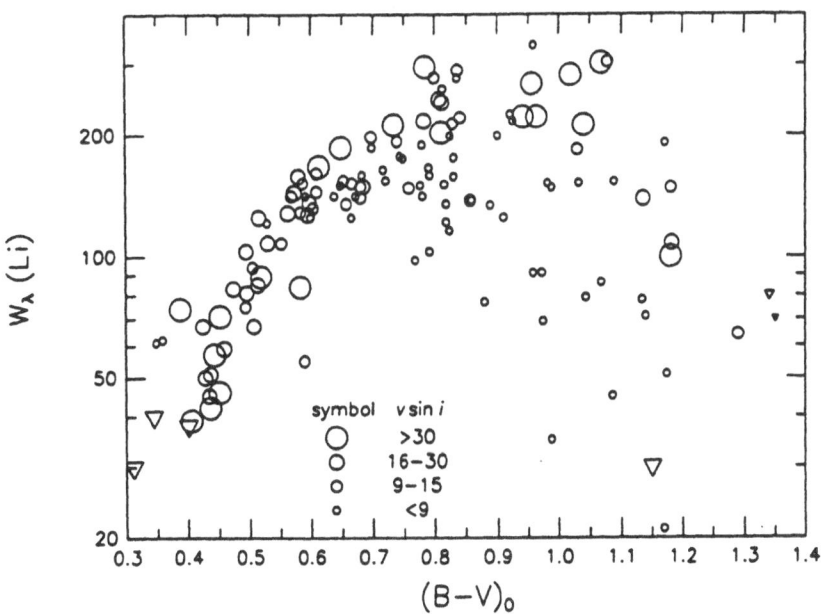

Fig. 2. Equivalent width of the Li 6708 Å feature versus dereddened $(B - V)$ color for stars of the Pleiades. The circles are in proportion to the $v \sin i$ of the star in the manner indicated.

2 Keck Observations

In November, 1993, we used the Hires spectrograph on the Keck telescope to observe about a dozen Pleiades K dwarfs. We chose to observe some K dwarfs that were too faint to observe at Mount Hamilton and which were known rapid rotators, with the expectation that these stars should show the most Li if the correlation of Figure 2 extends to cooler stars.

The results are shown in Figure 4, supplemented by some similar observations of García-López et al. (1994). For these mid-K dwarfs of the Pleiades, we see rapid rotators that clearly have much less Li than more slowly-rotating stars of the same colors. Why should this be?

First, we are confident that Figure 4 is an accurate representation of the observations and that we have not made major errors in estimating W_λ in the rapidly-rotating stars. To measure equivalent widths, we artificially broadened the spectra of slow rotators (for which the equivalent width or a firm upper limit can be established) to match the rest of the spectrum, and we then looked at the difference between the two.

Second, it seems unlikely that stars with $(B - V)_0 \gtrsim 1.15$ should fail to show a correlation between Li and rotation when bluer stars do for reasons related to the intrinsic physics of the stars. We say this because there is nothing significant that we know of that changes at that color in the structure of a star.

Third, the breakdown in the Li-rotation correlation does not appear to be just a statistical artifact. The sample involved are big enough to avoid small number problems, and the correlation for $(B - V)_0 \lesssim 1.15$ is very strong (the few stars with abundant Li but low $v \sin i$ are likely to be seen pole-on).

At this time we do not have an explanation for the breakdown in the Li-rotation correlation for $(B - V)_0 \gtrsim 1.15$ ($T_{\mathrm{eff}} \lesssim 4300\,\mathrm{K}$), but we intend to examine the phenomenon in greater detail.

References

Boesgaard, A.M., Budge, K.G., and Ramsay, M.E. ApJ **327** 389 (1988)

Duncan, D.K., & Jones, B.F. ApJ **271** 663 (1983)

García-López, R.J., Rebolo, R., & Martín, E.L. A&A **282** 518 (1994)

Soderblom, D.R., Jones, B.F., Balachandran, S., Stauffer, J.R., Duncan, D.K., Fedele, S.B., & Hudon, J.D. AJ **106** 1059 (1993)

Soderblom, D.R. Mem. Soc. Astr. Ital., in press (1994)

Thorburn, J.A., Hobbs, L.M., Deliyannis, C.P., & Pinsonneault, M.H. ApJ **415** 150 (1993)

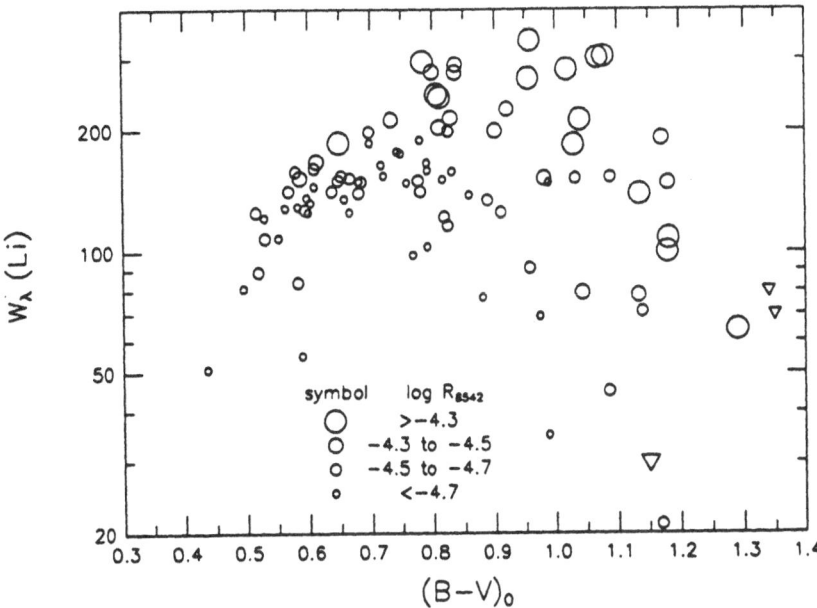

Fig. 3. Lithium equivalent width versus color, with symbols coded in proportion to activity.

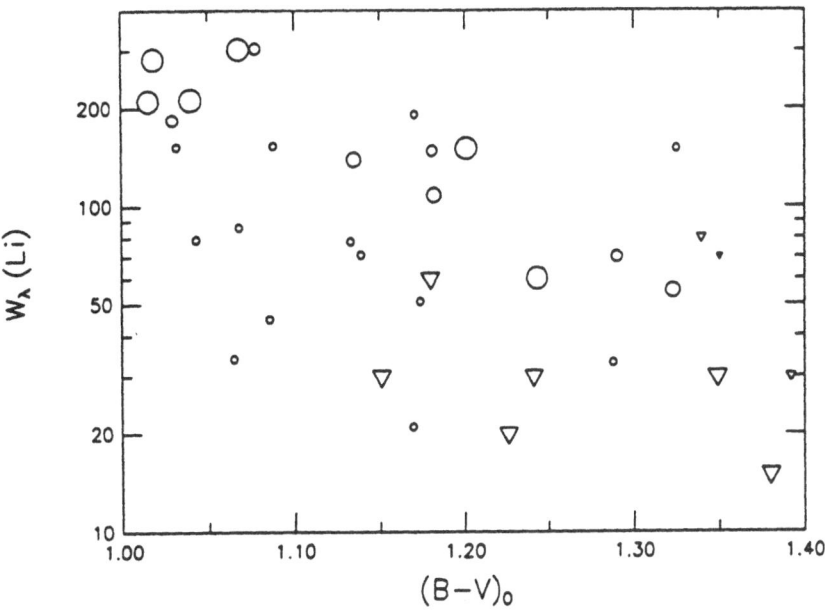

Fig. 4. Equivalent width of the Li 6708 Å feature versus dereddened $(B - V)$ color for K dwarfs in the Pleiades. The circles and triangles are in proportion to the $v \sin i$ of the star in the manner indicated in the Fig. 3.

Lithium Abundances of the Most Metal Poor Stars

Sean G. Ryan

Anglo–Australian Observatory, P O Box 296, Epping, NSW 2121, Australia

Abstract. Recent studies of halo main–sequence turnoff stars with [Fe/H] < -3.0 have yielded some lower Li abundances than are normally associated with Population II dwarfs, by 0.10 to 0.15 dex. Two possibilities are considered: (i) that the most metal poor stars have depleted their surface Li relative to higher metallicity halo dwarfs, or (ii) that some non–primordial production of Li is evident in halo dwarfs. It may be difficult to reconcile (ii) with galactic chemical evolution models, and there is no prediction in stellar evolution calculations that (i) would occur; the correct explanation for the observations remains unclear. It is unclear whether studying the most metal poor stars leads one closer towards or further away from the primordial Li abundance. Current NLTE calculations for the Li resonance doublet do not significantly affect the arguments.

1 Introduction

Population II stars warmer than $T_{\text{eff}} = 5600$ K have Li abundances A(Li) $=$ n(Li)/n(H)$+12.00$ generally within the range 2.0 to 2.2, over a range of metallicities spanning 2 dex. There is a widely held belief, following the pioneering work of Spite & Spite (1982), that the observed abundances are hardly altered from the Li abundance produced in the Big Bang. Since the lowest metallicity dwarfs preserve at their surfaces the most pristine material we can observe, one might expect to find the best indication of the primordial Li abundance by studying these stars. Three recent studies have measured Li in dwarfs with [Fe/H] < -3.0, viz. Spite & Spite (1993), Thorburn (1994) and Norris, Ryan & Stringfellow (1994). The Li abundances for many of these are lower than for more metal rich plateau stars, complicating efforts to discovering the primordial Li abundance.

2 Data

Many of the very metal poor stars appear in more than one study. Data for those with [Fe/H] < -3.0 (and $T_{\text{eff}} > 5600$ K) are given in Table 1; see the original publications for more details. Agreement on equivalent widths is generally good. This is partly because the lack of line absorption in this spectral region simplifies continuum placement and eliminates line blending. However, 815–43 is an exception for which four measurements fall into 2 divisions around 14 and 24 mA. The difference, considerably in excess of the formal errors of the measurements, is a good reminder that not all errors in astronomy are Gaussian or estimated reliably. Analyses of star–to–star scatter must allow for such non–Gaussian and/or underestimated errors.

Since there are systematic differences between authors in the derived abundances (primarily Thorburn on the one hand giving A(Li) approximately 0.1 dex higher than Spite & Spite and Norris et al for equivalent parameters), and some differences in the effective temperature scales, averages are given for the observed equivalent widths and for the effective temperatures, and these data are used to recompute abundances on a consistent abundance scale. [1] Lithium abundances were recomputed as described in Norris et al (1994), using models from Bell (1981) since these are comparable to those used for higher metallicity stars during the previous decade. Systematic differences would be introduced by adopting newer models for only the most metal poor stars; an attempt to bring diverse Li studies onto uniform effective temperature, stellar model atmosphere, and abundance scales will be made elsewhere. Gravities log g = 4.0 were used in all cases. The revised abundances and errors are given in columns (10) and (11) of the table.

As data have accumulated, it has become apparent that the lithium abundances even in plateau stars are higher at higher temperatures, Thorburn (1994) and Norris et al (1994) deriving slopes of 0.024 and 0.030 dex per 100 K respectively. The suggestion that this is due to the neglect of NLTE effects is NOT supported by existing NLTE calculations. Carlsson et al (1994) and Pavlenko (1994) derive NLTE corrections of 0.006 to 0.008 dex per 100 K and 0.000 to 0.004 dex per 100 K respectively for stellar parameters relevant to plateau dwarfs, at least a factor of 3 too small to eliminate the observed slope. Since depletion of the coolest plateau stars is not unexpected in standard stellar models, we should not be too surprised that the temperature dependence remains.

Irrespective of the cause of the slope, it is important not to ignore it. It would be inappropriate to take a blind average of the plateau abundances and call this the primordial value. Furthermore, merging the hot and the cold plateau stars without regard to the difference leaves unnecessarily large scatter in the combined sample. Norris et al and Thorburn compensated for the slope by renormalizing the abundances to a temperature towards the hot end of the plateau. The final column of the table gives abundances renormalized to 6200 K, by 0.03 dex per 100 K. (The effect on abundances for the very metal poor sample is minor since it is composed primarily of turnoff stars anyway, since these are the brightest dwarfs.)

---

[1] Blind averages of the effective temperatures lock systematic (non–Gaussian!) differences in the effective temperature scales into the data and would be improved upon in a more thorough investigation.

Table 1. Li data for stars with [Fe/H] < −3.0 and T_{eff} > 5600 K.

Star	[Fe/H]	W,ref mÅ	σ mÅ	T_{eff}	A(Li)	W_a mÅ	$σ_a$ mÅ	$T_{eff,a}$	A(Li)	σ	A6200
(1)	(2)	(3)	(4)	(5)	(6)	(7)	(8)	(9)	(10)	(11)	(12)
G64-12	-3.4	31,SS	b	6325	2.34	29	3	6260	2.31	.05	2.29
"	-3.52	28,T	4	6197	2.37						
G64-37	-2.70	14,T	2	6376	2.15	14	2	6360	2.03	.07	1.98
"	-3.38	14,NRS	2	6350	2.03						
G238-30	-3.03	29,T	3	5771	2.08	29	3	5770	1.95	.05	2.08
G268-32	-3.50	27,T	4	5841	2.09	27	4	5840	1.96	.07	2.07
-13:3442	-3.14	30,T	3	6168	2.39	30	3	6170	2.26	.05	2.27
-24:17504 *	-3.70	22,SS		6100	2.04	21	2	6130	2.07	.04	2.09
"		21,T	3	6168	2.21						
"		19,NRS	2	6075	1.97						
-33:1173	-3.29	17,SS		6400	2.11	13	2	6360	2.00	.07	1.95
"		12,T	2	6319	2.03						
"		10,NRS	1	6350	1.87						
22186-17	-3.14	17,T	4	5978	1.96	17	4	5980	1.85	.12	1.92
22876-32A	-3.80	15,T	·2	6319	2.14	15	2	6230	1.99	.06	1.98
"	-4.21	15,NRS	3	6150	1.91						
22884-108	-3.28	15,T	5	6319	2.14	15	5	6320	2.04	.18	2.00
22958-42	-3.34	<12,T		6217	<1.96	<12		6220	<1.87		<1.86
22964-214	-3.29	16,T	4	6370	2.21	16	4	6370	2.10	.12	2.05
29499-60	-3.31	16,T	4	6217	2.10	16	4	6220	2.00	.12	1.99
29506-7	-3.10	24,T	5	6217	2.31	24	5	6220	2.19	.10	2.18
29527-15	-3.50	<10,T		6319	<1.94	<10		6320	<1.84		<1.80
815-43	-3.20	27,SS		6350	2.32	:24	3	6390	2.31	.06	2.25
"		22,T	2	6423	2.40	:					
"		15,NRS	3	6350	2.06	:14	2	"	2.05	.07	1.99
"		13,NRS	2			:					
831-70 *	-3.4	26,SS		6000	2.06	24	2	6060	2.05	.04	2.09
"		23,T	3	6119	2.22						

NOTES:
a. adopted values
b. Spite & Spite give their accuracy as 3 to 6 mA
REFERENCES:
SS=Spite & Spite 1993   T=Thorburn 1994   NRS=Norris, Ryan & Stringfellow 1994

The mean lithium abundance, A6200 = 2.07, is 0.10 to 0.15 dex less than is typical of higher metallicity halo stars. Current stellar models do not predict that the most metal poor halo stars should deplete more Li than those which are more metal rich. However the alternative, that the more metal rich halo stars show the products of galactic nucleosynthesis of Li, has low appeal since the source would have to act only during formation of the lowest metallicity stars. For example, the production of Li between forming stars at [Fe/H] = −3.5 and at [Fe/H] = −2.5 must be made to cease in going to [Fe/H] = −1.5. It is not clear, then, what is responsible for the lower lithium abundances seen in the most metal poor stars: whether unexpected depletion has occurred in the most metal poor stars from a somewhat higher primordial value, or whether some galactic enrichment has occurred to raise the observed value from a lower primordial value (or worse, some combination of both).

Not all metal poor stars appear to have low Li abundances. The Li equivalent widths of G64-12 and G64-37 appear well determined and differ by a factor of two. Temperature differences of 400 K would be required to explain this with a uniform Li abundance; such a temperature difference stretches acceptable bounds. (Li abundances are not particularly sensitive to errors in the adopted gravity or microturbulence.) Then again, the four measurements of 815-43 are not in overall agreement; even when caution is exercised, consistent results are not guaranteed.

3 Conclusions

With the exception of a few highly Li–deficient stars, most plateau dwarfs have Li abundances within the range 1.9 to 2.3, but there is a trend towards higher Li abundances with increasing $T_{\rm eff}$, and stars at the lowest metallicities studied have lower lithium abundances. Blind averages of plateau star abundances to find the primordial value ignore these details and overlook possible stellar processing and/or galactic chemical evolution which may have altered the primordial Li abundance by more than one might naively wish.

References

Carlsson, M., Rutten, R. J., Bruls, J. H. M. J., & Shchukina, N. G. 1994, in press
Norris, J. E., Ryan, S. G., & Stringfellow, G. S. 1994, ApJ, 423, 386
Pavlenko, Y. V. 1994 preprint and poster, this volume
Spite, F., & Spite, M. 1982, A&A, 115, 357
Spite, F., & Spite, M. 1993, A&A, 279, L9
Thorburn, J. A. 1994, ApJ, 421, 318

Lithium Abundance in Pop II Stars: Influence of a Small Mass Loss

Sylvie Vauclair and Corinne Charbonnel

Observatoire Midi-Pyrénées, 14, av. E. Belin, 31400 Toulouse, France

Abstract. Lithium abundance computations including microscopic diffusion below the convection zone in halo stars lead to results in contradiction with the observations. Rotation-induced turbulence may prevent this settling but it leads to results which do not reproduce the plateau shape. Here we claim that a small stellar wind can prevent the lithium settling without bringing up to the convection zone the matter in which lithium has been destroyed by nuclear reactions. The results obtained with this wind model are in very good agreement with the observations.

1 Diffusion and rotation

The observations by Spite & Spite (1982), that the upper values of the lithium abundance in halo stars are one order of magnitude smaller than the ones observed in galactic stars, lead to a large debate in the astrophysical community about the primordial lithium abundance. The whole problem can be summarized by the following question: has lithium been depleted in the outer layers of halo stars, or has the original lithium been preserved at their surfaces since the beginning?

It seems difficult to maintain the original lithium abundance in halo stars during all their lifetime. Either lithium is depleted due to element separation, or it is destroyed by nuclear reactions. Computations by Profitt & Michaud (1989) showed that nowhere inside halo stars the lithium abundance could have remained at its original value.

It was suggested by Vauclair (1988) that rotation-induced turbulence could lead to a nuclear destruction of lithium in halo stars large enough to explain their present abundances, with an original abundance equal to the present galactic one. It seemed possible that the "plateau shape" of the abundances be preserved if the turbulent diffusion coefficient decreased rapidly with radius, as in Zahn (1987) (see also Pinsonneault et al. 1992). Recent computations with all parameters identical to those tested for the Sun and for galactic clusters (Gaigé 1994, Charbonnel et al. 1992, Charbonnel et al. 1994) show however that rotation-induced turbulence leads to a negative slope in the "plateau" which is not observed (Figure 2).

On the other hand, if no macroscopic motion is introduced below the convection zone, the theoretical lithium plateau shows a tendency of bending down for the hottest stars (Figure 2) (see also Vauclair & Charbonnel 1994).

2 Influence of a small mass loss

We suggest that the lithium dilemma for halo stars may be solved by taking into account a small mass loss, of the order or slightly larger than the solar wind. The idea is that a small wind may prevent the element separation without levitating up to the convection zone the layers where lithium has been largely destroyed by the nuclear reactions.

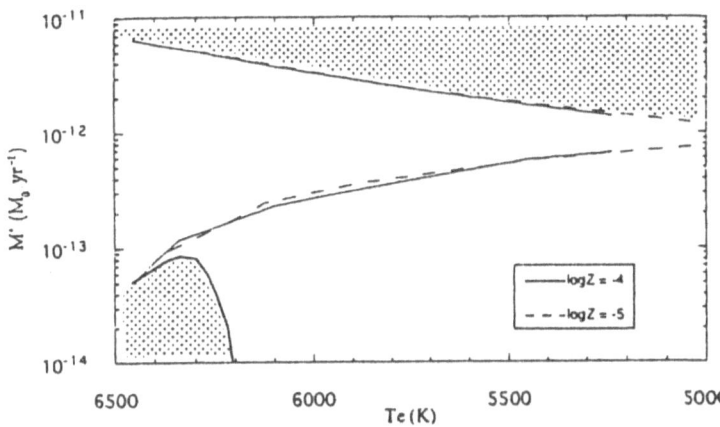

Fig. 1. Approximate values of the mass loss rates needed to prevent microscopic diffusion of lithium in halo stars (lower curves), and mass loss rates which bring up to the convection zone the matter in which lithium has been largely destroyed by nuclear reactions. The thick curve represents the mass loss rates which prevents lithium from being depleted by more than a factor 2

The influence of stellar mass loss to counteract the effect of gravitational diffusion was first introduced by Vauclair (1975) who suggested that this could be the reason for the existence of helium-rich main sequence stars. If a star is subject to a small wind, it slowly readjusts its internal structure according to the mass loss. For small mass fluxes (at most two orders of magnitude larger than the solar wind) the internal changes in the star are negligible except that the matter in the outer layers is slowly replaced by other matter coming from underneath, to satisfy mass conservation. This slow motion can prevent element gravitational settling below the convection zone in cool stars.

The order of magnitude of the minimum mass loss rate needed to prevent microscopic diffusion below the convection zone in halo stars may be evaluated with the assumption that the upward mass loss flux is opposite to the downward lithium diffusion flux below the convection zone. The results obtained for an age of 15 Gyr are given in Figure 1. There is no need however to prevent microscopic diffusion in all the stars, as lithium is depleted only in the hottest stars (Figure 2). The mass loss rate needed for lithium to be depleted by less than a factor 2 is also given in Figure 1.

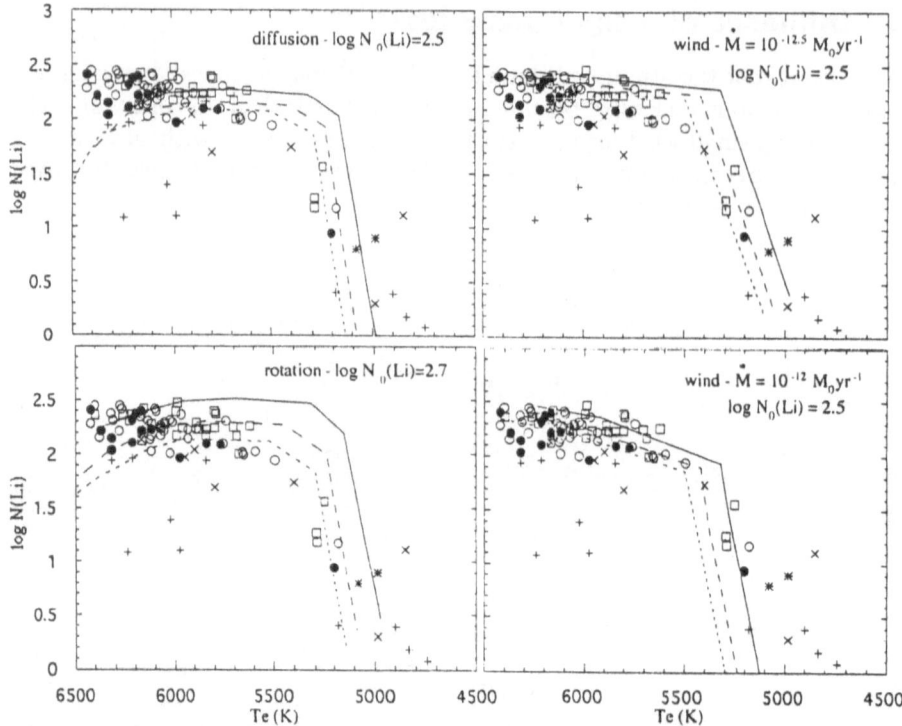

Fig. 2. Theoretical lithium depletion curves at three ages (solid line : 10Gyr; dashed line : 15Gyr; dotted line : 18Gyr) for a metallicity $Z=10^{-4}$. Upper left : pure diffusion; lower left : diffusion and rotation-induced turbulence, with all parameters identical to those tested for the Sun; upper right : diffusion and mass loss ($10^{-12.5}$ $M_\odot$ yr^{-1}); lower right : diffusion and mass loss (10^{-12} $M_\odot$ yr^{-1}). The observations are by Thornburn (1994) and Spite et al. (1994)

The order of magnitude of the mass loss rate which would bring up to the convection zone during the star's lifetime the layers in which lithium has been largely destroyed by nuclear reactions may also be evaluated from the models. This represents the maximum mass loss rate allowed for lithium to be preserved in the outer convection zone. The results are also given in Figure 1. This maximum value decreases for cooler stars, so that we may expect some lithium depletion increasing in stars with decreasing effective temperature. In any case the range of mass loss rates which may preserve lithium in halo stars and account for the "lithium plateau" seems quite comfortable from these preliminary computations.

Figure 2 compares the results obtained by introducing the mass loss rate as a parameter in the Toulouse-Geneva evolutionary code to the recent observations by Thornburn (1994) and Spite et al. (1994). A very good agreement is obtained.

3 Conclusion

The observations of lithium in halo stars can be accounted for by introducing a small stellar wind at least in the hottest stars of the "plateau". The mass loss rate must be larger than 10^{-13} $M_\odot yr^{-1}$ for stars with effective temperatures larger than 6250K. For all stars it must be smaller than 2 to 5 10^{-12} $M_\odot yr^{-1}$ for the lithium destroyed layers not to show up at the surface. The few stars with no observed lithium can be those which suffer larger mass loss rates.

These results suggest that the initial lithium abundances in halo stars are given by the largest abundance observed in the hottest stars of the "plateau". **The primordial lithium abundance should be derived from the upper envelope of the observations and not from their average value.**

References

Charbonnel C., Vauclair, S., Zahn, J.P. (1992): A&A 255, 191

Charbonnel C., Vauclair, S., Maeder, A., Meynet, G., Schaller, G., (1994): A&A 283, 155

Gaigé, Y. (1994): PhD Thesis

Pinsonneault, M.H., Deliyannis, C.P., Demarque P., (1992): ApJS 78, 181

Proffitt, C.R., Michaud, G., (1989): ApJ 346, 976

Spite, F., Spite, M., (1982): A&A 115, 357

Spite, M., Pasquini, L., Spite, F., (1994): preprint

Thorburn, J.A., (1994): ApJ 421, 318

Vauclair, S., (1975): A&A 45, 233

Vauclair, S., (1988): ApJ 335, 971

Vauclair, S., Charbonnel, C. (1994): preprint

Zahn, J.P., (1987): in Summer school in Astrophysical Fluid Dynamics (Les Houches, Elsevier Sci. Publ.)

Lithium in Late–Type Subgiants

S. Randich[1], R. Pallavicini[2], L. Pasquini[3], R. Gratton[4]

[1] MPI für Extraterrestrische Physik, D-85740 Garching, Germany
[2] Osservatorio Astrofisico di Arcetri, I-50125 Florence, Italy
[3] European Southern Observatory, Casilla 19001, Santiago 19, Chile
[4] Osservatorio Astronomico di Padova, Italy

Abstract. We present the preliminary results of a survey of Lithium in Pop. I subgiants. Our sample is compared with both main sequence stars and evolved giants. Our main conclusion is that Li-rich subgiants are likely to be the descendants of Li–rich old main sequence stars, rather than objects which have enriched their Li content in the early phases of post-main sequence evolution.

1 Introduction

Herbig & Wolff (1966) obtained Li abundances for 12 F6 to K1 subgiants, finding that at a given position in the M_V − (B-V) plane the Li abundance is not unique. In particular, they found a few solar–mass objects, among which the G2 IV star β Hyi, with a [Li/Ca] ratio substantially higher than expected. Although they suggested some speculative interpretations, a definitive explanation was not given.

Now, nearly 30 years later, we are still far from completely understanding both main sequence and post–main sequence evolution of Li abundances in solar–type stars. In particular, while new observations of β Hyi have confirmed that its Li is a factor ~ 30 higher than in the Sun, other solar–mass subgiants have been found with a Li content much larger than the Sun. Dravins et al. (1993). have hypothesized that Li could have diffused downward during main sequence lifetime, and temporarily stored below the convective zone. As the star evolves off the main sequence, Li could be dredged-up to the surface, thus leading to a temporary enhancement of Li abundance before appreciable dilution occurs. However, observations of subgiants in the old cluster M67 suggest that Li is destroyed, rather than stored in deeper layers (Balachandran 1990).

Another open question is the large spread in Lithium found for K-type giants which often have Li abundances much higher than dwarfs of the same spectral type (Brown et al. 1989). Except for very rare extremely Li-rich stars, the *maximum* Li abundances observed in giants are consistent with theoretical predictions; however, the precise relationship between mass and Li abundance in cool giants is still unclear.

A systematic study of Li in subgiants has not yet been reported in the literature apart from the one by Herbig & Wolff. We decided therefore to carry out such a survey with the purpose of following the evolution of Li in stars of different mass as they move from the main sequence to the base of the red giant branch. In particular we were hoping to test whether the diffusion mechanism proposed by Dravins et al. is operating or not. We present here the preliminary results of this survey.

2   Data Sample and Observations

Our sample consists of 102 southern stars with spectral type later than F7 taken from the Bright Star Catalogue. Here, we give the results for a subsample of 69 objects obtained by excluding known binaries and stars with an uncertain luminosity classification.

The observations were carried out at ESO using the 1.4m CAT telescope and the CES spectrograph, with the Short Camera and CCD # 9. The nominal resolving power was R=60,000. A spectral range of about 50 Å centered at 6708 Å was covered. Data reduction was performed in the standard way using the MIDAS package. A signal to noise ratio S/N > 150 was obtained for most of our spectra. Such a ratio corresponds to a minimum detectable Li EW of $\sim$ 3-4 mÅ. Abundances were derived from equivalent widths using the curves of growth of Soderblom et al. (1993). In this preliminary analysis we determined effective temperatures from $B - V$ colours and the polynomial relationship between $B-V$ and $\log T_{eff}$ given by Gray (1992).

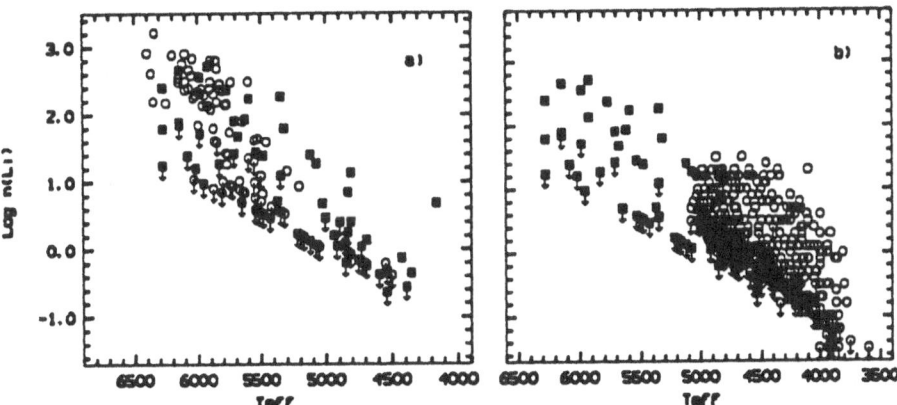

Fig. 1. a) Li abundances vs. T_{eff} for subgiants (filled squares) and dwarfs from the literature (open circles). b) Li abundances vs. T_{eff} for subgiants (filled squares) and giants from Brown et al. (open circles).

3   Results

In Figure 1a we plot Li abundances as a function of effective temperature for the subgiants in our sample (filled squares) and for a sample of dwarfs from various sources in the literature (open circles). The figure shows that, within the pattern of a decreasing Li with decreasing T_{eff}, subgiants show a large spread in Li at each spectral type, similar to what observed for main sequence stars. At temperatures greater than about 5600 K, main sequence stars show a slight tendency to have higher Li abundances than the more evolved subgiants, while the opposite occurs at $\sim$ 5000 K, where subgiants tend to have larger Li abundances than dwarfs of similar colour. In Figure 1b the same sample of subgiants is compared to the giants of Brown et al. Two main features are evident in the figure: first, giants show a larger spread in Li than subgiants especially at cooler temperatures; second, the subgiants are on average less Li rich than giants of similar temperatures. In particular, if we focus on stars with 4500 K $\leq T_{eff} \leq$ 5000 K, the mean Li abundances for the giants and the subgiants are, respectively, 0.39 and 0.06 (note that mean Li abundances have been computed considering upper limits as detections and the two surveys have approximately the same sensitivity to Li EW's). At temperatures < 4800 K, we do not find any subgiant, but one, with significant Li abundance.

4   Discussion

A) Warm subgiants: comparison with dwarfs

Although our sample is considerably larger than that of Herbig & Wolf, our results confirm their previous findings, i.e., the presence, among subgiants, of both stars which show Li depletion and stars which, on the contrary, are still rich in Li. This occurs also for objects of the same mass (like the Sun and β Hyi) and is similar to what observed for main sequence solar-type stars (e.g. Pasquini et al. 1994).

We believe that the high Li content observed in some G-type subgiants should be ascribed –whatever the mechanism is– to Li non–depletion on the main sequence, rather than to Li enrichment during post–main sequence evolution. In fact, if the Li abundance of β Hyi and similar stars were due to the dredging up mechanism suggested by Dravins et al., we should find that virtually *all* subgiants with T_{eff} similar to or higher than β Hyi are Li rich, either because they have about the same mass of β Hyi and are in the same evolutionary status, or because they are more massive and have progenitors which did not destroy Li on the main sequence. Fig. 1a clearly shows that this is not the case, unless the Li enrichment dredge-up phase lasts much less than the time spent by a star as a G-type subgiant. Therefore, we suggest that **Li rich, β Hyi like subgiants are just the early post–main sequence descendants of Li rich old dwarfs**, and that the spread in Li covered by subgiants at any given mass/temperature reflects the same spread observed among main sequence stars. Note that the slight shift of warm main-sequence stars toward larger Li abundances with respect to subgiants simply indicates the presence among main sequence stars of normal Li rich *young* objects, which, of course, are absent among subgiants.

B) Cool subgiants: comparison with giants

Unfortunately we know neither the mass nor the exact evolutionary status of our subgiants as well as of the giants in the Brown et al. sample. However, by looking at theoretical evolutionary tracks, we argue that the range of possible masses at a given temperature is much narrower for cool subgiants than for giants. In other words, a cool giant (at $T < 5000$ K) could have been a rather massive star on the main sequence, while a cool subgiant has necessarily a low–mass ($M \leq 1M_\odot$) progenitor.

We can expect, therefore, **a larger spead in Li abundances among giants than among subgiants as well as typically lower Li abundances among the coolest subgiants.** Fig. 1b shows evidence for both points. The larger spread in Li abundances among giants is likely due to the large dispersion in masses and in the distribution of Li abundances at the end of main-sequence evolution (note that a Li reach star like β Hyi will retain in the giant phase a larger Li abundance than a Li-poor star like the Sun, if Li dilution is the same –as expected– for both types of stars). The coolest subgiants, which have evolved from low-mass progenitors, have presumably depleted all of their Li on the main-sequence, while some of the subgiants at ~ 5000 K have more likely evolved from somewhat more massive progenitors, and thus have retained more Li before dilution occurs.

We intend to test these qualitative conclusions by putting the sample stars on the HR diagram for comparison with evolutionary tracks, as well as by measuring the $^{12}C/^{13}C$ ratio (for which observations have already been obtained for the stars in the sample).

References

Balachandran, S. (1990): in *Cool Stars, Stellar Systems, and the Sun*, ed. G Wallerstein, p. 257.

Brown, J.A., et al. (1989): *Astrophys. J. Suppl.* **71**, 293.

Dravins, D., et al. (1993): *Astrophys. J.* **403**, 385.

Gray, D.F. (1992): *The Observation and Analysis of Stellar Photospheres* (Cambridge University Press)

Herbig, G.H., and Wolff, R.J. (1966): *Ann. d'Astrophysique* **29**, 593.

Pasquini, L., Liu, Q. and Pallavicini, R. (1994): *Astron. Astrophys.* **287**, 205

Soderblom, D.R., Jones, B.F., and Balachandran, S. et al. (1993): *Astron. J.* **106**, 1059

Rotation and Lithium in Bright Giant Stars

A. Lèbre[1], J.-R. De Medeiros[2]

[1] GRAAL, CP 072, Université de Montpellier II, F-34095 Montpellier, France
[2] Departamento de física, Universidade Federal do Rio Grande do Norte, 59072–970 Natal, Brasil

Introduction

We present new measurements of high precision rotational velocities and new high resolution spectroscopic observations of the Lithium line 6707.81 Å for a large sample of bright giant stars with spectral types between F3II and G5II. We intend to study the link between the Lithium abundance and stellar rotation and in particular we search if in such spectral region lithium abundance could present a discontinuity paralleling the one found for the rotational velocity near the spectral type F8II–F9II.

1 Observations

1.1 Sample Data

Table 1 lists the program stars in order of increasing HD number. These bright giant stars are northern of −25o presenting visual magnitudes lower than 8.5. The spectral types are from Egret (1981). The rotational and radial velocities are from "a catalog of rotational and radial velocities for evolved stars" by De Medeiros and Mayor (1994). Such measurements were obtained between 1986 to 1993 with the **CORAVEL spectrometer** (Baranne et al., 1979) mounted at the Swiss 1m telescope at Observatoire de Haute Provence. These radial and rotational values present – at least for those stars showing rotation rates lower than about 50 km s^{-1} – an uncertainty of about 0.3 km s^{-1} and 2.0 km s^{-1} respectively.

1.2 Spectroscopic Data

These stars were observed in January 1992 at the 1.52m telescope at the Observatoire de Haute Provence with the AURELIE spectrometer :

a- 1200 lines/mm grating in the first order
b- Resolving power $R = \lambda/\Delta\lambda = 24{,}200$
c- Resolution 0.278 Å (central wavelength 6620 Å)
d- Signal to noise between 100 and 200.

The spectral region around Lithium line at 6707.81 Å was observed (Fig.1). Spectra are presented in the stellar rest frame wavelength and are normalized

Table 1. Program stars

star	Sp.Type	B - V	Vrad km s^{-1}	Vsini km s^{-1}	W(li) mÅ	spectro. binary
HD 371	G3II	1.04	- 6.43	7.4	30.0	
HD 9900	G5II	1.38	-10.81	5.5	107.0	
HD 12568	G1II	0.76	8.28	12.0	20.0	sb
HD 13122	F5II	0.34	- 5.08	66.0		
HD 13437	G5II	1.22	9.60	4.2	23.4	sb
HD 14173	G5II	0.95	2.96	2.0	25.0	
HD 15000	F5II	0.40	-15.60	44.0		
HD 15784	F4II	0.47	1.60	16.1	85.0	
HD 20084	G3II	0.92	32.20	2.3	11.5	sb
HD 20123	G5II	1.15	1.17	7.9	33.6	
HD 23010	F5II	0.38	30.27	75.0		
HD 23230	F5II	0.42	- 9.91	58.7		
HD 26673	G5II	1.01	-17.58	10.6	25.1	sb
HD 34658	F5II	0.42	10.37	72.0		
HD 38232	F5II	0.69	- 4.77	9.0		sb
HD 39455	F5II	0.46	-13.99	31.5		
HD 41994	G5II	1.02	5.56	6.5	21.9	
HD 43282	G5II	1.32	- 2.90	7.0	35.5	sb
HD 45207	F8II	0.59	-35.96	16.2	21.8	
HD 52497	G5II	0.95	- 9.57	8.0	15.7	
HD 57048	G5II	0.95	- 2.04	2.3	23.4	
HD 57728	G2II	0.87	- 0.96	2.0	23.3	sb
HD 67542	G0II	0.81	19.37	5.9	9.1	
HD 68752	G5II	1.08	16.39	6.9	21.0	
HD 84441	G0II	0.81	4.61	5.9	13.5	sb
HD 85015	F3II	0.40	- 6.22	8.3		sb
HD 92125	G0II	0.81	- 8.49	7.7	11.0	
HD 101828	G5II	0.89	-17.05	2.1	30.9	
HD 101841	F3II	0.34	-13.15	66.0		sb
HD 106556	G5II	1.01	-21.76	8.9	41.3	
HD 114988	G2II	0.77	- 2.20	2.8	26.0	
HD 254429	F8II	0.61	4.26	80.0		

to a pseudo continuum (cubic spline function). We have detected very weak LiI features except for HD 15784 and HD 9900 that show strong lithium features. When possible, the equivalent width (W in mÅ) on LiI line was measured (cf Table 1) with an accuracy of 7 mÅ.

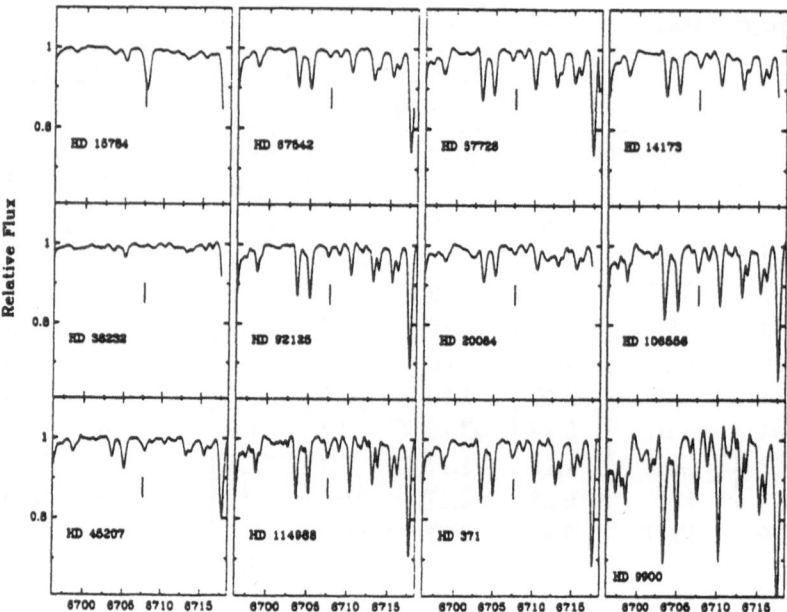

Fig. 1. spectroscopic observations around the Lithium line 6707.81 Å (straight line)

2 Preliminary Results

2.1 Rotational Discontinuity

Fig. 2 presents the behaviour of the distribution of projected equatorial rotational velocities as a function of (B - V) colour. The well defined rotational discontinuity near B - V = 0.61 (F8II), established by De Medeiros and Mayor (1989) is very clear, despite the limited sample used here. On the right side of such rotational discontinuity all single stars have vsini lower than about 10 km s^{-1}. To the left of such discontinuity the vsini values range from a few km s^{-1} to 100 km s^{-1}. Evolutionary effects and a mixing of populations seem to be the origin of such discontinuity. Indeed blue loops become important for stars with mass larger than about 4 Mo and then a fraction of slowly rotating brigh giants may be in blue loops evolutionary stage.

2.2 Lithium Behaviour

Fig. 3 presents the distribution of the measured equivalent width of the LiI line as a function of (B - V) colour. One sees no discontinuity paralleling the one found for the rotational velocity around F8II (cf Fig. 2). However such trend is not conclusive because of the difficulty in observing lithium region of high rotators. Only two stars HD 9900 (G5II) and HD 15784 (F4II) show important

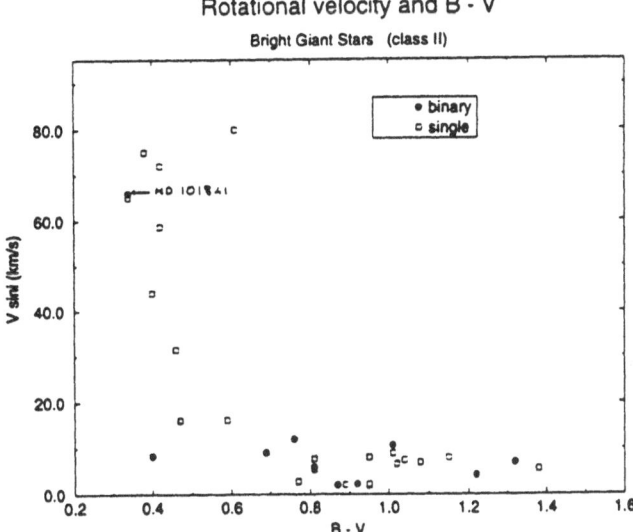

Fig. 2. Distribution of the projected equatorial rotational velocities as a function of (B - V) colour

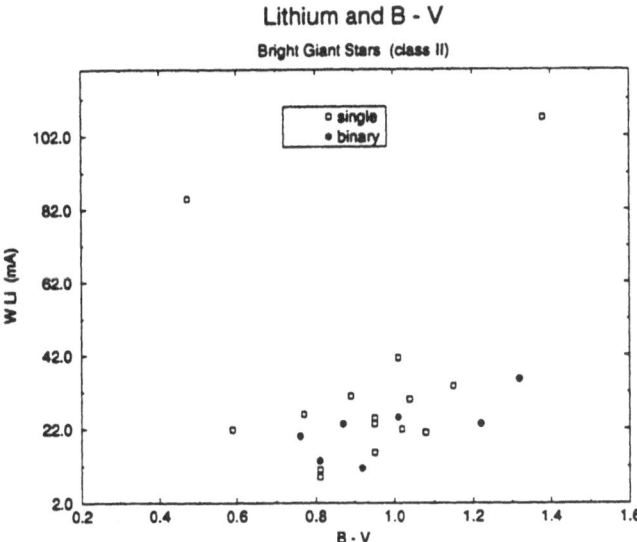

Fig. 3. Distribution of the measured equivalent width of the LiI line as a function of (B - V) colour

lithium features (cf Fig.1). One sees a tendency for an increase of the equivalent width of the LiI line with the (B - V) colour index. But caution is required because such trend may be just an artefact.

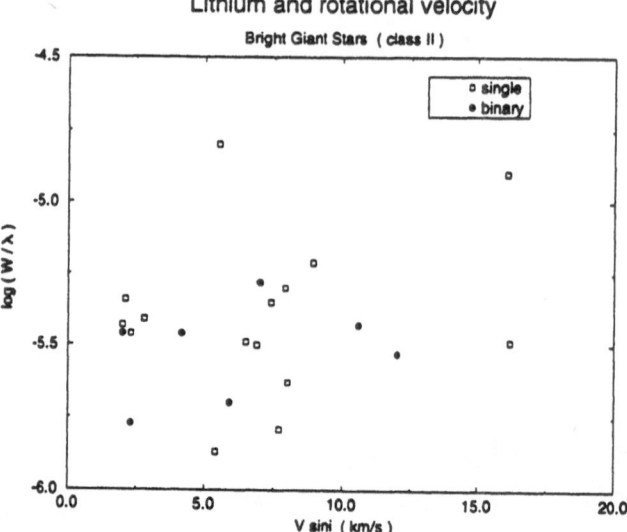

Fig. 4. Distribution of the measured equivalent width of the LiI line ($logW/\lambda$) as a function of the rotational velocity

Fig. 4 presents the distribution of the measured equivalent width of the LiI line ($logW/\lambda$) as a function of the rotational velocity. There is no obvious link between the equivalent width of lithium and rotation. Within the uncertainties there is no difference between the lithium behaviour of single and binary bright giants for a given rotational velocity.

2.3 HD 101841

HD 101841 is a binary star with a period of 3.11 days and an assumed circular orbit (De Medeiros et al, 1994, preprint Obs. Geneve 128). Its spectral type was known to be F3II (Harlan, 1969). ¿From our spectroscopic observations no lithium feature has been detected. De Medeiros et al. (1994) show that this star probably is a giant star with spectral type in the range F0–F3 III–IV, which is in better agreement with the derived orbital period, too short for a bright giant star.

References

Baranne, A., Mayor, M., Poncet, J.L. (1979): Vistas in Astronomy **23** 279
De Medeiros, J.R., Mayor, M. (1989): PASP Conf. Series Vol. 9 p404
De Medeiros, J.R., Mayor, M. (1994): A&A Sup. in print
De Medeiros, J.R., Duquennoy, A., Lèbre, A., Burki, G. (1994): A&A **286** 481
Egret, D. (1981): Bull. Inf. CDS **18** 82
Harlan, E.A. (1969): AJ **74** 916

^{7}Li Production in Luminous AGB Stars

Bertrand Plez[1], Verne V. Smith[2]

[1] Niels Bohr Institute, Blegdamsvej 17, DK-2100 Copenhagen, Denmark
[2] Astronomy Department, University of Texas, Austin, TX 78712, USA

Abstract.

We summarize current knowledge on Li abundances in AGB stars in the Magellanic Clouds and the Galaxy, insisting on the difficulties of abundance analyses.

1 Introduction

The existence of Li-rich red giants has been known since McKellar's discovery of a very strong 6707Å Li I feature in the spectrum of WZ Cas (1940). Since that time a handful of Galactic C and S stars have been shown to display a prominent 6707Å line. Recently, Smith and Lambert (1989, 1990) found that the most luminous thermally pulsing AGB stars (with M_{bol} from about -6 up to the tip of the AGB at -7.1) in the Magellanic Clouds had a strong 6707Å Li I line. They interpret this strong feature as being due to Li production on the AGB. The Be-transport mechanism of Cameron and Fowler (1971) is operating in the deep, hot-bottom convective envelope (HBCE) of these luminous AGB stars as has been found in certain models (e.g. Sackmann and Boothroyd 1992, and Mowlavi, this volume). As these stars might be a significant source of ^{7}Li for the interstellar medium, more detailed studies have been prompted by the discovery of Smith and Lambert.

2 ^{7}Li Abundance Determination in AGB Stars

2.1 Difficulties

The determination of chemical abundances in oxygen-rich AGB stars (and in cool red giants in general) is a difficult task, owing to several problems. The effective temperature of these objects being less than 4000 K, their spectra are dominated by molecular line absorption from TiO, ZrO and, for the latest spectral types, LaO and other oxides. It is impossible to define the continuum level on the observed spectra. Detailed modelling of the spectrum is necessary to disentangle the lines of interest from the surrounding molecular veiling. In most cases, only strong atomic lines can be used for abundance determinations. The case of the Li I resonance doublet at 6707Å is particularly critical as strong TiO bandheads obliterate the continuum in this region. In addition this doublet is very strong in the Li-rich AGB stars and thus sensitive to details of the modelling in the upper atmospheric layers. Very little is known about non-LTE effects in M and S-type giants (see however the study on the formation of Li lines in AGB stars by Kiselman and Plez, 1994).Dynamics and shock waves are likely to have an impact on some spectral features. Circumstellar components (in emission and absorption) may arise in resonance lines as a consequence of heavy mass-loss. Fortunately, some of these difficulties may be circumvented by a careful analysis and modelling, using information from a variety of spectral lines.

2.2 Our Method

In our studies of AGB stars in the Magellanic Clouds (Plez, Smith and Lambert 1993, hereafter PSL, and Smith et al. 1994a, SPLL), and in the Galaxy (Smith et al., 1994b), we use spherically symmetric or plane-parallel model atmospheres from the SOSMARCS code (Plez, Brett and Nordlund 1992). Up-to-date atomic and molecular opacities are included in the opacity sampling approximation. As an equivalent width approach is completely out of question (see previous section), we do a detailed synthesis of spectral regions around the lines of interest. Of special importance is the veiling due to the numerous TiO lines that we include in the synthesis (about 1000 lines per Å). Our computations cover about 100Å in each spectral region in order to check the overall agreement of the calculated and observed molecular veiling. In addition to the 6707Å Li I doublet, we use in some cases the subordinate 8126Å line, which is weaker and in a less blanketed region of the spectrum. The K I 7699Å resonance line is very useful in checking the adequacy of the stellar parameters. PSL also synthesized regions around the 7800Å Rb I line, the 7300-7600Å region for the abundance determination of various atomic species and the CO bandheads at 2.2 microns. The simultaneous fit off all these regions constrains the stellar parameters. Examples of observed and calculated spectra are given in PSL and in SPLL.

3 Samples Studied

PSL did a careful study of 7 S-type AGB stars and 1 M supergiant in the SMC at a spectral resolution of about 18000. In addition to Li they measured abundances of some iron-peak elements, s-process elements and C isotope ratios. They derive the Li abundance from both 6707Å and 8126Å Li lines and use the K I 7699Å line to check their model parameters. They find Li abundances ranging from 2.0 to 3.5 on the logarithmic scale where H is 12. The 8126Å line yields systematically higher abundances.

SPLL studied a much larger sample of 112 M, S and C type stars in the SMC and the LMC. Their spectra are a mixture of high resolution (18000) and low resolution (4000) data. The Li line is detected in 29 S stars and 6 C stars. All stars showing the Li feature are AGB stars, with a majority of them falling in the luminosity interval $-7.2 \lesssim M_{bol} \lesssim -6.0$, in good agreement with HBCE models (Sackmann and Boothroyd 1992). Abundance estimates (from the 6707Å line only) range from 1.0 to 4.0. A few of these Li-rich stars are at lower luminosity, showing that $M_{bol} = -6$ is not a strict limit for the appearance of Li. HV1645 at $M_{bol} \approx -4.7$, with one of the strongest Li line yet observed in an S star, challenges any explanation.

Smith et al. (1994b) are studying a sample of more than 200 Galactic S stars. Of the 187 examined so far, 30 show a 6707Å Li I line. Abundance estimates have been derived, based solely on the resonance doublet. Some of the spectra could not be fit very well, or the C/O ratio had to be taken very close to 1. The Li I doublet becomes then extremely strong even at low abundances (a logarithmic abundance of 0.0 still gives an appreciably strong 6707Å line). The knowledge of the C/O ratio is thus necessary to constrain the Li abundance. Of great help are then the 8126Å line and the K I resonance line. We are now acquiring more observations of these stars at these wavelengths. Eventually, a full analysis of the type done by PSL will be made on a subsample of these Galactic stars. Better studies at higher spectral resolution can be carried out that will provide useful information on the HBCE burning. One problem is that the luminosities of the objects are unknown. It seems, however, that some information can be gained (until parallaxes are obtained by Hipparcos) by looking at the galactic distribution of different classes of S-stars. Figure 1 shows the Galactic latitude distribution of no-Li and Li-strong stars of 187 stars of the sample. The Li-strong stars are more concentrated towards low Galactic latitude, presumably because of a higher average mass and luminosity than the no-Li stars. This is in accordance with HBCE theories.

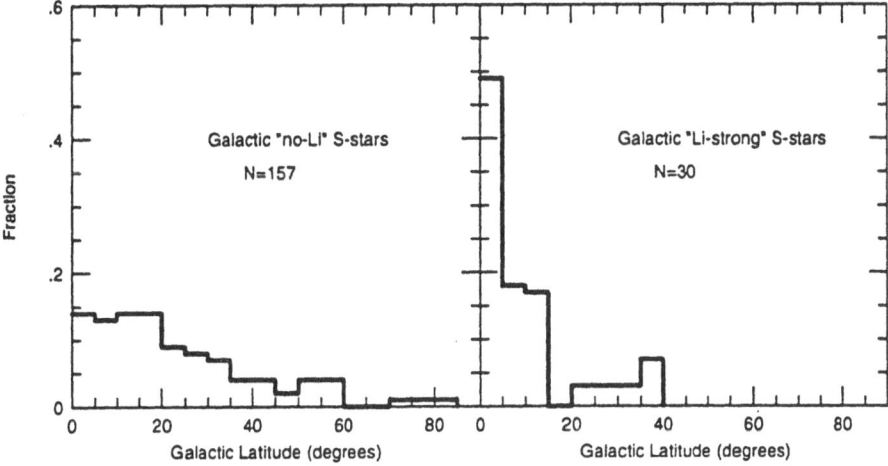

Fig. 1. Galactic latitude distribution of Li-strong and no-Li stars of Smith et al. (1994b)

4   Conclusions

Knowledge of Li abundances in AGB stars is now rapidly increasing. Large samples are being studied in the Magellanic Clouds and in the Galaxy using good models and synthetic spectroscopy of large portion of the spectrum. Non-LTE effects are under scrutiny. Large telescopes and efficient spectrographs open the prospect of doing similar studies in nearby galaxies at various metallicities. These data will provide new constraints for theories of HBCE-burning and nucleosynthesis on the AGB. And, maybe of more interest to participants to this meeting, we may soon have a definite answer on whether luminous AGB stars are an important contributor to the ^{7}Li enrichment of the Galaxy.

References

Cameron A.G.W., Fowler W.A., 1971, ApJ, 164, 111
Kiselman D., Plez B., 1994, Mem. S.A.It., in press
McKellar A., 1940, PASP, 52, 407
Mowlavi N., 1994, this volume
Plez B., Brett J.M., Nordlund Å., 1992, A&A, 256, 551
Plez B., Smith V.V., Lambert D.L., 1993, ApJ, 418, 812 (PSL)
Sackmann I.J., Boothroyd A.I., 1992, ApJ, 392, L71
Smith V.V., Lambert D.L., 1989, ApJ, 345, L75
Smith V.V., Lambert D.L., 1990, ApJ, 361, L69
Smith V.V., Plez B., Lambert D.L., Lubowich D.A., 1994a, ApJ, in press (SPLL)
Smith V.V., Plez B., Lambert D.L. et al., 1994b in preparation

Lithium Production and Hot Bottom Burning in AGB Stars

Nami Mowlavi

Institut d'Astronomie et d'Astrophysique
Bd. du Triomphe, 1050 Bruxelles, Belgium

Abstract. Lithium production in the envelope of massive asymptotic giant branch stars is studied through models of a $7\,M_\odot$, Z=0.02 star incoporating a time dependent convective diffusion algorithm coupled with the nuycleosynthesis. Lithium is found to be produced in our star before the first thermal instability develops. The evolution of the abundances of C, N, O and Al in the envelope is also followed. It is shown in particular that large amounts of ^{26}Al are expected to be produced in the envelope of super-rich lithium stars.

1 Introduction

Lithium is known to be present in the envelopes of some red giants in amounts far in excess of the solar or interstellar values (see for example Plez, 1994). A mechanism was suggested by Cameron & Fowler (1971) which would explain such high production of lithium *in the envelope* of Asymptotic Giant Branch (AGB) stars, and has been recently confirmed to occur in the most massive AGB stars through self-consistent AGB models (Sackmann & Boothroyd, 1992).

In order for the lithium to be efficiently produced through this mechanism, the temperatures at the base of the envelope must exceed $40\ 10^6$ K. This constraints the masses of the super-rich lithium stars to the most massive AGB stars ($M \geq 4\,M_\odot$). Besides, at these high temperatures, we expect other elements, such as C, N, O, and even Al, to be affected by nucleosynthesis (hot bottom burning, HBB). It is the purpose of this contribution to analyse the production of lithium in the envelope of super-rich lithium stars in relation with the evolution of these other elements. The comparison of these predictions with observations would give more clues in understanding the super-rich lithium stars. Section 2 presents the models and Section 3 the results. Section 4 makes a brief comparison with observations.

2 Models

A $7\,M_\odot$, Z=0.02 star has been evolved from the pre-main-sequence phase up to the first thermal pulses in the AGB phase. In order to follow the production of lithium through the Cameron-Fowler mechanism, a time dependent convective diffusion algorithm has been coupled with the nucleosynthesis (as in Sackmann & Boothroyd, 1992). 39 nuclides are followed from H to Si. OPAL radiative opacities are used, and we adopt α=2 for the mixing length parameter. Details of the code are given in Mowlavi & Forestini (1994). Most of the reaction rates of interest for our study are taken from the Caughlan & Fowler (1988) compilation. Exception is made for ^{25}Mg (p,γ) ^{26}Al and ^{26}Mg (p,γ) ^{27}Al (Illiadis et al., 1990), ^{26}Al^g (p,γ) ^{27}Si (Vogelaar 1989), and ^{27}Al (p,α) ^{24}Mg (Champagne et al., 1988).

3 Results

The evolution of the temperature at the base of the envelope of the $7\,M_\odot$ star is shown in figure 1a, together with the evolution of the abundances of ^{3}He and ^{7}Li. Lithium is produced when the temperature exceeds 40 10^6 K. This happens in our $7\,M_\odot$ star before the first thermal instability developes. The lithium abundance soon reaches an equilibrium value with ^{3}He, which decreases slowly as it burns. The lithium abundance reaches thus a maximum value of X(^{7}Li)=1.2 10^{-7} in mass fraction (log ϵ(^{7}Li)=4.5).

The abundance profiles ^{7}Li and ^{7}Be are *not* homogeneous in the convective envelope. The mean life time of ^{7}Li against p-capture is 10^{-8} yr at the bottom of the envelope, much shorter than the mixing time scale which is 10^{-4} yr. This explains why lithium is not homogenized instantaneously, requiring a time dependent convective diffusion coupled with the nucleosynthesis to follow its evolution. Berylium, produced by ^{3}He (α,γ) ^{7}Be, is transported to the outer layers, where it decays to ^{7}Li which is not destroyed because of the low temperatures in these regions.

As the temperature increases above 60 10^6 K, the CN cycle begins to operate, transforming ^{12}C to ^{14}N (see Figure 1b). Through this process, ^{13}C is first produced by p-capture of ^{12}C, and then destroyed as ^{12}C depletes. The ^{12}C/^{13}C ratio rapidly reaches its CN equilibrium value of 2-3.

At still higher temperatures, aluminium is produced at the base of the envelope through ^{25}Mg (p,γ) ^{26}Al^g. The mechanism is very efficient in our star as the temperatures reach values higher than 80 10^6 K. The abundance of ^{26}Al^g before the sixth pulse (our last model computed so far) already equals 6.5 10^{-7}, in mass fraction, and is still increasing (see Figure 1b). The ^{26}Al/^{27}Al ratio equals 0.01 at that time.

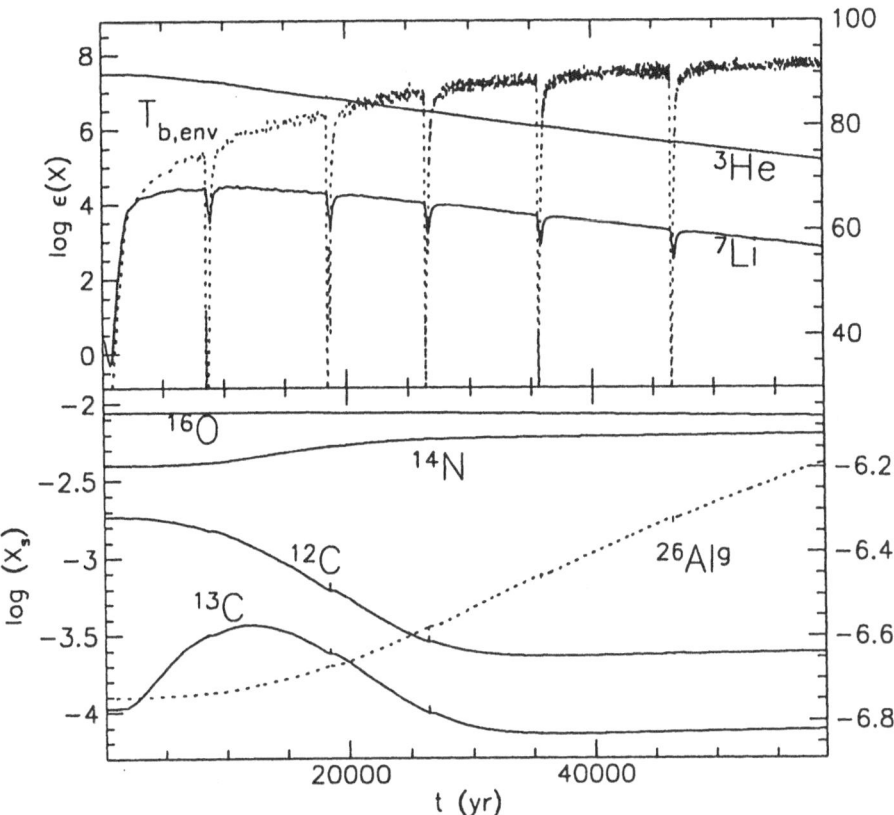

Fig. 1. *(a: upper figure)* Temperature at the base of the envelope (in 10^6 K, ride side scale) and surface abundances of ^{3}He and ^{7}Li (in $\log \epsilon(X)=12+\log \frac{N_X}{N_H}$, left side scale) as a function of time during the first 5 pulses of the $7\,M_\odot$, Z=0.02.
(b: lower figure) Evolution of the surface abundances of ^{12}C, ^{13}C, ^{14}N and ^{16}O (left side scale), and ^{26}Al^g (right side scale), in mass fractions, at the surface of the same star.

4 Discussion and Conclusions

- As mentionned in the introduction, the Cameron-Fowler mechanism can explain the production of lithium only in the most massive AGB stars. However, Abia et al. (1993) found that 2% of the galactic C stars are super-rich Li stars, for which the masses are estimated to lie between 1 and $2\,M_\odot$. The presence of Li in these stars, if the mass estimates are correct, is thus a mystery.

- Lithium is found to be produced in our $7\,M_\odot$, Z=0.02 star before the development of the first thermal instability on the AGB phase. This suggests that some luminous M star might be observed with high surface lithium abundance. No such observation has been reported yet.

- As a result of the HBB, ^{12}C is converted to ^{14}N through the CN cycle. A carbon star will thus turn back to a S star as lithium is produced in the envelope. This is in agreement with the super-rich lithium stars found in the Magellanic

Clouds which are all S stars (Smith & Lambert, 1989, 1990). It is however hard to explain the existence of galactic C stars which are super-rich lithium stars.

- Super-rich lithium stars should display low $^{12}C/^{13}C$ ratios. This seems to agree with the observations in the Magellanic Clouds (Plez et al., 1993). The analysis of this ratio in the galactic super-rich lithium stars is more controversial. Although most of them are J-type stars, Abial et al. (1993) do not find the expected theroretical correlations among lithium abundances and $^{12}C/^{13}C$ ratio.

- ^{18}O is completely destroyed by the HBB. $^{16}O/^{18}O$ increases from its first dredge-up value of $\sim$500 to more than 10^5! However, no observation for this isotopic ratio at the surface of super-rich lithium stars is reported yet.

- The production of $^{26}Al^g$ in the envelope of super-rich lithium stars offers another valuable prediction to test our understanding of these objects. The $^{26}Al/^{27}Al$ ratio increases above 0.01 in our $7\,M_\odot$ star. However, no observational data is available for this isotopic ratio either.

Finally, we have to mention that some clues can further be found in the recent study of meteorites. Nittler et al. (1994) find that some of the Tieschitz oxide grains have ^{26}Al enrichment ($^{26}Al/^{17}Al$ up to $8\ 10^{-3}$) and important ^{18}O depletions ($^{16}O/^{18}O > 5000$). On the other hand, Amari et al. (1993) find some anti-correlation in SiC grains between the $^{12}C/^{13}C$ and $^{26}Al/^{27}Al$ ratios. Both types of grains could have been formed around a massive AGB star.

References

Abia, C., Boffin, H.M.J., Isern, J., Rebolo, R.: 1993, *A&A* **272**, 455

Amari, S., Hoppe, P., Zinner, E., Lewis, R.: 1993, *Nat.* **365**, 806

Cameron, A.G.W., Fowler, W.A.: 1971, *Ap. J.* **164**, 111

Caughlan, G.R., Fowler, W.A.: 1988, *Atomic Data Nucl. Data Tables* **40**, 283

Champagne A.E., Cella, C.H., Kouzes, R.T., Lowry, M.M., Magnus, P.V., Smith, M.S., Mao, Z.Q.: 1988, *Nucl. Phys.* **A487**, 433

Iliadis, Ch., Schange, Th., Rolfs, C., Schröder, U., Somorjai, E., Trautvetter, H.P., Wolke, K., Endt, P.M., Kikstra, S.W., Champagne, A.E., Arnould, M., Paulus, G.: 1990, *Nucl. Phys.* **A512**, 509

Mowlavi, N., Forestini, M.: 1994, *A&A* **282**, 843

Nittler, L.R., Alexander, C.M., Gao, X., Walker, R.M., Zinner, E.K.: 1994, *Nat.* **370**, 443

Plez, B.: 1994, *these proceedings*

Plez, B., Smith, V.V., Lambert, D.L.: 1993, *Ap. J.* **418**, 812

Sackmann, I.-J., Boothroyd, A.I.: 1992, *Ap. J. Lett.* **392**, L71

Smith,V.V., Lambert,D.L.: 1989, *Ap. J. Lett.* **345**, L75

Smith,V.V., Lambert,D.L.: 1990, *Ap. J. Lett.* **361**, L69

Vogelaar, B.: 1989, *Ph. D. thesis*, California Institute of Technology (unpublished)

Resolution of the Classical Hyades Lithium Problem

John Faulkner[1], Fritz Swenson[2]

[1] Lick Observatory and Astronomy Board, UC Santa Cruz,
Santa Cruz, CA 95064
[2] Los Alamos National Laboratory, X-2 MS B220,
Los Alamos, NM 87545

Abstract. For the very first time, it has recently proved possible to make a well-motivated, physically plausible, and self-consistent *prediction* of the Hyades G- and K-dwarf (Li,T_{eff}) relationship that *matches the long-unexplained observations*. The method employs the latest Iglesias & Rogers (OPAL) interior opacities and Alexander surface opacities (whose respective values are now themselves close to empirical predictions or estimates made earlier in this Hyades project), King's recently discovered [O/Fe] enhancement (another prediction!) and utterly conventional PMS (pre-main-sequence) evolution unaided by arbitrary and ad hoc adjustable parameters.

Thus, the following assumptions form a self-consistent set explaining the Hyades G- and K-dwarf (Li,T_{eff}) observations:

(i) Pop. I interior opacities are now essentially correct.

(ii) Pop. I surface opacities are now essentially correct.

(iii) The Hyades distance is now essentially correct.

(iv) The Hyades [Fe/H] and [O/Fe] are now essentially correct, and possible changes in [Ne, Mg, or Si/Fe] are likely to have only a small effect.

(v) Convective envelope overshooting is *negligible*.

(vi) Quarreling with the above means finding at least two compensating errors in assumptions (i) through (v).

1 Introduction

For some time, we and colleagues have pursued possible resolutions of a now 30-year old puzzle – the classical Hyades lithium problem. We have recently reached a most satisfying conclusion, a well-motivated, physically plausible and parameter free *self-consistent prediction* of the Hyades G- and K-dwarf (Li,T_{eff}) relationship that *matches the long-unexplained observations*. Space precluding full discussion, we here present a guide to the main points in our papers. In logical (not published) order, relevant references are SF (Swenson & Faulkner 1992), SSF (Swenson, Stringfellow, & Faulkner 1990a), SFRI90 (Swenson et al 1990b), SFRI94 (Swenson et al 1994a), SFIRA (Swenson et al 1994b).

2 The Problem; Earlier Ideas

Wallerstein, Herbig, & Conti (1965) noted that [Li/Ca] dropped quite sharply in Hyades G- and K-dwarfs, correctly suspecting a reduction in Li with decreasing T_{eff}. Since standard main-sequence (MS) convective envelopes were too shallow to explain Li-depletion via MS-burning, other explanations were sought. Bodenheimer (1965) seemed to show that conventional pre-main-sequence (PMS) burning was also (marginally?) inadequate; subsequent readers ignored his optimistic summary.

As time elapsed, the problem only seemed to worsen. Successively improved observations showed even stronger Li-depletion (Cayrel et al 1984; see Fig. 1); while reductions in measured $[Fe/H]_{Hyad}$ implied even less theoretical PMS depletion. The widening gap led to an era of (ad hoc)[2] theories. Ad hoc parameters (e.g., a multiple of the deep envelope scale-height) were introduced to characterize non-standard features of convection (MS "overshoot" — Straus et al 1976; PMS "extra-mixing" — D'Antona & Mazzitelli 1984). When fixed parameter values produced (Li, T_{eff}) curves markedly at odds with those observed, appeals were made to ad hoc variations in the ad hoc parameters.

3 Our Own Approach: Deduction, Production, and Use of Improved Opacities

We first briefly discuss how we eliminated another earlier promising idea.

3.1 Main-Sequence Mass-Loss Doesn't Work

Very early on, Weymann & Sears (1965) briefly explored a suggestion by N.J. Woolf, that mass-loss might lead to envelope Li-depletion (currently Li-rich material being lost while previously hotter, Li-free material continuously dilutes the envelope). They sketched the Li consequences – an ultimately exponentially decaying dependence on mass lost – of assuming an unchanging envelope mass for a given model. Hobbs et al (1989) did much the same.

However, the key to ruling out mass-loss lies in the realization that MS convective envelope masses vary (almost linearly!) with stellar mass. This happy circumstance yields quasi-analytic predictions for Li-depletion that not only match properly computed results very closely (SF), but also run along or parallel to most of the G- and K-dwarf (Li, T_{eff}) observations. Explaining the observations then requires that those dwarfs all started out with very closely defined masses (e.g. $1.08 \pm 0.01 M_\odot$), before losing various amounts of mass; a circumstance so contrived that it rules out this kind of explanation.

3.2 The Importance and Influence of Opacities for PMS Li-Burning

While working on mass-loss, we became aware not only that extant PMS studies relied on older opacities, but also that a careful systematic examination of the separate influence of interior and surface opacities had never been made. As we were to show (SSF), increments in these separate opacities have oppositely signed effects on PMS envelope evolution and therefore ultimate Li-content. (Simply put, the surface opacity helps determine which adiabat the deep envelope attains, while the interior opacity determines where the radiative interior leaves the adiabat – Faulkner & Swenson, in preparation.)

At the time, upward historical opacity trends were clear, although, for example, Iglesias & Rogers were yet to turn their attention to opacities for $T > 10^6$ K. (For PMS burning, the base of the envelope reaches its maximum temperature for only a few $\times 10^6$ yrs as a $\sim 1 M_\odot$ star begins to develop a radiative core – and therefore a retreating convective envelope – and thus moves from a Hayashi to a radiative contracting track. This time-scale sets the important temperature for PMS Li-burning at $T \sim 4 \times 10^6$ K, not the $\sim 2.5 \times 10^6$ K commonly misquoted.)

SSF made an end run around uncertain and gradually changing theoretical opacities by asking what the internal opacities would need to become (temporarily assuming unchanged surface opacities), to match the Li observations. Figure 1 (SFRI94) essentially summarizes the dramatic consequences SSF found with a procedure involving a $\sim 37\%$ increment over Cox & Tabor (1976) opacities. While SSF did not address the full question of consistency with other Hyades data, one consequence of the approach was both startling and gratifying – the observations, "like beads on a string" (Burt Jones), now lay along our theoretical curve. Unlike all other attempted explanations this was the first time that a close match to the characteristic curve was obtained; it remained a robust feature of our results as work continued.

In unpublished work, we allowed for a further 50% increase in surface opacities, finding that internal increments of $\sim 45 - 48\%$ would then fit the data. Spot check opacities for $T \sim 3.5$ and 4×10^6 K, kindly computed by Iglesias & Rogers, indeed matched these latter requirements closely! However, part of the change could be due to revised heavy element mixtures. To learn what change could be attributed *solely* to a change in opacity-computing methodology, Iglesias & Rogers undertook to recompute interior opacities for a mixture (King IVa) employed by Cox & Tabor. The resulting differences were substantial, if not as large as before. In SFRI90 we showed the (Li,T_{eff}) consequences of this change in internal opacity code: a removal of some 75-80% of the discrepancy. When we permitted ourselves (unpublished) to add a little helium, to restore the original mass-luminosity (M-L) relationship – leaving aside whether it was absolutely correct – a further 10-15% of the discrepancy was removed. (Sequences analogous to all three here described were later shown in Fig. 4 of SFRI94.)

The stage was now set for a massive undertaking (SFRI94) – to require that the spectroscopic [Fe/H] and the M-L relationship with the best-determined distance should both be fitted, *predicting* the (Li,T_{eff}) consequences (and, incidentally, the H-R diagram) for a succession of changes in physical inputs: internal

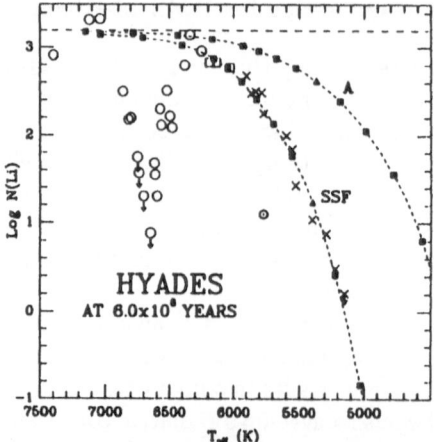

Fig. 1. The Hyades (Li,T$_{\text{eff}}$) plane. Observations (crosses, open squares, and circles) are from Cayrel et al (1984), Duncan & Jones (1983), and Boesgaard & Trippico (1986), respectively. Thorburn et al (1993) have recently published more observations that support and supplement the Cayrel et al data (see Fig. 2 of SFIRA). Arrows (one, at $\log N(\text{Li}) \sim 0.5$, obscured) indicate upper limits. Computed results are normalized to initial $\log N(^7\text{Li}_\circ) = 3.2$ (long dashed line) on a scale where $\log N(\text{H}) = 12.0$. Solid symbols give model results. Mass increments between models are $0.05 M_\odot$ ($M < 1.2 M_\odot$) and $0.10 M_\odot$ ($M > 1.2 M_\odot$). Triangles identify $1 \mathcal{M}\odot$ models. Sequence A shows the (Li,T$_{\text{eff}}$) predictions for models with older fiducial physics. Sequence SSF is the fit produced with a 37 percent maximum enhancement of Cox & Tabor (1976) interior opacities as described in SSF.

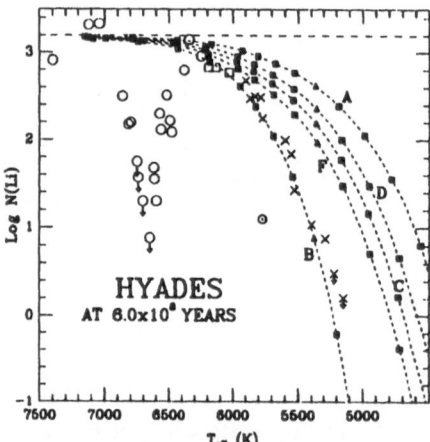

Fig. 2. Model predictions in the Hyades (Li,T$_{\text{eff}}$) plane. See Figure 1 for general description. Sequences, computed with different inputs or physics, are labeled A, B, C, D, and F and are described in SFRI94. They show the non-monotonic influence of successive changes on self-consistent predictions. Note that all $1 M_\odot$ models have roughly the same temperature, an empirical constraint resulting from the required consistency with the solar constraint and Hyades M-L information.

opacities, surface opacities, heavy element mixes, equation-of-state, etc. (As with our change in opacity code only – but not mixture – in SFRI90, such careful isolation of effects is not common in the field.) Whenever a physical input was changed, we re-calibrated the mixing-length parameter α by re-evolving a solar model with the same physics, and completed the [Fe/H] and M-L self-consistent procedure described in SFRI94. Interested readers can there follow the ebb and flow in (Li, T_{eff}) *predictions* as five major input improvements are pursued. All but one are shown in Fig. 2.

It is historically ironic that the *prediction* of an excellent fit to the H-R diagram (Fig. 2b of SFRI94) is an incidental consequence. One of us (Faulkner 1964) first pointed out that the Hyades M-L and H-R data could simply not be fitted simultaneously with the then distance modulus of 3.03 mag; he suggested moving the Hyades away! The present excellent H-R *prediction* using a distance modulus of 3.35 ± 0.02 mag (Peterson & Solensky 1988) serves as a remarkable confirmation of the general correctness of this much greater value.

Final SFRI94 Li predictions (sequence F; see Fig. 2) were much closer to the observations than sequence A, but still left room for improvement. We showed that four relatively modest input changes could each produce an essentially perfect fit. One such change, noting that oxygen was second only to iron in providing metallic opacities at $T \sim 4 \times 10^6$ K, was a modest relative oxygen enhancement. Shortly after SFRI94 was submitted, we heard that Jeremy King had quite independently found that the Hyades [O/H] was indeed somewhat higher than [Fe/H], at 0.265 ± 0.05 versus ~ 0.12 or 0.13 ± 0.03.

Thus, in our final letter, SFIRA, we examined the consequences of a "low" oxygen enhancement, OL, with $[O/H] \sim 0.2$, and a "high" enhancement, OH, with $[O/H] \sim 0.3$. Special sets of interior opacities (Iglesias & Rogers) and surface opacities (Alexander) were computed, with oxygen specifically enhanced. Prior to using these, we updated sequence F with several small improvements to various inputs. Sequence F, its update, and sequences OL and OH are shown in Figure 3. The OH prediction (within King's upper limit) essentially fits the observations. As a measure of how good the predictions are, "extra-mixing" by only $\sim 0.04 H_p$ would bring the OL results into perfect agreement (compared to D'Antona & Mazzitilli's $0.7 H_p$ for a poor quality shoulder region fit – see Fig. 7 of SFRI94). We also examined the separate influence of up to 18 elements on relevant opacities (Fig. 3 of SFIRA); only accurate abundances of key elements so identified can put our (or anyone's) predictions on their firmest footing.

Swenson has continued improving our evolutionary code's equation of state (to give closer agreement with that used in the opacity calculations). With the latest improvement, the revised OL prediction moves down to the position of the OH curve in Figure 3; thus, the revised OH and OL predictions straddle the very small spread of the data about a mean fitting curve.

To summarize: we feel that we and our colleagues have done the best job of *predicting* the consequences for the (Li, T_{eff}) Hyades relationship from standard models employing the best physical inputs we can, consistent with all avail-

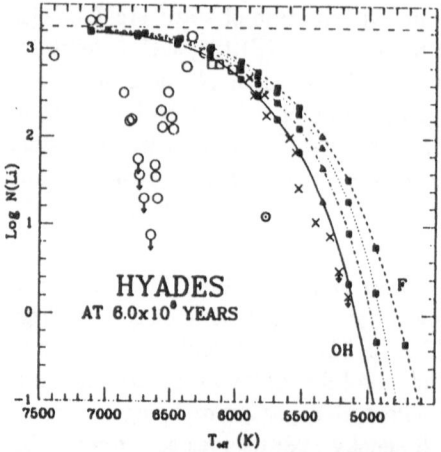

Fig. 3. Figure 1 of paper SFIRA. Computed results normalized to log N(^{7}Li$_o$)= 3.25 initially. *Dashed curve*, replotted sequence F ("best" model from SFRI94); *dotted curve*, updated sequence F (see text); *dash-dot curve*, sequence OL (lower of two oxygen enhancements); *solid line*, sequence OH (higher oxygen enhancement).

able and known constraints for this best observed and fundamentally important cluster. When we do so, lo and behold, the *prediction* matches the observations!

References

Bodenheimer, P. 1965, ApJ, 158, 419

Boesgaard, A.M., & Tripicco, M.J. 1986, ApJ, 302, L49

Cayrel, R., Cayrel de Strobel, G., Campbell, B., & Däppen, W. 1984, ApJ, 283, 205

Cox, A.N., & Tabor, J.E. 1976, ApJS, 31, 271

D'Antona, F., & Mazzitelli, I. 1984, A&A, 138, 431

Duncan, D.K., & Jones, B.F. 1983, ApJ, 271, 663

Faulkner, J. 1964, The Observatory, 84, 257

Hobbs, L.M., Iben, I., Jr., & Pilachowski, C. 1989, ApJ, 347, 817

Peterson, D.M., & Solensky, R. 1988, ApJ, 333, 256

Straus, J.M., Blake, J.B., & Schramm, D.N. 1976, ApJ, 204, 481

Swenson, F.J. & Faulkner, J. 1992, ApJ, 395, 654 (SF)

Swenson, F.J., Faulkner, J., Iglesias, C.A., Rogers, F.J., & Alexander, D.R. 1994b, ApJL, 422, L79 (SFIRA)

Swenson, F.J., Faulkner, J., Rogers, F.J, & Iglesias, C.A. 1990b, BAAS, 22, 1223 (S-FRI90)

Swenson, F.J., Faulkner, J., Rogers, F.J, & Iglesias, C.A. 1994a, ApJ, 425, 286 (SFRI94)

Swenson, F.J., Stringfellow, G.S., & Faulkner, J. 1990a, ApJL, 348, 33 (SSF)

Thorburn, J.A., Hobbs, L.M., Deliyannis, C.P., & Pinsonneault, M.H. 1993, ApJ, 415, 150

Wallerstein, G., Herbig, G.H., & Conti, P.S. 1965, ApJ, 141, 610

Weymann, R., & Sears, R.L. 1965, ApJ, 142, 174

Lithium in Old Binary Stars

L. Pasquini[1], M. Spite[2], F. Spite[2]

[1] European Southern Observatory, Casilla 19001, Santiago 19, Chile
[2] Observatoire de Paris, Section de Meudon, DASGAL, URA 335 du CNRS, 92195 Meudon Cedex, France

Abstract. A detailed analysis of high resolution, high signal to noise spectra of twenty metal deficient binaries ($-2.9 <$ [Fe/H] < -0.4) with periods ranging from 220d to 1.8d is presented. For most of the stars, in addition to the Li abundance determination, a careful determination of the stellar physical parameters (T_{eff} , log g, [Fe/H]) has been performed. For $T_{eff} \geq$ about 5000K the lithium abundance in long and short period binaries does not differ from that of single dwarfs in the same interval of temperature, gravity and metallicity. When T_{eff} is $\leq$ about 5000K the lithium abundance is larger in the short period binaries. These results may cast some doubts on theories predicting a strong Li depletion in Pop II dwarfs induced by strong angular momentum losses.

1 Observations and Data Analysis

The problem of how Li is depleted in the stellar interior is a difficult question which has not yet been completely understood. According to standard evolutionary theories, in fact, lithium abundance should be uniquely determined by the stellar age, mass and metallicity. On the other hand, recent studies have questioned this picture, suggesting that stellar rotation may play an important role in the history of Li depletion. In particular, rotational models were recently developed in which the lithium depletion is proportional to the loss of angular momentum (Pinsonneault et al. 1989, 1992, Zahn 1992). In some cases, the lithium depletion is predicted to be almost one order of magnitude in warm metal-deficient dwarfs implying that the Li abundance measured in Pop II "plateau" stars (Spite and Spite 1982) would not reflect their initial Li content. In this framework, the study of old binaries is very relevant, in particular of those systems having short orbital periods. In fact, according to the binary evolution theory, short period binaries are expected to be tidally locked and short rotational periods can be sustained for large part of the stellar lifetime without a significant braking (cf. e.g. Duquennoy et al. 1992 and reference therein). As a consequence, we may expect that depletion in old, short period binaries should be less than in the long period binaries or single stars of comparable age.

In this contribution we discuss the lithium abundance of 20 metal deficient binaries with periods ranging from 2200 to 1.8 days.

Most of the spectra were obtained at ESO using the NTT telescope and the échelle EMMI spectrograph, or the 3.6m telescope with the échelle CASPEC spectrograph. Bell-Gustafsson's models for metal deficient dwarfs, computed with the same hypotheses as in Gustafsson et al. 1975, were used.

As a first step, the parameters of the atmosphere of the stars have been taken from the literature, and as a first approximation we adopted log $g = 4$.

In a second step, the parameters of the model (including the abundances) have been carefully redetermined from our spectra. For most of the stars, we measured the equivalent widths of about 50 FeI lines. After the iron abundance was computed, we checked the adopted temperature by comparing the abundances derived from the low and high excitation potential lines, and the microturbulence velocity by comparing the abundances deduced from the weakest and the strongest lines. Spectroscopic gravities were determined by comparing the iron abundance deduced from FeI and FeII lines.

When the errors on temperature, gravity, metallicity and microturbulent velocity are included, the accuracy of the lithium abundance may be estimated to be about 0.20dex.

2 Discussion

If the lithium depletion in stars is the result of mixing due to rotational spin-down, a tidally locked system in which spin-down is curtailed must exhibit a larger lithium abundance than a single star. Following Zahn (1994) the longest period for a metal-deficient binary to be tidally-locked should lies at about 6 days (8 days for the Pop I stars). ¿From the tidal interaction theories, synchronization is achieved *before* circularization (Van't Veer & Maceroni 1992) and a system with null eccentricity may be supposed to be tidally locked. Mayor et al.(1992) found that the transition between circular and eccentric orbit is 10 days for Pop I stars, while Latham et al. (1992) found a period of about 19 days for EMD stars. We have therefore chosen for the representation of our sample two characteristics orbital periods: 6 and 19 days. Long period binaries have P >19 days, intermediate period have 6< P <19 days and short period P <6 days.

To have a proper comparison, we separated the extreme PopII stars (stars with [Fe/H] < −1.4) from the old disk stars (−0.4 > [Fe/H] > −1.2). PopII binaries will be compared to PopII single dwarfs (Fig. 1a) and old disk binaries will be compared to single old disk dwarfs (Fig. 1b).

Long period binaries: $P > 19$ **days** It is expected that in the long period binaries, the lithium depletion is the same as in the normal single stars since their angular momentum evolution is about the same. ¿From Fig. 1a,b, we can conclude that the lithium abundance in the long period binaries, in agreement with the prediction, does not significantly differ from that of single stars, regardless of the stellar effective temperatures and metallicity.

Intermediate and short period binaries: $P < 19$ **days**

a) PopII stars. Three of the observed stars have a period less than 19 days, but only one has a period shorter than 6 days: HD 89499. This star, has a higher lithium abundance than the corresponding single stars. With a log $g= 2$, however, HD 89499 may not be a good case for Li overabundance, since the Li content of cool subgiants seems to be larger than in cool dwarfs (Pilachowski et al. 1993). The star G 65-22 which has a longer period (18.8 days) and a

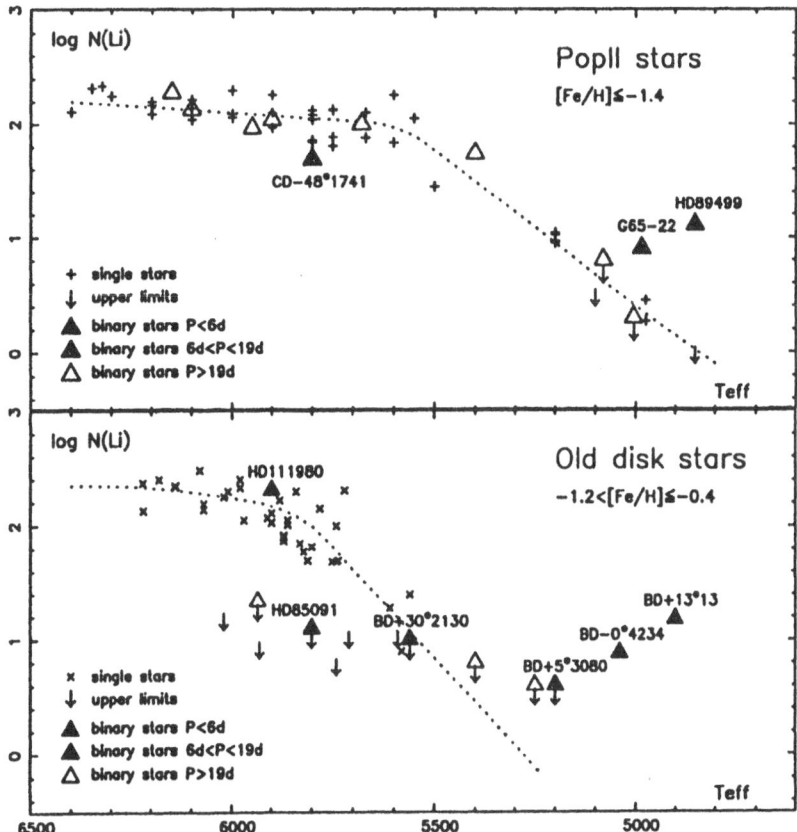

Fig. 1. Lithium abundance versus temperature for PopII stars and old disk stars. The dotted lines represent the mean lithium abundance versus temperature for PopII and old disk single stars.

circularized orbit, also shows a higher lithium content than single stars. On the contrary CD $-48°1741$, which is located in the region of the "plateau", has a normal lithium abundance. This star has a period P=7.6 days, shorter than the period of G 65-22, which is "lithium-rich". Its orbit is circular suggesting synchronization.

b) Old disk stars. Four cool short period binaries have been observed. Two of them, with periods less than 6 days and temperature less than 4900K (BD $-0°4234$ and BD $+13°13$) appear to be very lithium-rich. In the third star (BD $+5°3080$), which is slightly hotter and has a longer period (P=9.9d), lithium is not detectable. At this temperature, however, a low lithium content does not exclude an overabundance of lithium, since only upper limits are measured in normal stars. On the contrary BD $+30°2130$ is certainly not lithium rich. With a period of 6.6d and a metal deficiency of only -0.4 this binary is likely synchronized, but this does not bring any lithium enhancement. In the warmer part of Fig. 1b two other short period binaries have been observed. For HD 85091 (P=3.4d), only an upper limit can be set, thus its lithium abundance is lower than in the *majority* of the single stars. HD 111980 has a longer period (12.4 d) but a zero orbital eccentricity: its lithium content is the same as in normal single stars.

Our results on short and intermediate period binaries can be summarized as follows:

Cool short period binaries (T_{eff} *<~ 4900K) show larger Li abundances than the single or long period stars.*

Warm short period binaries (T_{eff} *>~ 5000K) do not show any significant overabundance of lithium.* One old disk short period binary is strongly Li depleted, a situation which is shared by about ~ 15 % of the single stars in the same temperature and metallicity range.

The fact that the lithium abundance of CD −48°1741 is not higher than the "plateau" level may cast doubts on the theories of Pinsonneault et al. (1992) and Zahn(1992). However Zahn (1994) computed that the cut off period for the inhibition of the lithium depletion in PopII stars is only about 6 days; it could be that the orbital period of CD −48°1741 is too long to test the theory. We note, on the other hand, that the *cool* 18.8 days period PopII star G65-22 does show enhanced Li abundance.

References

Duquennoy A., Mayor M., Mermillod J.-C., 1992, in Binaries as Tracers of Star Formation, A. Duquennoy and M. Mayor eds., Cambridge University Press, p. 52.
Gustafsson B., Bell R.A., Eriksson K., Nordlund A., 1975, A&A 42, 407
Latham, D.W., et al., 1992, AJ 104, 774
Mayor M. et al. 1992, in "Complementary Approach to Double and Multiple Star Research", ASP Conf. Ser. 32, H.A. McAlister & W.I. Hartkopf eds, p. 73
Pilachowski C., Sneden C., Booth J., 1993, ApJ 407, 699
Pinsonneault M.H., Kawaler S.D., Sofia S., Demarque P., 1989, ApJ 338, 424
Pinsonneault M.H., Deliyannis C.P., Demarque P., 1992, ApJS 78, 179
Spite F., Spite M., 1982, A&A 115, 357
Van't Veer F., Maceroni C., 1992, in Binaries as Tracers of Stellar Formation, eds. A. Duquennoy and M. Mayor, Cambridge University Press, p.237
Zahn, J.P., 1992, A&A 265, 115
Zahn J.P., 1994, A&A 288, 829

Lithium in Tidally Locked Binaries

Roberto Pallavicini[1], Sofia Randich[2]

[1] Osservatorio Astrofisico di Arcetri, I-50125 Florence, Italy
[2] MPI für Extraterrestrische Physik, D-85740 Garching, Germany

Abstract. We discuss the relationship between Li abundance and rotation rate for a large sample of tidally locked Pop I field binaries, in search of possible correlations between the presently observed Li abundance and the star rotational history. We conclude that such a connection, if exists, is largely masked by the influence of other parameters on the Li depletion process.

1 Observational and Theoretical Background

In the past few years there have been several reports in the literature that Lithium abundances in tidally locked binaries are enhanced with respect to single stars of the same effective temperature and age. First, there is the case of the Hyades, where the very few objects which deviate from the tight Li vs. T_{eff} curve of the cluster have been found to be short-period binaries (Soderblom et al. 1990; Thorburn et al. 1993). Second, there is the finding by several authors that RS CVn binaries and other chromospherically active binaries – in which chromospheric activity is the consequence of tidally enforced rapid rotation – have *on average* larger Li abundances than inactive stars of the same colour and in the same evolutionary status (Pallavicini et al. 1992; Randich et al. 1993, 1994a; Fekel and Balachandran 1993, 1994). Finally, there are recent reports by Balachandran et al. (1993), Spite et al. (1994) and Ryan and Deliyannis (1994) of enhanced Li abundances in Pop II and Old Disk stars that are members of short-period binaries.

A higher Li abundance in tidally locked binaries is predicted by some theoretical models which relate Li depletion to the loss of stellar angular momentum during main-sequence lifetime (Pinsonneault et al. 1989, 1990; Zahn 1992, 1994). Although these models differ in the details, they all predict that Li depletion should be larger for stars that have lost more angular momentum and have suffered more internal mixing due to angular momentum transport from the interior to the surface. Tidally locked binaries, which have been prevented from losing angular momentum by tidal interaction, should therefore have preserved more Li than single stars and long-period binaries *provided* they have not suffered appreciable Li depletion before synchronization occurs. Recent calculations by Zahn (1994) have shown that this is the case for binaries of sufficiently short period (≤ 8 days for Pop I stars and ≤ 6 days for Pop II stars), for which synchronization is reached during the pre-MS phase.

In this paper, we want to address the following specific question: is there any observational evidence for a relation between Li abundance and the *presently* observed rotation rate of a star ? And, if not, could other effects (like a dispersion in the initial rotation rates, different time scales for synchronization, different masses and/or metallicities, etc.) be so strong as to mask any such dependence on rotation ? In trying to answer these questions, we will use primarily Li observations of RS CVn binaries and related chromospherically active Pop I binaries and in particular data obtained by Randich et al. (1993, 1994a), which represent the largest homogeneous sample available at present for tidally locked binaries (81 individual components of 68 binary systems). Note that 19 of the 28 binary components discussed by Fekel and Balachandran (1993, 1994) are also in Randich et al., with good agreement between the two data sets. We refer to Thorburn et al. (1993), Zahn (1994), Spite et al. (1994) and Ryan and Deliyannis (1994) for a discussion of the Hyades binaries and Pop II binaries.

2 Li in Pop I Tidally Locked Binaries

Li observations of RS CVn binaries and other Pop I field binaries with active chromospheres have been discussed recently by Pallavicini (1994). In brief, the data show that there is a statistically significant tendency of these stars to have larger Li abundances than inactive Pop I stars of the same effective temperature and evolutionary state, not selected specifically on the basis of activity or close binarity (note that the latter sample, although including both single and binary systems, is not expected to contain a significant number of active short-period binaries). For instance, comparison of the tidally locked binaries of Randich et al. (1993, 1994a) –which are all listed in the catalog of chromospherically active binaries of Strassmeier et al. (1993)– with samples of "inactive" giants (Brown et al. 1989) and of "inactive" subgiants (Randich et al. 1994b), show that the active binaries have *on average* larger Li abundances. However, the same data also show that a large spread exists at each colour and not all cool (T $\leq$ 5000 K) active binaries have large Li abundances (see also similar conclusions by Fekel and Balachandran 1994).

In order to check whether the different Li abundances of active binaries at each colour could be due to different rotation periods, we have divided the sample in two subsamples, one for P $\leq$ 8 d, the other for P > 8 d (we have also done the same for P $\leq$ and, respectively, > 14 d, since 14 d is the canonical period used in the definition of the RS CVn class). We find no statistically significant difference in Li abundances for the short and long period binaries (Fig. 1), in spite of the fact that the short period binaries are predicted to have reached synchronization before the long period ones (Zahn 1994). Virtually all binaries in the sample (including the long period ones, except a couple) are synchronized *at the present epoch*, as shown by the approximate equality of orbital and photometric period, cf. Strassmeier et al. 1993. Yet, we do not find any significant dependence of Li abundance on period (Fig. 2), contrary to what one would have expected if the longer period systems have synchronized much later on the MS than short-period binaries. The only obvious trend we see in Fig. 2 is the dependence of Li abundance on T_{eff}. A similar plot of Li abundance vs. *vsini* shows basically the same lack of correlation, and is further affected by the usually unknown inclination angle and the different radii for subgiants and giants.

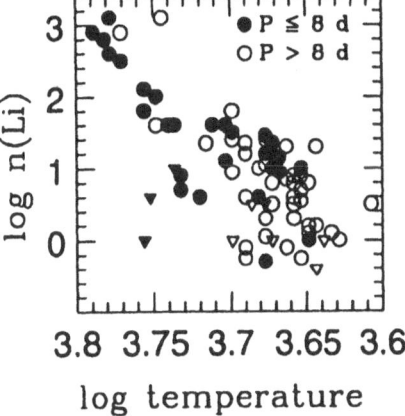

Fig. 1. Li abundances vs. T_{eff} for Pop I chromospherically active binaries. Different symbols indicate short and long period systems.

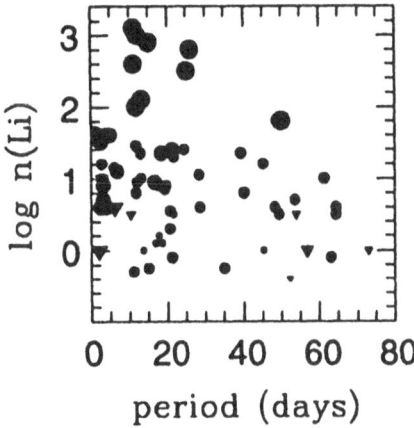

Fig. 2. Li abundances vs. period for Pop I chromospherically active binaries. Larger symbols indicate higher effective temperatures.

Since we do not find any obvious correlation with the rotation period and/or the present rotation rate, we have considered the alternative possibility that the larger (on average) Li abundances of active binaries might be due to systematically larger masses ($M \geq 1.5\ M_\odot$) and thus reduced Li depletion on the MS for stars that have thin or absent outer convective zones. This has been discussed by Randich et al. (1994a), who have shown that there is no significant dependence of Li abundance on mass for those binary stars for which masses are known from orbital solutions. Comparison with theoretical evolutionary tracks leads to the same conclusion, although in this case uncertainties on T_{eff} and distances, and the dependence of evolutionary tracks on metallicity, make the comparison more dubious. At any rate, it appears that a combination of factors (including mass, rotational history, metallicity, and possibly others) may contribute to the observed Li abundances of active binaries, and that these factors are likely to affect individual objects in a different way, thus contributing to the observed large scatter in Li abundances.

3 Conclusions

Pop I tidally locked binaries –which *on average* have larger Li abundances and have lost less angular momentum than single stars and non-locked binaries– are qualitatively in agreement with the Yale rotating models of Li depletion and with the tidal synchronization theory of Zahn. However, the number of parameters which may affect Li depletion in these stars is so high –and the sample is so inhomogeneous– as to prevent a meaningful comparison of theoretical predictions with observations. The lack of an obvious correlation between Li abundances and presently observed rotation rates (or periods) cannot be used as an argument against the Yale + Zahn theory, but, on the other hand, the qualitative agreement found is nor by itself a proof that the theory is correct. In any case, caution should be used when applying simple-minded theoretical predictions to individual cases, unless almost everything is known about the star and its past rotational history.

References

Balachandran, S., Carney, B.W., Fry, A.M., Fullton, L.K., and Peterson, R.C. (1993): *Astrophys. J.* **413**, 368.

Brown, J.A., Sneden, C., Lambert, D.L., and Dutchover, E.Jr. (1989): *Astrophys. J. Suppl.* **71**, 293.

Fekel, F.C., and Balachandran, S. (1993): *Astrophys. J.* **403**, 708.

Fekel, F.C., and Balachandran, S. (1994): in *Cool Stars, Stellar Systems, and the Sun*, ed. by J.-P. Caillault (Publ. Astron. Soc. Pacific: San Francisco), in press.

Pallavicini, R. (1994): in *Cool Stars, Stellar Systems, and the Sun*, ed. by J.-P. Caillault (Publ. Astron. Soc. Pacific: San Francisco), in press.

Pallavicini, R., Randich, S., and Giampapa, M.S. (1992): *Astron. Astrophys.* **253**, 195.

Pinsonneault, M.H., Kawaler, S.D., Sofia, S., and Demarque, P. (1989): *Astrophys. J.* **338**, 424.

Pinsonneault, M.H., Kawaler, S.D., and Demarque, P. (1990): *Astrophys. J. Suppl.* **74**, 501.

Randich, S., Gratton, R., and Pallavicini, R. (1993): *Astron. Astrophys.* **273**, 194.

Randich, S., Giampapa, M.S., and Pallavicini, R. (1994a): *Astron. Astrophys.* **283**, 893.

Randich, S., Pallavicini, R. Pasquini, L., and Gratton, R. (1994b): this conference.

Ryan, S.G., and Deliyannis, C.P. (1994): *Astrophys. J.*, submitted.

Soderblom, D.R., Oey, M.S., Johnson, D.R.H., and Stone, R.P.S. (1990): *Astron. J.* **99**, 595.

Spite, M., Pasquini, L., and Spite, F. (1994): *Astron. Astrophys.*, in press.

Strassmeier, K.G., Hall, D.S., Fekel, F.C., and Scheck, M. (1993): *Astron. Astrophys. Suppl.* **100**, 173.

Thorburn, J.A., Hobbs, L.M., Deliyannis, C.P., and Pinsonneault, M.H. (1993): *Astrophys. J.* **415**, 150.

Zahn, J.-P. (1992): *Astron. Astrophys.* **265**, 115.

Zahn, J.-P. (1994): *Astron. Astrophys.*, in press.

Lithium in Companions to Compact Objects

R. Rebolo[1], E. L. Martín[1,2], J. Casares[3], P.Charles[3]

[1] Instituto de Astrofisica de Canarias, Via Lactea s/n, 38200 La Laguna, Tenerife, Spain
[2] Astronomical Institute "Anton Pannekoek", Kruislaan 403, 1098 SJ, Amsterdam, The Netherlands
[3] Dept. of Astrophysics, Nuclear Physics Lab., Oxford OX1 3RH, UK

Abstract. We report the detection of lithium in the quiescent spectrum of the neutron star binary Cen X-4. We obtain, via spectral synthesis LTE analysis, a surface Li abundance in the K-type secondary of log N(Li)=3.3 ±0.4. Non-LTE effects could reduce this value by 0.2 dex. The Li abundance is also derived for the secondaries of two black hole binaries: A0620-00 and V404 Cyg. High Li abundances seem to be a common feature in stars orbiting around neutron stars and black hole candidates, a result totally unexpected mainly because the mass transfer history of these systems. The most plausible explanation appears to be that Li production takes place in the environment of the compact objects. The mechanism may be connected with the repeated strong outbursts that characterize transient X-ray binaries. This kind of objects may have produced a significant amount of Li in our Galaxy. We propose some observational tests to confirm it.

1 Introduction

The detection of a strong Li I $\lambda6708$ resonance line in the quiescent spectrum of V404 Cyg (Martín et al. 1992), currently considered the best candidate for a stellar mass black hole (Casares et al. 1992), prompted us to initiate an observational programme specifically aimed at searching for Li in other X-ray binaries. The next two obvious targets were A0620-00 because it also harbours a black hole candidate, and Cen X-4 because of its similarity with A0620-00 (McClintock & Remillard 1990) even though its compact primary is a neutron star. These three X-ray binaries are classified as X-ray transients (XTs), sometimes referred to as X-ray novae. They are a subclass of low mass X-ray binaries (LMXBs) which is characterized by strong outbursts lasting several weeks, followed by long quiescent periods of several years. In outburst the XTs can become the brightest X-ray sources in the sky, reaching peak X-ray luminosities of about 510^{38} ergs s^{-1}. On the other hand, while in quiescence, the optical luminosity of XTs is dominated by the low mass secondary star, thereby allowing to study its properties. Before attempting to search for Li in A0620-00 we learned of its detection by Marsh, Robinson & Wood (1994).

We report here the discovery of a strong Li I feature in the quiescent spectrum of Cen X-4, which is thus the third XT where Li has been found, and the first one with a known neutron star primary. A detailed description of the Li discovery in Cen X-4 is presented in Martín et al. 1994.

2 Lithium in X-ray Transients

In Fig. 1 we show the spectrum (FWHM~1.5 Å) of Cen X-4 obtained at the Anglo-Australian Telescope. Note the strong LiI absorption and H_α emission. The Li feature in Cen X-4 moves in phase with all the secondary's absorption lines, and thus it must be formed in the stellar photosphere. No equivalent width variations are seen at the level of accuracy of our measurements.

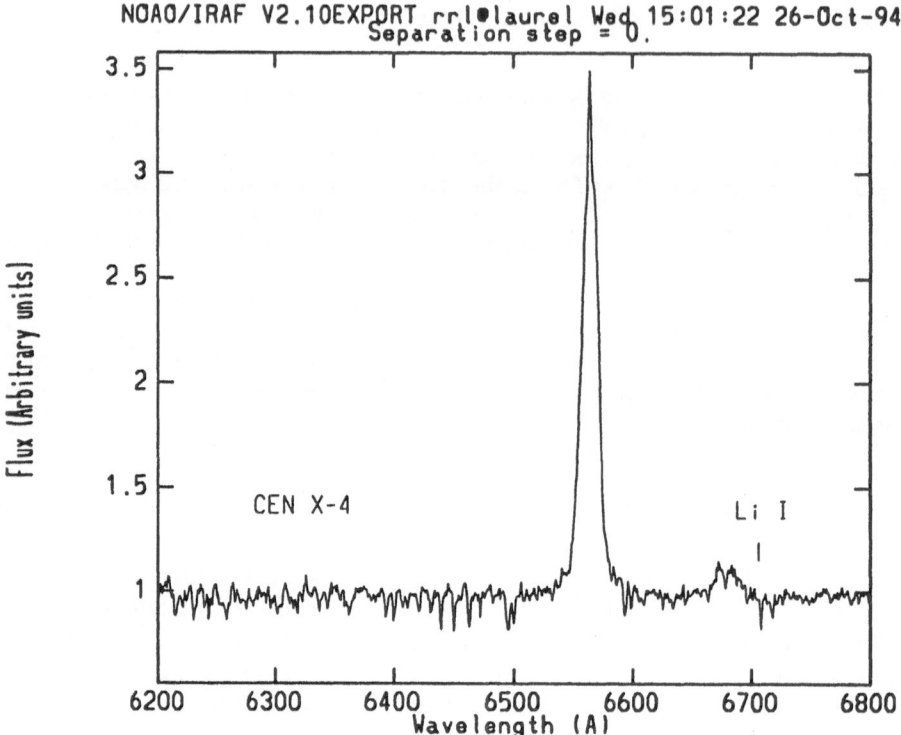

Fig. 1. Spectrum of Cen X-4 in quiescence (V=18.4) showing strong Li I λ6708 absorption.

We have performed LTE spectral synthesis and COG (including NLTE effects) analysis for the XTs where Li has been detected. We assumed *a priori* that we could use solar metallicity atmospheric models, an assumption that proved to be justified by the good fits to the Ca I λ6717 line obtained. The weak Fe I λ6707.4 line and all the known lines blended with the Li I doublet in our spectra were included together in the spectral synthesis, and solar iron abundance was used. A microturbulence value of 2 km s^{-1} was adopted in all computations. Spectral types and veiling factors for the secondaries were taken from the literature, but we checked that they were consistent with our data. The effective temperatures were assigned using calibrations of spectral type vs. T_{eff} and luminosity class. The gravities used were of log g=3.5 for V404 Cyg and log g=4.0 for the other two secondaries. We derived these values considering the estimates of radii and mass densities available in the literature and assuming that the secondaries fill their Roche lobes.

Table 1. Li in binary systems with compact objects

Name	Nature of Compact Obj.	Sp. T. Second.	Mass ($M_\odot$) Second.	P_{orb} days	W_λ (LiI) mÅ	W_λ(CaI) mÅ
A0620-00	BH (XT)	K3-K5	0.5	0.32	235	190
V404 Cyg	BH (XT)	G9-K1	1.0	6.47	285	210
Cen X-4	NS (XT)	K5-K7	0.2	0.63	380	270

Lithium is present at nearly cosmic levels in all 3 secondaries of XTs that have been searched for it so far. In marked contrast, it is not present in several secondaries of CVs that we have studied. In Table 1 we give Li I λ 6708 equivalent width values uncorrected for veiling effects (which are taken into account by Martín et al. 1994).

3 Discussion

The presence of Li cannot be explained with conventional ideas on the formation and evolution of X-ray transients. The e-folding time of Li depletion in MS K-type stars is 10^8 years (see Rebolo 1991 for a review), and the mass transfer from the secondaries of XTs entails further depletion via loss of surface layers and dilution. Hence, the absence of Li in the secondaries of CVs are the expected result, while the Li detections in the secondaries of XTs is the unexpected result which needs interpretation.

Li can be created by nuclear reactions (spallation of CNO nuclei and $\alpha\alpha$ collisions) during the energetic outbursts of XTs. Possible tests of this scenario are: 1) the detection of γ emission lines associated with de-excitation of ^{16}O (6.1 MeV), ^{12}C (4.4 MeV), D (2.2 MeV) and ^{7}Li (0.47 MeV). 2) more Li detections in LMXB secondaries, and ideally the discovery of a Li abundance much larger than cosmic in one of them. 3) measurement of the Li isotopic ratio, which should be in the range 1.1 to 1.4, i.e. much higher than in the interstellar medium, depending on the dominant reactions. 4) detection of high abundances of other light elements (D, Be, B).

Two possible scenarios for the Li enrichment of the secondary have been proposed; a disk wind model and a direct hit model. As discussed by Martín et al. (1995), the disk wind model seems more likely to work as there is evidence for substantial amounts of intervening matter between the compact object and the secondary. The wind will presumably enrich the interstellar medium with Li, as well as the secondary. Preliminary estimates indicate that if an average population of 1000 XTs (a number not ruled out by the presently observed rate of XT outbursts) has been present throughout the Galactic disk's lifetime, the contribution of Li from these systems may have been enough to explain the present Li abundance in the interstellar medium.

References

Casares, J., Charles, P.A. & Naylor, T. 1992, Nature, 355, 614

Marsh, T.R., Robinson, E.L. & Wood, J.H. 1994, MNRAS, 266, 137

Martín, E.L., Rebolo, R., Casares, J. & Charles, P.A. 1992, Nature, 358, 129

Martín, E.L., Rebolo, R., Casares, J. & Charles, P.A. 1994, ApJ, in press (Nov. 10)

Martín, E.L., Spruit, H. & van Paradijs, J. 1995, AA, in press

McClintock, J.E. & Remillard, R.A. 1986, ApJ, 308, 110

McClintock, J.E. & Remillard, R.A. 1990, ApJ, 350, 386

Rebolo, R. 1991, "Evolution of Stars: The Photospheric Abundance Connection", eds G. Michaud and A. Tutukov, IAU Symp. 145, Kluwer, 85.

Welsh, W.F., Horne, K. & Gomer, R. 1993, ApJ, 410, L39

Stellar Production of Lithium

F.Matteucci[1], F. D'Antona[2] and F.X. Timmes[3]

[1] Astronomy Department, University of Trieste, Italy
[2] Osservatorio Astronomico di Roma, Italy
[3] Laboratory for Astrophysics and Space Research, Enrico Fermi Institute, Chicago, U.S.A.

Abstract. Under the assumption that the abundance of ^7Li in population II stars represents the primordial ^7Li abundance (with perhaps a small contribution from GCR spallation) and the GCR spallation/fusion processes cannot contribute to more than 10% of the ^7Li abundance observed in population I stars and in the solar system, one must conclude that most of ^7Li in population I stars has a stellar origin. Possible stellar ^7Li sources are discussed: low mass AGB stars ($2\text{-}5M_\odot$), high mass AGB stars ($5\text{-}8M_\odot$), supernovae of type II ($M > 10M_\odot$) and novae. We include the yields from these different ^7Li sources in a model for the chemical evolution of the Galaxy and predict the behaviour of the log N(Li) vs. [Fe/H] relation. We conclude that, although a unique model cannot be found, due to the uncertainties still present in the stellar nucleosynthesis, the most likely scenario is that ^7Li is partly produced in type II supernovae (ν-induced nucleosynthesis) and partly in massive AGB stars.

The Low Primordial ^7Li Scenario

The observed log N(Li) vs. [Fe/H] shows that the ^7Li abundance in population I and TTauri stars is at least a factor of ten higher than in population II stars. The problem of ^7Li can be expressed as follows:

Either there is a "high" primordial ^7Li abundance observed in young matter in the Galaxy (logN(Li)$\simeq$ 3.0 dex), implying non-standard Big Bang nucleosynthesis, with the consequence that ^7Li is progressively destroyed inside stars,or

There is a "low" primordial ^7Li abundance (logN(Li)$\simeq$2.0 dex) as suggested by the uniformity of ^7Li abundance in Pop. II stars (the Spite plateau) with the consequence that an efficient production of ^7Li during the Galaxy lifetime is required to explain the ^7Li abundance in Pop. I stars. In the following we will assume the low primordial ^7Li scenario. In this framework a substantial stellar ^7Li production is required to explain the Pop. I ^7Li abundance. This is because cosmic-ray $\alpha\text{-}\alpha$ reactions can at most contribute to $10 - 20\%$ of the Pop. I ^7Li (Walker et al.1985; Prantzos et al.1993, but see Beckman, this conference). The similarity of the $^7Li/^6Li$ ratio between the solar system and the ISM (Lemoine et al.1993) also argues in favor of ^7Li stellar production (Reeves 1993). However, Lemoine (this conference) has shown that a ratio of ~ 2, lower than the solar system ratio, has been observed in one of the two absorbing components in the direction of ζ Oph. If this low value will be confirmed, new cosmic ray induced spallation mechanisms could be required.

Observational and Theoretical ^7Li -rich Stars

Massive AGB and M supergiants in the Magellanic Clouds and in the Galaxy are ^7Li -enriched. These stars present enhanced s-process elements from which one infers that they are in their Thermal Pulsing (TP) phase, but they are not C-stars. Smith and Lambert (1989, 1990) suggested log N(Li)$\simeq$ 2.0 − 3.8 dex for these stars in the Magellanic Clouds, more recently Plez et al.(1993) indicated an average abundance log N(Li)$\simeq$ 3.0 dex.

Other possible ^7Li producers are Galactic C-stars with luminosities ranging from $M_{bol} \simeq$ -6 to -3.5. These stars, in fact, show in some cases very high ^7Li abundances (log N(Li) $\simeq$ 4.5-5.5 dex, Abia et al.1993).

¿From the theoretical point of view, it is necessary to have a site in which the reaction $^3He(\alpha, \gamma)^7Be$ can occur, but ^7Be should also be rapidly transported (by convection) into low temperature regions where it can decay into ^7Li by k-capture (Cameron and Fowler, 1971).

Possible ^7Li stellar sources, suggested from a theoretical point of view are:
1) C-stars $(2 - 5M_\odot)$
2) high-mass AGB stars $(5 - 8M_\odot)$
3) supernovae of type II $(M > 10M_\odot)$
4) fast novae (binary systems with a WD plus a low mass MS star)

- **C-stars**

^7Li is produced during the TP phase and then destroyed: survival may last a few thermal pulses. D'Antona and Matteucci (1991) (DM) assumed that each C-star can produce logN(Li) =3.85 and concluded that they are not important ^7Li producers in the Galaxy, due to their low mass loss. Abia et al.(1993) estimated the ^7Li production from C-stars from observations and concluded also that C-stars are probably not very important ^7Li producers.

- **High mass AGB stars**

Sackmann and Boothroyd (1992) showed that stars of masses $4 - 6M_\odot$, in the luminosity range $-6 \leq M_{bol} \leq -7$, can reach log N(Li)=4.5 in the envelope as a consequence of hot-bottom burning. DM showed that these stars can be very important contributors to ^7Li Galactic enrichment by assuming that each star would produce logN(Li)=5.0 dex.

- **Type II SNe**

ν-induced nucleosynthesis can create a substantial amount of ^7Li : the flux of neutrinos following the core collapse to form a neutron star becomes so great that can induce substantial transmutation in the ejected material (Woosley et al.1990; Woosley and Weaver 1994).

- **Novae**

The explosive formation of ^7Li in novae has been revisited recently (Boffin et al.1993). These results seem to suggest that novae cannot be important ^7Li producers in the Galaxy. DM had included ^7Li production from novae following the prescriptions of Starrfield et al.(1978). They had also assumed that novae produce ^7Li with a substantial delay (of the order of 1 to 5 Gyr from the formation of the white dwarf). Therefore, the inclusion of nova production leads one to predict a strong increase in the ^7Li abundance from the time of formation of the solar system up to the present time. Such an increase has been claimed only by King (1993).

Galactic Chemical Evolution

The main characteristic of the plot log N(Li) vs. [Fe/H] (see Figure 1) is that Pop. II stars show a much smaller dispersion than Pop.I stars (although Thorburn's (1994) data show a larger spread in Pop. II stars than observed before). In the "low" primordial ^7Li scenario stars of the upper envelope should show the ^7Li abundance at their birth. Therefore, models of galactic chemical evolution, which predict the ^7Li abundance in the ISM, should fit the upper envelope of the log N(Li) vs. [Fe/H] data.

The chemical evolution model adopted here is the same as DM: the evolution of 20 chemical species is followed in detail both in the solar neighbourhood and in the whole Galaxy. Supernovae of type Ia, Ib and II and their nucleosynthesis products are taken into account. Infall of primordial material is considered. The nucleosynthesis prescriptions adopted (see Matteucci et al.1994 for more details) are:

- Renzini and Voli (1981) for the mass range $0.8 - M_{up}$. Tornambè and Chieffi's (1986) results on the variation of M_{up} (i.e. the limiting mass for the formation of a CO degenerate core) with the stellar metallicity Z are also taken into account.
- Woosley and Weaver's (1994) new results for ^7Li production due to ν-induced nucleosynthesis in type II SNe ($11 - 40M_\odot$).
- ^7Li production from novae is also included for comparison and the prescriptions are the same as in DM.

Results for log N(Li) vs [Fe/H] and Conclusions

The most recent data on ^7Li in dwarfs (Boesgaard and Tripicco, 1986; Rebolo et al., 1988; Spite and Spite, 1986; Deliyannis et al., 1990; Lambert et al., 1991; Thorburn, 1994; Norris et al., 1994) have been collected and compared with model predictions. Models including various combinations of C-stars, massive AGB, SNe II and novae have been computed, with an assumed primordial ^7Li abundance logN(Li)=2.2 dex. Our main results can be summarized as follows:

Models with massive AGB stars and variable M_{up} plus SNe II give the best agreement with the observed log N(Li) vs. [Fe/H], as it can be seen in Figure 1 where our best model is shown. C-stars are found to be negligible producers of ^7Li , mainly because of the small mass they can loose. The best values for ^7Li production in massive AGB is log N(Li)=4.15, although a value of 3.5 could be still acceptable. In this last case the production by SNe II would be predominant, whereas in the other case is roughly the same as that of AGB stars. The variation of M_{up} with the stellar metallicity improves the fit of the log N(Li) vs.

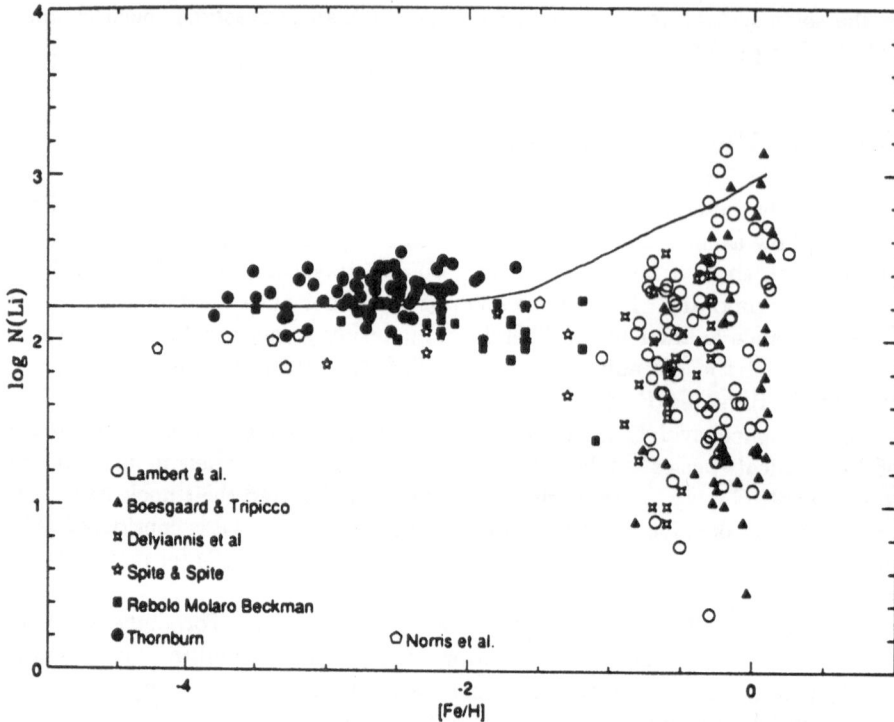

Fig. 1. Comparison between observational logN(Li) vs. [Fe/H] and our best model

[Fe/H] plot mainly in the region around [Fe/H]=-1.0. The theoretical predictions in this region tend, in fact, to give higher log N(Li) values than observed. However, it is not clear whether this is a lack of observational points in the critical region of the transition between halo and disk or is a real feature. Models where novae are included give far too high ^7Li abundances in the solar system if also all the other mentioned sources are considered. They give an acceptable solution if the SNe II are suppressed, although they predict a substantial growth of ^7Li between the solar system formation time and now. However, ^7Li production from novae has probably been overestimated by DM. Therefore, from this study we can conclude that the most likely scenario to explain the log N(Li) vs. [Fe/H] plot, under the assumption that most of ^7Li in Pop. I stars has a stellar origin, is that supernovae of type II and massive AGB stars contributed in roughly the same proportions to the ^7Li Galactic enrichment.

References

Abia, C., Isern, J., Canal, R.: Astron. Astrophys. **275**, 96 (1993)

Boesgaard, A.M., Tripicco, M.J.: Astrophys.J. **302**, L49 (1986)

Boffin, H.M.J., Paulus, G., Arnould, M., Mowlavi, N.: Astron. Astrophys. **279**, 173 (1993)

Cameron, A.G.W., Fowler, W.A.: Astrophys.J. **164**, 111 (1971)

D'Antona, F., Matteucci: Astron. Astrophys. **248**, 62 (1991)

Delyiannis, C.P., Demarque, P., Kawaler, S.: Astrophys.J.Suppl. **73**, 21 (1990)

King, J.R.: Astron.J. **105**, 1087 (1993)

Lambert, D.L., Heath, J.E., Edvardsson, B.: Mon.Not.R.Astr.Soc. **253**, 610 (1991)

Lemoine, M., Ferlet, R., Vidal-Majar, A., Emerich, C., Bertin, P.: Astron. Astrophys. **269**, 469 (1993)

Matteucci, F.,D'Antona, F., Timmes, F.X.: Astron. Astrophys. submitted

Norris, J.E., Ryan, S.G., Stringfellow, G.S.: Astrophys.J. **423**, 386 (1994)

Plez, B., Smith, V.V., Lambert, D.L.: Astrophys.J. **418**, 812 (1993)

Prantzos, N., Casse, M., Vangioni-Flam, E.: Astrophys.J. **403**, 630 (1993)

Rebolo, R., Molaro, P., Beckman, J.: Astron. Astrophys. **192**, 192 (1988)

Reeves, H.: Astron. Astrophys. **269**, 166 (1993)

Renzini, A., Voli, M.: Astron. Astrophys. **94**, 175 (1981)

Sackmann, I.J., Boothroyd, A.I.: Astrophys.J. **392**, L71 (1992)

Smith, V.V., Lambert, D.L.: Astrophys.J. **361**, L69 (1990)

Smith, V.V., Lambert, D.L.: Astrophys.J. **345**, L75 (1989)

Spite, F. Spite, M.: Astron. Astrophys. **163**, 140 (1986)

Starrfield, S., Truran, J.W., Sparks, W.M., Arnould, M.: Astrophys.J. **222**, 600 (1978)

Thorburn, J.A.: Astrophys.J. **421**, 318 (1994)

Tornambè, A., Chieffi, A.: Mon.Not.R.astr.Soc. **220**, 529 (1986)

Walker, T.P., Steigman, G., Schramm, D.N., Olive, K.A., Fields, B.: Astrophys.J. **413**, 562 (1993)

Woosley, S.E., Hartmann, D.H., Hoffman, R.D., Haxton, W.C.: Astrophys.J. **356**, 272 (1990)

Woosley, S.E., Weaver, T.: 1994, preprint

Li I Lines in POP II Dwarf Spectra: NLTE Effects

Yakiv V.Pavlenko

The Principal Astronomical Observatory, National Academy of Sciences of Ukraine, Golosiiv, Kyiv-22, 252650 UKRAINE

Abstract. Results of computations of the statistical equilibrium of lithium in dwarf atmospheres of T_{eff} = 6270, 5770, 5270 K, log g= 4.44 and $[m/H]$ = $-0, -1, -2, -3$ are discussed. It is shown that NLTE corrections are less than 0.1 dex for abundances of lithium observed in halo dwarfs.

1 The Procedure

Model atmospheres were computed using the **SAM71** program (Pavlenko 1992) which is a modification of **ATLAS5** (Kurucz 1975) Our main modification was to do opacity calculation using the opacity sampling technique (Peytremann 1976). Data for lines of atoms and ions were taken from Kurucz and Peytremann (1975) list. We assume that absorption lines are broadened by microturbulence v_t = 2 kms^{-1}. Each absorption line has a Voigt profile with damping equal to the sum of van der Waals, Stark and radiative values. In the model computation we assume the parameter of convection l/H = 1.0. In this paper we shall use appropriate designations for model atmospheres from our grid. We shall sign models as Mz, where M means X, Y, Z for models with T_{eff} = 5250, 5770, 6270 K and z means 0, 1, 2, 3 for $[\mu]$ = 0, $-1, -2, -3$ respectively. So Y0 means a solar model atmosphere, Z3 means model with T_{eff} = 6270, $[\mu]$ = -3. As it was noted above, all models were computed for log g= 4.44.

To solve the NLTE problem we followed the linearization method proposed by Auer and Heasley (1976). A lithium atom model used consists of 19 levels of *Li I* and the lowest state of *Li II*. 70 radiative transitions (19 bound-free (*bf*) and 51 bound-bound (*bb*)) were included into the rate matrix. The oscillator strengths of *bb* transitions were adopted from Wiese et al.(1966). The 10 strongest *bb* transitions were linearized. Other radiative transitions (*bb* and *bf*) were treated with fixed rates, because the absorption due to lithium atoms in their frequencies does not change the radiation field. The rates of collision excitation and photoinization cross sections were described elsewhere (Magazzu et al.1992; Pavlenko 1994).

Most of the *Li I* lines have multiplet structure which is accounted for by the absorption coefficient profiles of linerized lines (Pavlenko 1990). At the frequencies of all linerized *bb* radiative transitions of *Li I* , the absorption due to lines of other atoms, ions and molecules was taken into account. All possible transitions due to inelastic collisions with hydrogen atoms and electrons were included into a rate matrix. These rates were computed by formula $C_{ij} = C_{ij}^e + C_{ij}^h * q$, where C_{ij}^e and C_{ij}^h are the rates of $i \rightarrow j$ transition due to inelastic collisions with free electrons and hydrogen atoms respectively. To compute C_{ij}^h we followed the cross-sectins recipe proposed by Dravin and modified later by Steenbock and Holweger(1984). For the time parameter q is poorly defined (see Lambert 1992 for detail) and ranges from 0 to 1. In this paper we discuss results obtained for $q = 0$ and $q = 1$.

2   Results

I. The Sun. The model with $T_{eff} = 5770$ K, log g= 4.44 and $[\mu] = 0$ is a model of solar atmosphere. We performed NLTE computations also for two semiempirical model atmospheres: HOLMU (Holweger and Muller 1976), MACKKL (Maltby et al. 1984). More details for 5770/4.44/0.0 models atmosphere computations are given in the paper (Pavlenko 1994). A few results are given in the table 1.

Table 1. NLTE abundance corrections for lithium $\Delta\epsilon(Li) = \epsilon(Li)_{nlte} - \epsilon(Li)_{lte}$ in solar atmosphere. N_l - number of levels in model atom.

Authors	N_l	Model	q	$\Delta\epsilon(Li)$
Steenbock and Holveger(1984)	9	HOLMU	1.0	0.027
Pavlenko(1989)	6	MACKKL	1.0	0.13
This work	20	MACKKL	1.0	$> -0.001, < 0$
			0.0	0.017
	20	HOLMU	1.0	- 0.008
			0.0	0.006
	20	Z0	1.0	0.03
			0.0	0.05

We find that:
- for the Sun $abs(\Delta\epsilon(Li)) < 0.05 dex$;
- the NLTE abundance corrections are less than differences due to different temperature structures in the solar model atmospheres;
- semiempirical model atmospheres have shown less pronounced NLTE effects in comparison with theoretical ones; they "mimic" NLTE effects (see Rutten and Kostik 1982).

II. Dwarfs of halo and disc. In this paper we consider next problems:
- qualitative and quntitative regularities in NLTE of Li I lines in halo and disc dwarfs;
- dependence of NLTE on the q parameter;
- the impact of NLTE on results obtained in LTE;
- LTE and NLTE in resonance and subordinate lithium lines.

The computed LTE and NLTE equivalent widths of $Li\ I$ lines are given in Pavlenko(1994).

We find that:

- the sign and values of abundance correction due to NLTE effects in the resonance doublet 670.8 nm are small for halo dwarfs. So classical results of Spite & Spite (1982) related to discovery of "lithium plateau" in halo dwarfs cannot be changed by NLTE. - the abundance correction depends on the model atmosphere structure, transition rates due to inelastic collisions and the abundance of lithium. - for subordinate line 610.3 mn the sign and values of abundance corrections are always positive. - in a wide range of abundances $(1.0 < \epsilon(Li) < 2.8)$ the lithium abundance corrections due to NLTE effects are less than 0.1 dex for solar type dwarfs. - the dependence of our results on parameter q is not critically. - our computations show that the dependence of NLTE corrections for lithium abundances on the metallicity of solar like stars is rather weak.

To explain the last result we compare the temperature structures of models of the same T_{eff} and $\log g$ and different metallicities (for example, Z0 and Z3). It is a well known result, that temperature gradient in the model atmosphere depends on metallicity. In the Z3 model the gradient is lowered in comparison with the Z0 model. The difference between the radiative (and excitation !) temperature and electron temperature T_e in model Z3 is smaller than in model Z0. As a result the intensity of NLTE effects in model Z3 do not increase dramatically despite the decreasing of opacities in the frequencies of bound-bound and bound-free Li I transitions. More details about this are given in paper of Pavlenko (1994).

III. The dependence of $\Delta\epsilon$ on T_{eff} is more interesting. Results given in the table 2 may be used for explaining the slope of $\epsilon(Li)$ versus T_{eff} suggested by Rebolo *et al.*(1988).

Table 2. NLTE lithium abundance corrections as a function of metallicity.

Model atmosph.	μ	ϵ LTE	W_λ	$\epsilon(NLTE)$ $q = 0$	$\epsilon(NLTE)$ $q = 1$	$\Delta\epsilon(Li)$ $q = 0$	$\Delta\epsilon$ $q = 1$
5270/4.44	0	2.1	9.94	2.138	2.11	0.038	0.01
	-1	2.1	9.52	2.163	2.08	0.063	-0.02
	-2	2.1	8.88	2.189	2.075	0.089	-0.025
	-3	2.1	7.76	2.17	2.07	0.070	-0.03
6270/4.44	0	2.1	2.04	2.11	2.094	0.01	-0.006
	-3	2.1	1.976	2.13	2.072	0.03	-0.028

We find that:

- the NLTE correction factor increases from + 0.038 up to + 0.07 dex when T_{eff} drops from 6250 to 5250 K.

- the NLTE with $q = 1$ gives "better" results: in that case the slope $\epsilon(Li)$ versus T_{eff} decreases in comparison with LTE.

- the LTE abundance computations show an opposite tendency.

- in our analysis we ignore a dependence of equivalent widths on poorly known parameters of stellar atmospheres which are poor defined parameters:

e.g. gravity, microturbulence velocities, etc. It is possible, the dispersion of observed equivalents widths of Li I lines in dwarfs spectra may be explained by the dependence on these factors.

3 Conclusions

For the moment we can give qualitative and "zero approach quantative" estimation of the impact of NLTE on well known results. It is shown the NLTE has to be considered to resolve this problem. More refined results could be obtained for better defined model atmosphere, model atoms and homogeneity of observed data.

References

Auer,R.N., Heasley,I.N., 1976, Ap. J., **205**, 165.

Holweger,H., Muller, E. A. Solar physics, 1974, **9**, 19. N1

Kurucz R.L., 1975, SAO Special report N 309

Kurucz R.L., Peytremann, E.A, 1975, SAO Special report N 362.

Lambert, D.L., Physica Scripta, 1993, **T47**, 186

Maltby, P., Avrett, E.H., Carlsson, M.,Kjeldseth-Moe, O., Kurucz, R.,L., Loeser R. Ap. J., 1986, **306**, 284.

Magazzu, A., Rebolo, R., Pavlenko, Ya.V., 1993, Ap. J., **292**, 159.

Pavlenko Ya.V., 1990, Kinemat. i fizika nebes. tel, **6**, 58

Pavlenko Ya.V. Sov. Astron., 1994, in press.

Peytremann, 1974, A&A **203**, 33

Rebolo, R., Molaro P., Beckman, J.E A&A, 1988, **192**, 192.

Rutten R.J., Kostik R.I. A& A, 1982, **115**, 104.

Spite F., Spite M., 1982, A& A, **115**, 337.

Steenbock, W., Holweger H., 1984, A&A, **130**, 319

Wiese, W.L., Smith, M.W., Glennon,B.M., 1966, Atomic lines probabilities NSRD-NBS 4 1

Lithium Content Investigation at the 6-m Telescope

V.G. Klochkova, V.E. Panchuk

Special Astrophysical Observatory, N.Arkhyz, Stavropol Territory, 357147 Russia

Abstract. A program of spectroscopic investigation of Li content for metal–deficient s- tars has been fulfilled at the 6-m telescope since 1992. We used the echelle–spectrometer LYNX with a 530x580 CCD. The spectral interval registered simultaneously was 5500 – 8700 Å, the spectral resolution was equal to R=24000. Here we present the metal- licity [Fe/H], the Li content, the abundance of the light metals (Na, Si, Ca) and the measured radial velocity V_r for 5 first objects studied: HD 64090, G 29–20, G 122–57, G 182–7, and G 246–38.

1 Selection of Objects and Observations

Uniformity of lithium content in the atmosphere of unevolved metal poor sub- dwarfs with $T_e > 5600$ K at a level of about one tenth of the Li abundance in young disc stars is now a classical fact (Spite and Spite 1982). But the problem of origin (cosmological and galactic aspects) of this observed Li content is close- ly connected with the problem of efficiency of mechanisms destructed of such a fragile chemical element as Li in the envelopes of metal poor subdwarfs. One of the empirical approaches to solution of this problem is the investigation of the behavior of the Li content with the temperature for subdwarfs of different ages. For such a purpose we should perform a lot of Li, Fe abundance and radial velocity determinations for a sufficiently large sample of cold subdwarfs. Besides the problem of Li origin, the study of surface Li content in different type stars is important for sounding on stellar structure in the course of evolution of different mass stars.

Our program of spectroscopic investigation of Li content in the atmosphere of halo stars at the 6-m telescope with a high spectral resolution is based on the catalogue of Laird et al (1988). The northern objects with a metallicity $[Fe/H]_\odot < -1$ dex and with V < 11 mag were selected from the above mentioned catalogue. Mainly we prefer the stars with $[Fe/H]_\odot < -1.7$ dex.

Table 1. Initial data for the stars studied from the paper of Laird et al (1988)

Star	V	T_e, K	[Fe/H]
HD 64090	8.28	5380	−1.75
G 29–20	9.17	4988	−1.71
G 122–57	8.36	4744	−1.71
G 182–7	8.10	5023	−1.26
G 246–38	9.91	5234	−2.06

All observations were made at the Nasmyth–focus of the 6-m telescope with the echelle–spectrometer LYNX equipped with a CCD (Panchuk et al 1993). We obtained 1 spectrum for each star with S/N >100 per pixel in the spectral region 5500–8800 Å with the spectral resolution R=24000 (Figure 1). A Th–Ar lamp was used for wavelenghts calibration with a precision better than 0.01 Å. The hot quickly rotating stars, HR 4687 and HR 1770, were observed in order to identify telluric lines and to separate these lines from stellar ones.

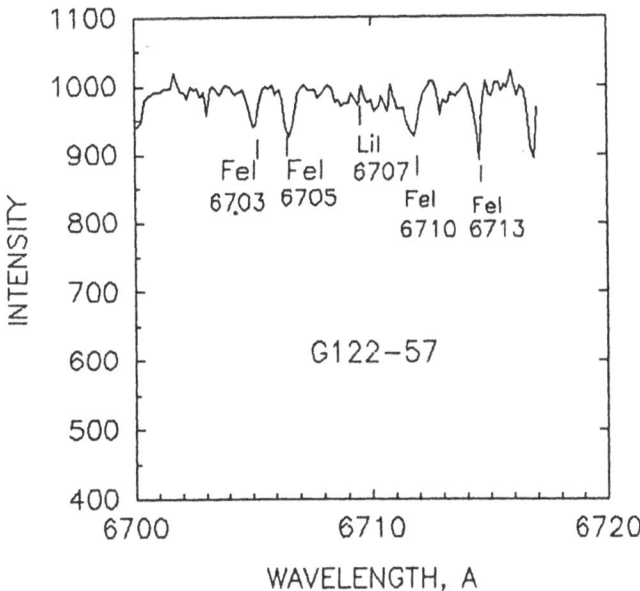

Fig. 1. The part of spectra of the star G 122–57 near Li I 6707 Å line

These spectra were reduced with the use of a special system of image processing and spectrum reduction DECH (Galazutdinov 1992). The generally accepted procedures of dark substraction, cosmic removing, optimal extraction of echelle orders were performed. A Gauss approximation was made for the measurement of equivalent widths W. To illustrate the accuracy of our observing data, a comparison of the measured W values with those from the paper by Rebolo et al (1988) for HD 64090 and from the paper by Pilachowski et al (1993) for G 122–57 is given in Figure 2. These above mentioned authors obtained the spectra with a spectral resolution and S/N close to ours.

According to the formula of Cayrel (1985) the accuracy of the measured W for the stars studied is better than 2–3.5 mÅ.

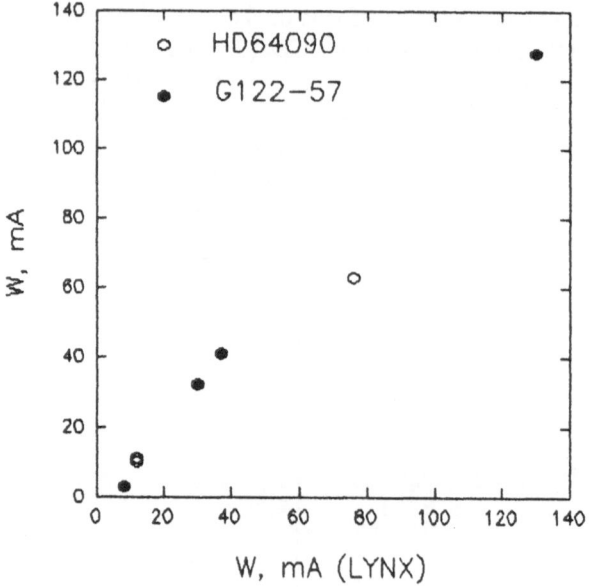

Fig. 2. The comparison of W obtained by authors with W from the papers published by Rebolo et al (1988) for HD 64090 and by Pilachowski et al (1993) for G 122-57

2 Determination of Model Parameters and Chemical Composition

Some necessary data for the stars under study (the magnitude V, effective temperature T_e and metallicity [Fe/H] from the paper by Laird et al (1988)) are given in Table 1. It should be recalled that these T_e and [Fe/H] values were determined from photometric indices.

In the Li content determination it is most important moment to define the effective temperature T_e because the Li doublet intensity is very sensitive to this parameter. As a first approach the T_e values from the catalogue of Laird et al (1988) were used. But thanks to the large registered spectral region we have a possibility of obtaining the main parameters of the model atmosphere T_e and surface gravity log g needed for the chemical composition calculation using the spectroscopic data free from the interstellar reddening. These model parameters were determined by the generally accepted procedure: T_e — by making the neutral iron abundance independent of the line excitation potential (see such a graphic for G 122-57 in Figure 3a), the surface gravity log g — by forcing the ionization equilibrium for FeI and FeII, the microturbulent velocity ξ_t — by forcing the independence of the abundance derived from individual FeI lines upon the equivalent width (see Figure 3b).

The adopted parameters for the model atmosphere of the stars under investigation together with values of these parameters from the earlier papers for comparison are listed in Table 2.

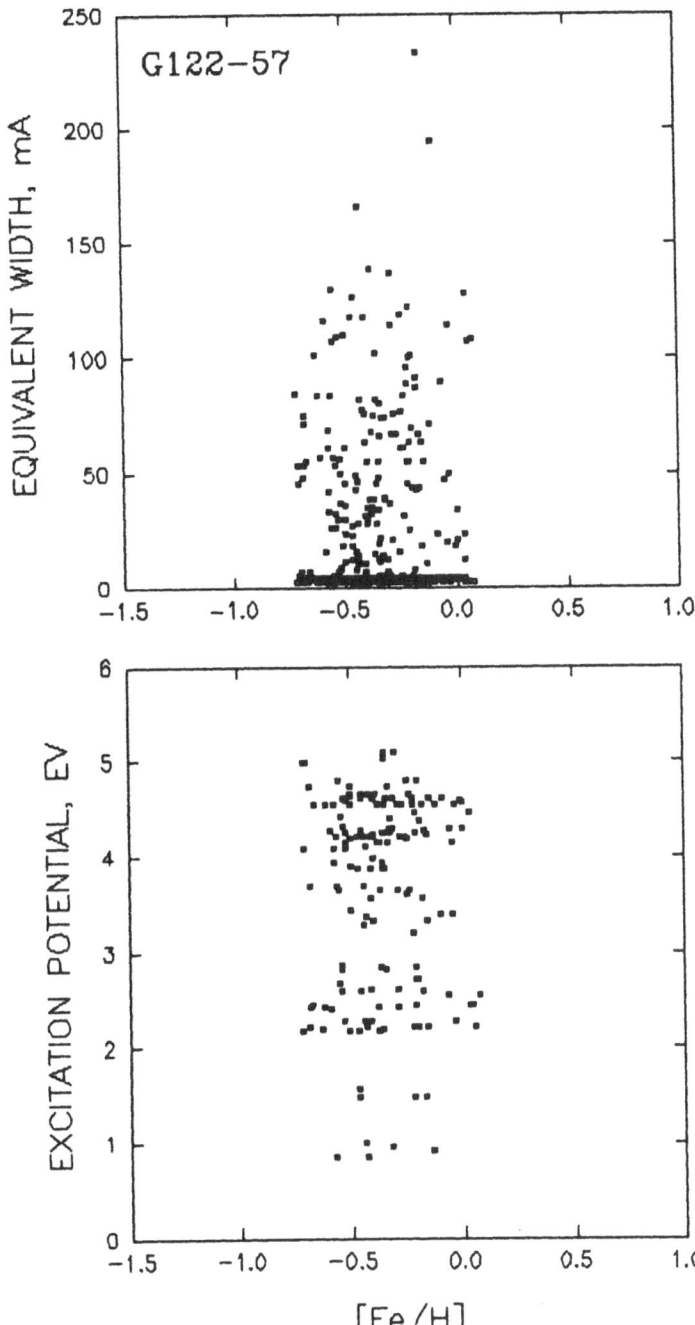

Fig. 3. The iron content for G 122-57 derived from FeI lines with different: b) equivalent width, b) low-excitation potential

Table 2. The parameters for the stars under study

Star	T_e, K	log g	ξ_t, km/s	$[Fe/H]_\odot$	$\log\epsilon$ (Li)	Ref
HD 64090	5361	3.54		-1.90		1
	5250	4.5		-1.60		2
	5400			-1.56		3
	5370	4.0		-1.90	1.17	4
	5500	4.5	2.0	-1.52		5
	5380			≤-1.60	≤0.90	6
					1.15	7
	5370	3.0	2.5	-1.76	1.24	8
G 29–20	5030	2.0	1.2	-0.91	1.03	8
G 122–57	4800			-0.92	≤-0.10	9
	5040	3.0	1.0	-0.33	0.76	8
G 182–7	5500	4.2	2.0	-0.14	1.22	8
G 246–38	5240	4.0		≤-2.50	≤1.10	4
	5240	3.5	3.5	-2.00	0.89	8

References to Table 2: 1 – Hearnshaw 1976; 2 – Peterson 1978; 3 – Peterson 1981; 4 – Rebolo et al 1988; 5 – Krishnaswamy-Gilroy et al 1988; 6 – Hobbs and Duncan 1987; 7 – Spite and Spite 1986; 8 – this paper; 9 – Pilachowski et al 1993.

Practically all T_e values published earlier were derived by the authors using (R–I) or (b–y) indices and different calibrations. Note that our spectroscopic T_e value for the well studied closely located (distance from the Sun d = 26 pc) and therefore unreddened star HD 64090 is equal to the averaged T_e values given for this star in Table 2.

The Kurucz's and Bell–Gustafsson's models and WIDTH6 program were used for a chemical composition calculation. The gf–values for the main bulk of spectral lines were given from the Thevenin's papers (1989, 1990), however for the lines of the neutral atom Fe we prefer the accurate experimental data of the Oxford group (most recent reference: Blackwell et al 1986). The calculated abundances of selected chemical elements are given in Tables 2 and 3 together with the earlier published results. It should be noted that the results are presented in the generally accepted form:

$$[X/Fe]_\odot = [\log\epsilon(X) - \log\epsilon(Fe)]_* - [\log\epsilon(X) - \log\epsilon(Fe)]_\odot$$

We used here the solar chemical composition data following Grevesse (1993).

The accuracy of abundance determination is due to the accuracy of our equivalent widths W, by the accuracy of model parameters (T_e, log g, ξ_t), and by the conformity of the adopted model atmosphere to the real stellar one. The influence of errors of W, T_e, log g, ξ_t determination for the cold stars was repeatedly analyzed and published by other authors (see, for example, Norris et al 1993). It is not necessary to repeat here such estimations. Note that our results on FeI abundance are based on numerous individual lines (from 60 lines for the G 246–38 to 180 ones for G 182–7 and G 122–57). The internal error of log ε (FeI) is $\delta < 0.02$dex. The typical number of lines is $n = 4$ for Na and $n > 20$ for Ca and Si.

Table 3. The abundance of light metals and the measured radial velocities for the stars studied

Star	$[Na/H]_\odot$	$[Si/H]_\odot$	$[Ca/H]_\odot$	$V_r \pm \sigma.$ km/s
HD 64090	-1.38	-1.21	-1.41	-236.6±0.4
G 29–20	-1.05	-0.61	-0.43	-191.0±0.1
G 122–57	-0.37	-0.18	0.00	58.2 ±0.1
G 182–7	-0.05	-0.15	0.13	-0.3 ±0.2
G 246–38	-2.28	-1.38	-1.73	-162.2 ±0.2

3 Radial Velocity

The essence and some details of V_r measurement procedure based on our CCD echelle spectra were published by Galazutdinov (1992). The determined radial velocities V_r for the studied objects are given in Table 3. Our result for HD 64090 coincides with V_r published by Carney (1978). The radial velocities based on high-resolution low S/N spectra for another 4 stars were measured by Carney and Latham (1986). Our data for these stars, except G 122–57, are in a good agreement with those from the cited paper. But for the star G 122–57 we found the value of V_r essentially different from the value $V_r = 47.06 \pm 0.51$ km/s published by Carney and Latham (1986). Such a variability of radial velocity permits us to suspect the binarity of the the subdwarf G 122–57.

4 Discussion and Conclusions

We have reported new Li I observations together with metallicity and light elements abundances for 5 objects from the survey of proper–motion stars of Laird et al (1988). The derived results concerning Li and Fe contents for the well studied halo subdwarf HD 64090 and for the metal poor star G 246–38 are in good agreement with the earlier published data (see Table 2). The light metals are overabundant for these two stars as would be expected for halo stars for which $[\alpha/Fe] \geq 0.3 - 0.5$dex (Wheleer et al 1989; Norris et al. 1993).

A large discrepancy between the value of metallicity determined in this work and the metallicity from the Laird et al (1988) catalogue for two stars G 122–57 and G 182–7 is found. A main cause of this disagreement is a difference in the adopted effective temperature for these stars. We are apt to think that our value of metallicity close to solar is more fit for G 182–7 because it is probably a disc star according to its radial velocity $V_r = -0.39$km/s. Such a suggestion is also supported by the absence of significant enhancement of the light elements Na, Si, Ca in its spectrum: $[Na/Fe] = 0.09$dex, $[Si/Fe] = -0.01$dex, $[Ca/Fe] = 0.27$dex; while the α–process elements content for metal–deficient halo stars must be essentially enhanced relative to the iron abundance.

Acknowledgement

This work was supported by the grant 94-02-03281-a from the Russian Foundation of Fundamental Researches.

References

Blackwell D.E., Booth A.J., Haddock D.J. et al (1986): Mon Notic Royal Astron Soc **220** 549

Carney B.W. (1978). In: Abia C., Rebolo R. Ap J **347** 186 (1989)

Carney B.W., Latham D.W. (1986) Astron J **92** 116

Cayrel de Strobel G. (1985): in Proc 111th IAU Symp., *Calibration of Fundamental Stellar Quantities*, eds. by D.S. Hayes, L.E. Pasinetti, and A.G.David Philip, Dordrecht et al, pp. 137–162

Hearnshaw J.B. (1976) A&A **51** 71

Hobbs L.M., Duncan D.K. (1987) Ap J **317** 796

Galazutdinov G.A. (1992) Preprint Special Astrophys. Observ. **92**

Grevesse N. (1993) The paper reported at d'Evry Schatzman Colloqiuim "Physical Processes in Astrophysics" held at Paris Observ., Sept. 1993.

Krishnaswamy-Gilroy K., Sneden C., Pilachowski C., Cowan J. (1988) Ap J **327** 298

Laird J.B., Carney B.W., and Latham D.W. (1988) Astron J **95** 1843

Norris J.E., Peterson R.C., Beers T.C. (1993) Ap J **415** 797

Panchuk V.E., Klochkova V.G., Galazutdinov G.A. et al (1993) Pis'ma Astron Zh **19** 1069

Peterson R.C. (1978) Ap J **222** 181

Peterson R.C. (1981) Ap J **244** 989

Pilachowski C.A., Sneden C., Booth J. (1993) Ap J **407** 699

Rebolo R., Molaro P., Beckman J.E. (1988) A&A **192** 192

Spite F., Spite M. (1982) A&A **115** 357

Spite F., Spite M. (1986) A&A **163** 140

Thevenin F. (1989) A&A Suppl **77** 137

Thevenin F. (1990) A&A Suppl **82** 179

Wheeler J.C., Sneden C., Truran J.W., Jr. (1989) Ann Rev A&A **27** 279

Part VII

Lithium Isotopes

The Lithium Isotope Ratio in Old Stars

P.E. Nissen

Institute of Physics and Astronomy, University of Aarhus, DK-8000 Aarhus C

1 Introduction

A number of important observational studies of the light element abundances in old, metal-poor stars have been published in recent years. Beryllium has been detected and found to evolve with a nearly constant ratio to iron (Gilmore et al. 1992, Ryan et al. 1992, Boesgaard & King 1993). Boron has been shown to be about 20 times more abundant than Be, when non-LTE effects in the formation of the B I 2496.7 Å line are taken into account (Duncan et al. 1992, Kiselman 1994, Edvardsson et al. 1994). Thorburn (1994) has made an extensive survey of lithium in very metal-poor, main-sequence stars, which suggests that the abundance of Li increases significantly with increasing metallicity. Finally, ^6Li has probably been detected in the metal-poor ($[Fe/H] \simeq -2.4$) turnoff star HD 84937 (Smith et al. 1993, Hobbs & Thorburn 1994) at a level of $^6Li/^7Li \simeq$ 0.05. All of these new data are very significant for our understanding of the Galactic evolution of the light element abundances and for the determination of the amount of ^7Li produced in the Big Bang phase of the Universe.

In the present paper we concentrate on the lithium isotope problem. The methods and accuracy by which the $^6Li/^7Li$ ratio can be determined are discussed. Earlier and recent works on old F and G stars are reviewed and some new observations obtained with the ESO Coudé Echelle Spectrometer (CES) are presented. Finally, the existing results on the lithium isotope ratio and the new data on Be, B and Li abundances are compared with models for cosmic ray production of the light elements as well as the possible depletion of these elements in stellar envelopes.

2 Analysis of the Li I 6707 Å line

The Li I resonance line at 6707 Å is a doublet with hyperfine structure. Accurate interferometric wavelength measurements have been carried out by Meissner et. al. (1948). The doublet splitting is 0.152 Å and the isotope shift is 0.158 Å with ^6Li having the longest wavelength (see Table 1 of Andersen et al. 1984). Hence, the stronger doublet component of ^6Li is superimposed on the weaker ^7Li component. Due to the various line broadening effects in a stellar atmosphere a skew but rather smooth profile results. Fig. 1 shows a model-atmosphere calculation of the Li line in HD 160617 for three values of $^6Li/^7Li$. As seen, a change of the ^6Li abundance has two effects: i) a shift of the center-of-gravity (cog) wavelength of the Li line amounting to 15 mÅ when $^6Li/^7Li$ is increased from 0.0 to 0.1, and ii) an increase of the FWHM of the line, which amounts to 18 mÅ for the same increase of $^6Li/^7Li$. Thus, we have essentially two methods to determine the $^6Li/^7Li$ ratio: The wavelength method and the profile method.

2.1 The Wavelength Method

When measuring the cog-wavelength of the Li line we must reduce the observed wavelength for the Doppler shift due to the radial velocity of the star and the gravitational redshift. This Doppler shift can be determined from nearby metallic absorption lines. For the most metal-poor stars ($[Fe/H] < -2.0$) only two lines are available: an Fe I line at 6677.987 Å and a Ca I line at 6717.677 Å. For stars with $[Fe/H] \simeq -1.0$ Fe I lines at 6703.567, 6705.102 and 6726.667 Å may also be used. In metal-poor stars these lines are practically unblended. Accurate wavelengths (± 2 mÅ) of the Fe I lines lines have recently been measured by Nave et al. (1994). Furthermore, M. Rosberg and S. Johansson, Lund, have kindly measured a similar accurate wavelength of the Ca I line. In principle it should therefore be possible to determine the cog-wavelength of the Li I line to an accuracy of say ± 2 mÅ, which would correspond to about ± 0.01 in $^6Li/^7Li$. However, as discussed in detail by Nissen et al. (1994a), the Doppler shift determined from lines with high and low excitation potential lines differ by up to 6 mÅ (0.3 $km\,s^{-1}$), which translates to an error of 0.04 in $^6Li/^7Li$. These small differences in the Doppler shift are probably caused by convective motions in the stellar atmosphere. Similar systematic differences are seen in the solar spectrum (Dravins et al. 1981), and have been explained by a hydrodynamical model in which hot, rising bright granules are balanced by a downflow in darker (cooler) inter-granular regions. The result is a convective blueshift of the lines, which is more pronounced for the weak high excitation lines because they are formed deep in the atmosphere where the convective motions are more vigorous.

We conclude that until the convective blueshift of lines in the spectra of metal-poor stars have been mapped as a function of line strength and excitation potential we cannot use the cog-wavelength of the Li I resonance line to determine very accurate values of the lithium isotope ratio. For the time being the profile method appears more accurate.

2.2 The Profile Method

In this method the observed profile of the Li I line is compared with synthetic, model-atmosphere spectra computed for various values of $^6Li/^7Li$ (see Fig. 1). The basic problem is that the broadening of the Li I line due to stellar rotation and macro-turbulent motions in the atmosphere must be determined from other lines in the spectrum. As an example we briefly discuss HD 160617, that was recently observed with the ESO CES (Nissen et al. 1994a) in the spectral region 6676 - 6734 Å with a resolution of $R = 115.000$ and a S/N = 400. Synthetic spectra were computed by the aid of a programme *BSYN* kindly made available by the stellar atmosphere group in Uppsala. The model atmosphere used has parameters, $T_{eff} = 5800\,K$, $\log g = 3.8$ and $[Fe/H] = -2.0$. The model atmosphere computation includes thermal and microturbulent broadening with $\xi_{turb} = 1.7\,km\,s^{-1}$ as determined by Nissen et al. (1994b). In order to determine the broadening due to the combined effects of the instrumental profile and stellar rotation and macro-turbulence the synthetic spectrum was convolved by various broadening functions and compared to the observed profile of the Fe I line at 6678 Å. It turns out that pure rotational broadening does not lead to a satisfactory agreement between the observed and the synthetic profile. The best fit is obtained with a so-called radial-tangential profile (Gray 1976), which corresponds to radial and tangential motions in the atmosphere each with a Gaussian distribution of the velocities. The radial-tangential profile is more V-shaped than the U-shaped profiles corresponding to pure rotation or isotropic Gaussian broadening.

Fig. 2 shows the resulting fit of the Fe I line for three values of the velocity broadening parameter $\zeta_{RT} = 4.3$, 5.0 and 5.7 $km\,s^{-1}$. The equivalent width is the same for the observed and the synthetic lines. ¿From the comparison we determine $\zeta_{RT} = 5.0\pm0.3\ km\,s^{-1}$. Very nearly the same value is determined from the Ca I line at 6717 Å. The fit is, however, not perfect. A slight asymmetry is seen in the wings of the Fe I line - probably caused by the convective motions in the stellar atmosphere.

In Fig. 1 the observed profile of the Li I line of HD 160617 is compared to three synthetic profiles corresponding to $^6Li/^7Li = 0.0$, 0.1 and 0.2, respectively, all folded with a radial-tangential broadening function having $\zeta_{RT} = 5.0\ km\,s^{-1}$ as determined from the analysis of the Fe I line. As seen, the profile corresponding to $^6Li/^7Li = 0.0$ gives a nearly perfect fit to the data. Hence, there is no evidence of the presence of ^6Li in HD 160617. ¿From Fig. 1 an upper limit $^6Li/^7Li < 0.02$ can be set.

3 Determinations of the Li Isotope Ratio

3.1 Previous Attempts

Herbig (1964) and Feast (1966 and 1970) were apparently the first to study the lithium isotope ratio in bright solar-type stars. High-dispersion, photographic spectra were analyzed with the wavelength method. Feast concluded that the results were consistent with an upper limit of about 0.5 for the $^6Li/^7Li$ ratio, and that this high value was in fact reached in subgiants. Cohen (1972), however, showed that the upper limit of $^6Li/^7Li$ was 0.1 in 14 bright F, G and K stars, which she observed with a Fabry-Perot scanner at a resolution of 70.000 and a S/N of about 50.

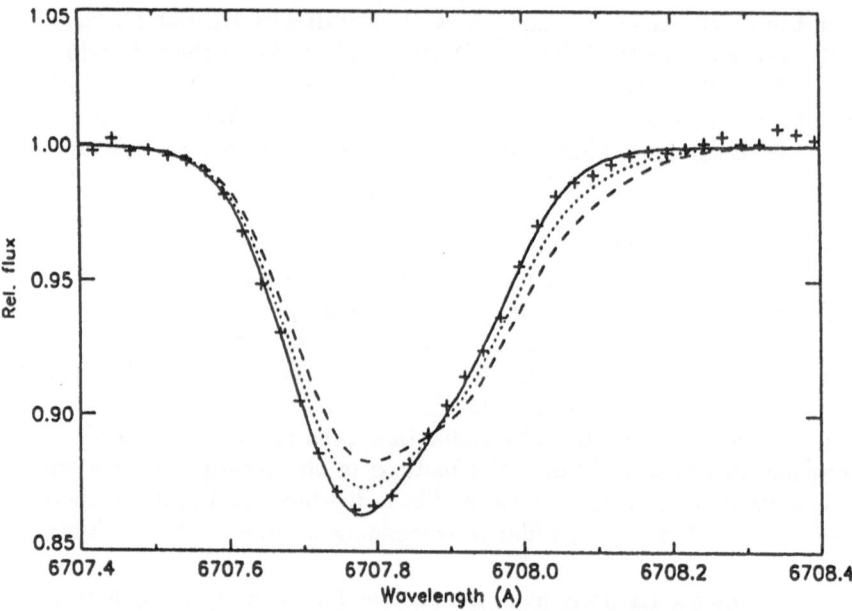

Fig. 1. The observed profile (+) of the Li I line of HD 160617 compared with synthetic model-atmosphere profiles convolved with a radial-tangential broadening function with $\zeta_{RT} = 5.0\ km\,s^{-1}$ and corresponding to $^6Li/^7Li = 0.0$, 0.1 and 0.2, respectively

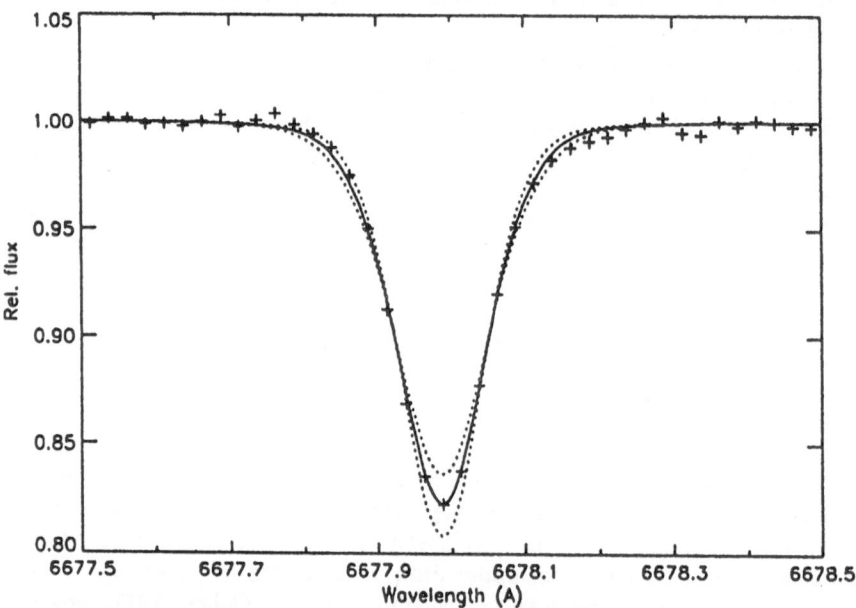

Fig. 2. The observed profile (+) of the Fe I line of HD 160617 compared with a synthetic model-atmosphere profile convolved with a radial-tangential broadening function with $\zeta_{RT} = 4.3$, 5.0 and 5.7 $km\,s^{-1}$, respectively

Following the construction of new effective, high resolution echelle spectrometers equipped with sensitive, linear detectors a number of important studies of the $^6Li/^7Li$ ratio appeared. Andersen et al. (1984) used the ESO CES to set an upper limit of $^6Li/^7Li < 0.1$ for 8 solar-type dwarfs. Maurice et al. (1984) used the same spectrometer to study four metal-deficient stars. The upper limit for two old disk stars were $^6Li/^7Li < 0.05$, whereas the limit for two halo stars (ν Ind and HD 76932) were $^6Li/^7Li < 0.1$. Hobbs (1985) using the coudé spectrograph of the 2.7 m telescope at the McDonald Observatory also found an upper limit $^6Li/^7Li < 0.1$ in a study of 5 F and G dwarfs. Finally, Pilachowski et al. (1989) made a careful analysis of the metal-poor, turnoff star HD 84937 and derived the limit $^6Li/^7Li < 0.1$. All of these results were based on the profile method.

3.2 Recent Studies of HD 19445 and HD 84937

The studies just mentioned were based on spectra of the Li I line having a resolution of about 100,000 and a S/N of 100 to 200, except in the case of Maurice et al. (1984), who used spectra with S/N $\simeq$ 300 but a resolution of about 50,000 only. In an attempt to reach a higher accuracy in the determination of the lithium isotope ratio Smith, Lambert & Nissen (1993) (hereafter SLN) used the McDonald Observatory's 2.7 m reflector and coudé spectrometer to obtain spectra with S/N $\simeq$ 400 and $R = 125,000$ for two stars, HD 19445 and HD 84937. The atmospheric parameters are $(T_{eff}, \log g, [Fe/H]) = (5820, 4.6, -2.2)$ for HD 19445 and $(6090, 4.0, -2.4)$ for HD 84937 according to SLN. Hence, as discussed by SLN, HD 19445 is a main-sequence star with a mass of about 0.7 $M_\odot$, for which one expects nearly complete depletion of ^6Li, whereas most of the ^6Li has survived in HD 84937 according to standard stellar models.

SLN derived the $^6Li/^7Li$ ratio in the two stars by the profile method. The velocity broadening was determined from the Ca I 6162 Å line. In Fig. 3 the fit to the Li I line is shown for the two stars. The total abundance of Li was chosen so that the equivalent widths of the synthetic and the observed lines are equal, and the Doppler shift was set to get the best overall fit. Using the sum of (observed − synthesis) residuals across the Li I line SLN got the following results:

$$HD\ 19445: {}^6Li/^7Li = 0.00 \pm 0.02$$

$$HD\ 84937: {}^6Li/^7Li = 0.05 \pm 0.02$$

The main contribution to the quoted (one sigma) error arises from the noise in the observed spectrum and the error in the determination of the velocity broadening parameter ($\pm 0.4\,km\,s^{-1}$). As discussed by SLN errors in the model atmosphere parameters and the microturbulence do not affect the derived Li isotope ratio significantly. The same conclusion was reached by Andersen et al. (1984) and Maurice et al. (1984).

Recently, Hobbs & Thorburn (1994) have studied the lithium isotope ratio in six halo stars including HD 19445 and HD 84937. New observations were obtained with the same equipment as used by SLN. This independent study by Hobbs & Thorburn resulted in $^6Li/^7Li \leq 0.04$ for HD 19445 and $^6Li/^7Li = 0.07 \pm 0.03$ for

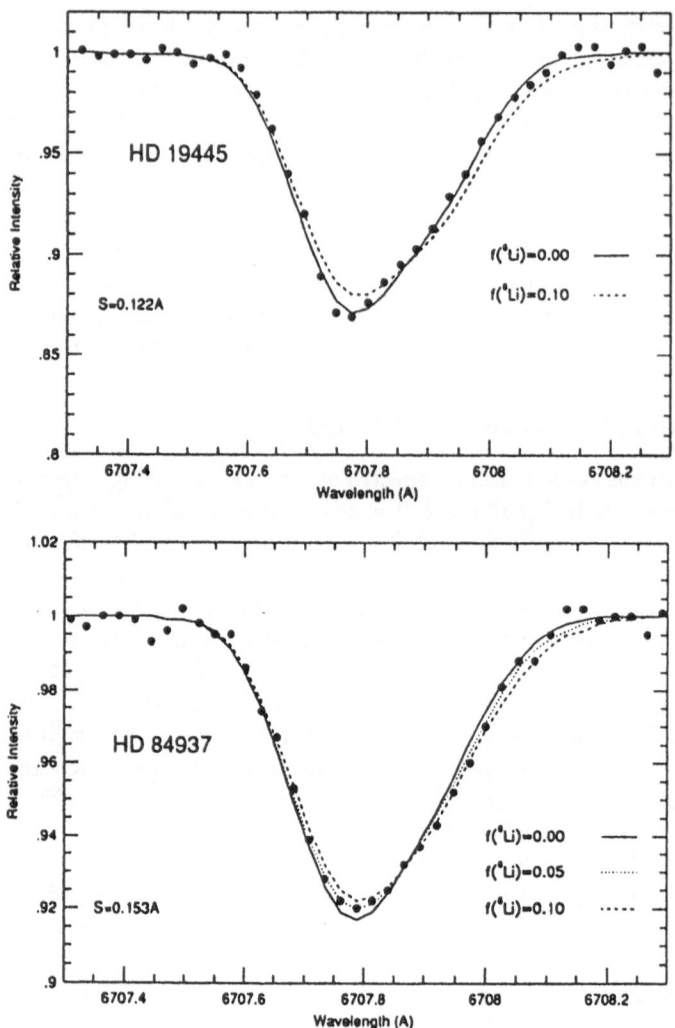

Fig. 3. The observed profiles (•) of the Li I line of HD 19445 and HD 84937 compared with synthetic model atmosphere profiles for various $^6Li/^7Li$ ratios. S is the FWHM of the Gaussian smoothing function as determined from the observed profile of the Ca I 6162 Å line. See Smith, Lambert & Nissen (1993) for details

HD 84937. When combined with SLN the detection of ^6Li for HD 84937 seems highly significant from a statistical point of view. The most critical assumption is that the velocity broadening profile determined from the Ca I line at 6162 Å is also valid for the Li I line. The Ca I line is slightly stronger and has a higher excitation potential than the Li I line. Hence, the depth of formation in the atmosphere is not exactly the same for the two lines. Clearly, this problem should be further studied by determining the velocity broadening profile for many lines with different strength and excitation potential.

The best argument for the reality of the detection of ^6Li in HD 84937 is that a difference $\triangle(^6Li/^7Li) = 0.05 \pm 0.02$ relative to HD 19445 is found. The two stars were observed and analyzed in exactly the same way and they have similar model atmosphere parameters. One would therefore expect that a possible systematic error in the derived $^6Li/^7Li$ would be the same for the two stars. HD 19445 so to say sets the zero point of the method.

4 Discussion and Conclusions

As discussed in detail by Steigman et al. (1993) the presence of ^6Li in the atmosphere of HD 84937 at a level of 5% of the ^7Li abundance is consistent with the measured Be abundance (Boesgaard & King 1993) in this star within the context of i) Standard Big Bang nucleosynthesis, ii) Pop. II cosmic ray nucleosynthesis and iii) standard (non-rotating) models for Li depletion. In particular, Steigman et al. derive $D_6 > 0.2$, where D_6 is the depletion factor for ^6Li. As shown by Chaboyer (1994) standard stellar evolution models with new opacities predict $D_6 \simeq 0.4$ for turnoff stars and subgiants with $T_{eff} > 5900\,K$, whereas $D_6 < 0.01$ for a main-sequence star like HD 19445 with $T_{eff} \simeq 5800\,K$. The same models predict $D_7 \simeq 1.0$, i.e. no ^7Li depletion for both main sequence and subgiants with $T_{eff} \geq 5800\,K$.

Non-standard models with rotational induced mixing predicting a strong ^7Li depletion ($D_7 \simeq 0.1$) (Pinsonneault et al. 1992) seem to be excluded by the detection of ^6Li in HD 84937, because the same models predict a very severe ^6Li depletion ($D_6 < 0.01$).

It remains to be seen if models predicting a mild ^7Li depletion ($D_7 \simeq 0.5$), e.g. the diffusion models of Chaboyer & Demarque (1994), are consistent with the presence of ^6Li in HD 84937, but in any case we conclude that the degree of ^7Li depletion in Pop. II stars is severely constrained by the amount of ^6Li detected in their atmospheres. This has important implications for Big Bang nucleosynthesis; e.g. inhomogeneous Big Bang models predicting $\log \varepsilon(Li) \simeq 3.0$ can probably be excluded.

In addition to HD 19445 and HD 84937, Hobbs & Thorburn (1994) studied four other Pop. II stars. For two of them an upper limit $^6Li/^7Li < 0.1$ could be set. For HD 140283, a subgiant with $T_{eff} = 5750\,K$ and $[Fe/H] = -2.5$, the limit is $^6Li/^7Li \leq 0.03$. For the fourth star, HD 201891, a detection ($^6Li/^7Li = 0.05 \pm 0.02$) is claimed. According to Hobbs and Thorburn this star is a subgiant with $T_{eff} = 5900\,K$ and $[Fe/H] = -1.5$. It is, however, puzzling that ^7Li in HD 201891 ($\log \varepsilon(Li) = 1.90$) appears to be depleted relative to the plateau value ($\log \varepsilon(Li) \simeq 2.2$). One would therefore expect ^6Li to be totally depleted. We also note that Edvardsson et al. (1993) find HD 201891 to be a main-sequence star with $[Fe/H] = -1.1$ for which one would expect a high degree of ^6Li depletion.

Smith, Lambert & Nissen have continued observations of the Li I line with the ESO CES and the McDonald spectrometer. An example of the spectra obtained was discussed in Sect. 2. Preliminary results indicate that the abundance of ^6Li

in subgiants with $T_{eff} < 5900$ is less than in the turnoff star HD 84937. This agrees with the limit $^6Li/^7Li \leq 0.03$ for HD 140283 found by Hobbs & Thorburn (1994). A rapid decline of ^6Li for $T_{eff} < 5900$ K due to dilution in connection with the deepening of the convection zone as a star evolves along the subgiant branch is in fact predicted by the standard models of Chaboyer (1994).

In connection with the discussion of depletion coefficients predicted from stellar evolution models it is important to know if the stars observed are subgiants or main sequence stars. According to their positions in the Strömgren $(b - y) - c_1$ diagram (Schuster & Nissen, 1989), HD 140283 and HD 160617 are subgiants and HD 84937 is very close to the turnoff point. As discussed by Chaboyer (1994), the parallax of HD 84937 places the star on the main sequence. However, the parallax is uncertain and the case of HD 140283 shows that parallaxes are sometimes affected by large systematic errors (see discussion by Nissen et al. 1994b). Clearly, accurate HIPPARCOS parallaxes will be very valuable for the discussion of the evolutionary status of the stars for which ^6Li has been studied.

We conclude that further studies of the ^6Li abundances in metal-poor Pop. II stars are much needed to confirm that turnoff stars and the hottest subgiants do indeed have a small amount of ^6Li in their atmospheres. Determinations of $^6Li/^7Li$ in somewhat more metal-rich Pop. II stars would also be very interesting. As discussed by Thorburn (1994), one would expect that $^6Li/^7Li$ increases with increasing metallicity. Pop. II turnoff stars are however rare. Only one (HD 84937) is brighter than $m_V = 9$. In order to reach the fainter turnoff stars with $R > 100,000$ and S/N > 400 very large telescopes equipped with high quality spectrometers are needed.

References

Andersen, J., Gustafsson, B., Lambert, D.L. 1984, A&A 136, 65

Boesgaard, A.M., King, J.R. 1993, AJ 106, 2309

Chaboyer, B.C. 1994, ApJ Letters (submitted)

Chaboyer, B.C., Demarque, P. 1994, ApJ (in press)

Cohen, J.G. 1972, ApJ 171, 71

Dravins, D., Lindegren, L., Nordlund, Å. 1981, A&A 96, 345

Duncan, D.K., Lambert, D.L., Lemke, M. 1992, ApJ 401, 584

Edvardsson, B., Andersen, J., Gustafsson, B., Lambert, D.L., Nissen, P.E., Tomkin, J. 1993, A&A 275, 101

Edvardsson, B., Gustafsson, B., Johansson, S.G., Kiselman, D., Lambert, D.L., Nissen, P.E., Gilmore, G. 1994, A&A (in press)

Feast, M.W. 1966, MNRAS 134, 321

Feast, M.W. 1970, MNRAS 148, 489

Gilmore, G., Gustafsson, B., Edvardsson, B., Nissen, P.E. 1992, Nature 357, 379

Gray, D.F. 1976, "The Observation and Analysis of Stellar Photospheres", John Wiley and Sons, p. 426

Herbig, G.H. 1964, ApJ 140, 702

Hobbs, L.M. 1985, ApJ 290, 284

Hobbs, L.M., Thorburn, J.A. 1994, ApJ 428, L25

Kiselman, D. 1994, A&A 286, 169

Maurice, E., Spite, F., Spite, M. 1984, A&A 132, 278

Meissner, K.W., Mundie, L.G., Stelson, P.H. 1948, Phys. Rev. 74, 932

Nave, G., Johansson, S., Learner, R.C.M., Thorne, A.P., Brault, J.W. 1994, ApJS (in press)

Nissen, P.E., Lambert, D.L., Smith, V.V. 1994a, The ESO Messenger 76, 36

Nissen, P.E., Gustafsson, B., Edvardsson, B., Gilmore, G. 1994b, A&A 285, 440

Pilachowski, C.A., Hobbs, L.M., De Young, D.S. 1989, ApJ 345, L39

Pinsonneault, M.H., Deliyannis, C.P., Demarque, P. 1992, ApJS 78, 179

Ryan, S., Norris, J., Bessell, M.S., Deliyannis, C.P. 1992, ApJ 388, 184

Schuster, W.J., Nissen, P.E. 1989, A&A 222, 69

Smith, V.V., Lambert, D.L., Nissen, P.E. 1993, ApJ 408, 262 (SLN)

Steigman, G., Fields, B.D., Olive, K.A., Schramm, D.N, Walker, T.P. 1993, ApJ 415, L35

Thorburn, J.A. 1994, ApJ 421, 318

Constraints on the Galactic Evolution of the Li Abundance from the ^{7}Li/^{6}Li Ratio

C. Abia[1], J. Isern[2], R. Canal[3]

[1] Dpto. Física Teórica y del Cosmos, Universidad de Granada (Spain).
[2] Centre d'Estudis Avançats CSIC, Blanes (Spain).
[3] Dpto. Física de la Atmósfera, Astronomía y Astrofísica, Universidad dei Barcelona (Spain).

1 Introduction

Our knowledge concerning the evolution of the lithium abundance in the galaxy has been extended in recent years. Smith et al. (1993), for the first time, claim the detection of ^{6}Li in the atmosphere of a metal-poor star (HD 84937, [Fe/H]= -2.4), ^{7}Li/^{6}Li= 20 ± 8. This limits the amount of Li produced by s-pallation reactions in the ISM during the early galaxy and, at the same time, strongly constrains the theories of Li depletion in main-sequence metal deficient stars. From this observational data one might conclude that Li probably has not been destroyed in the atmosphere of metal-poor dwarfs by more than a factor of 2 and that the contribution by spallation reactions to the Li abundance observed in halo stars (Li/H$\approx 2 \times 10^{-10}$) was no more than 10%. In consequence, the abundance level of the *lithium plateau* in halo stars probably represents the pregalactic abundance of this element. On the other hand, Meyer et al. (1993) and Lemoine et al. (1993) recently derived the ^{7}Li/^{6}Li ratio in the ISM toward the line of sight of several bright stars: 6.8(+1.4, -1.7) toward ζ Oph, 5.5(+1.3, -1.1) toward ζ Per and 12.5(+4.3, -3.4) toward ρ Oph, respectively. These results, taken together with the Solar System meteoritic ratio (^{7}Li/^{6}Li)$_\odot = 12.5 \pm 0.1$, imply that the ^{7}Li/^{6}Li ratio has remained nearly constant or even decreased in the last ~ 4.5 Gyr. This has crucial implications on the evolution of the lithium isotopic abundances, as we will try to show. The aim of this short contribution is to understand the observations quoted above on the grounds of the nucleosynthetic mechanism currently believed to be responsible for the origin of the Li production in the galaxy.

2 The Model

We constructed a simple model of chemical evolution for the solar neighbourhood in order to study the evolution of the lithium isotopic abundances. The model parameters and approximations used are essentially those made in Bravo et al. (1993). Concerning the lithium sources we assume that both isotopes are produced in the ISM mainly by $\alpha + \alpha$ fusion reactions. The production rate,

$Q^i(t)$, is calculated following the model of Meneguzzi et al. (1971) with a cosmic ray (CR) spectrum $\phi(E) \approx E^{-2.6}$ ($E > 100$ Mev/nucl). This rate is modulated in time by a factor $\sigma(t)/\sigma(T)$, where σ is the surface gas density and $T \equiv 13$ Gyr is the current age of the galaxy. For the stellar production of ^{7}Li we assume that AGB stars are the sole stellar source of lithium in the galaxy. Note that these stars constitute, as far as we know, the only observational evidence of Li production in stars. The production of ^{7}Li in novae and supernovae is ruled-out according to the lastest theoretical calculations (Boffin et al. 1993; Timmes et al. 1994). The production rate of ^{7}Li by AGB stars, $S_*^7(t)$, is calculated as by Abia et al. (1993) but considering their estimated yield as a lower limit. That is because: a) Only C-stars are considered as Li producers. However, some S-stars, which are also AGB stars, show strong Li lines in their spectra. b) The C-stars sample on which the calculations are based is probably not complete (see Abia et al. 1993 for details). Finally, we adopt a primordial abundance $(^7\text{Li/H})_p \approx 2 \times 10^{-10}$.

Figure 1 shows the evolution of the Li/H vs. [Fe/H] obtained assuming only CR production (solid line) and adding to that the ^{7}Li production by AGB stars in the hypothesis of a mass range $1.2 \leq M/M_\odot \leq 3$ for their progenitors (dashed line). Figure 2 shows the corresponding evolution of the ^{7}Li/^{6}Li ratio for these two assumptions. ¿From these figures we can conclude: i) A source of ^{7}Li (in this case AGB stars) is needed in order to explain the observed evolution of Li/H vs. [Fe/H] relationship and the $(^7\text{Li}/^6\text{Li})_\odot$ ratio. ii) However, as a consequence it is difficult to maintain constant (or disminish) the ^{7}Li/^{6}Li ratio at the level of ~ 12.5 during the last ~ 4.5 Gyr. The present ^{7}Li/^{6}Li ratio obtained in our best model is ~ 20. Several alternatives are possible in order to solve this problem: p 1) On the hypothesis of the existence of an average value of the ^{7}Li/^{6}Li ratio in the galaxy, the actual ratio in the ISM is unknown. Thus, the recent observations in the ISM are probably chemical inhomogeneities (see Lemoine in this volume).

2) The contrary solution is to believe that the Solar System $(^7\text{Li}^6\text{Li})_\odot$ ratio is not representative of the ISM ~ 4.5 Gyr ago. In this case, it is possible to fit the present observations by assuming, for instance, a lower range $(1.2 \leq M/M_\odot \leq 1.7)$ for the AGB progenitor's stellar mass (Figure 2, dotted line). p 3) A source of ^{6}Li besides CR exists in the galaxy. The observational fact of $d(^7Li/^6Li)/dt \approx 0$ during the last ~ 4.5 Gyr implies,

$$\frac{dX^7}{dt} \approx \left(\frac{X^7}{X^6}\right)\frac{dX^6}{dt} \tag{1}$$

where $X^{6,7}$ is the abundance of ^{6,7}Li by mass in the ISM respectively, thus $X^7/X^6 \approx 10.7$. Under this hypothesis, it is straightforward to estimate the current production rate of ^{6}Li , $S_*^6(t)$, needed in order to maintain $^7\text{Li}/^6\text{Li} \approx 12$. Stating the evolutionary equations which govern the evolution of $X^{6,7}$ in the ISM and taking into account that the relation between the production rate by spallation reactions of ^{6,7}Li is $Q^7 \approx 1.4Q^6$, from (1) it is easy to obtain an expression for S_*^6,

$$S_*^6(t) = \frac{X^6}{X^7}\left[S_*^7(t) + X_p^7 f(t)\right] + \sigma_o Q^6(t)\left[1.4A^7\frac{X^6}{X^7} - A^6\right] \quad M_\odot pc^{-2}Gyr^{-1} \tag{2}$$

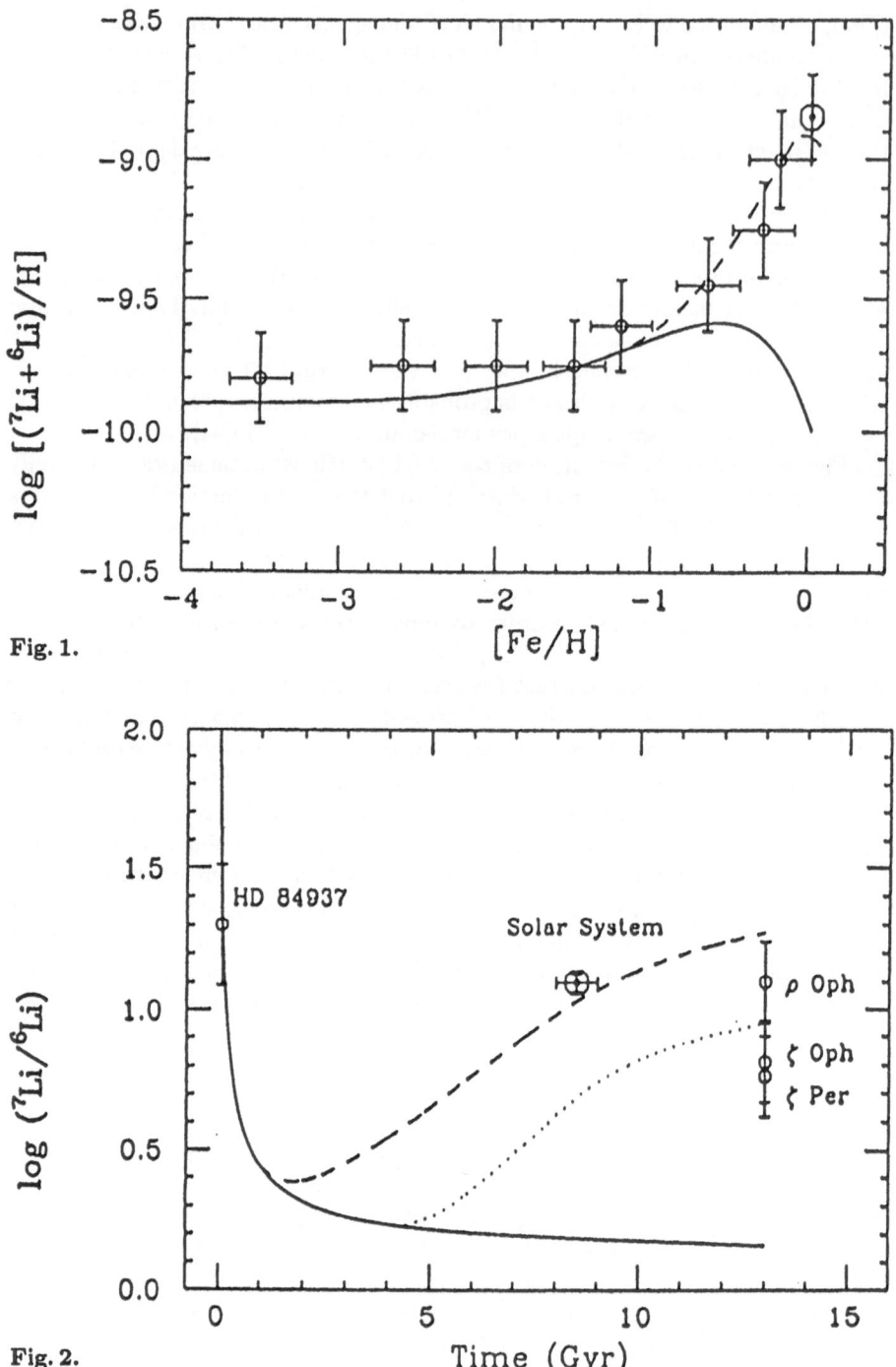

Fig. 1.

Fig. 2.

where X_p^7 is the primordial abundance by mass of ^{7}Li, $f(t)$ the infall rate, $\sigma_o \approx 6$ $M_\odot pc^{-2}$ the present surface gas density in the model and $A^{6,7}$ the atomic mass of both isotopes. Taking the present values for $S_*^7 \approx 5 \times 10^{-8}$ $M_\odot pc^{-2}$ Gyr^{-1} (Abia et al. 1994), $f \approx 0.1$ $M_\odot pc^{-2}Gyr^{-1}$ (Mirabel & Morras 1984) and $Q^6 \approx 8 \times 10^{-11}$ Gyr^{-1}, we obtain $S_*^6 \approx 2 \times 10^{-9}$ $M_\odot pc^{-2}Gyr^{-1}$. However, since it is difficult to understand a source of ^{6}Li without the simultaneous production of ^{7}Li, the value of S_*^6 estimated above should be a lower limit. Current models of AGB stars ruled out the production of ^{6}Li in these stars. Other scenarios must be sought: A low energy component ($E < 100$ Mev/nucl) in the CR spectrum? Spallation reactions in the accretion disk around compact objects? (see Reeves and Rebolo et al. in this volume for a discussion of this).

Abia, C., Isern, J., Canal, R. 1993, A&A 275, 96.

——————— 1994, A&A (submitted)

Boffin, H.M.J., Paulus, G., Arnould, M., Mowlavi, N. 1993, A&A 279, 173.

Bravo, E., Isern, J., Canal, R. 1993, A&A 270, 188.

Lemoine, M., Ferlet, R., Vidal-Madjar, A., Emerich, C., Bertin, P. 1993, A&A 269, 469.

Meneguzzi, M., Audouze, J., Reeves, H. 1971, A&A 15, 337.

Mirabel, I.F., Morras, R. 1984, ApJ 279, 86.

Meyer, D.M., Hawkins, I., Wright, E.L. 1993, ApJ 409, L62.

Smith, V.V., Lambert, D.L., Nissen, P.E. 1993, ApJ 408, 262.

Timmes, F.X., Woosley, S.E., Weaver, T.A. 1994, ApJ (preprint)

The Interstellar ^{7}Li/^{6}Li Ratio

Martin Lemoine

Institut d'Astrophysique de Paris, CNRS, 98 bis boulevard Arago, 75014 Paris, France

Abstract. I discuss the observational status of the interstellar lithium isotopic ratio and its significance with respect to the galactic evolution of lithium.

1   Introduction

This talk discusses the observational status of the lithium isotopic ratio in the interstellar medium (ISM). Although only four such measurements have been performed up to now, this discussion justifies itself in that the value of the interstellar ^{7}Li/^{6}Li ratio may have some drastic consequences on the galactic evolution of lithium. It will be shown furthermore that the published values of the interstellar ^{7}Li/^{6}Li ratio cannot be considered as representative values for the ISM. In a first part, I will therefore review, discuss, criticize and, where possible, correct each of those measurements in order to draw a preliminary observational status of this quantity. I will then present the analysis and the results of new high quality observations of this ratio on a previously observed line of sight, and discuss the extreme consequences of the atypical values derived. A complete observational status of the interstellar ^{7}Li/^{6}Li ratio and its significance with respect to the galactic evolution of lithium will be given as a conclusion. I will not discuss the beautiful measurements of the ^{7}Li/^{6}Li ratio performed in the photospheres of metal-deficient Population II stars, and I refer the reader to the talk of Nissen in these proceedings for further interest. Before proceeding with this review, let us outline briefly the purpose and basic significance of interstellar ^{7}Li/^{6}Li ratio measurements.

Lithium–7 is now generally accepted to originate in the hot Big Bang primordial nucleosynthesis (BBN), with a primordial abundance $(^7\text{Li/H}) \simeq 10^{-10}$, in excellent agreement with the observed uniformity of the ^{7}Li abundance in very metal deficient Pop II stars $(^7\text{Li/H}) = 1.4 \times 10^{-10}$ (Spite, these proceedings). During the galactic evolution, both lithium isotopes are created by spallation reactions of galactic cosmic rays (GCR) interacting with the ISM, that yield $(^7\text{Li/H}) \simeq 2. \times 10^{-10}$ in 10 Gyrs, with a production ratio $(^7\text{Li/}^6\text{Li})_{GCR} = 1.4$ (Meneguzzi et al., 1975; Reeves, 1994). The major problem in understanding the evolution of lithium in the Galaxy is to explain the observed Pop I abundance of ^{7}Li, $(^7\text{Li/H})_{PopI} \sim 10^{-9}$, of which only 30% is accounted for by BBN and GCR spallation mechanisms, as well as the high ^{7}Li/^{6}Li ratio measured in meteorites, representative of the solar system formation epoch 4.6 Gyrs ago, $(^7\text{Li/}^6\text{Li})_\odot = 12.3$, whereas the above mechanisms predict a ratio around 2. In order to account for these two quantities, the existence of an extra source of

^{7}Li of stellar origin has been suggested, AGB C and S stars being the best candidates as they are observed to be super Li–rich (see Abia, these proceedings, for the chemical evolution of lithium, and Mowlawi, these proceedings, for the production of ^{7}Li in AGB stars). It was argued by Reeves (1993) from the comparison of the meteoritic $(^7$Li$/^6$Li$)_\odot$ ratio and the production rates ratio in GCR spallation reactions that GCR spallation alone tend to decrease the ^{7}Li$/^6$Li ratio with time, and that, starting 4.6 Gyrs ago with a ratio ^{7}Li$/^6$Li $=12.3$, one should observe today an interstellar ratio ^{7}Li$/^6$Li $\simeq 5$–6 without production of ^{7}Li in stars, or ^{7}Li$/^6$Li $\gtrsim 6$ with stellar production of ^{7}Li. Measuring the interstellar ^{7}Li$/^6$Li ratio thus provides a key test for the model of galactic lithium evolution. If this ratio is found to be ^{7}Li$/^6$Li $\lesssim 5$, then another scenario would have to be considered; one way out would be to consider a primordial abundance $(^7$Li$/$H$) \gtrsim 10^{-9}$ together with some form of internal mixing that could very well reproduce the observed "Spite plateau" in Pop II stars (Pinseonneault et al., 1992) and some rotational mechanisms to reproduce via a gradual depletion of lithium in stars the observed abundances of ^{7}Li in stars of different metallicities (Vauclair, 1988). However, there is no obvious way of reproducing a primordial abundance as high as 10^{-9} since the inhomogeneous nucleosynthesis models involved up to a few years ago to yield such abundances no longer do (Reeves, 1994; Matthews et al., 1994).

2   Previous Measurements

The fact that there are so few measurements of such an important quantity as the interstellar ^{7}Li$/^6$Li ratio is simply due to the difficulty one has to face to detect the ^{6}Li isotope. The only accessible resonance lines of lithium form a transition doublet of ^{7}LiI at 6707.761–6707.912 Å, with a similar doublet for ^{6}LiI redshifted by 0.160 Å. In regards of the low abundance of lithium and the nearly complete ionisation of lithium to LiII in the ISM, the strongest equivalent widths observed for the 6708 Å line of ^{7}LiI are of a few mÅ on lines of sight already comprising a hydrogen column density well over 10^{20} cm^{-2}. The structure of the transition is such that the main component of the ^{6}LiI doublet is completely superimposed on the weaker component of the corresponding ^{7}LiI doublet. Therefore, the only absorbing line of ^{6}Li that one may hope to detect is the weaker one. Assuming a ratio ^{7}Li$/^6$Li $\simeq 10$, the oscillator strengths ratio of the doublet components being 2, the typical equivalent width of this line is $\sim 30 \,\mu$Å, which corresponds to a resolution element of 67 mÅ (i.e. a resolving power $\lambda/\Delta\lambda = 10^5$) lowered from the continuum by 0.04%... Moreover, on lines of sight showing a column density N(HI) $\gtrsim 10^{20}$ cm^{-2}, several interstellar components separated by $\sim$5–10 km.s^{-1} are often present so that the resulting absorption profile of the LiI transition doublet may be very complex (the isotopic shift of the doublets is 7.2 km.s^{-1}). A simulation of this resulting profile corresponding to our best fit solution for the ζ Oph line of sight is shown in Fig.1 as the solid line, the individual contributions for ^{7}Li and ^{6}Li of each of the two interstellar absorbing clouds being shown in dashed line. Therefore, in order to detect ^{6}Li and derive a

^{7}Li/^{6}Li ratio as accurate as possible, one needs to observe a bright target with a line of sight as dense as possible and whose velocity structure must be as trivial as possible. It is also necessary to observe hot stars so that the lithium region be not contaminated by stellar lines of other elements. Naturally, the number of such targets comes down to a few. One needs also to reduce the data with care in order to avoid any instrumental pollution of the profile. Finally, it is necessary to use sophisticated profile fitting methods in order to probe the observed profile for all the contributions shown in Fig.1, since neglecting one of them would mean neglecting a contribution of the order of the ^{6}Li absorption one is looking for.

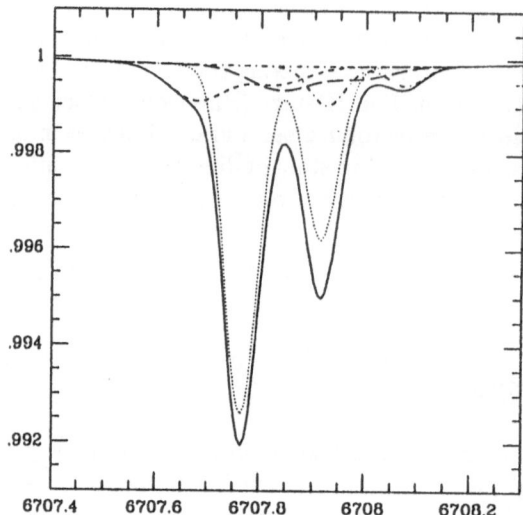

Fig. 1. Simulation of the observed profile of the λ6708 Å LiI line as taken from our best fit solution in the direction of ζ Oph (solid line). The individual contributions of the two absorbing clouds A and B are shown as: dotted line ^{7}Li$_A$, short dash ^{7}Li$_B$, dot–dash ^{6}Li$_A$, long dash ^{6}Li$_B$. Note that the profile in solid line would only be observed if the S/N and the sampling were already infinite.

Nevertheless, the first estimation of the interstellar ^{7}Li/^{6}Li ratio was tempted ten years ago by Ferlet & Dennefeld (1984, hereafter FD) in the direction of ζ Oph using the Coudé Echelle Spectrometer (CES) at the ESO 1.4m Coudé Auxiliary Telescope and a Reticon detector. Although they obtained high quality data with a resolving power $\lambda/\Delta\lambda=10^5$ and a signal-to-noise ratio per pixel S/N=4000, ^{6}Li was actually not detected for that time. I will therefore not discuss further the value obtained for the main absorbing cloud:

$$^7\text{Li}/^6\text{Li} \gtrsim 25 \ (\sim 38)$$

The first actual detection of ^{6}Li was reported by our group (Lemoine et al., 1993, hereafter L93) in the direction of ρ Oph. The observations were conducted at ESO using the 3.6m Telescope linked via fiber optics to the CES, providing a resolving power $\lambda/\Delta\lambda=10^5$ and a signal-to-noise ratio S/N~ 4000 per pixel of a CCD detector. The KI $\lambda7699\,\text{Å}$ line was also observed since LiI and KI are known to behave similarly in the ISM (White, 1986). The typical equivalent width of the KI line is also much stronger than that of the LiI line at $6708\,\text{Å}$, and thus allows to derive an accurate velocity structure of the line of sight. Two interstellar absorbing clouds, one main (A) and one weak (B) were detected, but only the $^7\text{Li}/^6\text{Li}$ ratio in cloud A was evaluated, yielding:

$$^7\text{Li}/^6\text{Li} = 12.5^{+4.3}_{-3.4} \ (2\sigma)$$

This value was immediately interpreted by Reeves (1993) as strong evidence for the existence of an extra source of ^7Li. The only critics I would address to this measurement is that, due to numerical complexity and to the weakness of the B component in ^7Li, the $^6\text{Li}_B$ contribution was neglected. Using a more sophisticated profile fitting algorithm (to be shortly discussed in 3.), we were able recently to re-analyze this line of sight, and the following ratios were derived taking into account all detected contributions:

$$\left(^7\text{Li}/^6\text{Li}\right)_A = 11.1$$

$$\left(^7\text{Li}/^6\text{Li}\right)_B \sim 3$$

The error bar on the $^7\text{Li}/^6\text{Li}$ $_A$ ratio is to remain as previously, i.e. ±2 at 1σ, but the $(^7\text{Li}/^6\text{Li})_B$ ratio is uncertain since $^6\text{Li}_B$ is not formally detected above the photon noise. The fit is shown in Fig.2, the corresponding χ^2 is 37.4/43 giving a level of confidence of 71%.

Meyer, Hawkins & Wright (1993, hereafter MHW) reported shortly thereafter $^7\text{Li}/^6\text{Li} =6.8^{+1.4}_{-1.7}$ toward ζ Oph, and $^7\text{Li}/^6\text{Li} =5.5^{+1.3}_{-1.1}$ toward ζ Per, values which cast severe doubt as to the existence of a stellar source of ^7Li, hence on the "canonical" scenario for the galactic evolution of lithium. The data were obtained at the KPNO 0.9m telescope at a high quality, with $\lambda/\Delta\lambda\simeq2$-$3\times10^5$ and S/N$\simeq$ 2000 per pixel. These results seem to us to be questionable for the following reason. It is obvious from the MHW spectra that at least two absorbing clouds are well detected in ^7Li toward each of the two targets, and only the main component was taken into account in the profile fitting in each case. It is in fact well known that toward ζ Oph and ζ Per, several interstellar components are indeed present. Recalling the discussion above, these values seem to be biased, and are at most average values of the $^7\text{Li}/^6\text{Li}$ ratio on the lines of sight, hence *a priori* not representative of the general ISM. It happens moreover that the new data we discuss in 3. were also obtained in the direction of ζ Oph. We derive a ratio $^7\text{Li}/^6\text{Li} \simeq9.8$ when looking for a one cloud solution, with a χ^2 of 303/13 and 103/58 for the KI line and the LiI line respectively, i.e. respective confidence levels of 0% and 0% because of the non-negligible presence of a second absorbing cloud. This result is in slight agreement with that derived by MHW albeit higher, suggesting that an average value for the $^7\text{Li}/^6\text{Li}$ ratio on this line sight would lie closer to 8-9 than to 6-7. For similar reasons, we are led to believe that the value derived toward ζ Per are not representative either of the $^7\text{Li}/^6\text{Li}$ ratios on

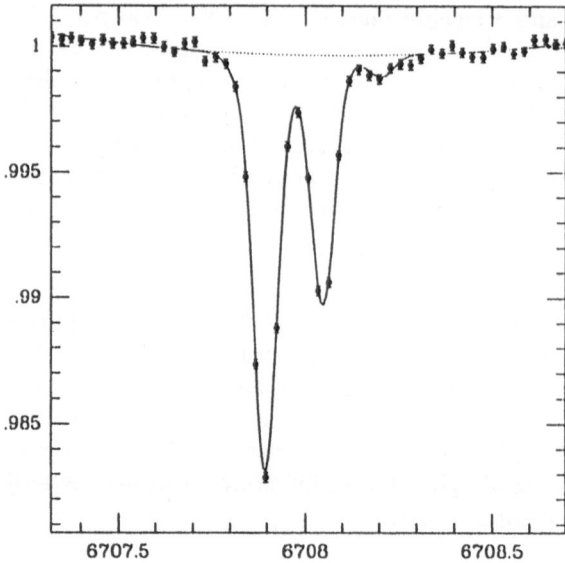

Fig. 2. Best fit solution for the ρ Oph line of sight including all ^{7}Li and ^{6}Li contributions for two interstellar absorbing components. Error bars are 1σ. The level of confidence of the fit is 71%.

this line of sight. These data are of a such quality that they would deserve being re-analyzed in more detail, as we did for ρ Oph.

In the end, among these four measurements of the interstellar ^{7}Li/^{6}Li ratio, it seems that only the values derived and corrected toward ρ Oph may be kept, i.e. ^{7}Li/^{6}Li $\simeq 11.1 \pm 2$, with an uncertain value ^{7}Li/^{6}Li ~ 3 to be confirmed or not. The values derived by MHW do not seem to be representative, and the data should be re-analyzed in more detail, as their high spectral resolution and high signal-to-noise ratio would for sure allow accurate estimations of probably two ^{7}Li/^{6}Li ratios on each of the two lines of sight.

3 New Data Toward ζ Oph

We obtained new data of the $\lambda 6708$ Å LiI line toward ζ Oph at the ESO 3.6m Telescope linked to the CES with fiber optics. ζ Oph had already been observed by FD and MHW, yielding two very different values for the ^{7}Li/^{6}Li ratio. Our data revealed to be of a higher quality, showing $\lambda/\Delta\lambda=10^5$ together with S/N=7500 per pixel. The data were carefully reduced using different approaches in order to estimate the importance of systematics on the LiI profile, hence on the ^{7}Li/^{6}Li ratios. Indeed, it was found that the noise is dominated by interference fringes remnants at a level of $\sim 10^{-4}$ *rms* and not by the photon noise, as it can be seen from the non-gaussian statistics of the continuum in the spectrum of Fig.3. The $\lambda 7699$ Å KI line was also observed in order to link our LiI observations with KI and with others. The spectra were normalized and the continua fitted using a cross-validation statistical criterion. In order to probe deeper the profiles for the different contributions (see Fig.1), a sophisticated profile fitting technique based on a simulated annealing minimization algorithm allowing to include all kinds of constraints on the minimization procedures, e.g. atomic and physical constraints on the parameters that define the profile, was developed. For further details on the reduction and analysis of this line of sight, the reader is referred to Lemoine et al. (1994). Two interstellar components were detected in KI and in LiI, one main (A) and one secondary (B). The limiting detectable equivalent width of the spectrum is $18\,\mu\text{Å}$, or $50\,\mu\text{Å}$ including systematics. We were thus able to derive two $^7\text{Li}/^6\text{Li}$ ratios:

$$\left(^7\text{Li}/^6\text{Li}\right)_A = 8.6 \pm 0.8\ (\pm 1.4)$$

$$\left(^7\text{Li}/^6\text{Li}\right)_B = 1.4\ ^{+1.2}_{-0.5}\ (\pm 0.6)$$

where the error bars within brackets are associated to systematics and were estimated by fitting different sets of spectra reduced using different techniques. In fact, these numbers constitute upper limits to the systematics as they somehow include statistical errors associated with the photon noise. The fit is shown in Fig.3, the corresponding χ^2 is $50.5/54$, or equivalently a level of confidence of 61%. In Fig.4 (Fig.5), I show the $^6\text{Li}_A$ ($^6\text{Li}_B$) doublet as calculated by subtracting to the observed profile all the calculated absorption except that of the $^6\text{Li}_A$ ($^6\text{Li}_B$) doublet. As toward ρ Oph, we find an extremely atypical $^7\text{Li}/^6\text{Li}$ ratio. Why is it that this ratio be so low?

It might be suggested for instance that the "strong" $^6\text{Li}_B$ contribution observed is due to an instrumental effect. A mimicking absorption feature, such as an interference fringe remnant as observed in emission on the blue and red sides of the doublet (see Fig.3), would indeed provide more $^6\text{Li}_B$ absorption than real and thus imply a low $^7\text{Li}/^6\text{Li}$ ratio. However, as it may be seen from Fig.5, the $^6\text{Li}_B$ doublet is detected above the noise, and appears at the right position with seemingly good oscillator strengths ratio and fine structure shift of the lines. This makes the presence of an extra instrumental feature very improbable. Other systematic effects, such as the contamination of the profile by cosmics, were carefully looked for in individual spectra, and none was found.

Another way to excuse this atypically low ratio would be to assume the presence of a high radial velocity cloud absorbing in ^7LiI at the position of the $^6\text{Li}_B$ ratio, which would mimic perfectly the doublet structure of the line. This may be checked by introducing in the observed KI profile an absorption redshifted so as to correspond to a $^7\text{LiI}_C$ doublet matching the $^6\text{LiI}_B$ absorption (see L93). This in fact partly possible from our KI spectrum since the redshifted absorption takes place in the red wing of the KI profile (see Fig.10b, Lemoine et al., 1994). This is not the case for cloud A, where a similar contamination of the $^6\text{Li}_A$ doublet is strictly ruled out. It is well known that many absorbing clouds are present on the ζ Oph line of sight, but as to whether these other absorbing clouds contribute in ^7LiI is another matter. Only extremely high resolution observations of the KI profile may help, and these were underway at the time of this workshop at the AAT; very preliminary results will be discussed in the epilogue. Although

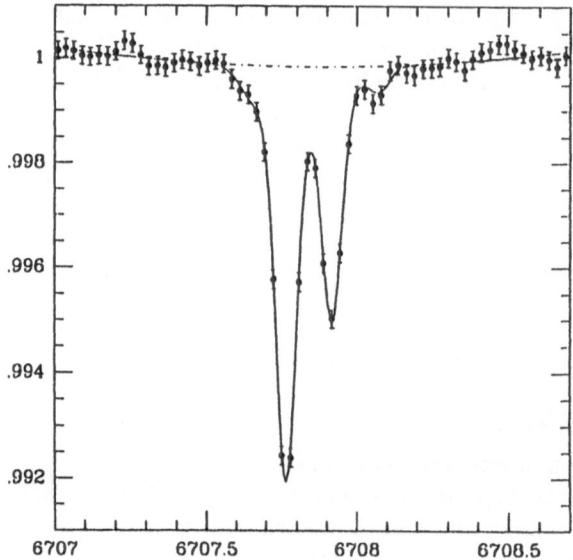

Fig. 3. Best fit solution for the ζ Oph line of sight, including two absorbing components. The level of confidence is now 61%.

this explanation is attractive, it would however not explain the atypical LiI/KI ratio observed in cloud B: it was shown that the LiI/KI ratio is rather constant around 3.10^{-3} in the ISM (White, 1986), as it is observed in cloud A for instance; the ratio measured in cloud B is $\simeq 0.03$. There is no reason why a third absorbing cloud would explain this abnormal ratio since the LiI column density in cloud B is derived mainly from the $^7\mathrm{Li}_B$ doublet, which could not be contaminated by the $^7\mathrm{Li}_C$ line. Finally, although our combination S/N–$\lambda/\Delta\lambda$ did not allow to look with confidence for more than two absorbing clouds, we nevertheless tried 3 cloud solutions in different configurations and in each case, there was no way to escape having one of the $^7\mathrm{Li}/^6\mathrm{Li}$ ratios around 2–3.

The only standard way to obtain a ratio as low as $\sim$2 in the ISM comes through a massive interaction of the GCR with the material of cloud B. Using the calculations of Reeves (1993), Steigman (1993), one may evaluate that a burst of GCR spallation over 10^6 yrs is required with an enhancement of the GCR flux by a factor $\zeta \sim 2 \times 10^4$ to reproduce a ratio of 2, and $\zeta \sim 2 \times 10^5$ for a ratio 1.5. Such an enhancement is so enormous that it would make ζ Oph a γ–ray source with a flux detectable above the present instrumental thresholds, which has not been reported as yet. Still, this would explain in a natural way the LiI/KI ratio measured in cloud B, as in this scenario one should expect to measure a ratio $(\mathrm{LiI/KI})_B \sim (1 + 10^{-4}\zeta)(\mathrm{LiI/KI})_A$, the factor 10^{-4} arising from the ratio of the duration of the burst to the age of the Galaxy. This scenario seems unrealistic as for now, but one should note that the calculations are extremely coarse in that they assume a magnification of the whole GCR proton flux spectrum, and not simply the presence of a low-energy excess for instance (Meneguzzi et al., 1975)

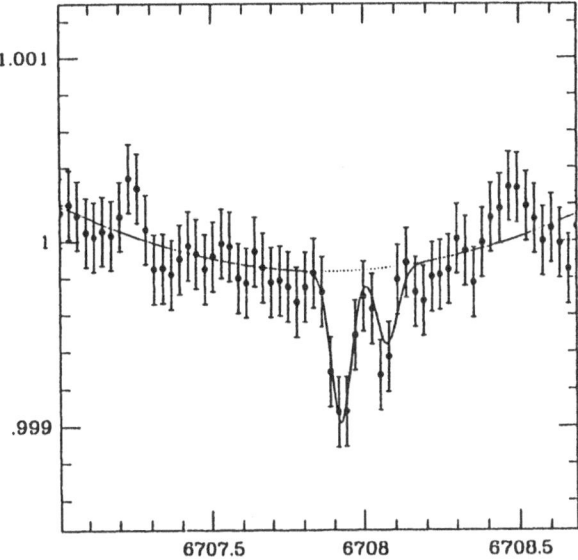

Fig. 4. After subtracting all calculated contributions to the observed profile of Fig.3 except that of the ^6Li_A doublet, this latter is shown here as residuals. The solid line shows the fit to this doublet calculated in Fig.3.

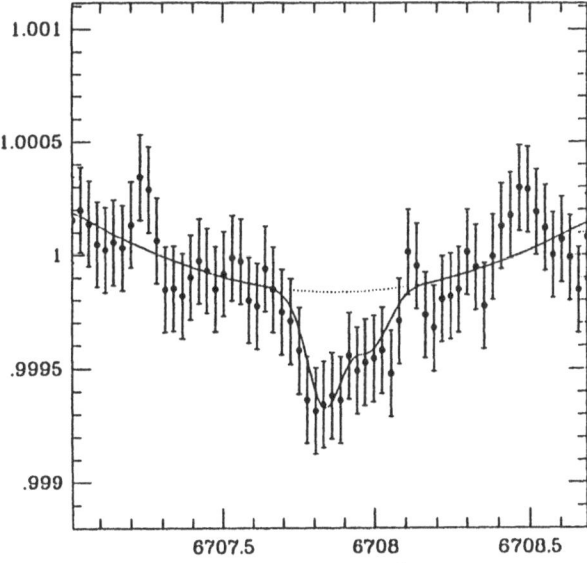

Fig. 5. Same as in Fig.5, but for the ^6Li_B doublet. Note the scale in ordinates.

Of course, the $(^7Li/^6Li)_A$ ratio now depends on the suggestions presented to account for the $(^7Li/^6Li)_B$ ratio, although none of them appears entirely satisfactory. In each case however, one should expect that the actual $(^7Li/^6Li)_A$ ratio be higher than what is measured, viz. $(^7Li/^6Li)_A \gtrsim 8.6$. Our final results toward ζ Oph are thus:

$$\left(^7\mathrm{Li}/^6\mathrm{Li}\right)_A \gtrsim 8.6$$

$$\left(^7\mathrm{Li}/^6\mathrm{Li}\right)_B \sim 2 \text{ (unexplained?)}$$

Very recently, Vangioni-Flam, Cassé & Lehoucq (these proceedings), Cassé, Vangioni-Flam & Lehoucq (1994) offered a very elegant explanation to the $(^7\mathrm{Li}/^6\mathrm{Li})_B$ ratio. They have interpreted the recent detection of an extremely high γ–ray flux in Orion as due to the interaction of accelerated α and $^{16}\mathrm{O}$ nuclei with the ISM. They showed that a SNII exploding inside an interstellar cloud would indeed accelerate such nuclei, which are typical yields of a massive star, and through interaction with the ISM would produce $^7\mathrm{Li}$ and $^6\mathrm{Li}$ isotopes, among others, with a production ratio $^7\mathrm{Li}/^6\mathrm{Li} \simeq 3$ depending on certain parameters. It happens that although this interaction process should last $\Delta t \simeq 10^5$ yrs, its efficiency in creating Li atoms is some 10^6 times that of usual GCR spallation, and this is precisely the $\zeta \times \Delta t$ factor we were looking for above. This efficiency results from the very low thresholds of α–α reactions, the shape of the energy spectrum, and the weak ionization losses of α nuclei when propagating in the ISM. Assuming now that cloud B is a fragment of an interstellar cloud irradiated *via* this process, one should expect to measure $(^7\mathrm{Li}/^6\mathrm{Li})_B \sim 3$, and $(\mathrm{LiI/KI})_B \simeq 11$, which is indeed very satisfying! Finally, I cannot leave this section without quoting the work of the workshop organizer, P. Crane. Crane & Lambert (private communication) obtained lithium data at the MacDonald 2.7m in the direction of ζ Oph, with $\mathrm{S/N} \simeq 2000$ and $\lambda/\Delta\lambda = 1.25 \times 10^5$. They report the presence of two interstellar components with $^7\mathrm{Li}/^6\mathrm{Li}$ ratios above 10 in each, though the secondary component is endowed with a thermal width $b \sim 13\,\mathrm{km.s^{-1}}$, i.e. far above what is typically observed (a few $\mathrm{km.s^{-1}}$). Our solution also provides a good fit to their data, which in turn argues against an instrumental contamination.

4 Conclusions & Epilogue

Previous measurements of the interstellar $^7\mathrm{Li}/^6\mathrm{Li}$ ratio were discussed and criticized in turn, and new data on an already observed line of sight were presented. At the end, the observational status of this important isotopic ratio restrains itself to the following:

$$\left(^7\mathrm{Li}/^6\mathrm{Li}\right) = 11.1 \pm 2 \quad (\rho\,\mathrm{Oph})$$

$$\left(^7\mathrm{Li}/^6\mathrm{Li}\right) \gtrsim 8.6 \quad (\zeta\,\mathrm{Oph})$$

$$\left(^7\mathrm{Li}/^6\mathrm{Li}\right) \sim 2 \quad (\rho, \zeta\,\mathrm{Oph}, \text{to be confirmed?})$$

The measurements of MHW in the directions of ζ Per were shown to be biased as all the detected ISM components were not taken into account in the profile fitting. The lower limit $^7\mathrm{Li}/^6\mathrm{Li} \gtrsim 25$ obtained by FD toward ζ Oph is unexplained. A $^7\mathrm{Li}/^6\mathrm{Li}$ ratio in the secondary cloud detected toward ρ Oph has to be further discussed as it was done for the $^7\mathrm{Li}/^6\mathrm{Li} = 2$ ratio toward ζ Oph. This latter is not readily explained, but might be the result of a new spallation process proposed by Vangioni-Flam, Cassé & Lehoucq (these proceedings), where α and ^{16}O nuclei accelerated in SNII would interact with the ISM to create LiBeB isotopes, among others. First calculations show that this process is much more efficient in creating the LiBeB isotopes than the usual GCR spallation and compensates largely for the short duration time of the interaction. In any case, this process is to be elaborated on further and cannot be neglected any longer in studies of the galactic evolution of the light elements.

New observations of typical $^7Li/^6Li$ target stars were conducted in June 1994 at the Anglo-Australian Telescope using the newly commissioned Ultra High Resolution Facilities spectrograph at $\lambda/\Delta\lambda = 6 \times 10^5$. Toward each of the observed targets ρ Oph, ζ Oph, and others, more than two interstellar components seem to be detected in KI, some of them separated by only $\simeq 1\,\mathrm{km.s}^{-1}$... This means that to derive an accurate description of the velocity structure of a studied line of sight, which is essential as shown in 2., it is in fact mandatory to go to a resolution $\lambda/\Delta\lambda \gtrsim 3.10^5$. What this entails for the $^7Li/^6Li$ ratios is far more depressing: there is no hope in LiI to resolve ISM components separated by $\Delta v \simeq 1\,\mathrm{km.s}^{-1}$ because LiI is precisely a light element, and the natural width of the LiI line is thus already $\simeq 1.2\,\mathrm{km.s}^{-1}$. Whenever the $^7Li/^6Li$ ratio is around 2 in one component and 20 in the other, the resulting profile will be indistinguishable from a profile obtained from a single absorbing cloud with a ratio $^7Li/^6Li \simeq 10$... This means as well that the above $^7Li/^6Li$ ratios are average values of $^7Li/^6Li$ ratios in clouds with very similar radial velocities $\Delta v \simeq 1\,\mathrm{km.s}^{-1} < \lambda/\Delta\lambda$. One may however give the following considerations. First of all, obtaining a representative value of the interstellar $^7Li/^6Li$ ratio is no longer a matter of a few measurements. Just as for the D/H ratio on QSO lines of sight, it is a matter of statistics at a long term, and it certainly deserves the effort. If variations of the $^7Li/^6Li$ ratio are detected, then they must be greater in reality than what is observed, as the values measured are already averages over different absorbing clouds. This would be the trace of an inhomogenous physical process of ^7Li creation, possibly the one proposed by Vangioni-Flam et al.. If no variations are detected, we may have got the right value of the interstellar $^7Li/^6Li$ ratio, the system is homogeneous, and after all, one may consider that two interstellar clouds separated by only $1\,\mathrm{km.s}^{-1}$ are somehow physically linked and that they should exhibit similar $^7Li/^6Li$ ratios. The epilogue of this review is thus that we only see the beginning of it. Stating a ratio $^7Li/^6Li \sim 10$ as representative of the ISM is an attractive possibility, but it is a bit too premature.

Acknowledgements: the work reported of above was done in collaboration with Roger Ferlet and Alfred Vidal-Madjar.

References

Abia, C.: these proceedings

Cassé, M., Vangioni–Flam, E., Lehoucq, R.: Nature, to appear

Ferlet, R., Dennefeld, M.: 1984, AA **138**, 303 (FD)

Lemoine, M., Ferlet, R., Vidal-Madjar, A., Emerich, C., Bertin, P.: 1993, A&A **269**, 469 (L93)

Lemoine, M., Ferlet, R., Vidal-Madjar, A.: 1994, to appear in A&A

Matthews et al. ?: 1994, ApJ

Meneguzzi, M., Audouze, J., Reeves, H.: 1975, A&A **40**, 99

Meyer, D. M., Hawkins, I., Wright, E. L.: 1993, ApJ **409**, L61 (MHW)

Mowlawi, B.: these proceedings

Pinseonneault, M. H., Deliyannis, C. P., Demarque, P.: 1992, ApJ Supp. **78**, 179

Reeves, H.: 1993, A&A **269**, 166

Reeves, H.: 1994, Rev. Mod. Phys. **66**, 193

Spite, F.: these proceedings

Vangioni–Flam, E., Cassé, M., Lehoucq, R.: these proceedings

Vauclair, S.: 1988, *in "Dark Matter"*, ed J. Audouze and J. Tran Thanh Van, p. 269

Part VIII

Beryllium and Boron

The Galactic Evolution of Beryllium

Ann Merchant Boesgaard

Institute for Astronomy, University of Hawaii, Honolulu, HI 96822, USA

Abstract. Observations of beryllium are presented and the derived Be abundances are interpreted in the context of the origin of Be and the evolution of its abundance during the evolution of the Galaxy. Recent Keck/HIRES observations in some very metal–poor stars are included.

1 Introduction

The possibility exists that under certain circumstances the element Be can be produced by nucleosynthesis during the Big Bang. For standard Big Bang Nucleosynthesis (BBN) the predicted amounts are $N(Be)/N(H) = 10^{-17}$ to 10^{-16}, but depending on the parameter values, inhomogeneous BBN could produce up to 10^{-13} (for example, see Mathews et al 1991). By searching for Be in the oldest stars, we should be able to determine if there has been production of Be in amounts greater than predicted by the standard BBN. In particular, if there is a plateau (which is above 10^{-16}) in the abundance of Be found in several very metal–poor stars over a range of low metallicities, we could distinguish between standard and non–standard BBN; the Be abundance thus found would provide important constraints to the large parameter space for inhomogeneous BBN models.

There will be production of Be in the early history of the Galaxy by spallation reactions also. Bombardment of high energy cosmic rays and O atoms will produce the rare light elements, Li, Be, and B. Massive stars produce oxygen during their evolution, and during supernovae explosions oxygen atoms and energetic cosmic rays are released. If the Be is created in the vicinity of the type II supernovae which produce both cosmic rays and O, then the Be abundance should be proportional to the O abundance. If production occurs in the general interstellar medium from encounters of energetic O atoms with protons, then the amount is proportional to the O abundance squared. (See Pagel 1991.) Boesgaard and King (1993) found that $N(Be)$ was proportional to $N(O)^{1.12}$, indicating that most Be was formed in close proximity to supernovae, but that a small part was not.

During the evolution of the galactic disk, massive stars will be producing O atoms and, during the supernovae phase, galactic cosmic rays. In addition, the intermediate mass stars will produce C, N, O, and Fe as well as other elements. For disk stars then, the Fe abundance provides a ready chronometer with which to compare the Be abundances. Abundances of Fe are thought to be more securely known than those of O.

As stars evolve, their light elements are subject to destruction if circulation currents take them to temperatures of a few million degress where they are destroyed by nuclear reactions. Consequently, there will be destruction of Be in a process called astration. We do see many F and G dwarfs where this effect is apparent. In addition, the halo star HD 221377 with·[Fe/H] = −1.1 is very deficient in Be for its metallicity at N(Be)/N(H) < 10^{-13} (Boesgaard and King 1993). The reduced amounts of Be in stellar atmospheres provide excellent clues to stellar structure and evolution, but the galactic evolution effect – reduction of Be with time in the Galaxy – is thought to be negligible.

2 Observational Challenges

The resonance lines of Be II are at 3130 and 3131 Å. There are two challenging aspects to observations of Be II. 1) The Earth's atmosphere absorbs much of the incoming stellar radiation at these wavelengths so near to the atmospheric cut–off. 2) In the sun and solar–type stars the spectra are very crowded with lines. A four angstrom region near the Be II lines is shown in Figure 1. This is a spectrum of the sky taken at Keck with the HIRES spectrograph. The dispersion is 0.0218 Å px^{-1}. Although the weaker of the two Be II lines at λ3131 is fairly well–separated, the line at λ3130 is still blended even at this dispersion.

In order to observe Be well, high resolution, large telescopes, and efficient detectors are all needed, and it helps to be at a site like Mauna Kea where the atmospheric transparency in the ultraviolet is relatively high. In the era when Merchant (1966) used photographic plates and the Lick 3 m telescope to make observations of Be, it took about two hours to obtain a suitable spectrum of an star at V ∼ 4.5. With the 3.6 m CFHT and a Ford/Photometrics CCD Boesgaard and King (1993) were able to reach V ∼ 8.3 in two hours. However, with Keck and HIRES the gain is about a factor of 100 over the CFHT work (this gain factor includes the much increased spectral resolution).

The consequences of the blending are two–fold. It is necessary to obtain high resolution spectra to be able to distinguish the lines and it is difficult to ascertain the location of the continuum. Both of these difficulties are happily reduced in the most metal–poor stars where the blending lines are much weaker and where the true position of the continuum can be found. In those stars, however, we are searching for very weak Be II features so the signal–to–noise ratio (S/N) must be very high in order to detect weak lines or to set a meaningfully low upper limit.

3 Recent Observational Studies of Be

Recent work on Be abundances in metal–poor stars has been done by Rebolo et al (1988), Ryan et al (1990, 1992) and Gilmore et al (1991, 1992). Testimony is given to the difficulty of detecting Be in such stars by the predominence of upper limits on the Be abundance that are given for most of the stars in those papers. The 1992 contribution by Gilmore et al presents Be detections and abundances for eight stars with [Fe/H] < −1.5. New work by Boesgaard and King (1993)

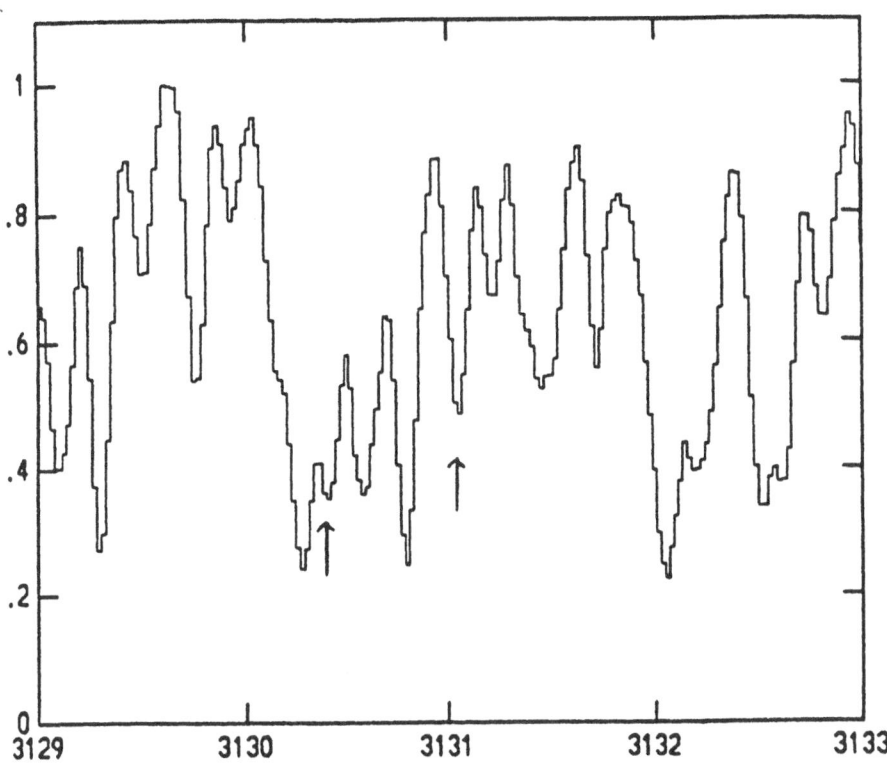

Fig. 1. This 4 Å region of a Keck/HIRES spectrum of the sky shows the spectral crowding in the region of the Be II lines in the sun and solar–type stars. The positions of the Be II lines are shown by the arrows.

contains observations of 14 halo stars, in two of which Be was undetected, and 27 (mostly old) disk stars. In order to investigate the galactic evolution of Be, it was necessary to make modern, i.e. CCD, observations of the disk stars. The previous disk star observations quoted by most of the above cited papers were from Boesgaard (1976) based on photographic spectra; however, Boesgaard and King did find remarkable agreement with the Boesgaard (1976) data for four of the five stars in common to the two studies.

The results of these investigations can be seen in Figure 2 from Boesgaard and King (1993) in the plot of log N(Be)/N(H) vs [Fe/H].

It is plausible, but far from conclusive from these data, that there is a plateau in the amount of Be present in the lowest metallicity stars near log N(Be/H) $\sim$ -12.8. As [Fe/H] increases from -2.2 to -1.0, log N(Be/H) increases; the slope is $1.2 - 1.3$, indicating a faster increase in the production of Be than in the production of Fe in the Galaxy. This is consistent with the production of Be from spallation reactions (between energetic cosmic rays and O atoms) from the products of massive stars and the production of Fe from intermediate mass stars.

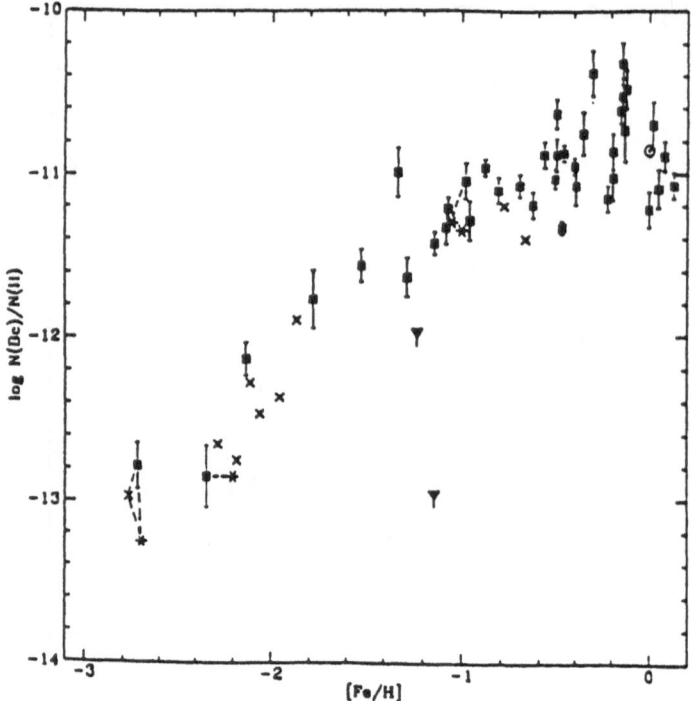

Fig. 2. Beryllium abundance results as a function of metallicity, [Fe/H]. The filled squares are from Boesgaard and King (1993), as are the two filled triangles which represent upper limits. The crosses are from Gilmore et al (1992) and the stars are from Ryan et al (1992). Points representing the same star from one or more studies are connected by dotted lines.

Gilmore et al (1993) have plotted the results of their Be abundances against their O abundances. For six stars with [O/H] < −1 they find a slope of 0.8. Boesgaard and King also determined O abundances for many of their Be stars; however, they found a steeper slope: 1.1. Since the slope of this relationship bears on the mechanism of galactic production of Be, additional effort should be made to determine its value.

4 The Keck/HIRES Studies – Be in the Early Universe

In order to reach even lower metallicity stars, which are rarer and thus fainter, it is necessary to use the Keck 10 m telescope and its high resolution spectrograph, HIRES. In fact, one of my goals, and thus one of the the science drivers for the design of some aspects of HIRES, was to observe Be in stars. Our team for this search for primordial Be also includes Beers, Deliyannis, Keane, King, Ryan, and Vogt – the HIRES "father."

We soon discovered that Keck + HIRES was about 100 times faster than what we were able to achieve with 4 m class telescopes. A comparison of spectra of the star, HD 19445 with [Fe/H] = -2.1, from CFHT and from Keck is shown in Figure 3.

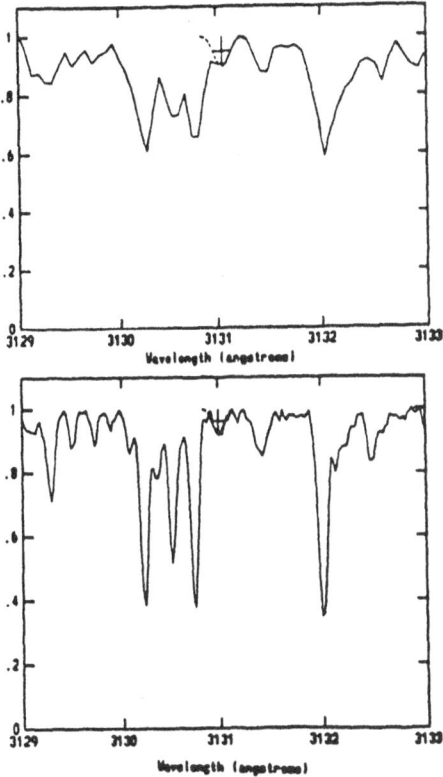

Fig. 3. Spectra in the Be II region of HD 19445. The top panel is a 2 hour exposure from CFHT at a dispersion of 0.067 Å px^{-1} and S/N = 78 per pixel. The lower panel is a 20 minute exposure from Keck at a dispersion of 0.022 Å px^{-1} and S/N = 156 per pixel. The Be II line at 3131 is fit by a gaussian profile in both spectra.

One of the brightest metal–poor stars is HD 140283 (V = 7.22), which has [Fe/H] = -2.8. This has been observed by Ryan et al (1990, 1992), Gilmore et al (1991, 1992) and Boesgaard and King (1993). Our Keck spectrum of this standard star is now *the* standard. It has S/N of 300 per resolution element (163 per pixel) and 12 pixels across the Be II 3131 feature. This is shown in Figure 4.

We have exposed for several hours on a star that is lower in metallicity than HD 140283 by a factor of more than two, BD +3 740. This star is about 400 K hotter than HD 140283 and has a slightly higher log g. Its spectrum is shown in Figure 5. This is a summation of four hours of exposure on two different observing runs and has a S/N of 132 per pixel. It is difficult to be sure that the Be II lines are detected; there is a feature with an equivalent width of 5.7 mÅ at

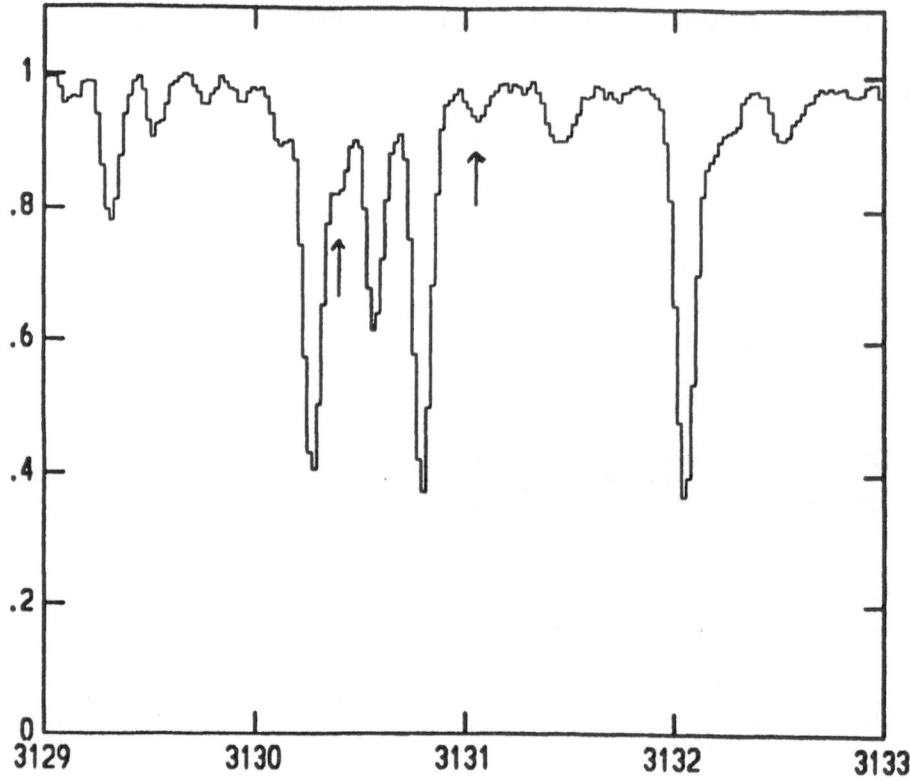

Fig. 4. The Keck/HIRES spectrum of the Be region of HD 140283.

the position of the stronger line of the doublet at λ3130. A detailed comparison of the features in HD 140283 and BD +3 740 shows that the two lines at 3129.75 and 3129.94 are present in both stars, but weaker in the more metal–poor star, BD +3 740. This lends credence to the existence of the Be II lines in BD +3 740.

New parameters have been determined in a self–consistent way for all the halo stars observed by Boesgaard and King, observed at CTIO and CFHT by Boesgaard, Deliyannis, and King, and observed at Keck by our team. In fact we have used two temperature scales, the one of King (1993) and the one of Carney (1983); the King scale is about 100 – 200 K hotter than the Carney scale. The effects of the various paramenters, log g, T_{eff}, [Fe/H], and ξ on the Be curves of growth can be seen in Figures 4 – 7 in Boesgaard and King (1993). The Be abundance results are very sensitive to log g. In addition the degree of ionization of Be I is sensitive to temperature, log g, and [Fe/H] in a complex way so the

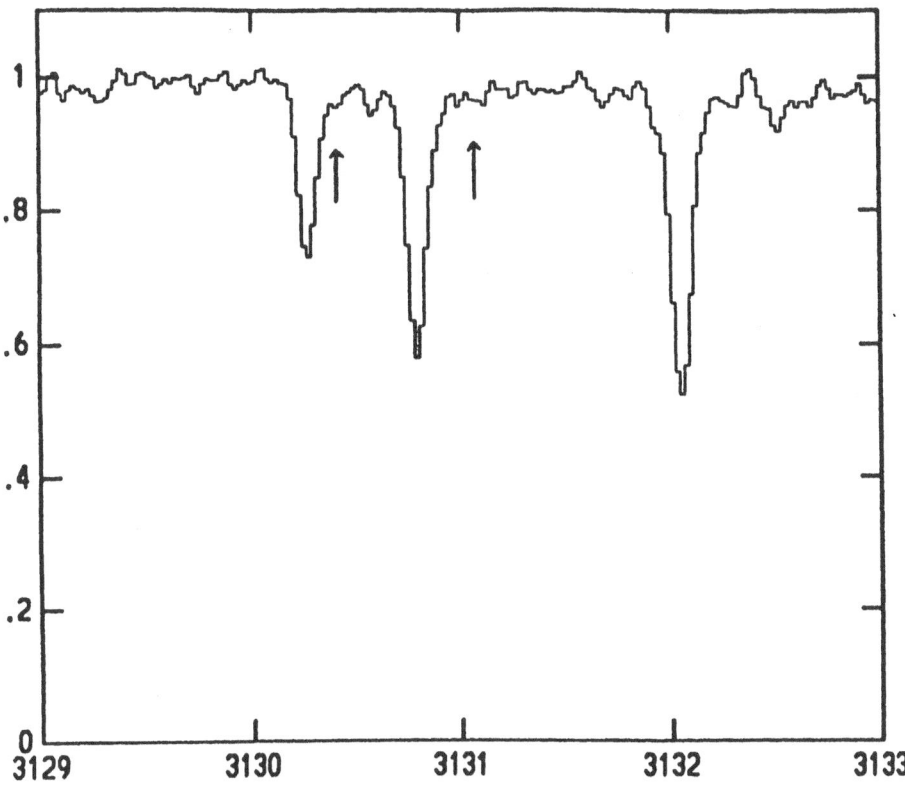

Fig. 5. The Keck/HIRES spectrum of Be II region of BD +3 740.

peak of Be II occurs at different temperatures for the various values of log g and [Fe/H].

Beryllium abundances have been determined or redetermined for all the stars. The results include the three stars whose spectra are shown in Figures 3, 4, and 5 – observed with Keck. These abundances are shown in Figure 6 plotted against [Fe/H] on the King (1993) temperature scale.

The slope of the Be vs Fe relation for the stars with [Fe/H] < −1.0 is 1.3. ¿From this figure all that can be said for now is that there may be a plateau for the lowest metallicity star near log N(Be/H) of −13.00. We have Keck observations of several more stars which we hope will provide information on the Be abundances in the most metal–poor stars.

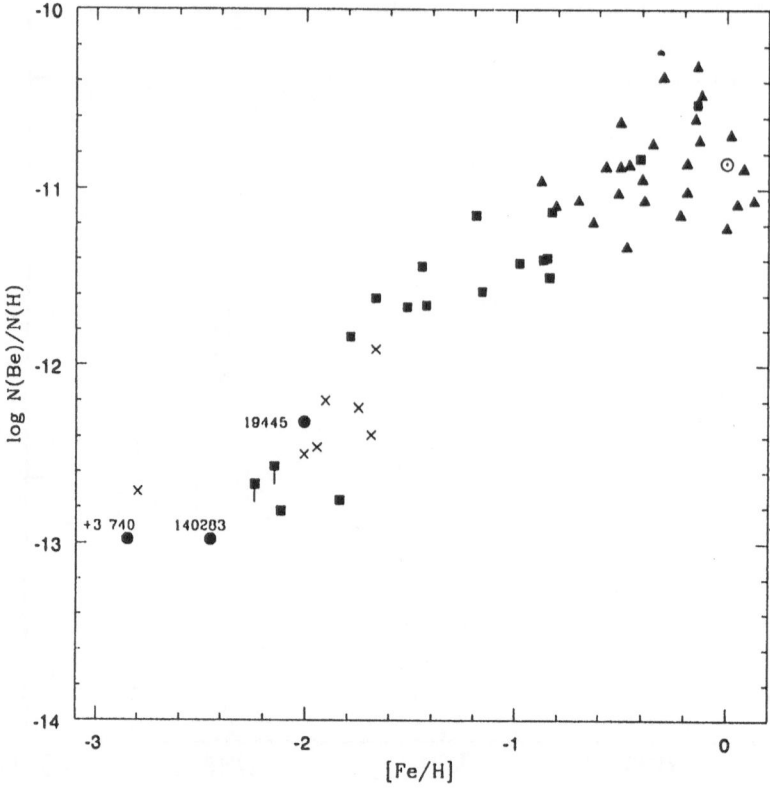

Fig. 6. Beryllium abundances vs [Fe/H] determined with new, self–consistently derived parameters. The filled triangles are the disk stars from Boesgaard and King (1993). The filled squares are the halo stars from observations at CFHT and CTIO. The crosses are the Gilmore et al. stars. And the large octagons are the Keck/HIRES data points.

5 Summary

We have reported on Be abundances in 23 halo stars and 27 disk stars. These abundances have been determined with newly derived sets of self-consistent stellar parameters. The global trends of Be with other abundances, [Fe/H] and [O/H], are thus based on a self-consistent data set.

As is seen in Figure 6, Be increases faster than Fe. More Be is produced in the early Galaxy than Fe. This is consistent with the basic ideas of galactic evolution wherein the early generations of massive stars and their supernovae produce Be from O and cosmic rays, while later generations of intermediate mass stars produce most of the Fe. Over the range $-2.5 < [Fe/H] < -1.0$ we find $N(Be) \propto N(Fe)^{1.3}$. At values of $[Fe/H] < -2.5$ there may be a plateau in the Be abundance at about log $N(Be/H) \sim -13.0$. Another simple model that could fit the data also would give a steeper slope for the most metal–poor stars, a slope of 1.0 for the stars of somewhat higher metal content, and a slope of 0 for the disk stars. Thus for $-3 < [Fe/H] < -2$, $N(Be) \propto N(Fe)^{2.0}$; for $-2 < [Fe/H] < -1$, $N(Be) \propto N(Fe)^{1.0}$; and for disk stars, $N(Be) \propto N(Fe)^{0.0}$.

From the work of Boesgaard and King (1993) we find $N(Be) \propto N(O)^{1.12}$. This indicates that Be is produced primarily near supernovae by short–lived cosmic rays, but that some is also produced in the interstellar medium by long–lived cosmic rays.

It is clear from Figure 6 that there is quite a large spread in the Be abundances of the higher metallicity stars near $[Fe/H] \sim 0$. This spread is probably real and can result from differing amounts of Be in the gas from which the different stars were formed and/or some depletion of Be in the stellar atmospheres by one or more of a number of mechanisms similar to those proposed for Li depletion, e.g. diffusion, rotationally–induced mixing, turbulent diffusion, meridonal circulation, overshoot.

There is much important work to be done on Be, both in the most metal-poor stars which contain information on cosmology and galactic Be production, and in solar–type stars to understand better the internal structure of stars and the depletion mechanisms for the light elements.

This work has been supported by grants from NSF: AST–9016778 and AST–9409793.

References

Boesgaard, A. M. 1976, ApJ 210, 466.

Boesgaard, A. M. and King, J. R. 1993, AJ 106, 2309.

Carney, B. 1983, AJ 83, 623.

Gilmore, G., Edvardsson, B. & Nissen, P. E. 1991, ApJ 378, 17.

Gilmore, G., Gustafsson, B., Edvardsson, B. & Nissen, P. E. 1992, Nature 357, 379.

King, J. K. 1993, AJ 106, 1206.

Mathews, G. J., Alcock, C., & Fuller, G. M. ApJ 349, 449.

Merchant, A. E. 1966, ApJ 143, 336.

Pagel, B. E. J. 1991, in "Lectures to Second IAC Winter School: Observational and Physical Cosmology," Tenerife, (Nordita Preprint No. 91/18 A).

Rebolo, R., Molaro, P., Abia, C., & Beckman, J. E. 1988, A&A 193, 193

Ryan, S. G., Bessell, M. S., Sutherland, R. S. & Norris, J. E. 1990, ApJ 348, L57.

Ryan, S. G., Norris, J. E., Bessell, M. S., & Deliyannis, C. P. 1992, ApJ 388, 184.

Non-LTE Effects on Be and B Abundance Determinations in Cool Stars

Dan Kiselman[1], Mats Carlsson[2]

[1] NORDITA, Blegdamsvej 17, DK-2100 Copenhagen, Denmark
[2] Institute of Theoretical Astrophysics, University of Oslo, Box 1029, Blindern, N-0315 Oslo, Norway

Abstract. We discuss the nature of non-LTE effects affecting abundance analysis of cool stars. The departures from LTE of importance for the B I lines in solar-type stars are described and some new results are presented. Boron abundances derived under the LTE assumption have significant systematic errors, especially for metal-poor stars. For beryllium, current results suggest that departures from LTE will not affect abundance analysis significantly.

1 Introduction

The interest in stellar abundances of boron and beryllium makes it imperative that the derivation of these abundances from observed spectra is as accurate as possible. The approximation of LTE, local thermodynamic equilibrium, is commonly used in the line-transfer computations needed for abundance analysis. We will discuss the shortcomings of the LTE and the systematic errors it introduces in boron and beryllium abundance determinations. The discussion will be restricted to cool stars, with effective temperatures from 5000 K up to 7000 K.

2 What Is a Non-LTE Effect?

The approximation of LTE, local thermodynamic equilibrium, enters twice in the standard procedure of deriving cool star elemental abundances from observations of absorption lines. LTE is first assumed in the computation of a model atmosphere for the star in question. LTE is then used in the treatment of line radiation transfer when calculating equivalent widths or synthetic spectra. The non-LTE effects discussed here are the errors introduced in the second stage. We will discuss the line-formation processes taking place in the standard LTE model atmospheres, but the treatment of the lines of interest is made without assuming LTE .

It should be noted that LTE is not an altogether unambiguous concept – there are several definitions around. Here, we will let LTE stand for the simplifying assumption that the atomic level populations in the stellar atmosphere under study are equal to what they would be in thermodynamic equilibrium given the local values of electron temperature (T_e), pressure, and chemical composition. In LTE we thus get the ionisation balance from the Saha equation and the excitation balance from the Boltzmann equation. The emission and absorption properties of the gas, commonly expressed as the absorption and scattering coefficients (κ_ν and σ_ν) and the source function (S_ν), can then easily be calculated and the radiative transfer equation solved for the desired frequencies. Note that with this definition of LTE, the total source function will not necessarily be equal to the Planck function since there may be a scattering contribution to the continuous opacities. The connection between these quantities and the emergent light intensity can be grasped from the formal solution of the transfer equation for outgoing vertical rays from a plane-parallel atmosphere

$$I_\nu = \int_0^\infty S_\nu e^{-\tau_\nu} d\tau_\nu,$$

where the optical depth is defined as $d\tau_\nu = (\kappa_\nu + \sigma_\nu)dz$.

The contributions to the total source function and opacity from a spectral line are the line source function (S_l) and the line opacity ($\kappa_{l,\nu}$). The line opacity is proportional to the population density of the line's lower atomic level, the line source function is (for the cases of interest here) proportional to the ratio between the population numbers of the upper and the lower levels. In LTE, the line source function is equal to the blackbody value at the frequency of the line: $S_l = B(T_e)$. These facts are useful to keep in mind when trying to understand the results of non-LTE calculations since it is convenient to discuss departures from LTE in the line opacity and in the line source function separately.

The reason that LTE is expected to fail in stellar atmospheres is the strong radiation field which will be important for setting the atomic level populations. Since the radiation is a *non-local* phenomenon we do not expect the population densities to be set by the *local* electron temperature only. In the non-LTE treatment of the line-formation problem we must solve the statistical equilibrium equations that describe the rates with which the atomic states are populated and depopulated. This must be done simultaneously with the solution of the radiative transfer equations.

Even when LTE fails, it is still useful to study the LTE case. It is practical for the abundance analyst to first make an LTE analysis and then apply non-LTE corrections to the "LTE abundances". It is also convenient to describe the line formation processes in terms of deviations from LTE. Thermodynamic equilibrium implies detailed balance, where each rate is exactly balanced by its inverse process. Departures from LTE are departures from detailed balance, caused by a radiation field that departs from the isotropic blackbody radiation characteristic of the local electron temperature. Collisional processes tend to push the system towards detailed balance. LTE is therefore a good approximation if collisional processes dominate in *all* the atomic transitions.

In the recent literature we find accounts of non-LTE mechanisms like *bound-bound pumping*, *ultraviolet overionisation*, *photon suction*, and *photon losses in resonance lines*. Refer to Carlsson et al. (1994) or Bruls et al. (1992) for detailed discussions. Note that these fancy names do not refer to any new basic physical processes, but rather to mechanisms that serve as useful tools for us to understand how the balance produced by the various ionisation, recombination, excitation, and deexcitation processes differs from what the Boltzmann and Saha formulae predict.

3 The Challenge of Non-LTE Calculations

Powerful methods now exist for solving the non-LTE line transfer problem in a given atmosphere. We use the operator perturbation technique of Scharmer & Carlsson (1985) as coded in the program MULTI by Carlsson (1986). We employ MULTI version 2.0, in which line blanketing is taken into account in the photoionization, the treatment of background opacities is improved and the many-level treatment is speeded up by using the local operator of Olson et al. (1986).

A major problem in non-LTE work is to assemble a sufficiently complex model atom. Accurate data is needed for many atomic levels, radiative, and collisional transitions, not just for the very line to be studied. Collisional cross sections are especially hard to come by. Large computational efforts aimed at producing high-accuracy atomic data, notably the Opacity Project (see Seaton et al. (1994) and references therein), have made this work easier.

4 Neutral Boron

The B I lines of interest for cool star abundance analysis are the resonance doublets at 250 nm and at 209 nm. The latter doublet has not yet been extensively observed, but it may be useful for isotopic ratio studies (Johansson et al. 1992).

4.1 Non-LTE Mechanisms for B I

Kiselman (1994) investigated the statistical equilibrium of neutral boron in three solar-type atmospheric models and found significant departures from LTE in stars hotter or more metal-poor than the Sun. This section relates to the mechanisms giving rise to non-LTE effects in the metal-poor stellar model investigated in that paper.

The resonance doublets have line source functions that are well described by a two-level atom approximation. This means that the line source functions are determined solely by processes in the line transition: radiative and collisional excitations and deexcitations. Furthermore, these transitions are dominated by the radiative transitions, and, since the lines are weak, their source functions are set by the background radiation field which in its turn is determined by the background opacities. Since the blue and ultraviolet radiation field in solar-type stars generally has a hotter radiation temperature than the local electron temperature we get the effect of *bound-bound pumping* and the line source function becomes greater than the Planckian value: $S_l > B$.

The line opacities are set by the ionisation balance since the overwhelming majority of neutral boron is in the ground state. The departure from the Saha equilibrium can be described as an overionisation caused by two mechanisms working in the same direction. The pumping in the ultraviolet resonance lines increases the ionisation rates since the radiation fields in the ionisation edges of the excited levels are so much richer than in the ground state ionisation edge (at $\lambda = 150$ nm). Furthermore, these radiation fields are stronger than the local Planckian value. Each of these effects is by itself sufficient to cause overionisation.

The pumping and the overionisation will both make absorption lines weaker than in LTE . This non-LTE effect is insignificant for the Sun, but increases in magnitude for hotter or more metal-poor stars. In these stars, the LTE assumption leads to underestimations of the boron abundance. For the well-known halo star HD140283 ([Fe/H] $= -2.6$), the non-LTE abundance correction amounts to $+0.54$ dex (Edvardsson et al. 1994).

4.2 New Results

We have made non-LTE calculations for a grid of model atmospheres using the B I model atom (31 levels and 114 line transitions) of Kiselman (1994). The model atmosphere code is that used by Edvardsson et al. (1993).

Kiselman (1994) did not include the effect of blending of other lines with the B I resonance lines – the "background" opacities were just the continuous ones. The effect of such blending can be expected to be a moderation of the pumping effect, thus a decrease of the line source function of the observed lines and a damping of the overionisation effect. Both effects will tend to decrease the non-LTE abundance corrections.

A full non-LTE treatment of blended lines of different atomic species is however a formidable task. In our calculations the "background" lines are assumed to be formed in LTE and pure absorption. It is a major task to assemble detailed linelists for the spectral regions of the B I resonance lines. Our treatment of blending is increasingly schematic when going from the 250 nm doublet to the shortest-wavelength resonance lines which are treated in the same way as photoionisations, i.e. the transitions are considered to be fixed with radiation fields computed by the atmospheric code, thus including the effect of line blanketing in an averaged way. It is not our aim to produce detailed synthetic spectra that can be directly compared with observations, but rather to study the departures from LTE and to acquire abundance corrections. The resulting spectra are, however, illustrative, and some are displayed in Fig. 1.

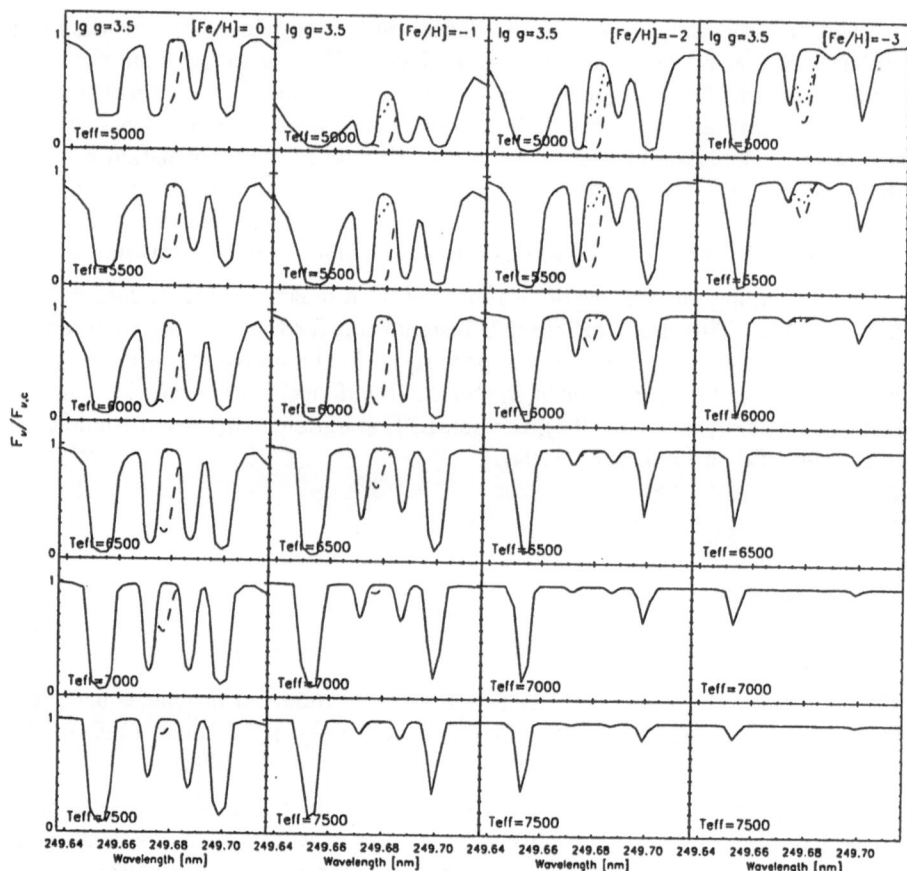

Fig. 1. Synthetic spectra of the 250 nm spectral region containing the B I line that has been used for abundance analysis. Spectra are shown for a section of the grid with $\lg g = 3.5$ [cgs]. Fulldrawn line: spectrum without boron. Long-dashed line: boron abundance = $\lg B/H + 12 = 0.3$. Short-dashed line: boron abundances = $0.6[Fe/H] + 2.4$. No macro-turbulent, rotational, or instrumental broadening have been applied. The apparent decrease of line strength in cooler stars when the metallicity is increased from 0 to −1 is due to the importance of metals for the continuous opacities – the spectra are normalised on the "continuum" flux

4.3 Non-LTE Abundance Corrections

The typical abundance analyst needs non-LTE corrections to improve LTE abundance estimates. The ordinary procedure is to calculate both LTE and non-LTE equivalent widths for a range of abundances. The two resulting curves of growth are then used to interpolate the LTE and the non-LTE abundances implied by a certain equivalent width. The non-LTE abundance corrections can be tabulated as functions of LTE abundance or line strength and atmospheric parameters.

There will, however, be complications if the lines are blended as in the boron case. The abundance analyst is here likely to use spectrum synthesis and not just measure equivalent widths. To provide abundance corrections we must then find the way of constructing curves of growth that gives the results most similar to those from the synthetic spectrum fitting procedure.

The definition of equivalent width is

$$W = \int (1 - \mathcal{F}_\lambda / \mathcal{F}_c) d\lambda.$$

When the continuum flux, $\mathcal{F}_c$, is a slowly varying function of wavelength, this definition is unambiguous. But the introduction of blending background lines makes two definitions possible:

A) $\mathcal{F}_c$ is the flux that would be present in the absence of the studied line. It is then a rapidly changing function of wavelength due to the presence of background lines. This normalisation is physically reasonable since W still measures the fraction of light that is blocked away. It gives, however, different weight to different parts of the measured line according to the variations of $\mathcal{F}_c$.

B) $\mathcal{F}_c$ is considered to be some constant or slowly varying function of wavelength, e.g. the flux that would be present if only "continuous" opacities were included. W is then equivalent to the line areas measured in typical spectrum plots like the ones in Fig. 1. In that particular case the definition B gives "equivalent widths" that are proportional to the area contained between the full-drawn and the dashed lines. This fact makes definition B seem a practical one, but the physical significance of W is now less clear than with A.

Numerical tests show that the difference between the two definitions, as measured by the the difference in curve-of-growth shapes and the difference in non-LTE abundance corrections that follows, becomes significant when the background lines are strong. We consider definition B to be the one that gives results most similar to the abundance analyst's spectrum fitting. The non-LTE abundance corrections presented in figures 2 and 3 are computed that way from our grid. The arrows displayed in the plots show the magnitude and the sign of the non-LTE abundance corrections by connecting LTE abundances (arrow base) and non-LTE abundances (arrow tip). Each panel shows the corrections for a range of boron abundances and metallicities. The four panels make it possible to see how the exclusion of background lines, increase of effective temperature, and increase of surface gravity affects the size of the corrections. It is clear that the corrections are most important for metal-poor stars and that they increase in size with increasing temperature and decreasing gravity. As expected, the importance of the background lines for the results are greatest in relatively metal-rich stars where they essentially remove the departures from LTE. It is, however, likely that our treatment of the background lines as formed in LTE and pure absorption is least accurate when these lines are strong. The effect of *photon escape* which will depress S_l and tend to strengthen the line relative to LTE is probably underestimated in those cases.

The plot in Fig. 4 shows the results of application of these results to literature LTE abundances of metal-poor solar-type stars.

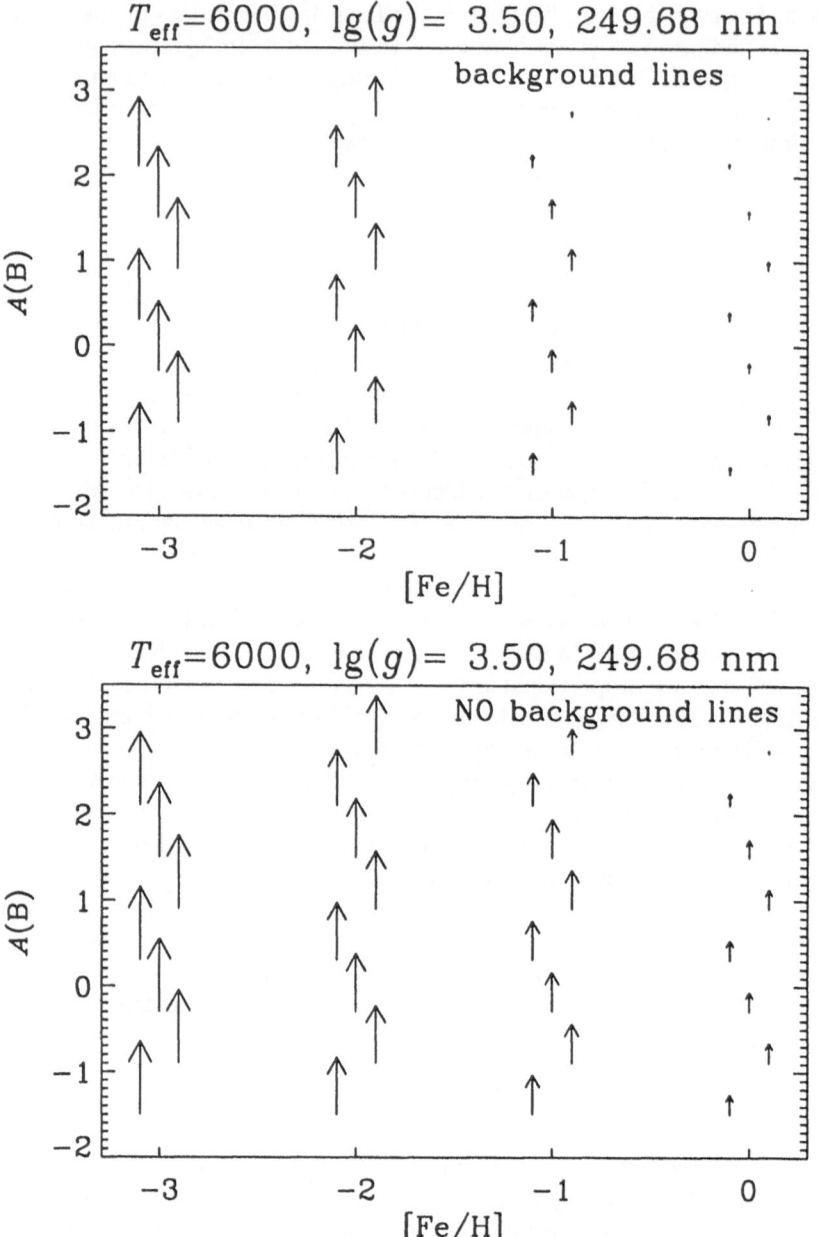

Fig. 2. Non-LTE corrections to boron abundances. Each arrow connects an LTE and a non-LTE abundance for one specific equivalent width. The spread in the position of the arrows for each metallicity (−3, −2, −1, and 0) has been introduced for clarity. The upper panel shows the correction when "background" lines are included in the calculations, in the lower panel these lines are excluded

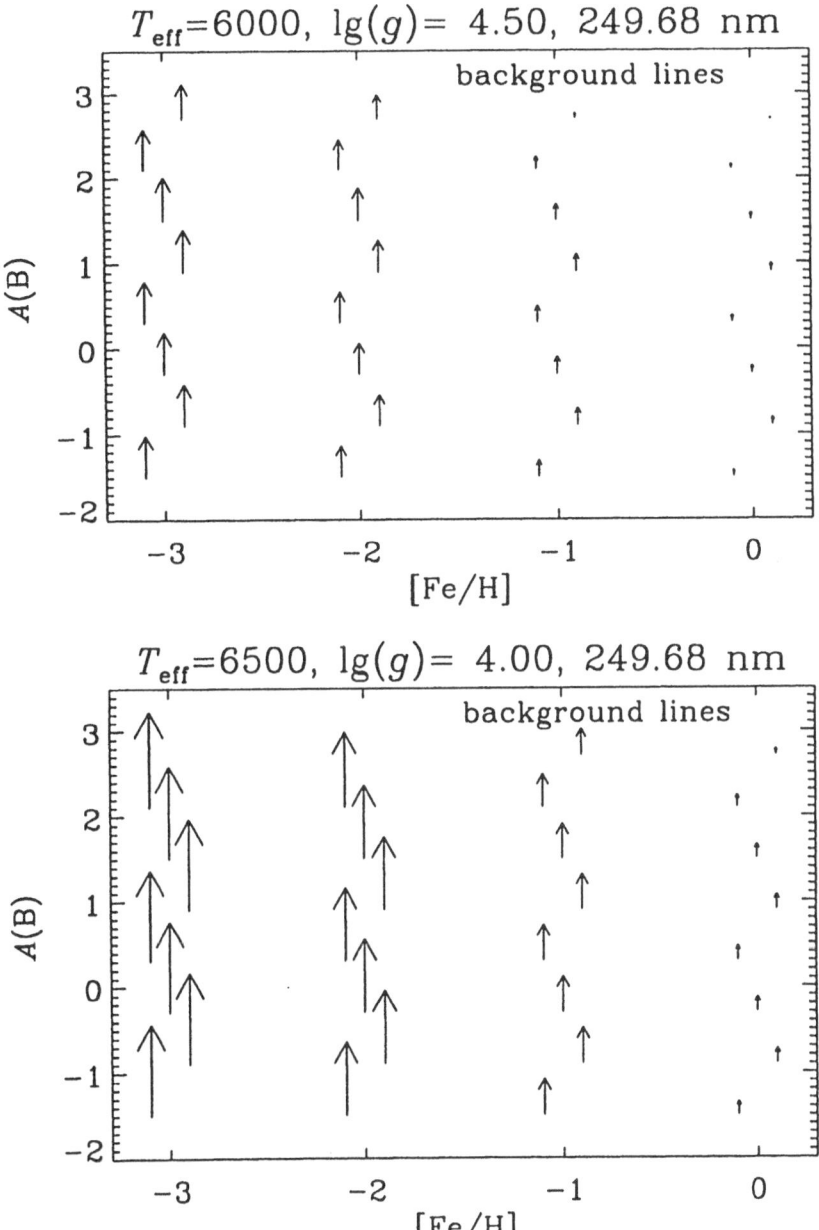

Fig. 3. Non-LTE corrections to boron abundances, similar to Fig. 2. Comparison to the upper panel of Fig. 2 shows how the non-LTE corrections vary with changing effective temperature and surface gravity

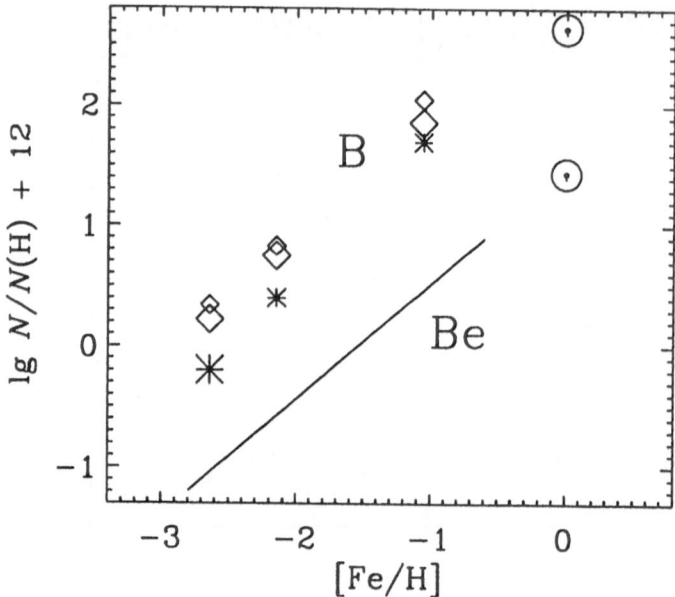

Fig. 4. Boron abundance as function of metallicity. The LTE results (stars) have been taken from Edvardsson et al. (1994) for HD140283 (big star) and from Duncan et al. (1992) (small stars). The non-LTE abundance corrections have been interpolated from the grid presented in this paper. Big diamonds are results with background line blanketing included, small diamonds without. The solar boron abundance is from Kohl et al. (1977). Indications of beryllium abundances are adopted from Duncan et al. (1992)

5 Beryllium

The spectral feature of interest for cool star abundance analysis is the Be II resonance doublet at 313 nm, just longwards of the atmospheric transmission cutoff. We have compiled an atomic model representing Be I and Be II (with 71 levels and 243 lines) and investigated the non-LTE formation of the beryllium lines in three atmospheric models representing the Sun, Procyon, and HD140283. The results should be considered as more preliminary than the boron results discussed above, since the collisional data still needs to be improved. For reasons to be discussed, however, it seems likely that the general results presented here will survive coming improvements.

For the HD140283 model, it seems that same general mechanisms as in the boron case just discussed cause the departures from LTE in beryllium. There is overionisation, apparently caused in the same way as in boron: bound-bound pumping in ultraviolet resonance lines plus a hotter-than-local-Planckian radiation field in the important ionisation edges. There is also pumping in the 313-nm doublet, causing its line source function to increase above the local Planckian value. The difference relative to the boron case is that the interesting lines now are lines arising from the ion: Be II This means that the overionisation will tend to make the lines stronger while the bound-bound pumping will weaken the lines. These effects effectively cancel each other. The lines *cannot* be said to be formed in LTE, but their equivalent widths do not differ much from the LTE -values! The resulting abundance correction is just +0.03 dex.

Since both the overionisation and the pumping effects are driven by the same phenomenon – the hot radiation field in the UV – they can be expected to cancel each other also if this should change by improvements in model atmospheres or background opacities. The non-LTE corrections for the Sun and Procyon are indeed also very small, less than 0.1 dex in both cases. It seems likely that non-LTE effects are not very significant for the 313-nm lines in solar-type stars. The largest effects are expected for very beryllium rich stars (if such exist) when the 313 nm lines are strong and affected by photon losses. The abundance corrections will in that case be negative.

The results for the Sun, in the sense of the line-strength in LTE being similar to non-LTE, is in agreement with earlier work by Chmielevski et al. (1975) and Shipman & Auer (1979).

6   Conclusions

Non-LTE effects should be taken into account when deriving abundances from B I lines in cool stars. So far, it seems that the assumption of LTE will not introduce significant errors in the derivation of Be abundances from Be II 313 nm in solar-type stars.

References

Bruls J.H.M.J., Rutten R.J., Shchukina N.G., 1992, A&A, 265, 237

Carlsson M., 1986, Uppsala Astronomical Observatory Report No. 33

Carlsson M., Rutten R.J., Bruls J.H.M.J., Shchukina N.G., 1994, AA, in press

Chmielewski Y., Müller E.A., Brault J.W., 1975, A&A, 42, 37

Duncan D.K., Lambert D.L., Lemke M., 1992, ApJ, 401, 584

Edvardsson B., Andersen J., Gustafsson B., et al., 1993, A&A, 275, 101

Edvardsson B., Gustafsson B., Johansson S.G., et al., 1994, A&A, in press

Johansson S.G., Litzén U., Kasten J., Kock M., 1993, ApJ, 403, L25

Kiselman, D., 1994, A&A, 286, 169

Kohl J.L., Parkinson W.H., Withbroe G.L., 1977, ApJ, 212, L101

Lemke M., Lambert D.L., Edvardsson B., 1993, PASP, 105, 468

Olson, G.L., Auer, L.H., Buchler, J.R.,1986, JQSRT, 35, 431

Scharmer G.B., Carlsson, M., 1985, J.Comput.Phys. 59, 56

Seaton M.J., Yu Yan, Mihalas D., Pradhan A.K., 1994, MNRAS, 266, 805

Shipman H.L., Auer L.H., 1979, ApJ, 84, 1756

LiBeB Production by Low Energy Galactic Cosmic Rays

Hubert Reeves[1] and Nikos Prantzos[1,2]

[1] SAp/DAPNIA, CE Saclay, 91191 Gif sur Yvette, France
[2] Institut d'Astrophysique, 98 bis boulevard Arago, F75014 Paris

Abstract. The recent detection of 4.4 MeV γ-ray lines from the Orion region suggests the presence of an important population of low-energy ($\sim$10 MeV) particles in that site. We explore the consequences of that observation for the synthesis of the light elements lithium, beryllium and boron (LiBeB).

1 Introduction

The discrepancy between the observed boron isotopic ratio in the solar system ($^{11}B/^{10}B$=4$\pm$0.4) and the computed formation ratio (2.3$\pm$0.5) in galactic cosmic rays (GCR) has been known for a long time (Meneguzzi, Audouze and Reeves 1971). It has been suggested that spallation reactions by low energy cosmic rays could resolve this problem. In the $\sim$10 MeV range larger boron isotopic ratios are expected, due to the $^{14}N(p,\alpha)^{11}B$ and to alpha-induced reactions. Excluded from the solar system by the modulation effects of the solar magnetic field, those hypothetical low-energy particles should be found close to their acceleration sources (Reeves and Meyer 1978), their short ionization range preventing them from propagating throughout the entire galactic volume. A recent observation (Bloemen et al. 1994) suggests that such low energy particles are indeed present in the Orion star forming region. In the following we attempt a preliminary exploration of the consequences of this observation for the nucleosynthesis of the LiBeB isotopes.

2 The Gamma Ray Emission of Orion

One clear sign of the existence of high fluxes of MeV particles would be the emission of nuclear gamma ray lines by excited nuclei and, in particular, the 4.4 MeV line of ^{12}C and the 6.1 MeV line of ^{16}O (Meneguzzi and Reeves 1975). In this respect, the reported discovery of these lines in the Orion Nebula is a most significant event. A high flux ($\sim$10^{-4} cm^{-2} s^{-1}) of gamma rays in the 3-7 MeV range has been detected, with wide resonances around 4 and 6 MeV. The resonance width ($\sim$0.5 MeV) is instrumental and does not allow to distinguish between the two possible mechanisms: 1) fast alphas and protons on interstellar C and O, with small expected width ($\sim$0.05 MeV) or 2) fast C and O on interstellar protons and alphas with a Doppler enlarged width ($\sim$0.5 MeV). The derived flux is found to be some 30 times larger than expected from the mere bombardment of standard galactic cosmic rays on the Orion clouds. The lack of enhancement of ~100 MeV gamma rays from pion decays indicates the absence of a corresponding enhanced GeV particle flux in this region. Another significant feature is the low reported value of the γ-ray flux at E<3 MeV, if it is not due to instrumental effects or data reduction (Bloemen, private communication). Meneguzzi and Reeves (1975) and Ramaty et al (1979) have computed the gamma ray fluxes resulting from the bombardment of high energy particles with GCR chemical abundances on interstellar gas. A large 1-3 MeV peak of Si,Mg,Ne and Fe γ-ray lines shows up in the calculations, several times larger than the 4-7 MeV peak of the corresponding C and O lines. Depending upon the (1-3 MeV)/(4-7 MeV) flux ratio in Orion, this feature seems to require either a reduced HHe /CNO ratio in the fast particle fluxes (to reduce the collisions of fast H and He on interstellar Si, Mg, Ne and Fe) and/or a reduced SiMgNeFe/CO ratio in the fast particles (to reduce the collisions of fast Si, Mg, Ne and Fe on interstellar H and He). If the latter alternative is confirmed, it could be attributed to a First Ionisation Potential (FIP) effect, reducing SiMgFe but not Ne.

3 LiBeB Production Yields and Galactic Activity

The Orion observation implies the need of enlarging our views on the problem of the origin of LiBeB. It should be noted that in the case of a reduced HHe/CO ratio, the energetics of the formation of these elements is considerably reduced compared with the case of standard GCR flux (Ryter et al 1970). For instance the price of one ^{6}Li atom can be reduced from ~3 ergs to as little as 10^{-2} ergs.

Computations have been performed to evaluate the formation ratios of the light isotopes in an Orion-type episode. Power-law kinetic energy spectra of fast particles are assumed, bounded between 8 and successively 20, 50 and 150 MeV/nucleon. CNO nuclei are in a solar relative composition while protons and α particles are reduced by factors of 30 or 300 over the solar composition. The bombarded cloud has a solar composition always and an escape length of 3 g cm^{-2} is assumed. Both H and He induced nuclear reactions are considered. Figure 1 shows the computed abundance ratios as a function of the exponent of the kinetic energy spectrum. It can be seen that the resulting ^{11}B/^{10}B ratio reaches values as high as ~7, while the ^{7}Li/^{6}Li ratio varies around 2. Similarly the B/Be and Li/Be ratios vary in the range 25-40 and 30-70, respectively, the latter being very much dependent on the assumed He abundance. The results for other values of the cut-off are qualitatively the same. Similar results have been obtained by Cassé et al. (1994).

Assuming that the Orion type activity is widespread in the Galaxy, could it have a substantial contribution to the overall synthesis of LiBeB? The observed 4.4. MeV Orion emissivity of ~2 10^{39} photons s^{-1} corresponds, from our calculations, to a ^{11}B production rate of ~ 4 10^{38} s^{-1}. The observed stellar ratio in Pop I stars is B/H~2 10^{-10} (Boesgaard and Heacox 1978; Lemke et al 1994) implying a total number of 10^{58} boron atoms in the Galaxy, or a mean pro-

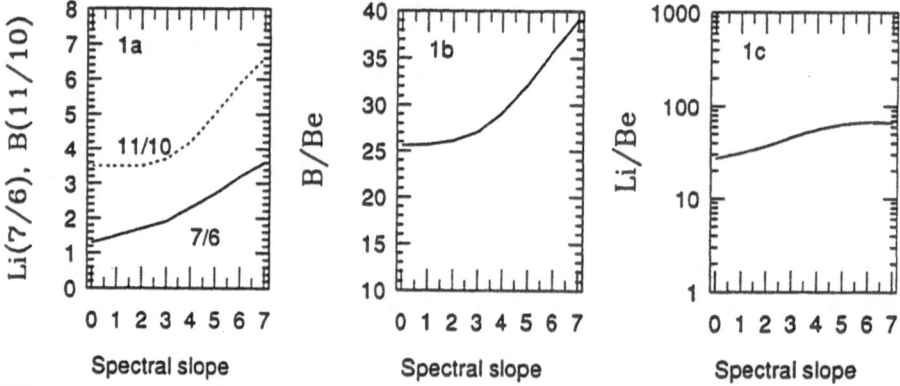

Fig. 1. Formation ratios of Li Be B as a function of the exponent of the power-law kinetic energy spectra of the bombarding particles, with energies between 8 and 50 MeV. The ratio of protons and alphas to heavier elements in the fast particles is reduced by 30 with respect to solar composition. The gas composition is solar.

duction rate of $\sim3\ 10^{40}\ s^{-1}$ integrated over the galactic life. Thus, ~100 Orion type sources are required to operate permanently in the Galaxy, i.e. a sizeable fraction of the ~500 existing giant molecular clouds. If these sources are spread evenly over the galactic plane, their average separation distance would be ~2 kpc. Very few of them should be detectable by the COMPTEL telescope and it may not be surprising that, so far, only the ~0.5 kpc distant Orion source has been seen individually. The absence of a corresponding signal in the direction of the galactic center is not surprising either, since it is expected to be $\sim10^{-5}$ photons s^{-1} cm^{-2}, some ten times weaker than the observed Orion flux.

4 Interpretations of the Abundances

To interpret correctly the elemental and isotopic observations of LiBeB in a given astrophysical setting, we must first integrate the calculated yields over the galactic lifetime, taking into account the fact that, while the effect of the high energy GCR flux is felt continuously and globally over the volume of the Glaxy, the low energy and short ionization range Orion-type fluxes are localized in space and most likely episodic.

Next we must sum up the effects of the two components. First a *global* "background" component, in which the yields of past Orion type episodes have diffused throughout the galactic volume adding their contributions to the GCR yields and also to the stellar originating component of ^{7}Li. Second a *local* component, generated by recent, perhaps still ongoing, Orion type episodes, not yet spatially diffused but creating local inhomogeneities with specific abundance ratios.

4.1 The Early Galaxy

One interesting property of the Pop II abundance of Be/H and B/H (Rebolo et al 1992; Ryan et al 1990, 1992; Gilmore et al 1991, 1992; Edvardsson et al 1994; Duncan et al 1992; Boesgaard and King 1993) is that they behave as primary elements (slope of ~ 1 in the Be or B vs. Fe plot). This is difficult to interpret in the framework of the standard scenario, where these nuclei are produced as secondaries, i.e. by GCR p and αs hitting the increasingly more abundant C-NO nuclei of the ISM. One straightforward solution is to assume that they are mostly produced by fast CNO nuclei having always the same composition (Duncan et al 1992) and this could indeed be the case if supernovæ accelerate their own ejecta. Observations do not favour this possibility for high energy GCR (>200 MeV/nucleon, their composition been essentially that of Pop I stars), but there are no observational constrains for ~ 10 MeV/nucleon particles. Orion-type sources could then be the site of such an activity. Assuming that the two LiBeB sources (i.e. >100 MeV/nucleon GCR with time-dependent CNO composition and ~ 10 MeV/nucleon CR with time-independent CNO composition) have comparable contributions today, it turns out that the low-energy component should dominate in the early Galaxy.

The B/Be ratio should be a an indicator of the relative importance of the two fluxes. However, in the only star where a consistent non-LTE analysis has been performed up to now (HD140273), the derived value B/Be=17^{+17}_{-8} (Edvardsson et al. 1994) can accomodate any scenario. *If* it turns out to be in the upper range (i.e. >20, say) then one may reasonably argue that Orion type fluxes dominated the yields in the early Galaxy: the slope of unity in the Be vs. Fe plot would indicate that fast CNO from SNII were the main contributors; the high B/Be would indicate that these fast CNO were mostly low energy (~ 10 MeV/nucleon) particles (see also Cassé et al. 1994).

One potential difficulty with this view come from the lack of dispersion of the B and Be abundances vs. Fe, which implies a successful mixing of these atoms throughout the halo, on a rather fast time scale ($\sim$a few 10^7 years). Indeed, it is difficult to understand why stars born in different star forming regions with (in principle) different levels of Orion-type activity should have the same Be and B abundances (Prantzos 1994).

An upper limit to the Li/ Be production ratio in the early Galaxy can be obtained from the flatness of the Li vs. Fe (Spite) plateau in hot halo stars. At [Fe/H]~ -1 we have Be/H$\sim 3 \ 10^{-12}$. The result depends, of course, upon the assumed depletion of Li on the surface of Pop II stars. From the flatness of the plateau, and from the presence of ^{6}Li in some halo stars (Smith et al 1993; Hobbs and Thorburn 1994) it can be safely assumed that the original abundance should not have been more than two or three times larger than the mean observed value $1.6 \ 10^{-10} < $ Li/H $< 3 \ 10^{-10}$. With these numbers one obtains for the formation ratio: Li/Be<50 to 100 at [Fe/H]~ -1. The calculated value (Fig. 1c) depends strongly upon the assumed He/CO ratio in the low-energy Orion-type fluxes, but a reduced (w.r.t. GCR) ratio is clearly required to satisfy the observations (alternatively, peculiar GCR spectra, flattened up to several hundreds of MeV/nucleon, could be invoked; e.g. Prantzos et al. 1993).

4.2   Later Galactic Life

The observed B/Be~14 ratio in Pop I stars (Lemke, Lambert and Edvardson 1994) may be taken as an indication that the GCR flux contribution later came to dominate the abundances, producing these species as secondaries. The observed primary-like behaviour of Be vs. Fe in disk stars can then be explained in the framework of a closed box model for the solar neighborhood, *but not if infall is assumed to solve the G-dwarf problem* (Pagel 1993; Prantzos 1994). On the contrary, there is no problem with the Be primary-like behaviour if Orion-type activity always dominates the Be and B production. Further observations and refined non-LTE analysis will help to clarify the role of the two candidate sources.

4.3   The Case of the Protosolar Nebula

We consider the possibility of an Orion episode in the protosolar OB association. The boron isotopic ratio in meteorites $^{11}B/^{10}B{\sim}4$ is too large to be explained solely in terms of the high energy GCR flux. Low energy cosmic rays have been suggested long ago (Meneguzzi et al 1971; Meneguzzi and Reeves 1975). Since we have no information on the value of this ratio outside the solar system we do not know if this is a "last minute" local effect or a global one.

Robert and Chaussudon (1994) have recently reported the detection of variations in the $^{11}B/^{10}B$ ratio at the level of five to ten percent. If these measurements are confirmed they could be assigned to the presence of the 1.5 million years radioctive ^{10}Be decaying in situ in the meteorites, thus confirming the occurence of an Orion-type episode in the protosolar nebula. A convincing signature of such a process would be a correlation between the $^{11}B/^{10}B$ ratio and the Be content of the mineral. The reported correlation, by the same authors, between the observed $^{11}B/^{10}B$ ratio and other chemical elements is a possible sign of such a process, if all this is confirmed.

The discrepancy between the mean Pop I stars and the meteoritic boron has been known for a while. The solar system B/Be~45 (Anders and Grevesse 1989) may, again, imply the occurence of an Orion-type episode. The difference between the stellar value and the meteoritic value would also imply that the Orion-type contribution does not dominate the background yields (at least in Pop. I stars).

On the other hand the lithium isotopic ratio in the solar system $^{7}Li/^{6}Li{=}12.5$ (Kranzovsky and Muller 1967), does not seem to support the hypothesis of a protosolar Orion episode. The interpretation of lithium abundances is however somewhat complicated by the existence of a stellar-born ^{7}Li component. A reconciliation of the lithium isotopic ratio of the solar system with the occurence of an Orion episode would require a preirradiation value larger than 12.5. For instance, if a preirrediation value of $^{7}Li/^{6}Li$ as large as 20 is assumed, an Orion episode could account for all the observed LiBeB isotopic abundances in the solar system.

4.4 The Case of the Ophiucus cloud

Up to very recently, the lithium isotopic ratio was known only in the solar system and, in absence of other measurements, this value was taken as " cosmic". This view is challenged by recent observations in the Ophiucus clouds (Meyer et al 1993; Lemoine et al 1993).

The low lithium ratio ($^7Li/^6Li\sim 2$) in cloud no 2 of Lemoine et al (1993) suggests the recent occurence of a strong Orion episode, rather localized in space since it did not affect strongly the cloud no 1. The rather modest X and gamma ray flux coming from this region (Montmerle 1977) indicates that the process is no more going on at the present time. This conclusion is corroborated by the expected large ionisation rate generated by an Orion-type flux, not detected in Ophiucus. If this episode is not older than a few million years it might still show sign of ^{26}Al gamma ray lines at 1.8 MeV (Clayton 1994). Overabudances of boron, together with a large boron isotopic ratio would also be good evidence in favour of this scenario.

5   Conclusions

Four different processes appear to have played a role in the production of the lithium isotopes in the Galaxy. Over and above the Big Bang and the stellar contribution to the abundance of ^7Li we have to consider both the high energy (GCR) and the low energy (Orion-type) cosmic ray contribution to ^6Li and ^7Li. In this paper, we have tried to evaluate the relative importance of the two cosmic ray contributions throughout the galactic life. The B/Be ratio and the linear growth of B/ H and Be/H vs Fe/H may indicate that the Orion-type contribution dominated the early days of the galactic life. However, in view of the sporadic character of these processes, one would expect a larger more dispersion in the data. The case of Pop I stars is not so clear. The B/Be ratio in PopI stars seems to indicate that the GCR contribution has taken over by the time the [Fe/H] has reached -1. Signs of the (undiffused) effects of Orion-type fluxes are probably found both in the protosolar nebula and in the Ophiucus clouds.

References

Anders, E., and N. Grevesse, 1989, *Geochim. Cosmochim. Acta* , **53**, 197
Bloemen, H., et al. 1993, A&A, **281**, L5
Boesgaard, A.M., and Heacox, W.D., 1978, ApJ, **226**, 888
Boesgaard, A.M. and King, J.R., 1993, AJ, , **106**, 2309
Bykov, A. and Bloemen, H., 1994, A&A, **283**, L1
Cassé, M., Lehoucq, R. and Vangioni-Flam, E., 1994, *Nature*, submitted

Clayton, D.D., 1994, *Nature,* **368**, 222

Duncan, D.K., D.L. Lambert and M. Lemke, 1992. ApJ, **401**,584

Edvardsson, B. et al. 1994, A&A, in press

Gilmore, G., B. Edvardsson and P.E. Nissen, 1991, ApJ, **378**, 17

Gilmore, G., B. Gustafsson, B. Edvardsson and P. Nissen, 1992, *Nature,* **357**, 379

Hobbs, L., and Thorburn, J., 1994, ApJ Let., , **428**, L25

Kiselman, D., 1994, A&A, in press

Kraznowski, D. and Muller O. 1967, *Geochim. Cosmochim. Acta,* **31**, 1833

Lemke, M., Lambert, D.L., and Edvardsson, B., 1994, *PASP,* in press

Lemoine, M., R. Ferlet, A. Vidal-Madjar, C. Emerich, Bertin, C., 1993, A&A, **269**, 469

Lemoine, M. and Ferlet, R., 1994, IAP preprint 477. ESO Workshop on "*Light Element Abundances*". Elba . May 1992

Meneguzzi, M., J. Audouze and H. Reeves, 1971, A&A, **15**, 337

Meneguzzi, M., and H. Reeves, A&A, 1975, 40, 91

Meyer, D.M., Hawkins, I., and Wright, E.,L. 1993, ApJ, **409**, L61

Montmerle, T., 1977, ApJ, **217**, 872

Pagel, B. 1993, in *Cosmical Magnetism,* Ed. D. Lynden-Bell (Kluwer), p. 113

Prantzos, N. 1994, in *Nuclei in the cosmos,* (Gran Sasso, July 1994), Eds. M. Busso, C. Raiteri and R. Gallino, in press

Prantzos, N., Cassé, M. and Vangioni-Flamm, E. 1993, ApJ, **403**,630

Ramaty, R., Kozlovsky, B. and Lingenfelter, R.E., 1979, Astrophys. J. Suppl., **40**, 487

Rebolo, R., et al. 1992, in *Origin and Evolution of the Elements*", Eds. N. Prantzos, E. Vangioni-Flam and M.Cassé (Cambridge University Press) p. 149

Reeves, H. and J.P. Meyer, 1978, ApJ, **226**, 613

Robert, F. and Chaussudon, D. 1994, Proceedings of the Meteoretical Society Meeting, Prague, in press

Ryan, S. G., M. S. Bessell, R. S. Sutherland and J. E. Norris, 1990, ApJ Let., **348**, L57

Ryan, S. G. et al., 1992, ApJ, **388**, 184

Ryter, C., H. Reeves, E. Gradstztajn, and J. Audouze, 1970, A&A, **8**, 387

Smith , V. V., D. L. Lambert, and P. E. Nissen, 1993 ApJ, **408**, 262

Genesis and Evolution of LiBeB Isotopes I: Production Rates

E. Vangioni–Flam[1], R. Lehoucq[2] and M. Cassé[2]

[1] Institut d'Astrophysique, 98 bis boulevard Arago, F75014 Paris
[2] CE–Saclay, DSM/DAPNIA/Service d'Astrophysique, F91191 Gif sur Yvette cedex

Abstract. In the light of recent observations of gamma-ray lines of carbon and oxygen originating from the Orion nebula, we propose that fast nuclei accelerated by type II supernovæ embedded in giant molecular clouds interact with the ambient medium to produce copious amounts of light elements. Thus, they offer a new mechanism of LiBeB production which has the advantage of being metallicity independent (i.e. "primary"). Our aim is to study its effect on the evolution of these species in the Galaxy. Galactic evolution will be described in a companion paper presented at the Gran–Sasso conference (Nuclei in the Cosmos). In this article we concentrate on production rates.

1 Introduction

Recent gamma-ray observations by COMPTEL(Bloemen *et al.*,1994) have revealed that the Orion nebula is internally irradiated (Bykov and Bloemen, 1994, Clayton, 1994, and Cameron, 1994). The interaction of fast nuclei with the ambient cloud medium offers an unsuspected source of LiBeB (Cassé *et al.*,1994a, Cassé *et al.*,1994b) and a natural explanation to the strong excess of low energy particles, undetectable in the solar cavity due to the repelling effect of the solar wind, hypothesized long time ago by Reeves and co-workers to match the $^{11}B/^{10}B$ ratio observed in meteorites. We are thus led to reassess the evolution of these light elements through a galactic evolutionary model. Other attempts, invoking for instance an overconfinement of GCR in the early galaxy (Prantzos *et al.*,1993), associated with a large ^{11}B production through SN neutrino spallation (Olive *et al.*,1993) are comparatively more speculative. In this paper we present the physical processes responsible for the production of the light elements. The integration of this mechanism in a galactic evolutionary model is treated in a companion paper (Cassé *et al.*,1994b).

The sources of LiBeB isotopes have been traditionally associated to :

- the big-bang (^7Li)
- the spallation of galactic cosmic rays ($^{7,6}Li$, ^9Be and ^{10}B and part of ^{11}B)
- the spallation of a low energy cosmic ray component (most of ^{11}B), up tonow hypothetical, or neutrino spallation of ^{12}C in supernovæ
- AGB stars and novae (^7Li).

The situation has been reviewed by Reeves(1994) and by Pagel(1994). Recently, however, the role of fast particules confined in molecular clouds has been highlighted Bloemen *et al.*,1994). In the light of the COMPTEL observation a significant fraction of the abundance of LiBeB isotopes should be imparted to the interaction of low energy nuclei (about 10 MeV/n) injected and accelerated by type II supernovæ embedded in giant molecular clouds. We are inclined to think that this process plays a significant role in the shaping of the abundance of mass 6 to 11 isotopes. In the following we will illustrate the rather immediate connection between LiBeB and formation (and death) of massive stars in giant molecular clouds exemplified by the Orion nebula.

2 Formalism and Parameters

The influential parameters are :

- the composition of the fast particules,
- the composition of the medium with which they interact,
- the spectral shape of the fast particules.

It is reasonable to assume that particles are injected by SN due to their high internal temperature and accelerated in the surrounding turbulent medium. Thus fast particles would share the SN composition before propagating in the cloud. Turbulence in Orion should be sustained by shock waves due to winds of O and WR stars and SN. In a first approach we adopt a source composition similar to that ejected by a supernovæ massive enough to explode in a giant molecular cloud, typically of 15 to 35 $M_\odot$, with a preference for the highest mass due to its relatively short lifetime. Moreover note that 35 $M_\odot$ is the typical mass of a SN leading to the metallicity pattern observed in globular clusters (Brown *et al.*,1991). The composition of the ejecta taken from Weaver and Woolsey(1993) is especially α and oxygen rich. The interest of this SN composition is that it does not produce an undesirable excess of gamma-ray lines in the 1–3 MeV region. Low energy nuclei produced in molecular cloud would remain trapped there due to their small gyroradius and their short ionization range, producing specific gamma-ray lines and LiBeB nuclei on their way to thermalization. Under these circumstances, ^{6,7}Li are copiously produced by $\alpha + \alpha$ reactions and the other isotopes mainly by $O + \alpha$ and $\alpha + O$ reactions, where the first symbol denotes the particle and the second the target. This scenario has the great advantage to offer a metallicity independent production of light elements of the kind invoked by Duncan *et al.*(1992). The target composition is taken as pure H and He to simulate the early Galaxy or similar to that of the solar system in the Orion case. For the sake of simplicity we adopt a two slope injection spectrum (Ramaty *et al.*,1979) (0 up to a critical energy E_c and n above) representative of particles accelerated in a dense medium (Forman *et al.*,1986). The different production rates are obtained after integration of the energy dependent cross section (Ramaty *et al.*,1979), Read and Viola, 1985) on the steady state spectrum solution of the propagation equation. Alpha particules are favoured in the process not only by the fact that they are abundant in the ejecta of type II supernovæ but also because they are less affected by ionisation energy losses than heavier nuclei in the course of their propagation in the surrounding medium.

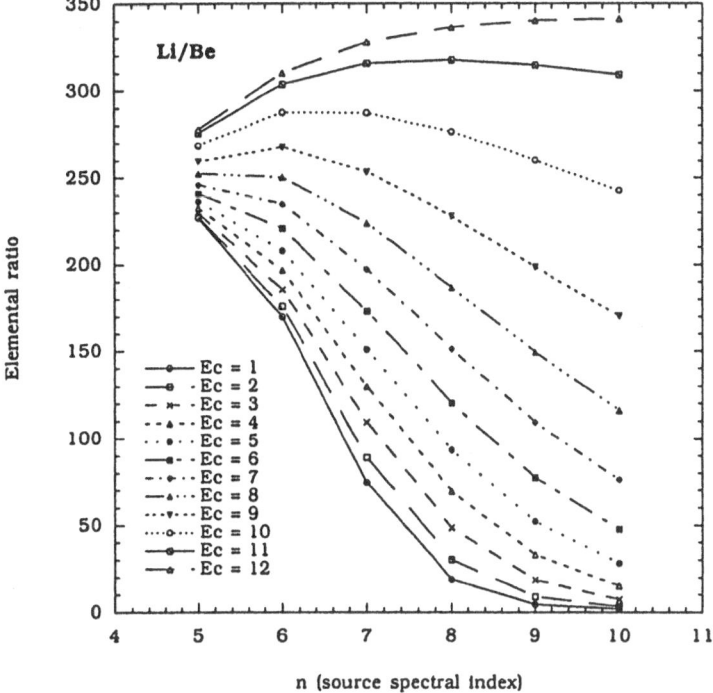

Fig. 1. Li/Be versus E_c and n.

3 Results

We have explored the parameter plane (E_c, n) for different beam compositions. Figures 1 to 4 present the elemental and isotopic ratios of interest. The results are obtained with $35\,M_\odot$ supernova composition, which is a representative case ; the target medium is pure H and He. Intending to apply the Orion mechanism to the early galaxy, we select (E_c, n) couples leading to Li/H< 100 (to avoid overproduction of Li) and $20 < B/Be < 35$ [19, 20]. Three kinds of solutions are left : (4,8), (5,8) and (7,9), but only the third one produces at solar metallicity a $^{12}C^*/^{16}O^*$ ratio compatible with the Orion observation. This selected solution, once integrated into a classical scheme of galactic evolution and complemented by standard galactic cosmic ray spallation and ^{7}Li production by stars, explains consistently the origin and evolution of all LiBeB isotopes (paper II).

In order to produce the total mass of Boron contained in the galaxy, of the order of $100\,M_\odot$, the total number of Orion-like regions integrated over the galactic lifetime, with an associated Boron production rate $= 10^{38}\,\mathrm{s}^{-1}$ [6] during 10^5 yr (Marti and Lingenfelter, 1994), would reach $3.5\,10^7$. This corresponds, in a steady state, to 350 objects in active Orion phase in the Galaxy on average. Taking into account the fact that the present star formation rate is about three times less than the average, we get today about 100 "Orions" in our Galaxy. As each one emits $3\,10^{39}$ photons s^{-1} (in the C, N and O lines) the total gamma-ray

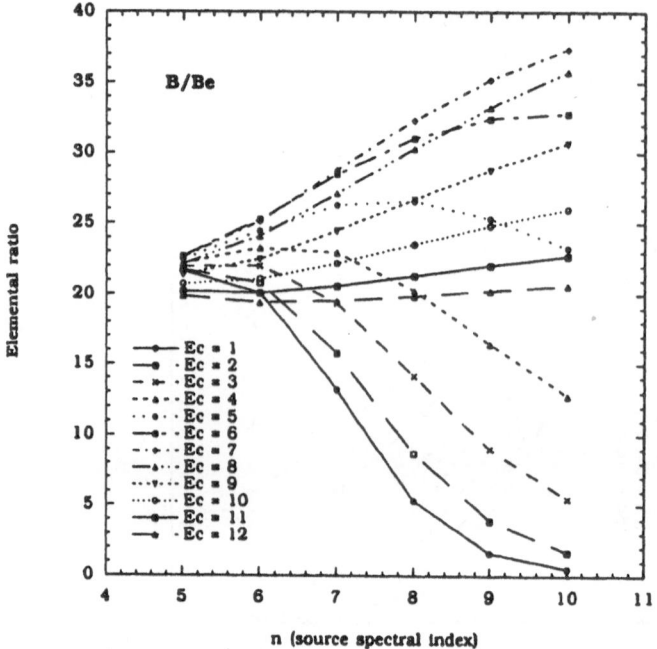

Fig. 2. B/Be versus E_c and n.

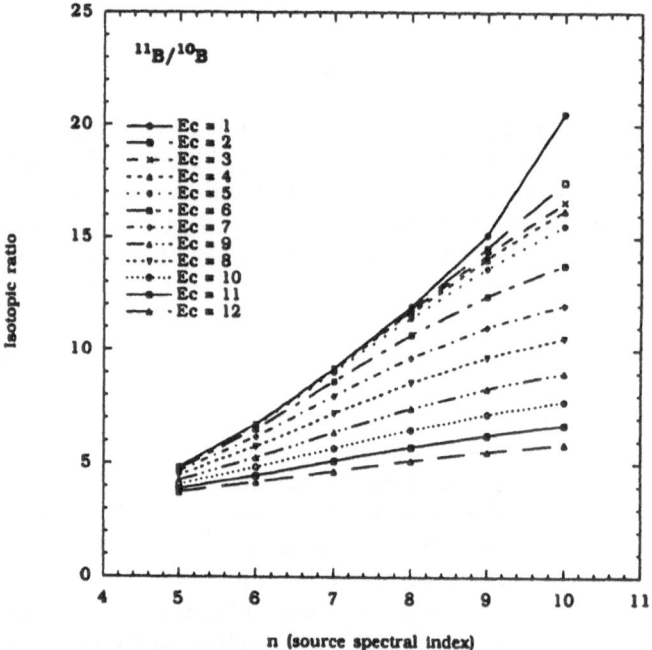

Fig. 3. ${}^{11}\text{B}/{}^{10}\text{B}$ versus E_c and n.

luminosity in this range is $3\,10^{41}$ photons s^{-1}. This is about 10 times less than the 1.8 MeV galactic emissivity associated to ^{26}Al decay as the sensitivity of the detectors are not different between 1.8 and 6 MeV. Thus the disk emission at 4.4 and 6.1 Mev, out of reach of the SMM satellite (Harris *et al.*,1990) should be detectable by the next generation of gamma-ray line telescopes like INTEGRAL.

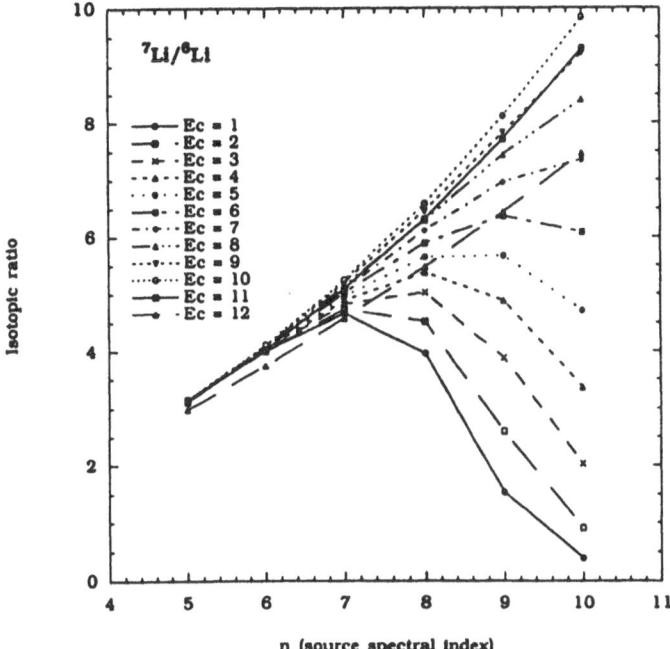

Fig. 4. ^{7}Li/^{6}Li versus E_c and n.

The Li production and hence the corresponding gamma-ray line emission due to the prompt deexcitation of ^{7}Li* and ^{7}Be* (Koslovsky and Ramaty, 1977) strongly depends on the adopted spectral shape since the threshold of $\alpha + \alpha$ fusion is located at 8 MeV/n. Therefore, the detection or lack of detection of the "Lithium feature" around 450 keV would impose stringent constraints on the spectrum of alpha particles. In the present case ($E_c = 7\,\text{MeV/n}$ and $n = 9$) the lithium feature would be much weaker than the combined carbon and oxygen signatures ($L_\gamma = 8.5\,10^{37}$ photons s^{-1} against $3\,10^{39}$ photons s^{-1}). Note that a ^{7}Li/^{6}Li ratio of the order of 2, as observed in a nearby cloud in the direction of ζ Ophiuchus (Lemoine, 1994 and Lemoine *et al.*,1994) would be reproduced with a low E_c (1 to 3 Mev/n) and a high n (8 to 10). This special case, however, would not reflect the average situation in time and space.

References

Bloemen H. et al.: A&A **281** L5 (1994)

Brown J.H., Burkert A. and Truran J.W.: ApJ **376** 115 (1991)

Bykov A. and Bloemen H.: A&A **283** L1 (1994)

Cameron A.G.W.: Nature **368** 192 (1994)

Cassé M., Lehoucq R. and Vangioni–Flam E.: Nature (1994), accepted

Cassé M., Vangioni–Flam E., Lehoucq R. and Oberto Y.: in "Nuclei in the Cosmos", L'Aquila, Italy (1994), to appear

Clayton D.D.: Nature **368** 222 (1994)

Duncan D.K., Lambert D.L. and Lemke M.: ApJ **401** 584 (1992)

Edwardson B. et al.: A&A (1994), in press

Fields B.D., Olive K.A. and Schramm D.N.: ApJ (1994), submitted

Forman M.A., Ramaty R. and Zweibel E.G.: in "The physics of the Sun" ed. P.A. Sturrock, Dordrecht:Reidel, p.249 (1986)

Harris M.J. et al.: ApJ **362** 135 (1990)

Koslovsky B. and Ramaty R.: ApJ Lett. **19** 19 (1977)

Lemoine M., Ferlet R. and Vidal–Madjar A.: A&A (1994), submitted

Lemoine M., this conference

Marti K. and Lingenfelter R.: in "Nuclei in the Cosmos", L'Aquila, Italy (1994), to appear

Pagel B.: in "Nuclei in the Cosmos", L'Aquila, Italy (1994), to appear

Prantzos N., Cassé M. and Vangioni–Flam E.: ApJ **403** 630 (1993)

Olive K.A., Prantzos N., Scully S. and Vangioni–Flam E.: ApJ **424** 666 (1993)

Ramaty R., Koslovsky B. and Lingenfelter R.: ApJ Suppl. **40** 487 (1979)

Read S.M. and Viola V.E.: Atomic Data and Nuclear Data Tables **31** 359 (1985)

Reeves H.: Rev. Mod. Phys. **66** 193 (1994)

Weaver T.A. and Woosley S.E.: Phys. Rep. **227** 65 (1993)

Using Beryllium to Explore Stellar Structure and Evolution

Constantine P. Deliyannis

Institute for Astronomy, University of Hawaii, 2680 Woodlawn, Honolulu, HI 96822, USA

1 Introduction

In order of increasing depth of survival, Lithium (Li), Beryllium (Be) and Boron (B) are all destroyed by (p,α) reactions, and survive only in the outermost layers of stars. *When used simultaneously,* these elements can provide strong clues about what physical processes might be important in stellar interiors. Yet, the literature on Li alone dominates by far. This in part due to the difficulties of observing Be and B : Be can only be observed near the atmospheric UV cutoff at 3130 Å, and B can only be observed from space, further in the UV. Fortunately, as telescopes become larger and the UV efficiencies of spectrographs and ccds increase, it becomes feasible to use Be (and B, using HST) effectively as probes of stellar structure. In this review I discuss several (though not all) contexts where Be, often together with other elements, has provided and continues to provide insight, and I point out potentially interesting areas for future research.

2 The Hg-Mn Stars

Abundance anomalies in Hg-Mn stars (late B stars with vsini $< 100\,\mathrm{km\,s^{-1}}$ and little or no magnetic fields) have found partial explanation in microscopic diffusion theory (Michaud 1970, Heacox 1979). Beryllium is often (but not always) found to be overabundant by 4–5 *orders of magnitude* (Sargent, Searle, & Jugaku 1962, Boesgaard et al. 1982). These Be observations lend further support to diffusion theory, in which large overabundances of Be can be created in Hg-Mn stars through radiative levitation of Be to the surface and its retention there (Borsenberger, Michaud, & Praderie 1984). However, the large range of Be abundances, the unexpected retention of B is some stars (Sadakane, Jugaku, & Takada-Hidai 1985), and other effects suggest that additional complications must also be important, perhaps including rotational mixing and mass loss.

3 The Hyades Giants

When stars evolve off the main sequence (MS), their surface convection zones (SCZ's) deepen considerably, diluting the surface abundances of Li, Be, and B. In standard stellar evolutionary models (which ignore microscopic diffusion, rotation, mass loss, overshoot, magnetic fields, etc.), structural changes are generally such that Li does not burn anew at the base of the SCZ, in spite of the fact that the SCZ contains more than half the stellar mass at maximum depth.

Boesgaard, Heacox, & Conti (1977) obtained Be upper limits in the Hyades red clump giants that were consistent with predicted standard dilution (Iben 1967). However, Li seemed to be overdepleted relative to dilution by a factor of 2–3, perhaps due to MS mass loss ($\sim 1\%$ of the mass). At this Workshop, Duncan has discussed HST B spectra for two Hyades giants that indicate probable B detections, depleted by ~ 1 dex relative to the turnoff, consistent with recent dilution models. Duncan suggested that non-LTE effects (Carlsson et al. 1994) raise the Li abundance by about a factor of 2. Thus, Li, Be, as well as B dilution, the $^{12}C/^{13}C$ ratio (§9), and the C/N ratio are all in remarkable agreement with standard physics alone of 2.25 $M_\odot$ models. Some MS mass loss ($\lesssim 1\%$ or so) and/or meridional circulation could be allowed within the errors.

4 The Boesgaard Li Gap

In a narrow T_{eff} range of only $\sim 300\,\mathrm{K}$, F stars show a severe Li depletion of at least 2 dex (e.g. Fig. 1, the Li "gap"). This gap's almost complete absence at the young age of the Pleiades (Fig. 1a) shows that the gap forms during the MS. Standard stellar models can deplete only a small amount of Li, do so only the pre-MS, and deplete more Li with decreasing T_{eff}, in sharp contradiction to the data. As a result, proposed mechanisms to explain the Li gap have proliferated, and include diffusion (Michaud 1986), mass loss (Schramm et al. 1990), and various mechanisms that result in slow mixing, including meridional circulation (Charbonneau & Michaud 1988), turbulence (Charbonnel & Vauclair 1992), gravity waves (García López & Spruit 1991), and instabilities that lead to angular momentum transport ("Yale" models, e.g. Pinsonneault et al. 1990). Some of these mechanisms may have broader implications (it has been suggested that diffusion and mass loss in stars with T_{eff} similar to the cool side of the Li gap may lower globular cluster age estimates; mixing could lead to high primodial Li with implications for cosmology), so it is important to ascertain which are important and which are not. Fortunately, the degeneracy of Li gap–producing mechanisms can potentially be lifted, at least in part, by differences in the predicted Be and B signatures.

Field stars show evidence of a Be gap, which presumably forms after the age of the Hyades, with interesting relative Li/Be depletion pattern :

a) To within the errors, θ UMa shows meteoritic abundances of Li and Be.

b) Almost all stars with detected Li but depleted by a factor of $10 - 50$ relative to meteoritic show Be abundances a factor of $2 - 4$ below meteoritic (Fig.

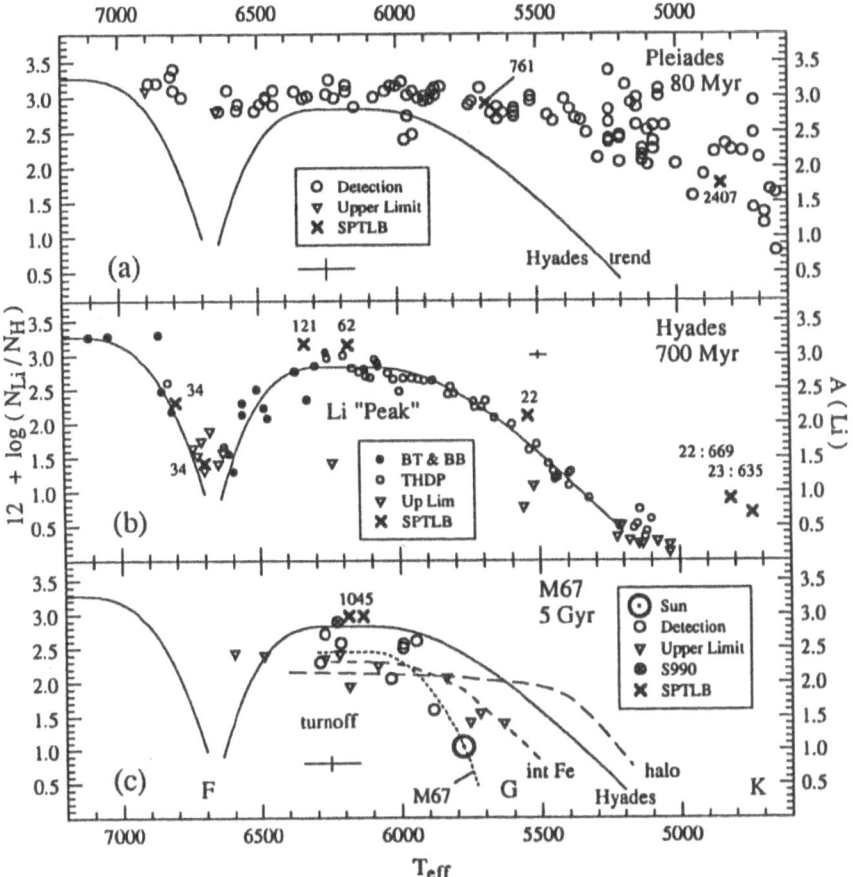

Fig. 1. The morphology of Li abundances in three open clusters, together with intermediate metallicity and halo dwarf trends. See Ryan & Deliyannis (1994) for sources.

2). Are these Be depletions? While the errors in these data may be as large as 0.3 dex in A(Be) for an individual star ($A(X) = 12 + \log (N_X/N_H)$), the average Be seems slightly smaller than meteoritic. Thus, unless the solar nebula was unusually rich in Be or there is some systematic error (§5), on average these stars seem depleted in Be.

c) 110 Her has detected Li, severely depleted by a factor of 100 – 200, and Be that is below meteoritic by a factor of ~ 10 . Its Be is likely depleted.

d) Stars with only upper limits in Be, below meteoritic by factors of 10 – 50, have only upper limits for Li, depleted by factors of more than 30 – 1000.

In short, it seems possible that, on their way to becoming severely Li depleted, *and while they still retain a detectable amount of Li,* Li gap stars already begin depleting Be and eventually deplete Be severely. If this striking Li/Be depletion pattern is indeed accurate and 110 Her represents a phase that normal Li gap stars undergo, some implications would be (Deliyannis & Pinsonneault 1993) :

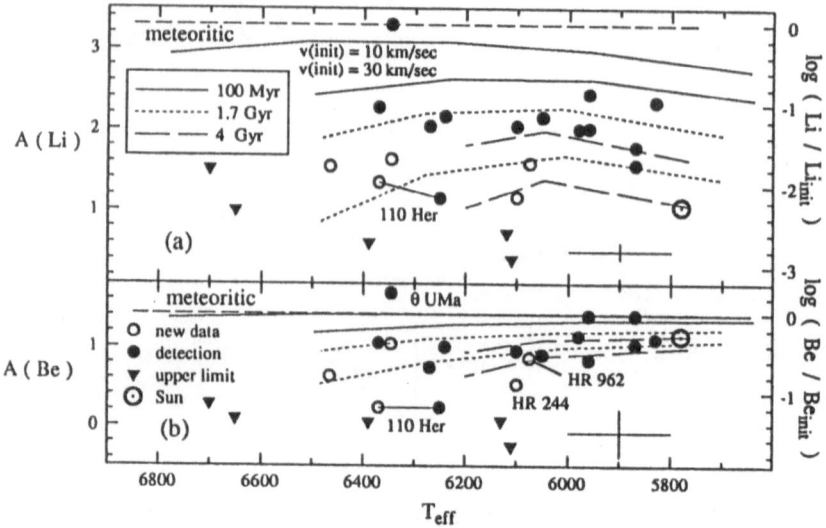

Fig. 2. Li and Be field star data, primarily from Boesgaard & Lavery (1986).

Mass Loss : When main sequence mass loss strips enough material to deplete Be by a factor of 5, all the observable Li would be lost as well. Thus, Li detections in Be depleted stars would provide strong evidence against the mass loss scenario.

Diffusion : Below 6400 K, predicted diffusion of Li and Be is approximately equal (e.g. Michaud & Charbonneau 1991), in sharp contradiction to the data (Fig. 2). Above 6400 K, Be and Li overabundances due to radiative levitation are predicted at different T_{eff} ; neither is observed. Diffusion is not likely to be the dominant cause of the Li gap (perhaps it is suppressed by turbulence).

Slow mixing : Qualitatively, slow mixing to sufficiently deep layers can create the observed Li/Be pattern. Figure 2 shows Yale model predictions against the Li and Be data. Note the difference in morphology between the predicted Li – and Be – T_{eff} curves, consistent with the data. Above 6000 K, increasing initial angular momenta with T_{eff} result in increased mixing. But below 6000 K, destruction of Li (but not Be) at the base of deeper SCZ's causes Li to decrease as T_{eff} decreases (as exemplified by the Sun), whereas Be remains nearly constant.

Given the potential significance of these implications, I have begun a program (with A. Boesgaard and J. King) to study a much larger number of stars. Some preliminary results are shown in Fig. 3. High resolution (R = 80000) and S/N (> 500) Li data enable detection of Li at lower abundances than before; higher resolution and S/N ccd data enable more accurate Be abundances. Figure 3 shows Li (88" telescope at the University of Hawaii) and Be spectra (Keck 10m telescope) in stars of similar parameters. HR 244 has weaker Be lines, correlated

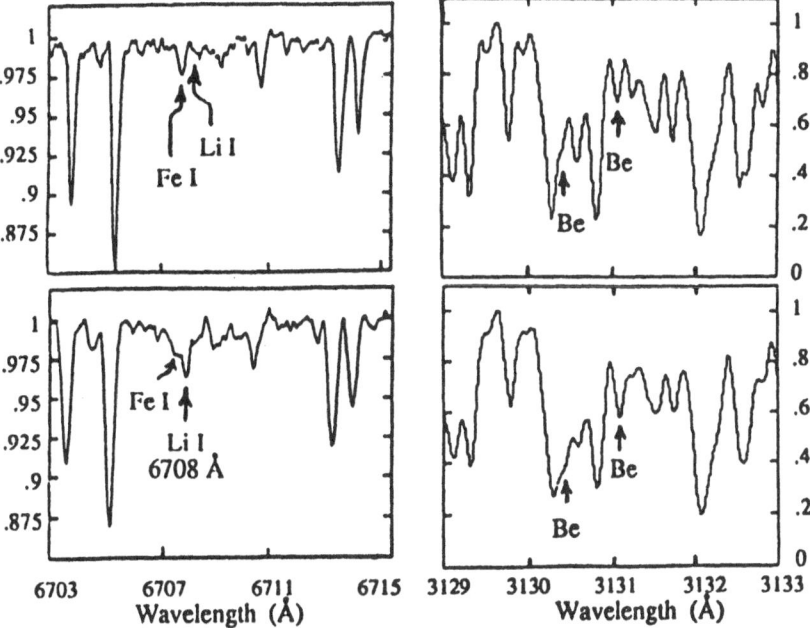

Fig. 3. Li (88", left) and Be (Keck, right) spectra for HR 244 (top) and 962 (bottom).

with weaker (and detected) Li lines, apparently supporting the idea that other stars besides 110 Her are able to deplete Be yet retain some Li. When analysis is complete we hope to have a statistical sample with which to determine whether or not the implications described above are correct.

5   Li and Be Depletion in the Sun, and in Cool Dwarfs

The depleted Li abundances observed in cool (T_{eff} < 6000 K) dwarfs, including the Sun (Fig. 1), have defied definitive explanation for approximately three decades. Comparison of open clusters of different ages shows that Li depletion increases with time, and in the case of G dwarfs, Li depletion continues during the main sequence (MS). This important point is illustrated in Fig. 1, which shows pre-MS Li depletion up to the age of the Pleiades, and progressively lower abundances in the older Hyades and still older M67 (and similarly aged Sun). MS Li depletion is also evident if we restrict attention only to clusters of solar metallicity : Ursa Major at ∼ 300 Myr (Soderblom et al. 1993), Coma at ∼700 Myr (Boesgaard 1987), and M67 show progressively lower Li abundances than in the Pleiades. The main difficulty with standard models is their general feature that the Li destruction occurs only during the pre-MS, at the base of the SCZ. Thus, while standard models may be able to explain the Li abundances in the Pleiades (and in younger clusters), they fail to explain the progressive Li depletion in older clusters. Increasing the depth of the rapidly mixed region, either by employing a larger mixing length or by overshoot, generally results in more Li depletion, but this too occurs in the pre-MS. Thus, some additional mechanism not included in standard models must act to deplete Li during the MS.

At this Workshop, J. Faulkner has claimed that standard stellar models explain the Hyades Li depletion. The higher Pleiades Li abundances would then have to result from less (pre-MS) burning in that cluster, perhaps because its lower metallicity implies shallower SCZ's. While this might at first glance seem plausible, standard models are nevertheless contradicted by the preponderance of data from other clusters. For example, keeping stellar input parameters fixed, according to standard models, G stars in all solar metallicity clusters should then resemble the Pleiades – they do not. Compared to the Hyades, the similarly aged but solar metallicity Coma has similar if not slightly *lower* Li as the Hyades, **not** the Pleiades. The slightly subsolar metallicity and older ($\sim$1.7 Gyr) cluster NGC 752 has lower Li, and the solar metallicity but still older M67 has still lower Li. It seems that age is a more important factor for Li depletion (in contrast to standard models), although metallicity may play a role.

In spite of the improved physics currently available and the attention to detail in Faulkner's models, there still remain uncertainties even in standard models that can result in significant differences in Li depletion. It is, in part, this freedom that allows one to match (even plausibly, sometimes) the Li depletion in any single cluster. (Faulkner's Hyades models present *a possible match* but not *the unique* solution to the Hyades Li data ; other equally plausible standard models can be constructed that *do not* match.) But it becomes much more difficult to match all cluster data and the Sun unless one varies parameters (mixing length, ^{4}He, still missing opacities, etc.) arbitrarily. Aside from lacking any physical motivation at present to do so, this would also ignore the rather strong clue of MS Li depletion with age provided by the totality of cluster data. Unless a convincing way is found for standard models to reproduce MS Li depletion, some additional mechanism will be required. I must emphasize that standard models also fail to explain the dispersion of Li observed at fixed T_{eff} , and the higher Li observed in short period tidally locked binaries (§8).

One proposed non-standard G dwarf MS Li depletion mechanism is mass loss (Hobbs et al. 1989) ; however, Swenson & Faulkner (1992) have convincingly shown this implies various absurdities, including in the IMF. Another proposed mechanism is slow mixing below the SCZ, such as due to turbulence (Schatzman 1969, Charbonnel & Vauclair 1992, Zahn 1994). The Yale models (§4) include a comprehensive treatment of instabilities that lead to angular momentum transport and mixing. As discussed above, an interesting feature of slow mixing is the ability to deplete some Be while still retaining detectable Li.

One unique and possibly very relevant feature of the Sun is the precision and confidence with which we know its initial Li and Be abundances, from meteoritic data. We thus know that its Li is depleted by about a factor of 200. If the data are taken at face value, with meteoritic A(Be) = 1.42 ± 0.04 (Anders & Grevesse 1988) and photospheric A(Be) = 1.15 ± 0.20 (Chmielewski, Müller, & Brault 1975), then Be is also depleted, by about a factor of 2. In standard models, any measurable depletion of surface Be resulting from its destruction at the base of the SCZ implies Li destruction by many more dex ; thus, the solar data would argue very strongly against standard models, and would strongly support slow mixing below the SCZ. However, I cannot entirely rule out the possibility that the apparent difference between the solar and meteoritic Be abundances is merely a reflection of some unsuspected systematic error. The general agreement between photospheric and meteoritic abundances of most elements (Anders & Grevesse 1988) might argue against such errors ; nevertheless, since slow mixing has broad implications, stronger evidence would be desirable.

6   Be in Cool Hyades Dwarfs

The prediction from standard cool dwarf models is unambiguous : Be depletion occurs at significantly lower T_{eff} than where Li is still detectable. Slow mixing, however, can in principle result in at least three Be patterns (with Li is still detectable) : as T_{eff} decreases, the Be abundance can increase slightly, stay constant, or decrease, depending on how the mixing is achieved. Data of García López et al. (1994) in Hyades dwarfs with detected Li seem to show a decline in Be at lower T_{eff} , in contrast to standard models but consistent with slow mixing. However, as they point out, large errors exist at lower T_{eff} due to increasing difficulty in separating Be from unknown blends, so the apparent trend may not be real. Better understanding of the Be line region at lower T_{eff} together with additional observations is clearly of considerable importance.

7 ^{7}Li and ^{6}Li in Halo Stars ; Groombridge 1830

Lithium abundances in halo dwarfs can in principle lead us to the primordial Li abundance, Li_p, *provided we can first understand the interior stellar evolution processes that might have affected the observed surface abundances.* Knowledge of Li_p in turn would constrain models of big bang nucleosynthesis (BBN ; i.e. the number of neutrino families and baryonic density in standard BBN). According to the current traditional view, the "Spite plateau", an almost constant Li abundance (at ~ 2.2) observed in halo dwarfs and subgiants spanning 5600 K – 6300 K, represents the nearly unaltered Li_p. My standard models have supported this view (Deliyannis et al. 1990, Fig. 4a), as they could reproduce the plateau, its apparent (small) slope, the cooler dwarf depletion, and subgiant morphology (§9). Then, ignoring recent observations of D in the direction of a quasar (Songaila et al. 1994), assuming ^{4}He$_p \le \sim 0.24$ (often taken as a 2σ upper limit) and traditional D+^{3}He constraints (e.g. Walker et al. 1991), standard BBN could be self- consistent in a density range of $\eta_{10} = \eta \times 10^{-10} \sim 2 - 4$ (Fig. 5).

However, recall that standard models fail to explain many of the Li and Be features observed in Pop I stars. Conceivably, their apparent success with halo stars could be coincidental. Indeed, I found that models with rotationally induced mixing could also be consistent with the data (Fig. 4b), but implied a

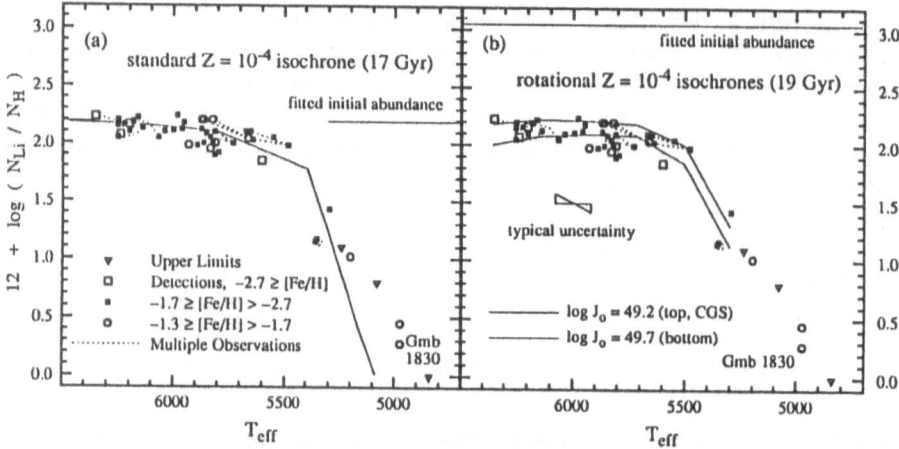

Fig. 4. Halo dwarf Li abundances (compiled in Deliyannis et al. 1990) plotted against (a) standard and (b) rotational stellar evolutionary models.

significantly higher Li_p (Deliyannis 1990 ; Pinsonneault, Deliyannis, & Demarque 1992). This splits the valid density ranges into two branches (Fig. 5). Consistency might be possible at $\eta_{10} \sim 6-7$ if one ignores the quasar D and believes that ^4He_p may be as high as ~ 0.245 (not entirely ruled out) or that incomplete knowledge of neutrino physics might allow a lower ^4He_p . Then it might be possible to account for dark matter (on scales of galactic halos and clusters) with baryons alone, especially at lower H_o. Alternatively, consistency might be possible at $\eta_{10} \sim 1$, with lower values of ^4He_p as preferred by some researchers, the quasar D, and no modification to neutrino physics. $D+^3He$ would be violated, but mixing suggests that this constraint might need modification anyway (§9). Implications for dark matter are strikingly different : non-baryonic dark matter would now be required. Finally, consistency is also possible in inhomogeneous BBN models that can accomodate high Li_p, low ^4He_p, high (quasar) D_p, and a variety of (the least certain) ^3He_p (e.g. Mathews et al. 1990), with unclear implications for dark matter. Additional possibilities will be discussed elsewhere.

I have argued that the constraints on BBN (and implications) are neither as straightforward nor as tight as is often suggested, in part on the basis of possible rotational depletion of Li. How realistic might this prediction be? It is worth digressing to place the rotational halo models in a broader perspective (e.g. see review by Sofia, Pinsonneault & Deliyannis 1991). I must emphasize that the halo rotational models are simply an application of the same method that can explain a variety of striking phenomena, including the Boesgaard Li gap, the relative Li/Be depletion ratio seen in F stars, the main sequence Li depletion in cool Pop I dwarfs, the Li and (possible) Be depletion in the Sun, and the observed Li dispersion at fixed T_{eff} seen in open clusters. They also predicted the existence of tidally locked binaries that preserve higher Li than normal stars, which is

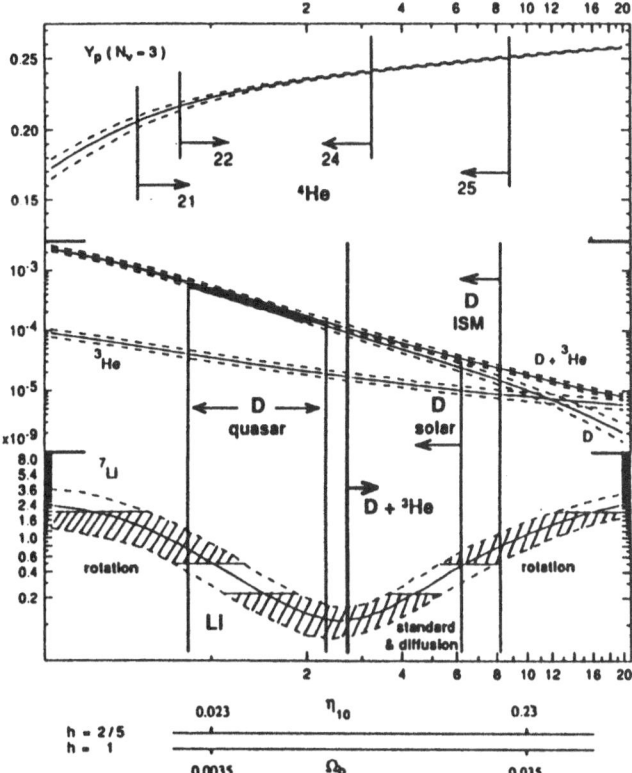

Fig. 5. BBN yields (small letters) of Kernan & Krauss (1994) against primordial abundance estimates (bold large letters).

observed in a variety of contexts, including in a halo subgiant (§8). Furthermore, they predict rapid stellar core rotation, as supported by observations of Pop I subgiants and halo horizontal branch stars, and yet are consistent with the very tight constraints imposed by solar seismology. The differences in Li morphology between halo stars and Pop I stars can be explained naturally in terms of the structural details in the models. Finally, more recent models in which some of the uncertainties have been significantly reduced (Chaboyer & Demarque 1994) give very similar results (significant Pop II depletion), underscoring the robustness of the models. In view of all these successes, the possibility of significant Li depletion in halo stars must therefore be examined quite seriously. At the same time, it must be realized that modelling rotation is far more complex than standard stellar evolution, and there remain a number of uncertainties. For this reason I give a generous error bar as to the predicted depletion for the Spite plateau – anywhere from a factor of 3 to 20. But it would be surprising if the true Spite plateau depletion turned out to be less than a factor of two. Confident determination of Li_p will thus ultimately be instructive regardless of its value : if high, we learn about cosmology ; if low, we learn about stellar evolution.

Either way, and regardless of the appeal of the arguments just listed, it is desirable to test the halo rotational model predictions as directly and as thoroughly as possible. One test is through their prediction that a small dispersion in Li abundances should exist as a result of differences in the initial angular momenta (J_o). Scrutiny of the data has found just such a dispersion (Deliyannis, Pinsonneault, & Duncan 1993), and a much larger sample has confirmed it (Thorburn 1994). However, Thorburn also found a slight slope in the Fe – Li relation, and proposed that Galactic Li enrichment combined with an age spread in the halo could account for the dispersion. Norris et al (1994) have argued that this would result in too much Li enrichment at higher metallicities. Further, the Li dispersion at [Fe/H] $= -3.5$ would require an age spread of several Gyr, which seems unpalatable at that metallicity. Rotational depletion may be the simpler alternative. But even if decision between these alternatives (which are not necessarily mutually exclusive) is difficult, the evidence strengthens the case that at least some Li processing of the Spite plateau has occured, up or down or both.

Boesgaard and King (1993) have found that the cool halo dwarf Gmb 1830 has slightly lower Be abundance than the mean trend. Yet Li has recently been detected in this star (Deliyannis et al. 1994a, Fig. 4). In analogy with the Sun, Be depletion coupled with a Li detection would argue strongly against standard models and strongly for slow mixing below the SCZ, possibly implying a high Li_p. But before such a conclusion can be made, it is desirable to re-observe Be in Gmb 1830 at higher resolution and S/N, and to better understand the Be line region. Further, it would be desirable to establish (or disprove) that any Be deficiency is indeed a depletion, and not an initial condition, which can potentially be done using the B/Be ratio.

The ^{6}Li isotope (generally believed to originate from cosmic rays interacting with the ISM, but can also be created in non-standard BBN models) is more fragile than ^{7}Li and can potentially provide very useful constraints for the models, both stellar and cosmological, if some obstacles come first be overcome. The presence of ^{6}Li would manifest itself as a very small assymetry superimposed on the ^{7}Li line, and is thus challenging to detect. It is even more challenging to identify such an assymetry unambiguously as ^{6}Li ; as Nissen has discussed at this Workshop, poorly understood assymetries due to convection have been observed. Stellar models predict increasing convective velocities toward the turnoff, and indeed Smith, Lambert, & Nissen (1993) report $5 - 6$ km s^{-1} worth of line broadening, the origin of which was not clarified, in the turnoff halo star HD 84937. It is a tiny assymetry on this line that they identify as ^{6}Li ; stronger proof of that assertion is required. If the assymetry is indeed ^{6}Li, then it must next be shown that it is protostellar in origin. Deliyannis and Malaney (1994) have found that it is energetically possible for the extremely shallow SCZ's of turnoff halo stars to be enriched by flare- produced ^{6}Li. Protostellar and flare-produced ^{6}Li can be distinguished observationally : the former should maintain a constant abundance in subgiants (as ^{7}Li is *observed* to do, §9) whereas the latter will begin diluting immediately after the turnoff. Finally, if protostellar ^{6}Li can be demonstrated to exist in HD 84937 and other stars, the models will be constrained, but not as severely as Nissen has claimed. The interesting question still remains, can significant ^{7}Li depletion occur in view of such ^{6}Li constraints? In analogy to the Li/Be situation in F stars like 110 Her (§4), the answer is yes ; one can envision plausible rotational models that deplete ^{7}Li by a factor of 5 – 10 and ^{6}Li by only a factor of 10 – 30, giving both high ^{7}Li$_p$ and plausible initial ^{6}Li.

8 Li in Short Period Tidally Locked Binaries (SPTLB's)

Another striking prediction comes from the combination of the Yale rotational models and the tidal circularization theory of Zahn & Bouchet (1989). In the Yale models, rotational mixing occurs primarily during the late pre-MS and MS. According to tidal theory, binaries observed today with sufficiently short period should have synchronized during the early pre-MS. They would thus avoid the subsequent spin-down, onset of instabilities, angular momentum transport and mixing that the Yale models predict for normal single stars. Main sequence SPTLB's should then exhibit *higher* Li abundances than the mean trend (Deliyannis 1990). This is indeed observed in a variety of contexts, for both dwarfs and subgiants of various metallicities (Ryan & Deliyannis 1994) :

a) Pleiades SPTLB's fall on the mean trend, as predicted (Fig. 1a),

b) cool Hyades SPTLB's are found above both moderately and severely depleted cool Hyades dwarfs (Fig. 1b)

c) two Hyades (marginal?) SPTLB's lie above the "Li peak" (Fig. 1b, below)

d) both stars in a much older M67 SPTLB lie above the Li peak (Fig. 1c)

e) the intermediate metallicity cool dwarf BD −0:4234 is higher than normal

f) the intermediate metallicity subgiant BD +13:13 is higher than normal

g) the halo subgiant HD 89499 lies above its corresponding mean trend, and,

h) −47:1741 and HD 111980 (halo) fall on the mean trend since their periods are too long for them to have synchronized in time to prevent Li depletion.

The region between the Boesgaard gap and the cool dwarfs, the "Li peak", is of particular interest. Open clusters show progressively lower Li with age, with intermediate metallicity stars slightly lower than M67 and the halo Li plateau just slightly lower still (Fig. 1). The SPTLB's lying above the Li peaks in the Hyades and especially in M67 argue that the decline of Li peak abundance with age is due to stellar depletion, **not** cluster Galactic Li enrichment (though some enrichment is allowed, Deliyannis et al. 1994b). Yale models deplete Li quite naturally, all the way to the intermediate metallicity stars and the halo Li plateau. *In this context, the halo Li plateau is completely analogous to the Li peak.* Evidence for a higher Li$_p$ might ultimately come from (or be argued against by) Li data in suitable halo SPTLB's. A word of caution : even SPTLB's can deplete Li, so they too may not give the actual value of Li$_p$ directly.

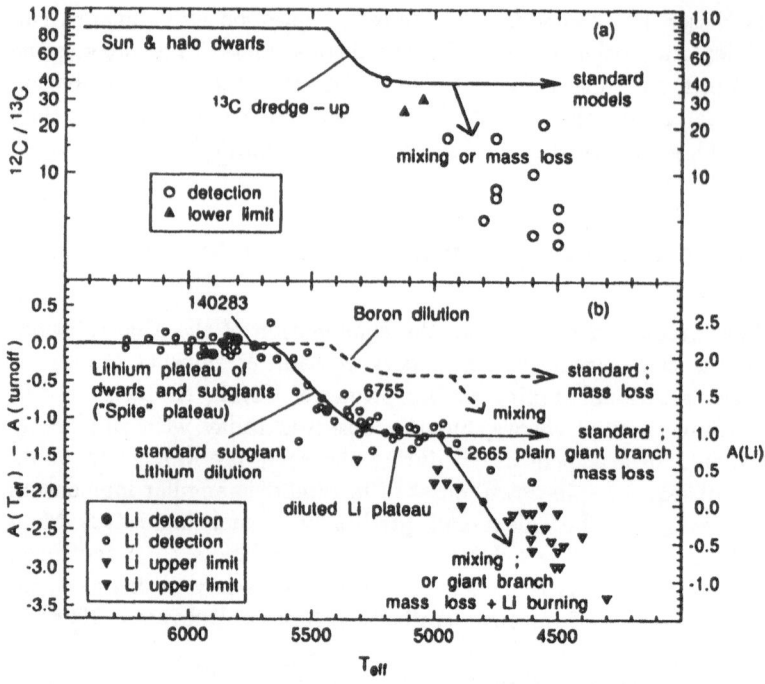

Fig. 6. Halo subgiant observations of (a) $^{12}C/^{13}C$ (Sneden et al. 1986) and (b) Li (Pilachowski et al. 1993) with theoretical standard dilution (Deliyannis et al. 1990).

9 Li and Be in Halo Subgiants

Few observations were available (Fig. 6b, solid symbols) when I first published a halo subgiant (standard) Li dilution curve (Deliyannis et al. 1990). Pilachowski et al. (1993) have augmented the sample considerably (open symbols), and agreement with the predictions is good down to 4900 K. Evidently :

a) The model SCZ deepens by orders of magnitude in terms of mass fraction from the turnoff (~ 6300 K) to the onset of dilution (~ 5600 K) ; the persistence of a constant observed Li abundance in that T_{eff} range indicates the preservation of a constant (though not necessarily unaltered) Li abundance throughout those depths

b) The onset of dilution evidences the Li preservation boundary

c) The slope of dilution from ~ 5600 to 5300 K illustrates the (rather sharp) profile of the Li preservation boundary

d) The model SCZ reaches a maximum depth (and maximum temperature, T_b, at its base) at ~ 5200 K near the base of the giant branch, after which it recedes back to the surface as the hydrogen burning shell moves outward. Concurrently, T_b slowly decreases. The establishment of an observed subgiant diluted Li plateau of constant Li supports this, and further suggests that, as deep as it is, the SCZ is indeed not hot enough to burn Li.

If the apparent dispersions in Li at fixed T_{eff} (e.g. near 5700 K, 5500 K, 5300 K) reflect true abundance spreads, they may be pointing to Li depletion processes, such as rotational mixing, that might have occured on the MS. Finally, note that, to be viable, models with additional physics must also be able to pass the detailed observational constraints listed above.

Together with A. Boesgaard and J. King, I have observed Be in a few halo subgiants (HD 2665, 6755) that have approximately diluted plateau Li abundances (Fig. 6) ; the predicted Be dilution is ~ 0.8 dex. Since Be evolves during the course of Galactic history, one must compare to the mean dwarf Be – Fe trend ; indeed, our upper limits lie approximately that much below the trend.

A related problem involves the $^{12}C/^{13}C$ ratio. Standard models predict a dwarf ratio of about 90 (as is observed in the Sun), which is reduced to about 30 – 60 in halo subgiants as their convection zones deepen and dredge up ^{13}C that was synthesized during the main sequence (e.g. Sneden et al. 1986). The observations are consistent with this (Fig. 6a) until ~ 4900 K, below which the C ratio declines rapidly and reaches, in some stars, the CN cycle equilibrium value of 3 – 4. This decline coincides with a rapid decline of the Li abundances below 4900 K ; both effects sharply contradict standard models. Contenders to explain the low C ratios include mass loss (Dearborn et al. 1976) or mixing on the giant branch (Sweigart and Mengel 1979). In the former scenario, as mass is lost, the MS synthesized ^{13}C peak finds itself closer to the SCZ and eventually gets dredged up. However, after maximum dilution the model SCZ contains more than half the mass, so the mass loss scenario cannot by itself significantly reduce the Li abundance any further, in sharp contrast to the observations. However, given current model uncertainties, it is not possible to completely exclude the possibility that below 4900 K, the SCZ becomes hot enough to burn Li. (It would be necessary to explain why this does not happen between 5300 and 4900 K.) One could then envision a scenario where mass loss explains the decline in the C ratio and burning explain the decline in Li.

However, mixing is a more natural explanation for several reasons. First, mixing will, by itself, naturally explain the decline in both Li and the C ratio. Li is mixed to regions where it is destroyed, whereas, ^{13}C becomes available from the increasing importance of the CNO cycle in the hydrogen burning shell. Second, initially the μ-gradient formed at the point of maximum penetration of the SCZ (near 5300 K) could prevent mixing between the receding SCZ and the layers below the m gradient, thus forming the diluted subgiant Li plateau. Third, after the hydrogen burning shell moves past the composition discontinuity (at about 4900 K in Fig. 6), meridional circulation (or other mixing) could extend all the way to the convection zone and lower $^{12}C/^{13}C$, consistent with observation. Horizontal branch stars do rotate rapidly (Peterson 1985), so the cores of halo giants could indeed be conducive to meridional circulation, strengthening the argument. It should be noted that additional instabilities related to rotation (e.g. see review by Sofia, Pinsonneault, & Deliyannis 1991) and other instabilities (Sneden et al. 1986) exist that could cause mixing. Fourth, mixing signatures continue to manifest themselves with evolution up the giant branch : for example,

Pilachowski (1988) has provided evidence of ON processing in evolved giants in M92, which is difficult to produce in the mass loss scenario.

A good test to distinguish between mixing and mass loss could be provided through observations of boron. B will not burn at the base of the SCZ even if Li does, so should stay almost flat with T_{eff} below 4900 K in the mass loss scenario. But if mixing is correct, a B decline should occur. Since B is sturdier than Li or Be, with the threshold for ^{3}He destruction being next, B (in tandem with the C ratio) is also an excellent probe for ^{3}He destruction.

If mixing is correct, then traditional D+^{3}He constraints on BBN may require modification. Mixing could possibly destroy not only the ^{3}He peak (synthesized during the MS) but also protostellar ^{3}He. But in deriving the D+^{3}He constraint, survival of ^{3}He has been estimated from standard models (Dearborn, Schramm, & Steigman 1986), and the contributions from low mass stars are given large weight in the Galactic chemical evolution. It is precisely these models that are contradicted by the Li and C ratio observations. If mixing significantly destroys the ^{3}He near the base of the giant branch, and then mass is indeed returned to the ISM by the time stars reach the horizontal branch (as is widely believed), then the D+^{3}He bound is moved to lower baryon density, perhaps by a large amount. This would more easily allow self-consistency with low ^{4}He$_p$, perhaps also with the high quasar D, and possibly also with high Li$_p$. For additional details and possibilties, see Deliyannis et al. (1994c).

10   Conclusions

Beryllium observations can enhance our understanding of stellar evolution and implications for cosmology, and can provide especially powerful constraints when used together with other light elements. In F stars, Be and Li together argue against diffusion or mass loss and for slow mixing as the primary cause of the Li gap. Be supports the predictions of standard subgiant dilution in the Hyades giants and in halo subgiants. Future Be observations may give insight as to what role (if any) slow mixing plays in cool Pop I dwarfs such as the Sun, halo dwarfs, and halo giants with unexpectedly low Li and $^{12}C/^{13}C$. Yet these and the other examples discussed here may well be only the beginning of what Be observations might potentially teach us about stellar interiors.

References

Anders, E. & Grevesse, N. 1989, Geo&Cosmo Acta, 53, 197

Boesgaard, A. M. 1987, ApJ, 321, 967

Boesgaard, A. M., Heacox, W. D., & Conti, P. S. 1977, ApJ, 214, 124

Boesgaard, A. M., Heacox, W. D., Wolff, S. C., Borsenberger, J., & Praderie, F. 1982, ApJ, 259, 723

Boesgaard, A. M. & King, J. R. 1993, AJ, 106, 2309

Boesgaard, A. M. & Lavery, R. J. 1986, ApJ, 309, 762

Borsenberger, J., Michaud, G., & Praderie, F. 1984, A&A, 139, 147

Carlsson, M., Rutten, R.J., Bruhls, J.H.M.J., & Shchukina, N.G. 1994, A&A, in press

Chaboyer, B. & Demarque, P. 1994, ApJ, 433, 510

Charbonneau, P. & Michaud, G. 1988, ApJ, 334, 746

Charbonnel, C. & Vauclair, S. 1992, A&A, 265, 55

Chmielewski, Y., Müller, E. A., & Brault, J. W. 1975, A&A, 42, 37

Dearborn, D. S. P., Eggleton, P. P., & Schramm, D. N. 1976, ApJ, 203, 455

Dearborn, D. S. P., Schramm, D. N., & Steigman, G. 1986, ApJ, 302, 35

Deliyannis, C. P. 1990, PhD Dissertation, Yale University

Deliyannis, C. P., Demarque, P., & Kawaler, S. D. 1990, ApJS, 73, 21

Deliyannis, C. P. & Pinsonneault, M. H. 1992, in *IAU Colloquium # 137 : Inside the Stars* in Vienna, Austria, ed. W.W. Weiss, ASP Conference Series, 40, 174

Deliyannis, C. P., Pinsonneault, M. H., & Duncan, D. 1993, ApJ, 414, 740

Deliyannis, C. P., Ryan, S. G., Beers, T. C., & Thorburn, J. A. 1994a, ApJ, 425, L21

Deliyannis, C. P., King, J. R., Boesgaard, A. M., & Ryan, S. G. 1994b, ApJ, 434, L81

Deliyannis, C. P. & Malaney, R. A. 1994, ApJ, submitted

Deliyannis, C. P., Pinsonneault, M. H., & Demarque, P. 1994c, in prep

García López, R. J. & Spruit, H. C. 1991, ApJ, 377, 268

García López, R. J., Rebolo, R., & Perez de Taoro, M. R. 1994, in *Cool Stars, Stellar Systems, and the Sun,* in Athens, Georgia, ed. J.-P. Caillault, ASPCS, 64, 282

Heacox, W. D. 1979, ApJS,41, 675

Hobbs, L. M., Iben, I., & Pilachowski, C. 1989, ApJ, 347, 817

Iben, I. 1967, ApJ, 147, 650

Kernan, P. J. & Krauss, L. M. 1994, preprint.

Mathews, G. J., Meyer, B. S., Alcock, C. R., & Fuller, G. M. 1990, ApJ, 358, 36

Michaud, G. 1970, ApJ, 160, 641

Michaud, G. 1986, ApJ, 302, 650

Michaud, G., & Charbonneau, P. 1991, Space Sci. Rev., 57, 1

Norris, J. E., Ryan, S. G., & Stringfellow, G. S. 1994, ApJ, 423, 386

Peterson, R. 1985, ApJ, 294, L35

Pilachowski, C. A. 1988, ApJ, 326, L57

Pilachowski, C. A., Sneden, C., & Booth, J. 1993, ApJ, 407, 699

Pinsonneault, M.H., Kawaler, S.D. & Demarque, P. 1990, ApJS, 74, 501

Pinsonneault, M. H., Deliyannis, C. P., & Demarque, P. 1992, ApJS, 78, 181

Ryan, S. G. & Deliyannis, C. P. 1994, ApJ, submitted

Sadakane, K., Jugaku, J., & Takada-Hidai, M. 1985, ApJ, 297, 240

Sargent, W. L., Searle, L., & Jugaku, J. 1962, PASP, 74, 408

Schatzman, E. 1969, A&A, 3, 331

Smith, V. V., Lambert, D. L., & Nissen, P. E. 1993, 408, 262

Sneden, C., Pilachowski, C. A., & Vandenberg, D. A. 1986, ApJ, 311, 826

Soderblom, D. R., Pilachowski, C. A., Fedele, S. B., & Jones, B. F. 1993, AJ, 105, 2299

Sofia, S., Pinsonneault, M.H., & Deliyannis, C.P. 1991, in *Angular Momentum Evolution of Young Stars,* Noto, Italy, eds. S. Catalano & J.R. Stauffer, (Kluwer), p.333

Songaila, A., Cowie, L. L., Hogan, C. J., & Rugers, M. 1994, Nature, in press

Sweigart, A. V. & Mengel, J. G. 1979, ApJ, 229, 624

Swenson, F. J. & Faulkner, J. 1992, ApJ, 395, 654

Thorburn, J. A. 1994, ApJ, 421, 318

Walker, T.P., Steigman, G., Schramm, D., Olive, K.A., Kang, H.S. 1991, ApJ, 376, 51

Zahn, J. - P., & Bouchet, L. 1989, A&A, 223, 112

Zahn, J. - P. 1994, A&A, in press

Implications of the B/Be and ^{11}B/^{10}B Ratios

Keith A. Olive[1]

[1] School of Physics and Astronomy, University of Minnesota Minneapolis, MN, 55455, USA

Abstract. Beryllium and boron are widely believed to be produced primarily by cosmic-ray nucleosynthesis. However the observed abundance ratios of B/Be and ^{11}B/^{10}B may indicate that an additional source for boron (presumably ^{11}B) is needed. Neutrino-process nucleosynthesis in supernovae may be such a source. When the ^{11}B yield from the neutrino process is normalized to obtain the correct ^{11}B/^{10}B ratio, one obtains a predictions that B/Be is large (much larger than the nearly model independent prediction of cosmic-ray nucleosynthesis), around 40 for [Fe/H] = -2.6. Large values of B/Be may be supported by recent NLTE analyses of boron observations.

A comparison of observed abundance ratios and their theoretical predictions is a good test of models of galactic cosmic-ray nucleosynthesis (Steigman & Walker, 1992; Prantzos, Cassé, & Vangioni-Flam, 1993; Walker et al., 1993; Steigman et al., 1993; Fields, Olive, & Schramm, 1994) and galactic chemical evolution (Prantzos, Cassé, & Vangioni-Flam, 1993); it may have implications for big bang nucleosynthesis as well (Walker et al., 1993; Olive & Schramm, 1993). The ratios of interest are ^{6}Li/^{7}Li, Li/Be, B/Be, and potentially ^{11}B/^{10}B. In the case of ^{6}Li/^{7}Li where the theoretical prediction of about 0.9 (from cosmic-ray nucleosynthesis) is robust, the observation of ^{6}Li (Smith, Lambert, & Nissen, 1992; Hobbs & Thorburn, 1994) is a good indication that Li is not strongly depleted in stars (at least not by nuclear burning) (Steigman, et al., 1993). The Li/Be ratio which can be used to probe the compatibility between cosmic-ray and big bang nucleosynthesis (Walker et al., 1993; Olive & Schramm, 1993) is much more model-dependent (Fields, Olive, & Schramm, 1994) and may take values anywhere from 10 to 1000. There is as yet no data on the ^{11}B/^{10}B ratio in Pop II objects, but such data would be very interesting. Standard cosmic ray nucleosynthesis models predict ^{11}B/^{10}B $\approx$ 2.5. Finally that brings us to the B/Be ratio which like the ^{6}Li/^{7}Li ratio is robust (Walker et al., 1993; Fields, Olive, & Schramm, 1994) and is an excellent tool to probe theoretical models. The standard model prediction for B/Be is 12 to 14.

In the table below, the available data for B is shown with the corresponding observed abundances of Be. The boron data is taken from Duncan, Lambert & Lemke (1992) for the three stars shown. In addition, for HD140283, the data reflects the measurement by Edvardsson et al. (1994) as well. The source of the Be data is given in the final column. As one can see from the B/Be ratios in table, some of the LTE ratios are in agreement with standard cosmic-ray nucleosynthesis model predictions, but most of them are on the low side of the prediction. For example, the overall average in the case of HD140283, gives B/Be = 6 ± 2. Effort has been concentrated for the most part in determining how low the B/Be can be made within the context of cosmic-ray nucleosynthesis. In

Walker et al. (1993), it was argued on the basis of spallation cross-sections that the extreme lower limit is B/Be > 7.

Table 1. Observed Pop II abundances of Be and B

Star([Fe/H])	[Be]	[B]	LTE B/Be	B/Be	Source*
HD 19445(-2.1)	-0.14 ± 0.1	0.4 ± 0.2	3.4 ± 1.8	~ 8	BK
HD 140283(-2.7)	-1.25 ± 0.4	-0.16 ± 0.14	12 ± 12	$34 - 50$.	R
HD 140283(-2.8)	-0.97 ± 0.25	-0.16 ± 0.14	7 ± 5	$23 - 33$	G
HD 140283(-2.7)	-0.78 ± 0.14	-0.16 ± 0.14	5 ± 2	$14 - 21$	BK
HD 140283(-2.5)	< -0.90	-0.16 ± 0.14	> 7	$> 21 - 30$	M
HD 201891(-1.3)	0.65 ± 0.1	1.7 ± 0.4	10 ± 10	~ 14.5	Re,BK

BK = Bosegaard & King, 1993; R = Ryan et al., 1990, 1992; G = Gilmore et al., 1991, 1992; M = Molaro et al. 1993; Re = Rebolo et al. 1988

Recently, Kiselman (1994) has performed a re-analysis of the inferred B abundance from the data. In the original analysis of Duncan, Lambert, & Lemke (1992), abundances based on the BI and BeII spectral lines were extracted using the assumption of local thermodynamic equilibrium (LTE). The beryllium abundance is believed to be relatively insensitive to this approximation but it was recognized in Duncan, Lambert, & Lemke (1992) that a non-LTE (NLTE) analysis could be a potentially important correction to the boron abundance. An overall upward correction to the boron abundance of 0.46 - 0.62 dex or a factor ~ 3 was found (Kiselman, 1994). The table, also reflects the NLTE corrected B/Be ratio (the second column of B/Be). The weighted average of the three positive observations of Be for HD140283 (corrected for differing surface gravities) is [Be] = -0.93 ± .12. After the Kiselman correction we find that B/Be = 17 - 25 (Fields, Olive, & Schramm, 1995), using the central value of Be.

In Fields, Olive, & Schramm (1995), we considered various cosmic-ray source spectra which are power law in momentum and total energy with escape pathlengths, Λ, between 10 - 1000 g/cm^2. The resulting ratios for Li/Be and B/Be are shown in Figures 1 and 2 for the momentum spectrum. The solid lines correspond to $\Lambda = 10$ g/cm^2 and the broken lines to $\Lambda = 1000$ g/cm^2. Assuming that Be is produced solely by cosmic ray nucleosynthesis and that Li is predominantly produced in the big bang, the B/Be ratio can be constrained to Li/Be < 1000; a very conservative bound based on the consistency of the Be observations (Walker et al. 1993). ¿From the Figures it is clear that B/Be < 14. A similar result is obtained for the energy spectrum as well.

Given the upper limit B/Be < 14, and the NLTE values in the table, we can pose the question as to whether an additional source of B is needed. The apparent lack of boron production in standard comic ray models is compounded when one notes that the observed isotopic ratio in the solar neighborhood ^{11}B/^{10}B = 4 relative to the prediction of 2.5.

A potentially important source for ^{11}B production has been found to result from neutrino induced nucleosynthesis in type II supernovae (Woosley et al., 1990). The core collapse of a massive star into a neutron star creates a flux of

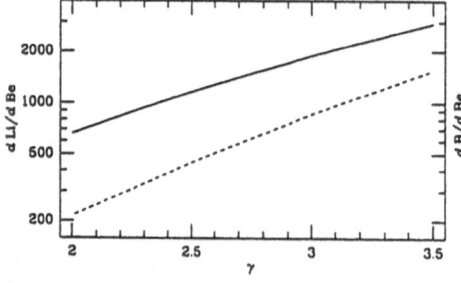

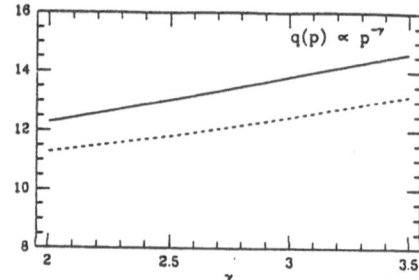

Fig. 1. Li/Be production rates for a source spectrum $q(p) \propto p^{-\gamma}$, plotted as a function of spectral index γ

Fig. 2. As in Figure 1, for the ratio of B/Be

neutrinos so great that in spite of the small cross sections involved, it may still induce substantial nucleosynthesis. It was found that considerable ^{11}B production can result as the flux of neutrinos passes through the He, C, and Si shells of the stellar envelope, primarily by neutrino spallation of ^{12}C. The dominant product is ^{11}B since it is favored for ν-spallation to knock out a single nucleon. In addition, some synthesis of ^{7}Li and ^{10}B takes place by this process but the production rate seems quite low.

An important aspect of the calculation in Woosley et al. (1990) is the full treatment of pre- and post-shock nucleosynthesis. Since the duration of the neutrino burst exceeds the time scale for the passage of the shock through the inner layers of the exploding star, ν-process nucleosynthesis can continue after the passage of the shock. In the outer layers, however, the destruction of fragile isotopes is a significant effect and is, for example, responsible for the destruction of ^{9}Be. Uncertainties in the calculation arise primarily from the assumed neutrino temperature and the cross sections for boron production. This process is attractive however as it naturally creates ^{11}B without much ^{10}B, and so provides the needed source of ^{11}B to augment the cosmic ray production and so reproduce the ^{11}B/^{10}B ratio (Olive et al., 1994).

Indeed, the ^{11}B yields from these processes (Woosley, Timmes, & Weaver, 1993) were incorporated in a chemical evolution model (Olive et al., 1994). Respecting the overall constraints imposed by the LiBeB observations in halo stars, we were able to obtain a solar isotopic ratio ^{11}B/^{10}B ≈ 4. Using the boron isotopic ratio to normalize the ν-process yields, it was shown that neutrino process nucleosynthesis leads to a relatively model independent prediction that the B/Be elemental ratio is large (> 50) at low metallicities ([Fe/H] < -3.0), assuming still that Be is produced as a secondary element as is the case in the conventional scenario of galactic cosmic-ray nucleosynthesis. In particular, at the metallicity corresponding to that of HD140283, [Fe/H] $\lesssim$ -2.6, it was predicted that the B/Be ratio should be close to 40. Though still on the high side, this is in overall good agreement with the NLTE corrected values shown in the data Table. In the Figure 3 the B/Be ratio is shown as a function of [Fe/H]. The theoretical curves are compared to available data. As one can readily see, the primary component of ^{11}B greatly enhances the B/Be ratio at low metallicities.

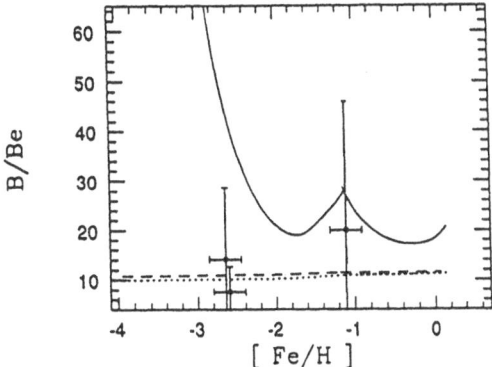

Fig. 3. The B/Be ratio as a function of [Fe/H] with (solid) and without (dashed) ν-process nucleosynthesis included. Also shown is the B/Be ratio in an enhanced model (Prantzos Cassé, & Vangioni-Flam, 1993) (dotted)

References

Boesgaard, A.M., & King, J. 1993, AJ, 106 2309

Duncan, D.K., Lambert, D.L., & Lemke, M. 1992, ApJ, 401, 58

Edvardsson, B., Gustafsson, B., Johansson, S.G., Kiselman, D., Lambert, D.L., Nissen, P.E., and Gilmore,G. 1994, A & A (in press)

Fields, B.D., Olive, K.A., & Schramm, D.N. 1994, ApJ (in press)

Fields, B.D., Olive, K.A., & Schramm, D.N. 1995, ApJ (in press)

Gilmore, G., Edvardsson, B., & Nissen, P.E. 1991, AJ, 378, 17

Gilmore, G., Gustafsson, B., Edvardsson, B., & Nissen, P.E. 1992, Nature, 357, 379

Hobbs, L., & Thorburn, J. 1994, ApJ, 428, L25

Kiselman, D. 1994, A&A, 286, 169; and these proceedings

Molaro, P., Castelli, F., & Pasquini, L. 1993, Origin and Evolution of the Light Elements, ed. N. Prantzos, E. Vangioni-Flam, & M. Cassé (Cambridge: Cambridge University Press), 153

Olive, K.A., Prantzos, N., Scully, S., & Vangioni-Flam, E. 1993, ApJ, 424, 666

Olive, K. A., & Schramm, D. N. 1993, Nature, 360, 439

Prantzos, N., Cassé, M., & Vangioni-Flam, E. 1993, ApJ, 403, 630

Rebolo, R., Molaro, P., Abia, C., & Beckman, J.E. 1988, A&A, 193, 193

Ryan, S., Bessel, M., Sutherland, R., & Norris, J. 1990, ApJ, 348, L57

Ryan, S., Norris, J., Bessel, M., & Deliyannis, C. 1992, ApJ, 388, 184

Smith, V.V., Lambert, D.L., & Nissen, P.E. 1992, ApJ, 408, 262

Steigman, G., Fields, B. D., Olive, K. A., Schramm, D. N., & Walker, T. P. 1993, ApJ, 415, L35

Steigman, G., & Walker, T. P. 1992, ApJ, 385, L13

Walker, T. P., Steigman, G., Schramm, D. N., Olive, K. A., & Fields, B. D. 1993, ApJ, 413, 562

Woosley, S. E., Timmes, F. X., & Weaver, T. A. 1993, in *Nuclei in the Cosmos*, eds. F. Käppeler & K. Wisshak, (London:Institute of Physics Publishing)

Woosley, S.E., Hartmann, D.H., Hoffman, R.D., & Haxton, W.C. 1990, ApJ, 356, 272

Beryllium Abundances in a New Set of Halo Stars

P. Molaro[1], P. Bonifacio[1], F. Castelli[2], L. Pasquini[3], F. Primas[4]

[1] Osservatorio Astronomico di Trieste, Italy
[2] CNR-GNA Unitá di ricerca di Trieste, Italy
[3] European Southern Observatory, La Silla, Chile
[4] Dipartimento di Astronomia, Universitá of Trieste, Italy

Abstract. We present observations of the Be II 3130 Å resonance doublet in 15 halo stars with metallicities ranging from [Fe/H]=-0.4 to -3.2 obtained with the CASPEC spectrograph with a FWHM≈0.09 Å   resolution at the ESO 3.6m telescope. Abundances are derived by means of the synthetic spectra technique employing Kurucz (1993) atmospheric models, with enhanced α-elements and no overshooting. The derived abundances show that Be correlates linearly with iron, giving strength to previous results. Three stars are found to be Be deficient. It is also shown that Be observations are useful to discriminate strongly Li-depleted stars and to follow the lithium galactic evolution.

Introduction

Beryllium is the lightest nuclide which is not synthesized in the standard big bang nucleosynthesis though traces are predicted in inhomogeneous models. It is destroyed in stellar interiors and most likely produced via spallation of heavy nuclei by energetic cosmic rays in the interstellar medium or close to supernovae. Thus, in particular for halo stars, Be observations are a probe for cosmological models, Galactic evolution, and stellar structure. Observations of Be in halo dwarfs began in the eighties (Molaro and Beckman 1984, Molaro, Beckman and Castelli 1984, Molaro 1987, Rebolo et al 1988, RMAB), but only recently the technology has become adequate for providing the required sensitivity and resolution at 3130 Å so that accurate observations of the BeII lines became possible (Ryan et al 1992, RNBD, Gilmore et al 1992, GGEN, Boesgaard and King 1993, BK). In the following we use the notation: $[Be]=\log(N_{Be}/N_H)+12$.

Observations

Data are obtained with the CASPEC spectrograph at the ESO 3.6m telescope along two observing runs in February 1992 and March 1993. The resolving power measured from the Th-Ar calibration lamp is of ≈ 35000 (FWHM=0.09 Å). Some examples of spectra showing the BeII lines at 3130.420 and 3131.065 Å are given in Fig 1. Be lines are detected in HD 140283 where the BeII 3131.07 Å has an EW of 8 mÅ. Be lines are also seen in HD 200654, which is particularly interesting

Table 1. Stellar parameters

Star	V	[Fe/H]	T_{eff}	logg	[Be]	[Li]	R
HD 3795	6.1	-0.8	5400	4.0	-0.60		
HD 6434	7.7	-0.6	5800	4.5	0.92		
HD 16784	8.0	-0.9	5500	4.0	0.40		
HD 25704	8.1	-1.1	5800	4.0	0.34	2.35	
HD 64096	5.2	-0.3	5900	4.3	1.03	1.80	
HD 64606	7.4	-0.9	5100	4.0			
HD 76932	5.8	-1.0	5800	3.5	0.40	1.96	
HD 106516	5.8	-0.9	5800	3.5	<-0.70	< 1.80	
HD 128279	8.0	-2.5	5100	2.7	-1.1	<<1.00	
HD 140283	7.2	-2.7	5650	3.4	-0.95	2.00	
HD 160617	8.7	-2.0	5800	3.8	-0.65	2.20	
HD 166913	8.2	-1.8	5900	3.7	0.06	2.17	
HD 200654	9.1	-3.2	5200	2.7	-1.26	0.94	
HD 211998	5.3	-1.5	5200	3.5	<-0.91	1.1	
HD 218502	8.5	-2.0	6000	3.8	-0.60	2.16	
HD 219617	8.2	-1.5	5800	4.0	-0.25	2.20	

because [Fe/H]=-3.2. However, Be is not seen by GGEN in this star so that our detection requires further confirmation.

Be abundances have been derived through spectral synthesis of the Be region obtained with the SYNTHE code (Kurucz 1993). The spectral synthesis is required by the heavy blending of the BeII resonance lines (Molaro et al 1993). By using the ATLAS9 code and the opacity distribution functions of Kurucz (1993), we computed the atmospheric models for the sample stars. BeII lines are not saturated in PopII stars. The parameters T_{eff}, log g and [Fe/H] are taken from the literature (cfr Table 1). The abundances are solar scaled for all elements except for the α-elements for which an abundance enhanced by 0.4 dex was assumed to represent better the chemical composition of the halo. Solar abundances used for reference are those of Anders and Grevesse (1989), except for iron up-dated to log N_{Fe}/N_{tot}=-4.53. The atomic and hydrides line lists are taken from Kurucz (1993). The mixing length parameter was 1.25 but we dropped the overshooting option which is instead effective in the Kurucz (1993) grid of models. The overshooting treatment affects the elemental abundances in the metal poor stars as shown in Molaro et al (1994). Be abundance is strongly dependent on gravity (Δlogg of +0.5 dex induce a Δ[Be] $\approx$ +0.4 dex) moderately dependent on T_{eff} and weakly dependent on the metallicity. We adopted a microturbulence of 1.5 km s^{-1}.

Results

Our [Be] abundances are shown in Fig. 2 together with all the Be observations available in literature. The stars with multiple determinations are connected with a line. For HD 76932 and HD 140283, which are the most studied halo stars, our determinations are in good agreement with those of other authors. When we restrict only to the high quality data (without HD 200654 and HD 128279) and take the average of multiple determinations, the best fit in the interval $-2.7 < [Fe/H] < -1.0$ gives [Be]=1.60+0.94[Fe/H]. This is in good agreement with the [Be] $\propto 0.8$[O/H] originally found by GGEN from only 6 determinations and is more gentle than the [Be]$\propto 1.256$[Fe/H] derived by BK. As we can see in correspondence of the most metal-poor objects, there is no evidence for a *plateau* in the Be abundances analogous to that observed for Li abundances. It is worth noting that the fit obtained for [Fe/H]< -1.0 provides a good fit for overall metallicities up to [Fe/H]$\approx$ 0, where it intercepts the meteoritic value of [Be]=1.56 (Reeves and Meyer, 1978). Be tracks iron closely and [Be/Fe] remains always at the meteoritic value, although some dispersion is present in particular at solar metallicities.

Standard models predict negligible Be depletion in dwarfs with $T_{eff} \geq$ 4900 K (Deliyannis et al 1990), but no detectable Be is found in the stars HD 3795, HD 106516 and HD 211998. The spectrum of HD 106516 is shown in Fig. 1. The atmospheric properties of this star are very close to those of HD 76932 (cfr Table 1) and when their spectra are divided all the spectral features cancel out with the exception of the BeII lines of HD 76932 which are shown at the bottom of Fig. 1. Two of these stars show also no Li, but HD 211998 has Li measured with [Li]=1.0, which is $\approx$ 1 dex below the plateau value, while Be appears even more depleted.

Li burns at lower temperatures than Be and simultaneous observations of the two elements provide information about the relative depletion factors. The amount of Li produced by spallation of high energy cosmic rays can be inferred from the observed Be, which is produced uniquely by these processes. Considering the whole sample of stars for which the two elements are measured, it turns out that the amount of Li produced by spallation of high energy cosmic rays is negligible for low [Fe/H], but for [Fe/H]> -1.0 it is often greater than the Li observed. These are stars where Li has been strongly depleted and occupy a well specified region in the [Li]-[Fe/H] diagramme (see Fig. 3). This behaviour strongly suggests that the envelope of the data points in Fig. 3 certainly follows the increase of Li by Galactic evolution.

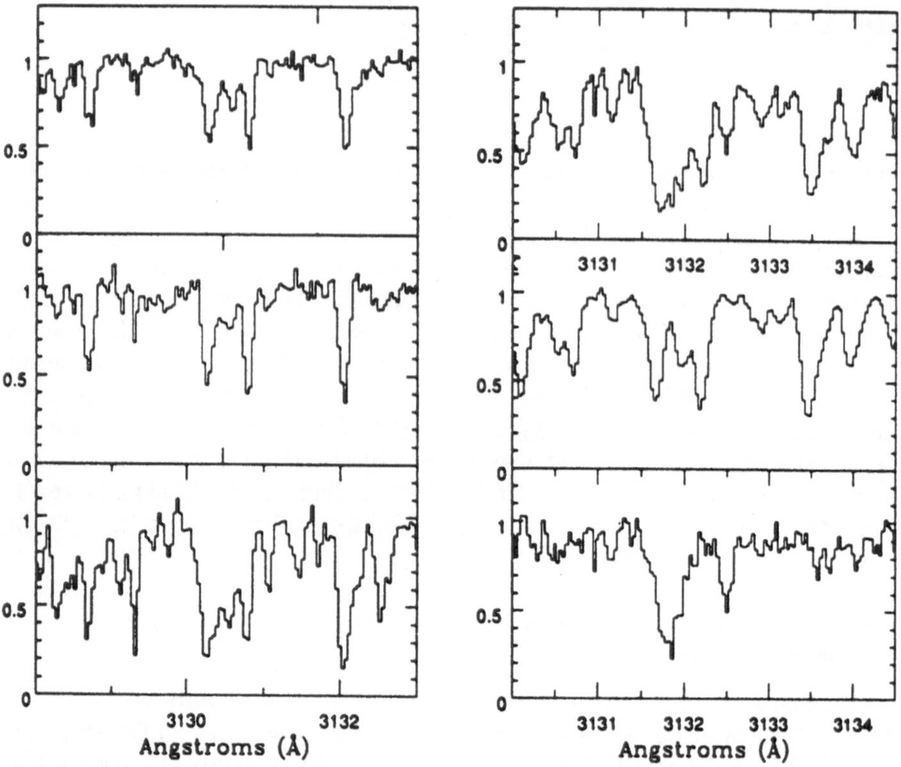

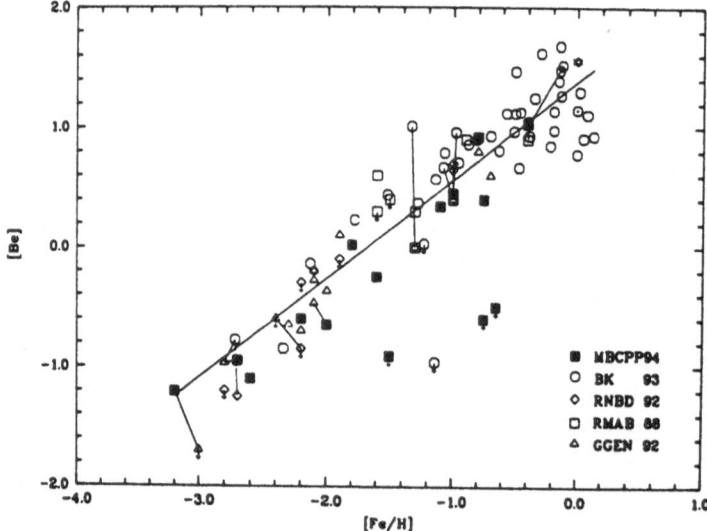

Fig. 1. Left panel. From the top: HD 140283, HD 218502, HD 25704. Right panel. From the top: HD 76932 and HD 106516, the bottom panel shows the spectrum resulting after the division of HD 76932 with the Be-weak star HD 106516.

Fig. 2. [Be] versus [Fe/H]. Filled squares are this paper, ($\star$) is the [Be] meteoritic value.

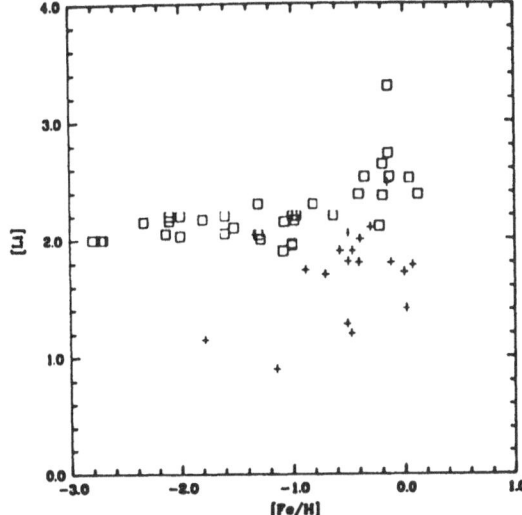

Fig. 3. [Li] versus [Fe/H]. (+) show stars which are likely Li depleted

References

Anders, E., Grevesse, N.: 1989 Geochim. Cosmochim. Acta, **53**, 197.

Boesgaard A., and King J. R.: 1993 Astron. J **106**, 2309.

Deliyannis, C.P. and Pinsonneault M. H.: 1990 Ap. J. 365, L67.

Gilmore G., Gustafsson, B., Edvardsson, B. and Nissen P.E.: 1992 Nature **357**, 379.

Kurucz, B.: 1993 CDROM.

Molaro, P., Beckman J. E., and Castelli F.: 1984 ESA SP-219, 197.

Molaro, P.: 1987 PHD Thesis SISSA Trieste.

Molaro P., and Beckman J.E.: 1984 A&A **139**, 394.

Molaro P., Castelli F., Pasquini L.: 1993 *Origin and Evolution of the Elements* 153.

Molaro, P., Primas, F., Bonifacio, P.: 1994 A&A submitted.

Norris, J.E., Ryan, S.G., Stringfellow, G.S.: 1994 Ap.J. **423**, 386.

Prantzos, N., Casse, M., Vangioni-Flam, E.: 1993 Ap. J. **403**, 630.

Reeves, H., and Meyer, J.-P.: 1978 Ap. J. **226** 613.

Rebolo R., Molaro, P., Abia, C., and Beckman, J.E.: 1988 A&A **193**, 193.

Ryan S., G., Norris, J.E., Bessel, M.S., and Deliyannis, C.P.: 1992 Ap. J. **388**, 184.

Beryllium in Metal-Poor Stars

R. Rebolo[1], R. J. García López[1,2], M. R. Pérez de Taoro[3]

[1] Instituto de Astrofísica de Canarias, E-38200 La Laguna, Tenerife, Spain
[2] Department of Astronomy, The University of Texas at Austin, Texas 78712-1083, USA
[3] Museo de la Ciencia y el Cosmos de Tenerife, E-38200 La Laguna, Tenerife, Spain

Abstract. We present new beryllium abundances for 20 metal-poor stars with metallicities ([Fe/H]) in the range -3.1 to -0.3, most of them with [Fe/H] below -1.3. Beryllium abundances were derived from 0.06 Å effective resolution spectra via spectral synthesis. Our results show evidence for a strong Be enrichment in the early Galaxy. Any primordial Be "plateau" is constrained to be at [Fe/H]≤ -3.

1 Introduction

Beryllium abundances in metal-poor stars are important for tracing the interaction of cosmic rays with interstellar matter at very early epochs in the life of our Galaxy and for probing non-standard Big Bang nucleosynthesis. It is expected that collisions between cosmic rays and intestellar gas nuclei (C,N,O) produce ^{9}Be (hereafter referred to as Be) as well as other light element isotopes (^{6,7}Li, ^{10,11}B). Measurement of Li, Be and B ratios in metal-poor stars provides a unique test for the occurrence of Galactic cosmic ray nucleosynthesis (e.g. Fields et al. 1993; Prantzos et al. 1993). Recently, there has been considerable interest in the possibility that non-standard Big Bang nucleosynthesis may produce Be (Kajino & Boyd 1990; Terasawa & Sato 1990; Kawano et al. 1991) at quantities measurable in very metal-poor stars. Observation of a Be "plateau" at very low metallicities, as found for Li, could be a strong indication of such primordial production.

We are conducting a programme aimed at determining beryllium abundances in a large sample of metal-poor stars (halo, thick-, and thin-disc populations). The observations presented here have been carried out at the 4.2-m William Herschel Telescope (WHT) with the Utrecht Echelle Spectrograph (UES). Additional observations have been obtained at the 2.5-m Nordic Optical Telescope using the IAC-Univ. of Belfast (IACUB) echelle spectrograph, and at the Anglo-Australian Telescope. The WHT spectra have an effective resolution of 0.06 Å. Several exposures were recorded for the faintest objects in the sample ($V \sim 9-10$), and each one was individually reduced. The extracted spectra were co-added after inspection of possible spikes in the Be spectral region, and then wavelength calibrated. The S/N was typically 30, always higher than 16, and for several stars (such as HD140283 and HD19455) it reached 65.

2 Stellar Parameters

Effective temperatures were estimated from the calibrations of the $V - K$ and $b-y$ indices by Carney (1983), Magain (1987a), and Alonso, Arribas, & Martínez-Roger (1994). In addition, the new Kurucz (1992) models for metal-poor stars were used to determine T_{eff} and $\log g$ for our sample in a systematic way. Using an interpolation method and the grid of colours provided by Kurucz, we obtained the "best" model describing the observed $V - K$, $b - y$, and c_1 indices for each star (stellar metallicities were taken from published values). We found that the combination $(V - K, c_1)$ provided a good discrimination between models, much more restrictive than $(b - y, c_1)$. The effective temperatures of the "best models" were then averaged with those from other independent calibrations to produce the final T_{eff} used in the analysis. We adopted the gravity of the "best" Kurucz models in our spectral analysis. The accuracy of the photometry permitted an estimate of $\log g$ to within ± 0.25 dex.

It is well known that Be abundance determinations (like oxygen abundances from OH lines) depend critically on the assumed stellar gravity. As a check on our gravity estimates, we compared our values for two stars in the sample with precise measurements available in the literature:

- HD103095. Smith, Lambert, & Ruck (1992) determined $\log g = 4.5 \pm 0.1$ and $T_{eff} = 5170 \pm 70$ K from a fine analysis of calcium and iron lines. It is in excellent agreement with our value of $\log g = 4.6$.

- HD140283. It has a precise parallax reported (accuracy 0."0019) by Harrington, from which Nissen et al. (1994) estimated $\log g = 3.39 \pm 0.15$. Our value $\log g = 2.96 \pm 0.2$ is clearly below the uncertainties associated with the parallax method. We also note that Magain (1987b) obtained $\log g = 3.2 \pm 0.2$ from a fine spectroscopic analysis. The value obtained from the Kurucz models appears to be too low in this case.

We cannot yet explain why the agreement is good for one star and poor for the other. More work on gravities of metal-poor stars is needed to address the problems concerning the nucleosynthesis and evolution of beryllium correctly.

Since there are no accurate parallaxes or detailed spectroscopic studies for the whole sample, we decided to use the gravities derived from the Kurucz models. They may be affected by unknown systematics, but will make our determinations internally consistent and will facilitate the comparison of our results with those from other groups obtained using, for instance, VandenBerg & Bell (1985) isochrones or other model atmospheres (i.e. MARCS of Gustafsson et al. 1975). Taking this into account, our abundances should be considered as preliminary.

3 Spectral Synthesis

The abundance analysis has been performed via spectral synthesis fitting of the weaker line, which is clearly resolved in our spectra. The presence of the stronger Be II line also provides a constraint to the derived abundances (given the uncertainties related to its blend with other lines). Synthetic spectra were computed using the code WITA2, a UNIX-based version of ABEL6 code (Pavlenko 1991), which allows to compute LTE atomic and molecular synthetic profiles for a given model atmosphere. We employed a grid of model atmospheres provided by Kurucz (1992), which were interpolated to get the models required for each one of the stars.

The $\log gf$ values for the Be II lines were taken from Wiese & Martin (1980). For the other lines we varied the values given in the line list of Kurucz (1992) until a good fit was obtained to the integrated flux solar atlas of Kurucz et al. (1984). We employed a model with $T_{\text{eff}} = 5750$ K, $\log g = 4.5$, [Fe/H]= 0.0, $\xi = 1.5$ km s^{-1}, and solar abundances for the different elements (including beryllium) which where taken from Anders & Grevesse (1989). Our fit shows a similar quality to those existing in the literature for the same purpose, and the agreement with the observed Be II weaker line is excellent.

A microturbulence of 1.5 km s^{-1} was adopted for all the stars. The spectral syntheses were not significantly sensitive to this assumption. On the other hand, García López, Severino, & Gomez (1994) have shown that NLTE corrections to the Be abundances in metal-poor stars are < 0.1 dex, suggesting that the LTE abundances are basically correct.

It should be noted that our present, as well as earlier, Be measurements may be affected by inadequate gravity estimations, inducing possible systematic errors as large as 0.25 dex. The uncertainty of present Be abundances in metal-poor stars, except for a few cases where detailed studies are available, may easily be larger than 0.25 dex.

4 Discussion

We plot in Fig. 1 our Be abundances versus stellar metallicities. Results from previous works are also included. There is a general good agreement between independent sets of measurements in the sense that all of them strongly suggest an early Be enrichment in our Galaxy. Our abundances are located slightly below those from others, which we mainly attribute to the lower gravities obtained with Kurucz models for our sample of stars. The slope of the relationship Be vs. Fe is well determined from our data.

A detailed comment on each object is not possible here, but let us make the following remarks on several individual stars:

(i) HD103095. Is Be depleted? Recently, Deliyannis et al. (1994) have claimed the detection of Li in this rather cool metal-poor star (log N(Li)= 0.27 ± 0.06, where log N(Li)≡log (Li/H)+12), which Boesgaard & King (1993)

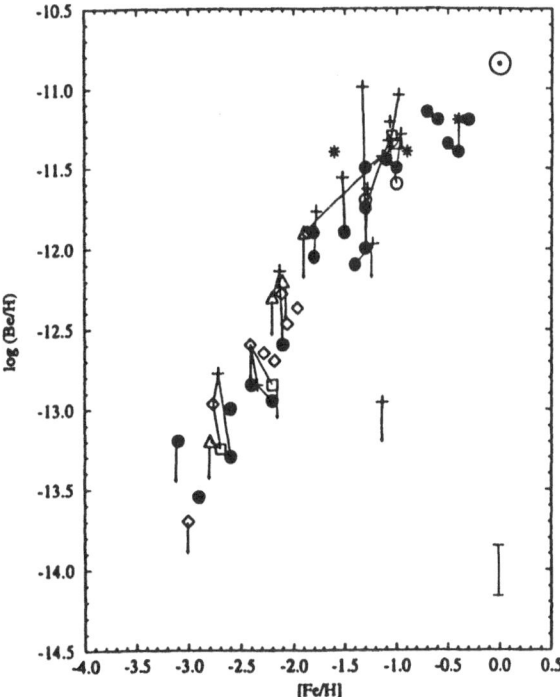

Fig. 1. Beryllium as a function of metallicity. We plot our results and others from the literature. Measurements for the same object are interconnected. Filled circles: this work; open circles: Rebolo et al. (1988); triangles: Ryan et al. (1990); open squares: Ryan et al. (1992); diamonds: Gilmore et al. (1992); asterisks: Rebolo et al. (1993); crosses: Boesgaard & King (1993).

suggested could be Be deficient. In our higher resolution spectrum we were able to detect the weak line of the Be II doublet in HD103095, and we measured a Be abundance log N(Be)=-0.1 ± 0.15 (note that this is one of the few stars where gravity is well determined). This abundance is lower than those of stars with similar [Fe/H], suggesting a mild deficiency of $0.2 - 0.3$ dex. Additional information is needed to establish whether this is due to a depletion process or to a lower initial abundance.

(ii) G 170-47 (BD+23 3130): the most metal-poor star with Be. We remark our detection of beryllium in G 170-47, a subgiant star with [Fe/H]= -2.9 and log N(Li)= 1.17 (Thorburn 1994). To our knowledge it is the lowest metallicity star where Be has been detected. Its Be abudance is lower by 0.25 dex than the Be abundance in HD140283, possibly indicating that the process of Be enrichment was active at such very early epochs in the life of our Galaxy. The low gravity suggests that the object is slightly evolved, and if dilution has taken place the initial Be content could have been higher. We have observed an even more metal deficient star in our sample, BD+3 740, but we could impose only an upper limit to its Be abundance.

References

Alonso, A. Arribas, S., & Martínez-Roger, C., 1994, A&A, in press

Anders, R., & Grevesse, N. 1989, Geochim. Cosmochim. Acta, 53, 197

Boesgaard, A. M., & King, J. R. 1993, AJ, 106, 2309

Carney, B. W. 1983, AJ, 88, 610

Deliyannis, C. P., Ryan, S. G., Beers, T. C., & Thorburn, J. A. 1994, ApJ 425, L21

Fields, B. D., Schramm, D. N., & Truran, J. W. 1993, ApJ, 405, 559

García López, R. J., Severino, G., & Gomez, M. T. A&A submitted

Gilmore, G., Gustafsson, B., Edvardsson, B., & Nissen, P. E. 1992, Nature, 357, 379

Gustafsson B., Bell R. A., Eriksson K., & Nordlund Å. 1975, A&A, 42, 407

Kawano, L. H., Fowler, W. A., Kavanagh, R. W., & Malaney, R. A. 1991, ApJ, 372, 1

Kajino, T., & Boyd, R. N. 1990, ApJ, 359, 267

Kurucz, R. L. 1992, private communication

Kurucz, R. L., Furenlid, I., Brault, J., & Testerman, L. 1984, Solar Flux Atlas from 296 to 1300 nm, NOAO Atlas No. 1

Magain, P. 1987a, A&A, 181, 323

Magain, P. 1987b, A&A, 179, 176

Malaney, R. A., & Fowler, W. A. 1989, ApJ, 345, L5

Nissen, P. E., Gustafsson, B., Edvardsson, B., & Gilmore, G. 1994, A&A, 285, 440

Pavlenko, Ya. V. 1991, SvA, 35, 212

Prantzos, N., Cassé, M, & Vangioni-Flam, E. 1993, ApJ, 403, 630

Rebolo, R., García López, R. J., Martín, E. L., Beckman, J. E., McKeith, C. D., Webb, J. K., & Pavlenko Ya. V. 1993, in Origin and Evolution of the Elements, N. Prantzos, E. Vangioni-Flam, & M. Cassé (eds.), (Cambridge: Cambridge Univ. Press), p. 149

Rebolo, R., Molaro, P., Abia, C., & Beckman, J. E. 1988, A&A, 193, 193

Ryan, S. G., Bessell, M. S., Sutherland, R. S., & Norris, J. E. 1990, ApJ, 348, L57

Ryan, S. G., Norris, J. E., Bessell, M. S., & Deliyannis, C. P. 1992, ApJ, 388, 184

Smith, G., Lambert, D. L., & Ruck, M. J. 1992, A&A, 263, 249

Terasawa, N., & Sato K. 1990, ApJ, 362, L47

Thorburn, J. A. 1994, ApJ, 421, 318

VandenBerg, D. A., & Bell, R. A. 1985, ApJS, 58, 561

Wiese W. L., & Martin, G. A. 1980, Wavelengths and Transition Probabilities for Atoms and Atomic Ions (NSRDS-NBS 68)

Boron in the Hyades Giants

D.K. Duncan[1], R.C. Peterson[2], J.A. Thorburn[1],
M.H. Pinsonneault[3], C.P. Deliyannis[4]

[1] University of Chicago, Dept. of Astronomy and Astrophysics,
5640 S. Ellis Ave., Chicago, IL, 60637, USA
[2] Board of Studies in Astronomy and Astrophysics, University of California, Santa Cruz, CA, 95064, USA
[3] Ohio State University, Dept. of Astronomy, 5040 Smith Lab, 174 W. 18th Ave., Columbus, OH, 43210, USA
[4] Institute for Astronomy, University of Hawaii, 2680 Woodlawn Dr., Honolulu, HI 96822, USA

Abstract. Spectra of two of the Hyades giants ϵ Tau (HD 28305) and δ^1 Tau (HD 27697), the reference K0II giant β Gem (HD 62509), and the Hyades main-sequence F star 45 Tau (vB 14) have been obtained with the *Hubble Space Telescope*. Boron abundances were determined from spectrum synthesis of the region near 2500Å. Since B is destroyed by (p,α) reactions at a temperature near 5 million degrees, it is expected to be preserved in only an outer shell on the main-sequence, and diluted by B-free interior material on the giant branch. We find the B abundance in each giant is reduced below solar by just the factor of 10 predicted by standard stellar evolution models. Li, which is more fragile than B, is reduced in the same stars by a factor of 160. The prediction is 60 – a long standing discrepancy. However, when NLTE effects are accounted for, the observed Li dilution factor becomes 80, equaling the predicted value within measurement error. No unusual mass loss or non-standard internal mixing is required to match the B or Li observations.

1 Introduction

The light elements lithium, beryllium, and boron are of particular use in probing the structure of late-type stars, since these elements undergo nuclear reactions at relatively low temperatures. The critical temperatures for destruction are approximately 2.5, 3.5, and 5×10^6K respectively at densities similar to those in the sun. Since these temperatures are reached near the bottom of the convection zone and well outside the core in solar-type stars, circulation and destruction of the light elements can result in observable abundance changes.

In giants, the largest observable changes are expected to be produced by the process of *dilution*. While the star is still on the main-sequence, each light element is depleted in the interior, where the critical temperature for destruction is reached, but present in a surface layer where the initial main-sequence abundance remains unchanged. As the star ascends the giant branch the convection zone deepens, incorporating interior material which is free of the light elements. The observed abundances should be reduced by the ratio of mass in the convection zone to mass in the surface region where the light elements were preserved

– a large ratio. A test of this prediction was first performed by Boesgaard, Heacox, & Conti (1977; see the discussion section), who discussed Li and Be. They concluded that dilution by the expanding convective envelope is *not* sufficient by itself to explain the abundances.

Furthermore, even in the Sun the destruction of light elements is not well understood. Standard solar models predict a temperature at the base of the solar convection zone very close to 2.0×10^6K, too low to destroy Li. However, the solar Li abundance is approximately 100 times less than the protostellar value determined from lunar or meteoritic samples (Nichiporuk & Moore, 1974; Anders & Grevesse, 1989), or in very young 1 M$_\odot$stars (Thorburn *et al.* 1993, Soderblom *et al.* 1994). This has led to the suggestion of a large number of possible additional mechanisms which might reduce Li/H, including convective overshoot, meridional circulation, rotationally-induced sheer, diffusion, and main-sequence mass loss (VandenBerg & Poll 1989, Michaud & Charbonneau 1991, Swenson & Faulkner 1992, etc.). Although many stellar evolution calculations have now been performed with "additional" mixing (e.g. Pinsonneault, Deliyannis, & Demarque 1992), no proposed mechanism fully explains the Li data (Chaboyer 1994). The Be and B in the sun appear close to the protostellar values (Chmielewski *et al.* 1975; Anders & Grevesse 1989). Li has been widely observed in stars; B and Be have not.

Additional data, in the form of Be and B observations, should be helpful in discriminating between the large number of mechanisms proposed to explain the Li observations, and in general improve our understanding of the internal structure of cool Population I stars. The present investigation represents a first step in that direction.

2 Observations

Spectra were obtained for the two Hyades giants ϵ Tau (HD 28305) and δ^1 Tau (HD 27697), the early-F Hyades dwarf 45 Tau (HD 26462; van Bueren 14), and the reference field K0II giant β Gem (HD 62509) using the G270M grating of the Goddard High Resolution Spectrograph (GHRS) of the *Hubble Space Telescope* (hereafter HST). Dates of observations in 1993 were August 27, February 5, October 9, and February 9 respectively. The standard FP-SPLIT procedure was followed which divides the spectrum into four sub-spectra, each recorded on slightly different parts of the photocathode to reduce instrumental noise. Exposure times of 9248, 9792, 1197, and 2067 s. respectively were required to produce 6000 counts per diode for the Hyades giants, 8000 for 45 Tau, and 12000 counts for β Gem. Note that one diode equals 4 pixels in this "quarter-stepped" data. The grating was centered at 2500 Å for a spectral coverage of approximately 45 Å with nominal dispersion of 0.094 Å per diode. (We use air wavelengths throughout this paper.) Nominal FWHM spectral resolution is 25,000.

Data reduction followed standard HST procedures. The S/N achieved for the Hyades giants was about 75 per diode, equal to the photon statistics limit. This was determined using the STSDAS reduction program "Specalign" which makes a least-squares determination of the spectrum and photocathode granularity of FP-SPLIT data. Wavelength calibration from the HST data was shifted onto a terrestrial (air) scale and velocity displacements due to stellar and Earth motion were removed by requiring that wavelengths of the strongest stellar features matched those of the line list of Kurucz (1993a).

3 Spectrum Synthesis

Boron determinations in cool stars are usually made of the pair of resonance lines of B I in the ultraviolet near 2500 Å. In cool giants as well as in the Sun that spectral region is very crowded, and a reliable analysis depends upon spectral synthesis. Knowledge of line identifications and their associated parameters is somewhat lacking, since UV laboratory measurements of atomic and molecular line transitions are exceedingly difficult. Consequently the analysis of such spectra poses significant difficulties.

We attempted to deal with these by first reanalyzing existing spectra of cool stars. Particularly useful in this regard were the three F stars previously studied by Lemke, Lambert, & Edvardsson (1993). Their F-star data, which includes the star Procyon, was obtained with the GHRS echelle grating, as described by those authors. This results in significantly higher spectral resolution, nominally 90000, and proportionately smaller wavelength coverage. We also examined our own solar and α Cen spectra, and the echelle HST spectra of HD 140283 (Edvardsson et al. 1994). The HD140283 spectrum was particularly helpful for determining gf values of lines which are saturated in the Lemke et al. stars. We insisted that the line list we developed provide an adequate fit to all these observations.

The synthesis program we used is the code SYNTHE (Kurucz 1993a), as modified to run on Unix SPARCstations by Steve Allen of the University of California at Santa Cruz. Kurucz provides lists of atomic lines including both lines measured in the laboratory and lines whose wavelengths are predicted from semi-empirical fitting to known energy levels, and the model atmospheres we used were also taken from Kurucz (1993a). Molecules were included in the calculations of species equilibria, but no molecular lines appear in the Kurucz lists of diatomic molecules within this region. Accurate gf values and wavelengths for the B I resonance doublet were taken from Johansson et al. (1993). Boron in the solar system has an abundance $\log \epsilon(B) = 2.88 \pm 0.04$ in carbonaceous chondrites (Anders & Grevesse 1989), and $\log \epsilon(B) = 2.6 \pm 0.3$ (Kohl, Parkinson, & Withbroe 1977) derived from the photospheric resonance lines. The measurements of the present paper refer to changes in B, not its absolute value. Our scale is set to the Kurucz default value, a solar B abundance of 2.56.

Our initial syntheses resulted in many lines in disagreement with that of the observed spectrum. One difficulty, not always appreciated, is that for predicted lines, *both* the wavelengths and the gf values are uncertain. Since wavelengths depend on differences of energy levels, and since at least one energy level of each predicted line is uncertain, having been extrapolated or interpolated from known energy levels, the positions of predicted lines can be in error by several angstroms in the ultraviolet. The procedure sometimes employed of taking predicted lines at their predicted wavelengths and adjusting their gf values is not valid.

We then chose to start using only lines whose positions have been measured in the laboratory. These have accurate wavelengths, but gf values with uncertainties of typically a factor of two in Kurucz's latest list. This produces a much better match, although many lines are too strong and a considerable number of lines appear to be missing from the synthesis, several of which are in the proximity of the boron lines. At this point we began to modify the Kurucz line list, changing the gf values of lines which were too strong, and referring to the spectra mentioned above.

A model with T_{eff} = 5000K, log g = 2.5, and [Fe/H] = +0.1 (Boesgaard et $al.$ 1977; Cayrel et $al.$ 1985) is shown fit to the sum of the data for the two Hyades giants in Fig. 1. (The individual giant spectra are extremely similar over this spectral range, so we fit their sum.) Positions of the B resonance are indicated. The longer wavelength component of the B doublet is highly blended, and the fit is made to the shorter wavelength component. Error bars on the points indicate ±1 std. deviation. Microturbulent velocity was set at v_t = 1.0 km s^{-1}. The resolution of 25000 is set by the instrumental broadening. A model was also run with T_{eff} = 4750K and log g = 2.5. This gave features that are too strong, particularly for low-excitation lines which are more temperature-sensitive. Boron abundances of [B/H] = −0.4, −1.1, −2, and −3 (essentially no B) are shown. The B I lines are considerably weaker than the model with [B/H] = −0.4; the Hyades giants have approximately one order of magnitude less than the solar boron abundance.

In Fig. 1 there are still discrepancies between the calculated and observed spectra. Most occur where Procyon and the other stars are also unmatched by the lab line list, and most do not affect the derived B abundances. We roughly estimate the uncertainty of the B fit to be 0.2-0.3 dex. Further details of the fit and the final line list we adopted will be published in a forthcoming Astrophysical Journal paper.

The main-sequence Hyades dwarf 45 Tau (= vB 14) was fit using the same line list as above and stellar parameters T_{eff} = 7000K, log g = 4.0, v_t = 1.0, and [Fe/H] = +0.1. Resolution of 15,000 matches the broadening set by stellar rotation. A very good fit is obtained with a solar B abundance, [B/H] = 0.0.

4 Discussion

The main point of the present paper is to test the growth of the convection zone by following the course of abundance changes in the light elements Li, Be, and B. The principle is straightforward. Each element is destroyed at rather low temperatures – Li most easily, then Be and B. While the star is still on the main-sequence each element is depleted in the interior, but present in a surface layer where the initial main-sequence abundance remains unchanged. This layer is thinnest for Li and thickest for B. Subsequent expansion of the convection zone

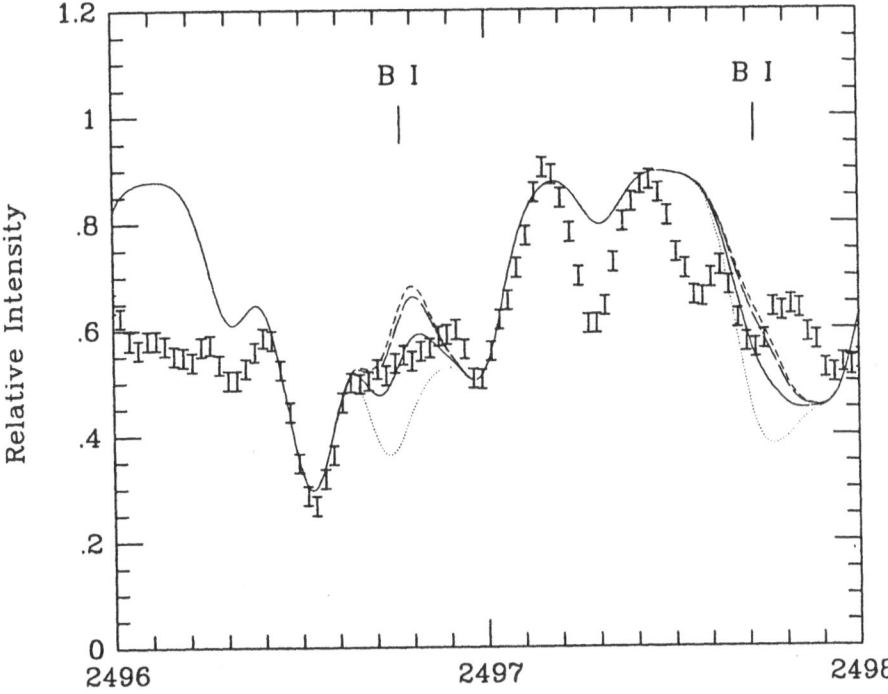

Fig. 1. Hyades giants spectra and syntheses with [B/H] = −0.4,−1.1 (best fit), −2, and −3. Error bars show ±1σ.

during giant-branch evolution dilutes the mass of Li, Be, and B in the surface layers where they were preserved by the total mass of the convection zone (into which they are mixed). Consequently underabundances occur which should be largest for lithium and least for boron.

This test was first performed by Boesgaard, Heacox, & Conti (1977) who looked for Be in six main-sequence Hyades dwarfs and the four Hyades giants. They did not detect Be but determined upper limits to its abundance, and also discussed the lithium abundance determinations of Zappala (1972) for dwarfs and Bonsack (1959) for giants. Boesgaard *et al.* concluded that dilution by the expanding convective envelope is not sufficient by itself to explain the abundances. This conclusion was based on a discrepancy between predicted Li dilution of about 60X and the observed dilution of 110-170X in the giants. They suggested that relatively small amounts of mass loss could "peel off" Li-containing material and reduce the abundance. To be effective, such mass loss must occur well before the giant branch is reached, while Li is still confined to a relatively thin surface layer. This amount of mass loss would also be consistent with the Be observations, but since they are only upper limits they do not provide a strong constraint themselves. The mass loss required, $\approx 10^{-11}$ M$_\odot$ yr^{-1} would not be detectable. However, this is a rate two orders of magnitude in excess of the solar wind, and there is no known reason for it to occur.

The B data of the present paper provides an independent check of predicted dilution. The main-sequence precursors of the Hyades giants are late A stars which do not have surface convection zones and are not expected to have depleted boron. This expectation is substantiated by our observation of the early F star 45 Tau, which shows solar [B/H] to within the error of measurement. The observed dilution between main sequence and giants in the Hyades is therefore -1.1 ± 0.3 dex.

Before obtaining the abundance results described above we made entirely new calculations of predicted Li, Be, and B depletion factors. This was done to see how calculations done using recent stellar models incorporating better opacities would compare to the calculations of Boesgaard et al. (1977), and also to calculate the predicted depletion factors for all the light elements in the same way for consistency. The calculations were done with the Yale stellar evolution code (Pinsonneault et al. 1992). Differences from the Boesgaard et al. predictions were not large. Our calculations also followed the $^{12}C/^{13}C$ ratio, which is observed to change in stars which have ascended the giant branch. Further details may be found in the forthcoming Ap. J. paper.

Measurements of $^{12}C/^{13}C$ in the Hyades giants are presented by Lambert, Dominy, and Siversten (1980) who also obtained more accurate Li abundances for the giants than were available to Boesgaard et al. 1977. We adopt $\log \varepsilon(\text{Li})$ = 0.9 ± 0.1 from their work. To measure Li dilution we need a value for the main-sequence precursors of the giants. For this we use the Hyades early-F star data of Boesgaard and Tripicco (1986), which gives $\log \varepsilon(\text{Li}) = 3.15 \pm 0.15$. We then derive a measured dilution of -2.20 ± 0.2 dex. This Li dilution factor is within the errors identical to that derived by Boesgaard et al..

All of the measured values and predictions are brought together in Table 1. The agreement for boron dilution and $^{12}C/^{13}C$ ratio is very good. We draw the conclusion that standard evolutionary models explain these abundances well. Be, too, is in reasonable agreement but since there is no firm detection in the giants the dilution is quite uncertain. Better spectra of the Hyades giants should soon make this number more accurate. The Li dilution presents the same discrepancy it did to Boesgaard et al., and greater accuracy has only made the problem more clear.

Table 1. Measured and Predicted Abundances

Species	Survival zone	Predicted Dilution	Observed
^{6}Li	0.8%	2.03 [dex]	
^{7}Li	1.4%	1.77	2.2 ± 0.2
^{7}Li (NLTE)			1.92 ± 0.2

^{9}Be	3.1%	1.44	> 1.5 (uncertain)
^{10}B	11.5%	0.87	
^{11}B	8.4%	1.01	
B combined in 1:4 ratio		0.98	1.1

The solution proposed by Boesgaard *et al.* of stripping 0.02 $M_\odot$ from the surface of the Hyades stars still could be invoked to reduce the Li abundance without altering the B abundance enough to cause problems. However, we believe that there is a more straightforward explanation which does not change the other numbers. Carlsson *et al.* (1994) have made an extensive study of NLTE effects in Li spectra in a range of giant and dwarf stars of different abundances and temperatures. They find effects that vary strongly with the strength of the Li feature and the temperature of the star. In fact, the sign of NLTE effects reverses across the temperature range between K and F stars. Applying the Carlsson *et al.* results to early-F stars which show an (LTE) abundance of $\log \varepsilon(\text{Li}) = 3.15$, one finds a correction of about -0.10. A giant with $\log \varepsilon(\text{Li}) = 0.95$ has a correction of +0.18. The dilution factor is consequently reduced to a value of 83 ± 40, whose errorbar overlaps with the predicted value of 60. It should also be noted that the uncertainty in the predicted dilution factor may be as much as a factor of two.

5 Conclusion

HST measurements of boron abundances of two of the Hyades giants and one dwarf have permitted a test of one of the basic predictions of stellar evolution theory: the growth of the convection zone as a star evolves up the giant branch. The observations are entirely consistent with prediction. The B abundance remains solar on the main-sequence, and then decreases by about a factor of 10 as the convection zone grows in mass and mixes in boron-free material from the stellar interior. The same test is performed with existing Li observations. Li survives in a thinner layer than B does while on the main-sequence, and its expected dilution is much greater; this is observed. A factor of two difference between prediction and observation first noted by Boesgaard *et al.* (1977) remains in our data. However, it disappears when the NLTE effects of Carlsson *et al.* (1994) are taken into account. There is no need to introduce any unusual amount of main-sequence mass loss to explain the light element abundance results. Recent observations of Be in the Hyades will provide one more test of stellar evolution predictions, and NLTE calculations should be done for B in the Hyades giants. At the present time, however, entirely standard models of stellar evolution explain the observations.

References

Anders, E., & Grevesse, N. (1989): *Geochim. Cosmochim. Acta*, **53**, 197.

Boesgaard, A.M., Heacox, W. D., & Conti, P.S. (1977): Astrophys. J., **214**, 124.

Boesgaard, A.M., & Trippico, M.J. (1986): Astrophys. J. Lett., **302**, L49.

Bonsack, W.K. (1959): Astrophys. J., **130**, 843.

Carlsson, M., Rutten, R.J., Bruhls, J.H.M.J., and Shchukina, N.G. (1994): Astron. Astrophys., **288**, 860.

Cayrel, R., Cayrel de Strobel, G., & Campbell, B(1985): Astron. Astrophys., **146**, 249.

Chaboyer, B. (1994): Ph.D. Thesis, Yale University

Chmielewski, Y., Müller, E.A., & Brault, J. (1975): Astron. Astrophys., **42**, 37.

Edvardsson, B., Gustafsson, B., Johansson, S.G., Kiselman, D., Lambert, D.L., Nissen, P.E., and Gilmore, G., (1994): Astron. Astrophys., **290**, 176.

Hobbs, L.M., & Pilachowski, C. (1988): Astrophys. J., **347**, 817.

Johansson, S.G., & Cowley (1988): J. Opt. Soc. Am. B, **5**, 2264.

Johansson, S.G. and Leckrone, D. (1992): private communication.

Johansson, S.G., Litzen, U., Kasten, J., and Kock, M. 1993, Astrophys. J. Lett., **403**, L25.

Kiselman, D. (1994): Astron. Astrophys., in press.

Kohl, J.L., Parkinson, W.H., and Withbroe, G.L. (1977): Astrophys. J. Lett, **212**, L101.

Kurucz, R. (1993a): CD-ROM.

Kurucz, R. (1993b): private communication.

Lemke, M., Lambert, D.L., and Edvardsson, B. (1993): Publ. Astron. Soc. Pacific, **105**, 468.

Lambert, D.L., Dominy, J.F., & Siversten, S. (1980): Astrophys. J., **235**, 114.

Meneguzzi, M., Audouze, J., & Reeves, H. 1971, Astron. Astrophys., **132**, 278.

Michaud, G., & Charbonneau, P. (1991): Space Sci. Rev., **57**, 1.

Nichiporuk, W., & Moore, C.B. (1974): Geochim. Cosmochim. Acta, **8**, 1691.

Pinsonneault, M.H., Deliyannis, C.P., and Demarque, P. (1992): Astrophys. J. Supp., **78**, 179.

Soderblom, D.R., Jones, B.F., Balachandran, S., Stauffer, J.R., Duncan, D.K., Fedele, S.B., & Hudon, D.H., (1993): Astron. J., **106**, 1059.

Swenson, F.J., & Faulkner, J., (1992): Astrophys. J., **395**, 654.

Thomas, D., Schramm, D.N., Olive, K.A., and Fields, B.D., (1993): Astrophys. J., **406**, 569.

Thorburn, J.A., Hobbs, L.M., Deliyannis, C.P., & Pinsonneault, M.H. (1993): Astrophys. J., **415**, 150.

VandenBerg, D.A., & Poll, H.E. (1989): Astron. J., **98**, 1451.

Zappala, R.R. (1972): Astrophys. J., **172**, 57.